Foundation Mathemati

SECOND EDITION

Modern Applications of Mathematics

Series editors:

D J G James Coventry University
R R Clements University of Bristol

Other titles in the series

Modern Engineering Mathematics *Glyn James*
Advanced Modern Engineering Mathematics *Glyn James*
Transforms in Signals and Systems *Peter Kraniauskas*
Engineering Mathematics: A Modern Foundation for Electronic, Electrical and Control Engineers *Anthony Croft, Robert Davison and Martin Hargreaves*

...Foundation Mathematics

SECOND EDITION

Dexter J. Booth
The University of Huddersfield

ADDISON-WESLEY PUBLISHING COMPANY

Wokingham, England · Reading, Massachusetts · Menlo Park, California
New York · Don Mills, Ontario · Amsterdam · Bonn · Sydney
Singapore · Tokyo · Madrid · San Juan · Milan · Paris · Mexico City
Seoul · Taipei

Cover designed by Designers & Partners of Oxford
and printed by The Riverside Printing Co. (Reading) Ltd.
Typeset by P & R Typesetters Ltd. (Salisbury, UK).
Printed in Great Britain at the University Press, Cambridge

First edition published 1991. Reprinted 1991.
Second edition printed 1993.

ISBN 0-201-62419-2

British Library Cataloguing in Publication Data
A catalogue record for this book is available from the British Library.

Library of Congress Cataloging in Publication Data applied for.

To my wife Liz

Preface to the Second Edition

The reception to the first edition of this book has been most gratifying. It would appear that it has touched a nerve, among teachers and students alike. In response to the many kind and useful comments that I have received about the first edition, I have made a number of changes – most minor but a few substantial. It is very tempting to include more and more material, but for the book to be usable it must also be manageable. While many of the suggestions made have been incorporated, a number have not because of space limitations.

The most significant change is the addition of a seventh chapter, dealing with the elements of sets and probability. Although there was no intent to introduce a statistics component to the book, it is felt that armed with the background presented in this chapter a student could confidently embark upon a fully fledged statistics course.

Minor changes include notational amendments and the correction of my earlier blunders. These have by now, hopefully, been cleared up and I apologize for any inconvenience caused as a consequence of them. However, in the best of schemes there is bound to be a glitch, and any comments or suggestions regarding the contents or presentation would be most gratefully received. The book is, after all, intended for you to use.

■ ■ ■ Preface to the First Edition

The tertiary education sector in Science and Engineering is facing a two-sided problem. On the one side, Science and Engineering departments are expected at least to maintain their student numbers, if not to increase them. On the other side, as the demographic curve falls, the total number of students adequately qualified to previously demanded standards continues to dwindle.

One consequence is that departments are reducing their qualification demands so as to maintain course viability. This is especially evidenced in the substantially increased ratio of BTech to A-level candidates. A major casualty of this situation is Mathematics. Increasingly, students are entering Science and Engineering courses inadequately prepared for the mathematical demands to be made of them in becoming graduate scientists and engineers.

This book is aimed at those students who need to bridge the gap. Such students may be enrolled on a foundation course and this text contains what I believe to be the necessary material for the mathematics component of such a course. The division of the book into 27 modules containing a total of 85 units makes it logistically ideal for the typical foundation year.

Other students may be granted exemption from such a foundation course, but may still be weak in their mathematics. This text can then be used as a self-help manual to enable them to build up their mathematics background while simultaneously studying mathematics in a more formal manner as part of their undergraduate studies.

The text is subdivided into 7 chapters, each chapter containing a number of modules covering a specific aspect of mathematics. Each module comprises a number of units, and each unit is designed so that the material it contains can reasonably be covered in one or two sittings at most. Every unit is accompanied by a set of worked examples and a set of part-worked exercises. The student is expected to read through the worked examples and then complete the part-worked exercises. This not only reinforces the material within the unit, but also serves as a continuing model of how to lay out solutions to problems, thereby generating confidence in the student's own abilities. While it may be tempting for a lecturer to complete the part-worked exercises, it is recommended that the students be permitted to attempt to work them for themselves first. They are designed to be an integral part of the learning process, and many of the comments I have received confirm that they do work.

At the beginning of each unit is a set of test questions. They are placed at the beginning rather than at the end for two reasons:

- Students who think they are familiar with the material contained in the unit can test themselves before reading the material. In this way any problems can be identified.
- Students who are unfamiliar with the material in the unit can see the nature of the expected outcome from studying the unit. This helps to focus the student's attention by giving an attainable goal from the beginning.

At the end of each module is a collection of exercises that range through all the material contained in the units in that module.

At the end of each chapter there is a collection of miscellaneous exercises. The solutions to these exercises require a knowledge of the material contained in that and previous chapters, alongside a little imagination.

It must be recognized that it is impossible to write a text book that covers all aspects of every foundation course in existence, and this book is no exception. You may find that certain topics have been omitted that you would have liked to have seen included. Indeed, you may find topics included that you may feel need not be there. However, in defence of the contents, I see the fundamental problem to be one of a lack of knowledge coupled with a lack of confidence and I have endeavoured to produce a book that provides a balance between these two aspects.

Acknowledgements

I should like to express my gratitude to my editors, Tim Pitts and Sarah Mallen of Addison-Wesley, for their professionalism and patience, to Professor Glyn James of Coventry University and to Dr John Turner of the University of Huddersfield for their many useful comments during the preparation of the first edition and to the many reviewers who have kindly and anonymously contributed their comments and suggestions. My special thanks are extended to the numerous readers who have responded to the first edition with their critical reactions, and particular mention must be made of Dr Joe Whitaker of Lancaster University, whose close editorial scrutiny has been greatly appreciated during the preparation of this second edition. Many amendments are due to his efforts, not least the change in the presentation of the Binomial Theorem. I also wish to thank the Mathematical Association of America for their kind permission to reproduce a number of questions that have appeared in their *Contest Problem Book II*. Also, I should like to thank all those students whom it has been my privilege to teach over the years, for teaching me so much.

Dexter J. Booth
Huddersfield 1993

■ ■ ■ To the Student

Do you have difficulties with mathematics? Difficulties with mathematics arise from two main sources – a lack of knowledge and a lack of confidence. It is very rare for difficulties to arise from a lack of innate ability.

Mathematics is a progressive subject where facts depend upon earlier facts, and if there are gaps in your knowledge it can become very difficult to understand the development of the subject. If you cannot see the development from one idea to the next, if you cannot see a pattern, then mathematics becomes a jumble of unrelated facts and your confidence in your own ability to understand the subject takes a nosedive. To address the problem of a lack of knowledge, we begin the book with arithmetic because, after all, that is where the subject starts. Also, a lot of people have a hazy knowledge of arithmetic and that makes an ideal breeding ground for later difficulties. You may feel that the early parts of the book are elementary, but different people have gaps at different places in their knowledge and a thorough competence in arithmetic is essential for future development. It may be that you are quite happy with your knowledge of arithmetic, but it is best to be sure. There may be gaps in your knowledge of which you are unaware.

To overcome a lack of confidence, the material in the book is highly structured, taking the form of a collection of self-contained units. Each unit contains worked examples, part-worked exercises for you to complete and a short test. Studying in this manner, by appealing to your short-term memory, your knowledge is developed and your confidence reinforced at each step along the way.

Using the Book

Before reading the material contained in a unit, you are encouraged to complete a short test on it. This may seem to be putting the cart before the horse, but there is a good reason for it. The last thing I want to do is to bore you – your continued interest is essential if you are to learn. There may be areas of mathematics where your knowledge is adequate and to cover old ground may be unnecessary. To cater for this, the test is there as a guide to tell you whether or not you need to read the unit. If you complete the test satisfactorily then proceed to the next unit.

If you decide to read the unit you will find that it ends with a collection of worked examples. You are expected to read through these carefully to make sure you follow the reasoning behind their solutions. The worked examples are as valuable and important as the unit material they exemplify. These are followed by a collection of similar, part-worked exercises. You are expected to complete these exercises and so reinforce your understanding of the material in the unit and the reasoning behind the worked examples. When you have completed the part-worked exercises, go back to the test at the beginning of the unit and try that. Before you leave a unit and move to the next one, make sure that you are completely satisfied with what you have done in that unit as it is essential that you understand the unit material. If you find you have difficulty remembering some of the material do not over-worry, we are looking for understanding at this stage – your memory will develop as you go along.

Having completed all the units in a module try the module exercises. These are very similar to the exercises and test questions that you have done in each unit of the module, and their purpose is to test your longer term memory. With long-term memory retention comes the confidence to tackle new areas of knowledge. Again, master one module before moving on to the next. By now you will find that you are getting into the swing of things and that your memory of past work is improving.

At the end of each chapter you will find a further collection of exercises that consolidate the work done in that and previous chapters. These exercises you will find challenging, so do not panic if you find that you are experiencing difficulties with them – you are supposed to. In all human endeavours we come across problems that we have not met before and to solve these problems the only tools we have available are our past knowledge and experiences. The chapter exercises are aimed at trying to encourage you to find a solution by linking your newly acquired mathematical knowledge to your confidence in unpractised situations. If you find that the solution to a particular problem continues to elude you then look at the back of the book where some solutions are given. Try to understand the reasoning behind these solutions and so develop both your ability to handle new problems and confidence in your ability to 'leave the nest' as it were and reason independently.

In conclusion, always remember that understanding is not beyond you – if you do come across something that refuses to make sense to you then stop. Leave it for a day or so and then go back to it. Even if you are not consciously thinking about it your brain will be beavering away trying to make sense of the problem and likely as not the answer to your problem will come when you are thinking about something entirely different. Above all, remember the thrill of learning how to do something that you could never do before and enjoy yourself.

Dexter J. Booth
Huddersfield 1993

▪ ▪ ▪ Contents

1
■ ■ ■ Handling Numbers

The aims of this chapter are to:

1. Review the elements of arithmetic and thereby introduce the real and imaginary number systems.
2. Demonstrate how to manipulate numerical data from its collection, through its organization to its analysis.
3. Exhibit linear relationships between two sets of numbers and model such relationships in the form of equations.
4. Display the linear relationships between two sets of numbers in graphical form and link linear equations and linear inequalities to their graphs.

This chapter contains four modules:

Module 1: Arithmetic
The elements of arithmetic are reviewed alongside simple uses of a calculator as preparation for using numbers to generate information.

Module 2: Data Handling
Information is extracted from numerical data by using what has been learned in Module 1. This includes the collection of numerical data by measurement and calculation, the organization of data into tabular format and the derivation of summary statistics from the tabulation.

Module 3: Relationships
The idea of a relationship between two sets of numbers is introduced. This relationship can be modelled in the form of an equation which is then used to extract information not contained in the original data.

Module 4: Displaying Relationships
Linear relationships between two sets of numbers are displayed on a cartesian graph. The links between equations and their graphs are established.

Module 1: Arithmetic

Aim: To review the elements of arithmetic and thereby introduce the real and imaginary number systems.

Objectives: When you have read this module you will be able to:

- ▶▶ Distinguish between integer, rational and irrational numbers and recognize the difference between the decimal form of a rational number and the decimal form of an irrational number.
- ▶▶ Manipulate numbers using brackets and the operations of addition, subtraction, multiplication, division and raising to a power using the precedence rules.
- ▶▶ Use rounding to estimate the possible outcome of a calculation prior to performing it exactly.
- ▶▶ Perform the arithmetic operations between imaginary numbers.

Unit 1 The Integers

Test yourself with the following:

1 Evaluate (a) $25 - 8 \times 4 + 6 \div 3$
(b) $(12 - 3) \times 5 - 4 \div (7 - 3)$
(c) $5 - \{2 + (3 \times [5 - 2]) - 9\}$
(d) $(-8) - (-4) + (-2) \times 3 - (-6)/(-2)$

2 Find (a) the prime factors and prime factorization
(b) all the factors
(c) the HCF and LCM

of 164 and 1025.

3 Round to the nearest 10, 100 and 1000: (a) 123 (b) 549 (c) -949

4 Estimate the value of $29 \times 11 - 48 \div 9$ and check the accuracy by evaluating the exact value.

The Natural Numbers

Numerals and Place Value

The **natural numbers** are written using the decimal numerals 0 to 9 where the position of a numeral in a number dictates the value it represents. For example:

123 stands for 1 hundred and 2 tens and 3 units

Here, the numerals 3, 2 and 1 are called the unity, tens and hundreds **coefficients** respectively. Every calculator contains a keypad with these numerals on the keys. Using these keys natural numbers can be displayed on the calculator screen up to the maximum number of numerals permitted. This concept of **place value** is crucial to our number system. Imagine the problem a Roman legionnaire would have had in working out a bakery order for IV loaves between III men per day to cater for CIX men for XIV days.

Points on a Line and Order

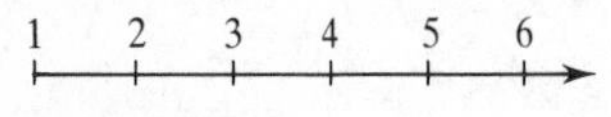

Figure 1.1

The natural numbers can be graphically represented by equally spaced points on a straight line. The natural numbers are **ordered** – they progress from small to large. On the line, numbers to the left of a given number are less than $(<)$ the given number and numbers to the right are greater than $(>)$ the given number. If two numbers are neither less than nor greater than each other they are equal $(=)$.

The Integers

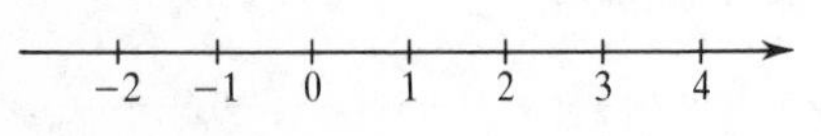

Figure 1.2

If the straight line containing the natural numbers is extended to the left we can plot equally spaced points to the left of zero. These points represent negative numbers which are written as a natural number preceded by a minus sign, for example -5. The natural numbers and the negative numbers shown here are collectively called the **integers**. The notion of order still applies: for example, $-3 < 2$ and $-4 > -6$. On a calculator a positive number can be changed into a negative number and vice versa by pressing the $+/-$ key.

The Arithmetic Operations

Brackets

Brackets can be used around the negative numbers to separate the minus sign attached to the numeral from the arithmetic operation signs. For example, $6 - -2$ should be written as $6 - (-2)$.

Addition and Subtraction

Adding a negative number is the same as subtracting a positive number. For example, $5 + (-3)$ is the same as $5 - 3$. The result is 2.

Subtracting a negative number is the same as adding a positive number. For example, $5 - (-3)$ is the same as $5 + 3$. The result is 8.

On a calculator addition and subtraction are performed using the $+$ and $-$ keys respectively.

Multiplication and Division

Multiplying or dividing two positive or two negative numbers results in a positive number. For example, 12×6 gives the same result as $(-12) \times (-6)$, they are both equal to 72. Also $12 \div 6$ gives the same result as $(-12) \div (-6)$, they are both equal to 2.

Multiplying or dividing a positive and a negative number gives a negative number. For example, $(-12) \times 6$ equals -72 as does $12 \times (-6)$. Also $12 \div (-6)$ and $(-12) \div 6$ both equal -2.

On a calculator, multiplication and division are performed using the $\times$ and $\div$ keys respectively. Alternative notations for the division operator are the forward slash (/) or the bar. For example, $12 \div 6$ can also be written as 12/6 or $\frac{12}{6}$.

Brackets and Precedence Rules

Brackets are used to separate the minus sign of negative numbers from the arithmetic operations. They can also be used to remove ambiguity by grouping numbers together in a calculation. For example, $5 \times 6 - 4$ could be either

$$30 - 4 = 26 \qquad \text{or} \qquad 5 \times 2 = 10$$

To remove the ambiguity we rely on the precedence rules. The precedence rules state that in any calculation we evaluate 'Brackets' first followed by 'Of' (which is the same as multiplication), 'Division', 'Multiplication', 'Addition' and 'Subtraction' in that order – the so-called BODMAS rule. This means that:

$$\begin{aligned} 5 \times 6 - 4 &= 30 - 4 \\ &= 26 \end{aligned}$$

Brackets can be used to produce the alternative result. That is:

$$\begin{aligned} 5 \times (6 - 4) &= 5 \times 2 \\ &= 10 \end{aligned}$$

because the precedence rules state that brackets are evaluated first.

When evaluating expressions containing **nested brackets** then the innermost brackets are evaluated first. For example:

$$\begin{aligned} 3 \times (4 - 2 \times [5 - 1]) &= 3 \times (4 - 2 \times 4) \\ &= 3 \times (4 - 8) \\ &= 3 \times (-4) \\ &= -12 \end{aligned}$$

Notation

When brackets are used we can omit multiplication signs and replace the division sign by a line, so that:

$$5 \times (6 - 4) \text{ becomes } 5(6 - 5)$$

Also,

$$(24 - 11) \div 5 \text{ becomes } \frac{24 - 11}{5}$$

Notice further that:

$$5(6 - 4) = (5 \times 6) - (5 \times 4) = (6 - 4)5 = 2 \times 5 = 10$$

and

$$\frac{24-11}{5}=\frac{24}{5}-\frac{11}{5}=\frac{13}{5}$$

Factors and Prime Numbers

Factors

Any pair of numbers are called **factors** of their product. For example, 3 and 5 are factors of 15. All natural numbers can be written in terms of products of their factors. For example, 12 has the factors 1, 2, 3, 4, 6 and 12. Consequently,

$$\begin{aligned}12 &= 12 \times 1\\ &= 6 \times 2\\ &= 3 \times 4\end{aligned}$$

Prime Numbers

If a number possesses only two factors, namely itself and unity, then it is called a **prime number**. The first six prime numbers are 2, 3, 5, 7, 11 and 13. Unity is not considered to be a prime number.

Prime Factorization

Any natural number can be written as a product of its prime factors. For example, 60 has the prime factors 2, 3 and 5 and its prime factorization is

$$60 = 2 \times 2 \times 3 \times 5$$

Highest Common Factor (HCF)

If two numbers have factors in common then the largest of these common factors is called their **Highest Common Factor** (HCF). For example: 18 has the factors 1, 2, 3, 6, 9 and 18; 30 has the factors 1, 2, 3, 5, 6, 10, 15 and 30. Consequently, the numbers 1, 2, 3 and 6 are their common factors, the largest of which is 6. The HCF of 18 and 30 is 6.

Lowest Common Multiple (LCM)

The smallest whole number that each one of a pair of numbers divides into exactly a whole number of times is called their **Lowest Common Multiple** (LCM). It is found by factorizing each number into its prime factors. For example, 6 has the prime factorization 2×3 and 10 has the prime factorization 2×5. The LCM is, therefore, $2 \times 3 \times 5 = 30$. Notice that the common factor 2 is only used once to find the LCM. If it were used twice then a common multiple would be obtained but it would not be the lowest one.

Estimation and Rounding

Calculations involving the natural numbers can be performed using a calculator. However, as a check that the result of a calculation is reasonable, an estimation of the result can be made using the idea of rounding. For example, using a calculator the addition of 19 and 32 yields the result 51. However, if 19 is rounded up to 20 and 32 is rounded down to 30 then the estimated result is given by adding 20 to 30 to give 50 which is a reasonable approximation to the exact result. The rules for rounding to the nearest 10 are as follows:

- If the unit coefficient of a number is 4 or less then round the number down by replacing the unit coefficient by zero. For example, 54 rounds down to 50 to the nearest 10.
- If the unit coefficient of a number is 5 or more then round the number up by increasing the tens coefficient by one and replacing the unit coefficient by zero. For example, 76 rounds up to 80 to the nearest 10.

The principle is that if the number is less than halfway to the next multiple of 10 then round down. If it is halfway or more then round up. This principle also applies when rounding to the nearest 100, 1000, 10 000 or more. For example, 549 rounds down to 500 to the nearest 100 and 1501 rounds up to 2000 to the nearest thousand.

EXAMPLES

1 Evaluate: **(a)** $36 + 12 \div 2 - 5 \times 4$
(b) $(17 + 3) \div 4 - 6 \div (8 - 5)$
(c) $3 + \{5 - (6 \times [7 - 4]) + 8\}$
(d) $(-2) - (-3) + (-4) \times 8 - (-18)/(-6)$

(a) $36 + 12 \div 2 - 5 \times 4 = 36 + 6 - 20$
$= 22$

(b) $(17 + 3) \div 4 - 6 \div (8 - 5) = 20 \div 4 - 6 \div 3$
$= 5 - 2$
$= 3$

(c) $3 + \{5 - (6 \times [7 - 4]) + 8\} = 3 + [5 - (6 \times 3) + 8]$
$= 3 + (5 - 18 + 8)$
$= 3 + (-5)$
$= -2$

(d) $(-2) - (-3) + (-4) \times 8 - (-18)/(-6) = -2 + 3 - 32 - 3$
$= -34$

2 Find: **(a)** the prime factors and the prime factorization
(b) all the factors
(c) the HCF and LCM
of 84 and 512.

(a) The prime factors of a number are found by successively dividing the number by the prime numbers starting from 2 and increasing. For the number 84 we have:

2	84
2	42
3	21
7	7
	1

The prime factors are, therefore 2, 3 and 7 and the prime factorization is $84 = 2 \times 2 \times 3 \times 7$.

The only prime factor of 512 is 2 and the prime factorization is $512 = 2 \times 2 \times 2 \times 2 \times 2 \times 2 \times 2 \times 2 \times 2$.

(b) All the factors of a number are found from the various products of the prime factors. The factors of 84 are:

1, 2, 3, 4, 6, 7, 12, 14, 21, 28, 42, 84

The factors of 512 are

1, 2, 4, 8, 16, 32, 64, 128, 256, 512

(c) The HCF is 4. The LCM is found from the prime factorizations thus:

$$\begin{aligned} 512 &= 2 \times 2 \times 2 \times 2 \times 2 \times 2 \times 2 \times 2 \times 2 \\ 84 &= 2 \times 2 \qquad\qquad\qquad\qquad\qquad\qquad\quad \times 3 \times 7 \\ 10\,752 &= 2 \times 2 \times 2 \times 2 \times 2 \times 2 \times 2 \times 2 \times 2 \times 3 \times 7 \end{aligned}$$

10 752 is the required LCM.

3 Round to the nearest 10, 100 and 1000: **(a)** 345 **(b)** 654 **(c)** -1099

(a) 345 is rounded to 350 to the nearest 10, 300 to the nearest 100 and 0 to the nearest 1000.
(b) 654 is rounded to 650 to the nearest 10, 700 to the nearest 100 and 1000 to the nearest 1000.
(c) -1099 is rounded to -1100 to the nearest 10, -1100 to the nearest 100 and -1000 to the nearest 1000.

4 Estimate the value of $18 \times 21 - 99 \div 11$ and check the accuracy by evaluating the exact value.

By rounding the sum can be approximated to:

$$\begin{aligned} 20 \times 20 - 100 \div 10 &= 400 - 10 \\ &= 390 \end{aligned}$$

The exact value is 369.

EXERCISES

1 Evaluate: **(a)** $18 - 3 \times 5 + 10 \div 2$
(b) $(18 - 9) \div 3 - 9 \div (6 - 3)$
(c) $12 - \{4 + (8 \times [12 - 4]) - 5\}$
(d) $(-6) - (-12) + (-4) \times (-2) - (-18)/(-6)$

(a) $18 - 3 \times 5 + 10 \div 2 = 18 - *$
$= 8$

(b) $(18 - 9) \div 3 - 9 \div (6 - 3) = * - *$
$= 0$

(c) $12 - \{4 + (8 \times [12 - 4]) - 5\} = 12 - *$
$= *$

(d) $(-6) - (-12) + (-4) \times (-2) - (-18)/(-6) = -6 * 12 + * - *$
$= 11$

2 Find: **(a)** the prime factors and prime factorization
(b) the factors
(c) the HCF and LCM
of 144 and 66.

(a) By successive division by the prime numbers it is found that the prime factors of 66 are:

2, * and *

with a prime factorization of *. The prime factors of 144 are:

2 and *

with a prime factorization of *.
(b) The factors of 66 are:

1, 2, *, 6, *, 22, *, 66

and of 144 they are:

1, 2, *, 4, *, 8, *, 12, *, 18, *, 36, *, *, 144

(c) The HCF is * and the LCM is found from the prime factorizations:

$$\begin{aligned} 144 &= 2 \times * \times * \times * \times * \times * \\ 66 &= 2 \times \qquad\qquad\quad * \times * \\ 1584 &= 2 \times * \times * \times * \times * \times * \times * \end{aligned}$$

3 Round to the nearest 10, 100 and 1000: **(a)** 1044 **(b)** 815 **(c)** −549

(a) 1044 rounds to * to the nearest 10, 1000 to the nearest 100 and * to the nearest 1000.
(b) 815 rounds to 820 to the nearest 10, * to the nearest 100 and * to the nearest 1000.
(c) −549 rounds to −* to the nearest 10, −* to the nearest 100 and −1000 to the nearest 1000.

4 Estimate the value of $99 \times 101 - 49 \div 6$ and check the accuracy by evaluating the exact value.

This sum rounds to:

$$* \times * - * \div * = 9990$$

The exact value is 9990.8 to one dec. pl.

Unit 2 Decimal Numbers

Test yourself with the following:

1 Evaluate: (a) $1.25 - 3.04 \times 4.16 + 8.23 \div 9.45$
(b) $(12.56 - 3.52) \times 8.91 - 17.23 \div (17.42 - 3.94)$
(c) $(-13.45) \times (17.26) - (-4.01)/(-3.26)$

2 Write each of the following numbers to (i) three significant figures and (ii) two decimal places:
(a) 13.451 (b) −7.056 (c) 19.995

3 Evaluate: (a) $\left(\frac{5^3 4^{-2}}{5^2}\right) \times \left(\frac{1}{2^{-1}}\right)$ without a calculator
(b) $0.296^{-1.34}$ with a calculator

4 Multiply and divide each of the following numbers by 10, 100 and 1000:
(a) 1.001 (b) 0.0502 (c) −193.1124

5 Write each of the following numbers in scientific notation:
(a) 115.23 (b) 0.00123

6 In the following assume that the numbers have been obtained by measurement. Estimate a value and then use a calculator to find the value to the correct level of accuracy:

$$\frac{(0.11)(15.0156)}{(9.9938)^2}$$

Division of Integers

If an integer is divided by an integer that is not one of its factors the result will not be another integer. Instead, the result will lie between two integers. For example, using a calculator it is seen that

$$89 \div 8 = 11.125$$

which is a number greater than 11 but less than 12.

As with natural numbers the position of a numeral within the number indicates its value. Here the number 11.125 represents

1 ten + 1 unit + 1 tenth + 2 hundredths + 5 thousandths

where the decimal point shows the separation of the units from the tenths. Numbers written in this format are called **decimal numbers.**

Rounding

All the operations of arithmetic that we have used with the integers apply to decimal numbers. However, when performing calculations involving decimal numbers it is not uncommon for the end result to be a number with a large quantity of numerals after the decimal point. To make such numbers more manageable they can be rounded either to a specified number of **significant figures** or to a specified number of **decimal places.**

Significant Figures

Significant figures are counted from the first non-zero numeral on the left of the number. When the required number of significant figures have been counted off, the remaining numerals are deleted with the following proviso: if the first number deleted is a five or more the last significant numeral is increased by 1, otherwise it is left unaltered.

For example, 11.125 to three significant figures is 11.1 and to four significant figures is 11.13. As a further example, 1234 to two significant figures is 1200. Here the first two figures of 1234 are significant but the remainder are not and so are replaced by zeros. Again, rounding must be employed so 1253 to two significant figures is 1300.

Decimal Places

The same rule applies for rounding to a specified number of decimal places. The only difference is that counting off the numerals starts after the decimal point. For example, 123.4467 to one decimal place is 123.4 and to two decimal places it is 123.45.

Trailing Zeros

Sometimes zeros must be inserted within a number to satisfy a condition for a specified number of either significant figures or decimal places. For example, 12 645 to two significant figures is 13 000 and 13.1 to three decimal places is 13.100. These zeros are referred to as **trailing zeros.**

Raising a Number to a Power

Raising to a power is devised from repetitive multiplication. The power is also called an **index** and the number to be raised to a power is called the **base**. For example

$$10 \times 10 \times 10 = 10^3$$

Here the number 3 is the power (index) and 10 is the base.

The Laws of Powers

The arithmetic of powers is contained in the following set of rules:

- *Power Unity*
  Any number raised to power 1 equals itself.

  $$7^1 = 7$$

- *Multiplication of numbers and the addition of powers*
  If two numbers are each written as a given base raised to some power then the **product of the two numbers** is equal to the same base raised to the **sum of the powers.** For example, $81 = 3^4$ and $9 = 3^2$ so

  $$\begin{aligned} 81 \times 9 &= 3^4 \times 3^2 \\ &= (3 \times 3 \times 3 \times 3) \times (3 \times 3) \\ &= 3 \times 3 \times 3 \times 3 \times 3 \times 3 \\ &= 3^6 \\ &= 729 \end{aligned}$$

  So $3^4 \times 3^2 = 3^6 = 3^{4+2}$
  To multiply the powers are added.

- *Division of numbers and the subtraction of powers*
  If two numbers are each written as a given base raised to some power then the **quotient of the two numbers** is equal to the same base raised to the **difference of the powers.** For

example, $7776 = 6^5$ and $36 = 6^2$ so

$$\begin{aligned}7776 \div 36 &= 6^5 \div 6^2 \\ &= (6 \times 6 \times 6 \times 6 \times 6) \div (6 \times 6) \\ &= \frac{(6 \times 6 \times 6 \times 6 \times 6)}{6 \times 6} \\ &= (6 \times 6 \times 6) \times \frac{(6 \times 6)}{(6 \times 6)} \\ &= 6 \times 6 \times 6 \times 1 \\ &= 6 \times 6 \times 6 \\ &= 6^3 \\ &= 216\end{aligned}$$

So $6^5 \div 6^2 = 6^{5-2}$
To divide, the powers are subtracted.

- *Power zero*
Any number raised to the power 0 equals unity. For example

$$\begin{aligned}1 &= 5^1 \div 5^1 \\ &= 5^{1-1} \\ &= 5^0\end{aligned}$$

So $5^0 = 1$

- *Multiplication of powers*
If a number is written as a given base raised to some power then that number **raised to a further power** is equal to the base raised to the **product of the powers**. For example, $16 = 4^2$ so

$$\begin{aligned}(16)^3 &= (4^2)^3 \\ &= 4^2 \times 4^2 \times 4^2 \\ &= 4 \times 4 \times 4 \times 4 \times 4 \times 4 \\ &= 4^6 \\ &= 4096\end{aligned}$$

So $(4^2)^3 = 4^6 = 4^{2 \times 3}$
To raise to a power, the powers are multiplied.

- *Negative powers*
A number raised to a negative power denotes the reciprocal. For example

$$\begin{aligned}2^{3-3} &= 2^0 \\ &= 1\end{aligned}$$

and

$$2^{3-3} = 2^3 \times 2^{-3}$$

Therefore

$$2^{-3} = \frac{1}{2^3}$$

Notice that in any operation with different indices *the base must be the same*. We cannot use these laws to combine different powers with different bases. For example

$2^3 \times 4^5$ cannot be written as 8^8

We can, however, combine different bases with the same power. For example

$2^3 \times 4^3$ can be written as 8^3

Powers on a Calculator

Powers can be evaluated on a calculator by using the x^y key. For example: enter the number 3, press the x^y key, enter the number 2 and press =. The result is 9 which is 3^2.

Decimal Powers

Decimal powers can be handled with a calculator. For example, using the x^y key it can be shown that:

$1.2^{1.2} = 1.2446$ to four decimal places

Multiplication and Division by 10

If a decimal number is multiplied by 10 the decimal point is moved one place to the right. If a decimal number is divided by 10 the decimal point is moved one place to the left. For example, $123.45 \times 10 = 1234.5$ and $123.45 \div 10 = 12.345$. These facts enable decimal numbers to be written in a standardized format called **scientific notation**.

Scientific Notation, Estimation and Accuracy

Scientific Notation

Any decimal number can be written as a decimal number between 1 and 10 (called the **mantissa**) multiplied by the number 10 raised to an appropriate power (called the **exponent**). For example

623.4367 can be written as $6.234\,367 \times 10^2$

and

0.002 34 can be written as 2.34×10^{-3}

Estimation

When performing a calculation involving decimal numbers it is always a good idea to check that your result is reasonable and that an arithmetic blunder or an error in using a calculator has not been made. This is done using scientific notation. For example,

$$69.845 \times 196.574 = 6.9845 \times 10^1 \times 1.965\,74 \times 10^2$$
$$= 6.9845 \times 1.965\,74 \times 10^3$$

This product can be estimated for reasonableness as:

$$7 \times 2 \times 1000 = 14\,000$$

The answer obtained using a calculator is 13 729.71 to two decimal places, which is 14 000 when rounded to the nearest 1000 indicating that the exact result could be reasonably expected.

Accuracy

If a set of measurements is made with an accuracy to a given number of significant figures then any calculation involving those measurements will only be accurate to one significant figure more than the least number of significant figures in any measurement. For example, the edges of a rectangular piece of cardboard are measured as 12.5 cm and 33.43 cm respectively. The area of the rectangle is given as the product of these two lengths. Using a calculator this product is

$$417.875 \text{ cm}^2$$

Because one of the edges is only measured to three significant figures of accuracy this result is only accurate to four significant figures. It should therefore be read as 417.9 cm^2.

EXAMPLES

1 Evaluate: **(a)** $0.86 + 5.23 \times 9.12 - 3.27 \div 2.75$
(b) $(18.34 + 5.26) \times 9.84 - 12.04 \div (5.23 - 1.05)$
(c) $(-8.75)/(-2.43) - (-5.22) \times (-0.14)$

(a) $0.86 + 5.23 \times 9.12 - 3.27 \div 2.75 = 0.86 + 47.698 - 1.189$
$= 47.37$ to two dec. pl.

(b) $(18.34 + 5.26) \times 9.84 - 12.04 \div (5.23 - 1.05)$
$= 23.60 \times 9.84 - 12.04 \div 4.18$
$= 232.224 - 2.880$
$= 229.34$ to two dec. pl.

(c) $(-8.75)/(-2.43) - (-5.22) \times (-0.14) = 3.601 - 0.731$
$= 2.87$ to two dec. pl.

2 Write each of the following numbers to (i) three significant figures and (ii) two decimal places:
(a) 17.545 **(b)** 0.050 27 **(c)** −10 032.5

(a) 17.545 is written as 17.5 to three sig. fig. and as 17.55 to two dec. pl.
(b) 0.050 27 is written as 0.0503 to three sig. fig. and as 0.05 to two dec. pl.
(c) −10 032.5 is written as −10 000 to three sig. fig. and as −10 032.50 to two dec. pl.

3 Evaluate: **(a)** $\left(\dfrac{3^2 2^{-4}}{16^{-1}}\right) \div \left(\dfrac{3^3}{2^5}\right)$ without a calculator

(b) $1.02^{-3.21}$ with a calculator

(a)
$$\left(\frac{3^2 2^{-4}}{16^{-1}}\right) \div \left(\frac{3^3}{2^5}\right) = \frac{3^2 \times 16 \times 2^5}{2^4 \times 3^3}$$
$$= \frac{2^5}{3}$$
$$= \frac{32}{3}$$
$$= 10.67 \text{ to two dec. pl.}$$

(b) $1.02^{-3.21}$ with a calculator enter the number 1.02. Now press the x^y key followed by 3.21. Press the $\pm$ key to change the display to -3.21 and press $=$ to produce the result: 0.938 411 775....

4 Multiply and divide each of the following numbers by 10, 100 and 1000:

(a) 100.0101 **(b)** 0.002 01 **(c)** -1.1

(a) Multiplying 100.0101 by 10, 100 and 1000 produces 1000.101, 10 001.01 and 100 010.1 respectively. Dividing produces 10.001 01, 1.000 101 and 0.100 010 1 respectively.
(b) Multiplying 0.002 01 by 10, 100 and 1000 produces 0.0201, 0.201 and 2.01 respectively. Dividing produces 0.000 201, 0.000 020 1 and 0.000 002 01 respectively.
(c) Multiplying -1.1 by 10, 100 and 1000 produces -11, -110 and -1100 respectively. Dividing produces -0.11, -0.011 and -0.0011 respectively.

5 Write each of the following numbers in scientific notation:

(a) 144.3 **(b)** 0.000 01

(a) $144.3 = 1.443 \times 10^2$ **(b)** $0.000\,01 = 1.0 \times 10^{-5}$

6 Assuming that the following contains numbers obtained by measurement, use a calculator to find the value to the correct level of accuracy:

$$\frac{(13.261)^{0.5}(1.2)}{(5.632)^3}$$

Using a calculator the result is obtained as follows:

Calculate $(5.632)^3$ and store the result, 178.643 796 in the memory.
Clear the display and calculate the numerator to give 4.369 878 717.
Now press $\div$, retrieve the number stored in memory and press $=$ to produce the result: 0.024 461 407 which is 0.0245 to three sig. fig.

We round to three significant figures because this is one more than the number with the least significant figures in the calculation.

EXERCISES

1 Evaluate: **(a)** $2.51 - 4.30 \times 6.14 + 3.28 \div 5.49$

(b) $(65.26 - 2.53) \times 1.98 - 37.21 \div (27.41 - 4.93)$

(c) $(-53.41) \times (27.21) - (-1.42)/(-6.23)$

(a) $2.51 - 4.30 \times 6.14 + 3.28 \div 5.49 = 2.51 - *$
$= -23.29$ to two dec. pl.

(b) $(65.26 - 2.53) \times 1.98 - 37.21 \div (27.41 - 4.93) = * - *$
$= 122.55$ to two dec. pl.

(c) $(-53.41) \times (27.21) - (-1.42)/(-6.23) = * - *$
$= -1453.51$ to two dec. pl.

2 Write each of the following numbers to (i) three significant figures and (ii) two decimal places:
(a) 13.664 **(b)** −3.156 **(c)** 0.091 45

(a) 13.664 written to three sig. fig. is 13.* and to two dec. pl. is 13.6*.
(b) −3.156 written to three sig. fig. is −3.1* and to two dec. pl. is −3.1*.
(c) 0.091 45 to three sig. fig. is 0.091* and to two dec. pl. is 0.*.

3 Multiply and divide each of the following numbers by 10, 100 and 1000:
(a) 0.011 01 **(b)** 6.0234 **(c)** −25.010 23

(a) 0.011 01 multiplied by 10, 100 and 1000 is *, * and 11.01 respectively. Divided it is 0.001 101, * and * respectively.
(b) 6.0234 multiplied by 10, 100 and 1000 is *, 602.34 and * respectively. Divided it is *, 0.060 234 and * respectively.
(c) −25.010 23 multiplied by 10, 100 and 1000 is −250.1023, * and * respectively. Divided it is *, * and −0.025 010 23 respectively.

4 Evaluate: **(a)** $\left(\frac{6^2 3^{-2}}{36}\right) \times \left(\frac{3^2}{2^{-2}}\right)$ without a calculator
(b) $0.012^{-5.26}$ with a calculator

(a) The result is 4.
(b) The result is 1.269×10^{10} to four sig. fig.

5 Write each of the following numbers in scientific notation:
(a) 2734.12 **(b)** 0.000 01

(a) $2734.12 = 2.734\,12 \times 10^{*}$ **(b)** $0.000\,01 = * \times 10^{*}$

6 Assuming that the following calculation contains numbers obtained by measurement, use a calculator to find the value of the following to the correct level of accuracy:

$$\frac{(8.342)(-9.456)^3}{(3.25)^4}$$

the result is −63.220 675 55 which, to the required level of accuracy is *.

Unit 3 Rational Numbers

Test yourself with the following:

1 Reduce each of the following fractions to their lowest terms:
(a) $\frac{125}{625}$ (b) $\frac{52}{78}$ (c) $\frac{98}{42}$

2 Evaluate: (a) $\frac{2}{5}+\frac{1}{3}$ (c) $\frac{7}{5}\times\frac{15}{14}$ (e) $\frac{4}{5}/\frac{6}{7}$ (g) $8^{1/3}$
(b) $\frac{3}{4}-\frac{4}{5}$ (d) $\frac{2}{3}\div\frac{2}{5}$ (f) $\frac{1}{4}$ of $\frac{1}{2}$ (h) $16^{-3/4}$

3 Convert to decimals: (a) $\frac{2}{9}$ (b) $\frac{4}{7}$ (c) $\frac{53}{11}$

4 A compound of sand, cement and water is mixed in the proportions $\frac{1}{3}$, $\frac{1}{6}$ and $\frac{1}{2}$ respectively. Give the components in ratio form.

5 What is: (a) 16% of 125 (b) 13 as a percentage of 52

Fractions

A fraction is a part of a whole and is represented by one integer – the **numerator** – divided by another integer – the **denominator**. For example, $\frac{3}{5}$ represents three fifths of a whole. Because fractions are written as one integer divided by another – a **ratio** – they are called **rational numbers**. Fractions are either **proper**, **improper** or **mixed**.

- In a proper fraction the numerator is less than the denominator.
- In an improper fraction the numerator is greater than the denominator.
- A mixed fraction is in the form of an integer and a fraction. For example $3\frac{4}{7}$.

Equivalent Fractions

Fractions which represent the same fractional part of a whole but which have different numerators and denominators are said to be **equivalent**. For example, $\frac{2}{3}$ is equivalent to $\frac{20}{30}$ and $\frac{10}{15}$. Equivalent fractions can be formed from any fraction by multiplying or dividing both the numerator and denominator by the same number. For example,

$$\frac{7}{9}=\frac{7\times 5}{9\times 5}=\frac{35}{45}$$

Here we have multiplied the numerator and denominator by the same number, namely 5, to produce an equivalent fraction.

Reduction to Lowest Terms

When the numerator and the denominator of a fraction have no factors in common the fraction is said to be in its lowest terms. A fraction can be reduced to its lowest terms by dividing both numerator and denominator by their common factors to produce an equivalent fraction. For example,

$$\frac{6}{24}=\frac{2\times 3}{2\times 2\times 2\times 3}=\frac{1}{2\times 2}=\frac{1}{4}$$

dividing top and bottom by the common factors 2 and 3.

This is also referred to as **cancelling** – identical factors in the numerator and denominator cancel each other out.

Addition and Subtraction of Fractions

To add or subtract two fractions each must be converted to equivalent fractions with the same denominator. For example,

$$\frac{1}{2}+\frac{2}{3}=\frac{3}{6}+\frac{4}{6}=\frac{7}{6}$$

Notice that the smallest common denominator is the LCM of the original denominators.

Multiplication and Division of Fractions

Two fractions are multiplied by multiplying their numerators and denominators independently. For example,

$$\frac{5}{7}\times\frac{9}{2}=\frac{5\times 9}{7\times 2}=\frac{45}{14}$$

To divide one fraction by another invert the divisor and multiply. For example,

$$\frac{3}{4}\div\frac{5}{6}=\frac{3}{4}\times\frac{6}{5}=\frac{18}{20}=\frac{9}{10}$$

Using 'of'

The use of the word 'of' is very common when dealing with fractions. For example, half of 6 means 3. This can be written as:

$$\tfrac{1}{2} \text{ of } 6 = 3$$

From this it can be seen that the word 'of' can be substituted by the **multiplication** sign, so

$$\tfrac{1}{2} \text{ of } 6 = \tfrac{1}{2}\times 6 = 3$$

Roots and Fractional Powers

Just as plants generate from roots so a number can be generated by the repetitive multiplication of a root with itself. For example,

$$4\times 4\times 4\times 4\times 4\times 4 = 4^6 = 4096$$

and we call 4 the sixth root of 4096 – six 4s are multiplied together to produce 4096. Similarly,

$$16\times 16\times 16 = 16^3 = 4096$$

so we call 16 the third root – or cube root – of 4096. Fractional powers denote roots. For example, the notation for the sixth root of 4096 is

$$4096^{1/6}$$

because

$$(4096^{1/6})^6 = 4096^{6/6} = 4096^1 = 4096$$

Square Roots

A square root is a second root. For example $2 = 4^{1/2}$ is a second (or square) root of 4 because $2 \times 2 = 4$. An alternative notation for the square root is

$$\sqrt{4}$$

Uniqueness

Not all roots are unique. For example $+2$ and -2 are both square roots of 4 because $(-2) \times (-2) = 4$. Indeed, all even roots lack uniqueness. For example,

$$81^{1/4} = \pm 3$$

Throughout the remainder of this book, unless indicated otherwise, the even root of a number is taken to be the positive value only. For example,

$$\sqrt{81} = 9 \text{ and } 16^{1/4} = 2$$

Notice also that we cannot find the square root of a negative number amongst the numbers we have discussed so far. This is because the product of any number with itself is always a positive number.

Converting Fractions to Decimals

A fraction is converted to a decimal format by performing the division. For example,

$$\begin{aligned} \tfrac{1}{2} &= (1.0) \div 2 \\ &= 0.5 \end{aligned}$$

All fractions have a decimal form that contains a sequence of numerals that are repeated indefinitely. For example,

$$\begin{aligned} \tfrac{1}{3} &= 0.333\,333\,3\ldots \\ \tfrac{1}{7} &= 0.142\,857\,142\,857\ldots \\ \tfrac{2}{5} &= 0.400\,000\,00\ldots \end{aligned}$$

Repeated sequences of numerals are denoted by placing a dot over them. For example,

$$\tfrac{1}{3} = 0.\dot{3}$$

means that the 3 repeats indefinitely, and

$$\tfrac{1}{7} = 0.\dot{1}4285\dot{7}$$

means that the sequence 142857 repeats indefinitely. However, we write

$$\tfrac{2}{5} = 0.4$$

In this case we ignore the repeated zeros, write 0.4 and refer to it as a **terminating decimal**.

Ratios and Percentages

Ratios

If a whole is separated into two fractional parts where each fraction has the same denominator then the numerators of the two fractions form a **ratio**. For example, if a quantity of salad dressing contains $\frac{1}{3}$ oil and $\frac{2}{3}$ vinegar the oil and vinegar are said to be in the ratio 'one to two' – written as 1:2. If a compound is formed from $\frac{3}{4}$ of substance A, $\frac{1}{6}$ of substance B and $\frac{1}{12}$ of substance C then the fractional parts can be written in equivalent fractional form as:

$$\tfrac{9}{12} \text{ of A, } \tfrac{2}{12} \text{ of B and } \tfrac{1}{12} \text{ of C}$$

The components A, B and C are then in the ratio 9:2:1 – 9 parts of A to 2 parts of B to 1 part of C. Notice that the sum of the parts $9 + 2 + 1 = 12$ is the same as the denominator of the fractional parts.

Percentages

A percentage is a fractional part of a whole where the denominator of the fraction is equal to 100. For example, if 20 out of 100 eggs are brown then the fraction of brown eggs is $\frac{20}{100}$ or 20%. To find a percentage part of a quantity we multiply the quantity by the percentage written as a fraction. For example, 24% of 75 is

$$\tfrac{24}{100} \times 75 = 18$$

Percentages can also be changed to decimals. For example,

$$15\% = \tfrac{15}{100} = 0.15$$

EXAMPLES

1 Reduce each of the following fractions to their lowest terms:

(a) $\frac{36}{108}$ **(b)** $\frac{100}{125}$ **(c)** $\frac{72}{18}$

(a) $\dfrac{36}{108} = \dfrac{3 \times 12}{3 \times 3 \times 12} = \dfrac{1}{3}$

(b) $\dfrac{100}{125} = \dfrac{4 \times 25}{5 \times 25} = \dfrac{4}{5}$

(c) $\dfrac{72}{18} = \dfrac{6 \times 4 \times 3}{6 \times 3} = 4$

2 Evaluate: **(a)** $\frac{1}{4} + \frac{1}{6}$ **(b)** $\frac{2}{3} - \frac{1}{2}$ **(c)** $\frac{8}{3} \times \frac{9}{11}$ **(d)** $\frac{3}{4} \div \frac{2}{5}$ **(e)** $\frac{7}{11}/\frac{5}{22}$ **(f)** $\frac{1}{3}$ of $\frac{1}{5}$ **(g)** $16^{1/4}$ **(h)** $8^{-2/3}$

(a) $\frac{1}{4} + \frac{1}{6} = \frac{3}{12} + \frac{2}{12} = \frac{5}{12}$

(b) $\frac{2}{3} - \frac{1}{2} = \frac{4}{6} - \frac{3}{6} = \frac{1}{6}$

(c) $\frac{8}{3} \times \frac{9}{11} = \frac{8 \times 9}{3 \times 11} = \frac{72}{33} = \frac{24}{11}$

(d) $\frac{3}{4} \div \frac{2}{5} = \frac{3}{4} \times \frac{5}{2} = \frac{15}{8}$

(e) $\frac{7}{11}/\frac{5}{22} = \frac{7}{11} \times \frac{22}{5} = \frac{154}{55} = \frac{14}{5}$

(f) $\frac{1}{3}$ of $\frac{1}{5} = \frac{1}{3} \times \frac{1}{5} = \frac{1}{15}$

(g) $16^{1/4} = \pm 2$

(h) $8^{-2/3} = (8^{1/3})^{-2} = 2^{-2} = \frac{1}{4}$

3 Convert to decimals: **(a)** $\frac{7}{9}$ **(b)** $\frac{5}{13}$ **(c)** $\frac{88}{15}$

(a) $\frac{7}{9} = 0.777\,777\,7\ldots = 0.\dot{7}$
(b) $\frac{5}{13} = 0.384\,615\,384\,615\ldots = 0.\dot{3}8461\dot{5}$
(c) $\frac{88}{15} = 5.866\,666\,66\ldots = 5.8\dot{6}$

4 Three sections A, B and C of a book have 52, 26 and 13 pages respectively. Give the number of pages per section in ratio form.

The total number of pages is 91 and sections A, B and C form $\frac{52}{91}$, $\frac{26}{91}$ and $\frac{13}{91}$ respectively of the book. These fractions can be reduced to the equivalent fractions $\frac{4}{7}$, $\frac{2}{7}$ and $\frac{1}{7}$ respectively. The pages of sections A, B and C are therefore in the ratio 4:2:1.

5 What is: **(a)** 48% of 250 **(b)** 15 as a percentage of 20

(a) 48% of 250 $= \frac{48}{100} \times 250$
$= 120$
(b) 15 as a percentage of 20 is $(\frac{15}{20}) \times 100 = 75\%$

EXERCISES

1 Reduce each of the following fractions to their lowest terms:
(a) $\frac{162}{243}$ **(b)** $\frac{94}{235}$ **(c)** $\frac{27}{18}$

(a) $\dfrac{162}{243} = \dfrac{2 \times * \times *}{3 \times * \times *} = \dfrac{2}{3}$

(b) $\dfrac{94}{235} = \dfrac{* \times *}{* \times *} = \dfrac{2}{5}$

(c) $\dfrac{27}{18} = \dfrac{* \times * \times *}{* \times * \times *} = \dfrac{3}{2}$

2 Evaluate: **(a)** $\frac{1}{7} + \frac{2}{3}$ **(c)** $\frac{3}{7} \times \frac{2}{5}$ **(e)** $\frac{3}{7}/\frac{2}{5}$ **(g)** $243^{-1/5}$
(b) $\frac{4}{9} - \frac{2}{3}$ **(d)** $\frac{5}{8} \div \frac{3}{4}$ **(f)** $\frac{1}{5}$ of $\frac{2}{3}$ **(h)** $9^{-5/2}$

(a) $\frac{1}{7} + \frac{2}{3} = \frac{*}{21} + \frac{*}{21} = \frac{17}{21}$
(b) $\frac{4}{9} - \frac{2}{3} = \frac{4}{9} - * = -\frac{2}{9}$
(c) $\frac{3}{7} \times \frac{2}{5} = \frac{* \times *}{*} = \frac{6}{35}$
(d) $\frac{5}{8} \div \frac{3}{4} = \frac{5}{8} \times * = \frac{5}{6}$
(e) $\frac{3}{7}/\frac{2}{5} = * \times * = \frac{15}{14}$
(f) $\frac{1}{5}$ of $\frac{2}{3} = * \times * = \frac{2}{15}$
(g) $243^{-1/5} = (243^*)^{-1} = *^{-1} = \frac{1}{3}$
(h) $9^{-5/2} = (9^*)^{-*} = \pm *^{-*} = \pm\frac{1}{243}$

3 Convert to decimals: **(a)** $\frac{8}{11}$ **(b)** $\frac{7}{3}$ **(c)** $\frac{9}{7}$

(a) $\frac{8}{11} = 0.***\,***\ldots = 0.\dot{7}\dot{2}$
(b) $\frac{7}{3} = 2.***\,***\ldots = 2.\dot{3}$
(c) $\frac{9}{7} = 1.***\,***\ldots = 1.\dot{2}8571\dot{4}$

4 A mixture of three chemical compounds A, B and C consists of $\frac{1}{5}$ of A, $\frac{2}{3}$ of B and the remainder of C. Give the component compounds in ratio form.

The compounds A, B and C are in the fractional parts

$\frac{1}{5}$, $\frac{2}{3}$ and $\frac{*}{*}$ respectively

In terms of common denominators these fractions are

*, * and *

Accordingly the compounds A, B and C are in the ratio

3:10:2

5 What is: **(a)** 12.5% of 32 **(b)** 9 as a percentage of 27

(a) 12.5% of 32 = $\frac{*}{100}$ * 32 = 4
(b) 9 as a percentage of 27 is $(\frac{*}{*}) \times * = 33.\dot{3}\%$

Unit 4 Real and Imaginary Numbers

Test yourself with the following:

1 What is the difference between the decimal form of a rational number and the decimal form of an irrational number?

2 What is a real number?

3 What is an imaginary number?

4 Simplify: (a) $\sqrt{-9}$ (b) j^4 (c) $(-81)^{1/2}$ (d) $\left(\frac{1}{j}\right)^3$ (e) $(-2)^{-3/2}$

Irrational Numbers

All rational numbers have a decimal format that either terminates or contains an infinitely repeated, finite sequence of numerals. By inference, any decimal number that neither terminates nor contains such a repeated sequence of numerals is not a rational number. It is called an **irrational number** – it is a number that cannot be written as one integer divided by another integer. Irrational numbers are not rare, indeed there are many more irrational than rational numbers. They do, however, present a problem. Because of their nature we cannot write them as a fraction: neither can we write down their exact decimal form as there is an infinite number of numerals after the decimal point. Instead, we devise other notations such as π, e and $\sqrt{2}$. The irrational number π (pi) is a number we shall meet when we measure the circumference of a circle. This number is on your calculator and is given as:

$\pi = 3.1416$ to four decimal places

The irrational number e is called the **exponential number**. It also is on your calculator and is given as:

e = 2.7183 to four decimal places

The Real Line

All the irrational numbers combined with all the rational numbers form the **real numbers.** When all the real numbers are plotted on a line every point on the line corresponds to a number – there are no gaps and no overlaps, the line is complete and the numbers are said to be continuous. This continuous line is referred to as the **real line.**

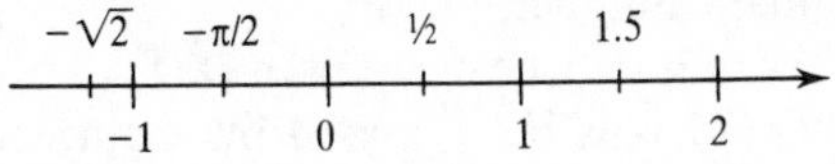

$-\sqrt{2}$ $-\pi/2$ ½ 1.5

−1 0 1 2

Figure 1.3

Using Real Numbers

An important point to realize is that all our calculations, all our measurements and all the numbers we use in our calculator are rational numbers. When we attempt to use an irrational number in a calculation we must, of necessity, round it off to a given number of decimal places – we approximate the irrational number to a rational number. While we may say that the circumference of a circle is πd where d is the diameter of the circle it is impossible either to calculate it or to measure it exactly.

Imaginary Numbers

The arithmetic of the real numbers poses two fundamental problems. The first problem concerns division by zero. This problem cannot be satisfactorily resolved by defining what is meant by such a division and so we resolve the problem by saying that division by zero is just not defined. In effect, we cannot do it. The second problem concerns the square root of a negative number. This problem can be resolved, but in doing so we open up a whole new panoply of numbers.

To reduce the problem to its essentials we consider the square root of minus unity:

$$(-1)^{1/2}$$

To resolve the problem we state that such a number exists and we give it the numeral j, where:

$$j^2 = -1$$

so that

$$j = (-1)^{1/2}$$

Now we find that we can define the square root of any negative real number. For example,

$$(-4)^{1/2} = (4 \times (-1))^{1/2} = (4j^2)^{1/2} = 4^{1/2}(j^2)^{1/2} = \pm 2j$$

Notice also that:

$$j^3 = j^2 j = -j$$

$$j^4 = (j^2)^2 = (-1)^2 = 1$$

Also, because

$$j^2 = -1 \text{ then } j = \frac{-1}{j}$$

Having solved the problem of defining the square root of a negative real number we now have a further problem. What sort of number is j? It is not a real number – the real line is complete – there are no points available on the real line where we could put j. The conclusion is that it is a new, different type of number. We call it an **imaginary number**. Indeed, any number of the form aj where a is a real number is called an imaginary number.

Note that in other mathematics texts this imaginary unit may be denoted by i; $i^2 = -1$. The notation i is the traditional symbol used by mathematicians but j is used by engineers and others to avoid confusion when the symbol i is used to denote electric current.

Terminology

The words real and imaginary as applied to these two different types of number are unfortunate choices of terminology. The word real has a connotation in the English language of solid reality, whereas the word imaginary connotes something elusive and unreal. There is nothing real in this sense about the real numbers, they are every bit as much an invention of the human imagination as the imaginary numbers are. It might be argued that real numbers can be used to count real objects whereas imaginary numbers cannot. But this is not really true. We have realized that we cannot quantify irrational numbers so we cannot use them in any counting procedure either. Any problem you may have in trying to understand what sort of number j is can best be resolved by just thinking of it as a new type of number that falls out of a problem concerning the real numbers but which is not itself a real number. Indeed, at the end of the next chapter we shall see yet another type of number emerging called a **complex number**.

EXAMPLES

1 Which of the following are rational numbers?

(a) 0.125 **(b)** $0.12\dot{5}$ **(c)** $\sqrt{5}$

(a) $0.125 = \frac{125}{1000}$ therefore it is rational.
(b) $0.12\dot{5} = 0.125\,555\,555\ldots$ a decimal with an infinitely repeating sequence of 5s, therefore it is rational.
(c) $\sqrt{5}$ cannot be written as one integer divided by another therefore it is not rational – it is irrational.

2 Simplify: **(a)** $\sqrt{-16}$ **(b)** $(-3)^{-3/2}$ **(c)** j^5 **(d)** $\left(\frac{2}{\text{j}}\right)^3$

(a) $\sqrt{-16} = \sqrt{[16 \times (-1)]} = (\sqrt{16})(\sqrt{-1}) = \pm 4\text{j}$
(b)
$$\begin{aligned}(-3)^{-3/2} &= [(-3)^{1/2}]^{-3}\\ &= [\pm(\sqrt{3})\text{j}]^{-3}\\ &= [\pm 3\sqrt{3}\text{j}^3]^{-1} \text{ because } (\sqrt{3})^3 = (\sqrt{3})^2\sqrt{3} = 3\sqrt{3}\\ &= \frac{\pm 1}{3\sqrt{3}\text{j}^3}\\ &= \frac{\pm 1}{3\sqrt{3}(-\text{j})} \text{ because } \text{j}^3 = \text{j}^2\text{j} = -\text{j}\\ &= \frac{\pm \text{j}}{3\sqrt{3}} \text{ because } \text{j}^2 = -1 \text{ and so } \text{j} = \frac{1}{-\text{j}}\end{aligned}$$

(c) $j^5 = j^2 \times j^2 \times j$
$= (-1) \times (-1) \times j$
$= j$

(d) $\left(\frac{2}{j}\right)^3 = \frac{8}{j^3} = \frac{-8}{j} = 8j$

EXERCISES

1 Which of the following are rational numbers?
(a) 3.2424 **(b)** $3.\dot{2}\dot{4}$ **(c)** $\sqrt{7}$

(a) $3.2424 = \frac{324\,24}{*}$ therefore it is *.
(b) $3.\dot{2}\dot{4}$ = * a decimal with an infinitely repeating sequence of *s, therefore it is *.
(c) $\sqrt{7}$ can/cannot be written as one integer divided by another therefore it is *.

2 Simplify: **(a)** $\sqrt{-81}$ **(b)** $(-2)^{-5/2}$ **(c)** j^7 **(d)** $\left(\frac{-3}{j}\right)^5$

(a) $\sqrt{-81} = \sqrt{[* \times (-1)]} = (\sqrt{*})(\sqrt{-1}) = \pm *$
(b) $(-2)^{-5/2} = [(-2)^{1/2}]^*$
$= [\pm(\sqrt{*})*]^*$
$= \pm *$
(c) $j^7 = j^* \times j^* \times j^* \times *$
$= *$
(d) $\left(\frac{-3}{j}\right)^5 = \frac{*}{j^*} = *$

Module 1: Further exercises

1 Evaluate: (a) $13 - 5 \times 3 + 8 \div 2$
(b) $(23 - 2) \times 8 - 16 \div (11 - 3)$
(c) $7 - \{9 + (2 \times [8 - 1]) - 3\}$
(d) $(-12) - (-8) + (-7) \times 4 - (-2)/(-1)$

2 Find: (a) the factors
(b) the prime factors
(c) the HCF and LCM
of 255 and 1020.

3 Round to the nearest 10, 100 and 1000: (a) 234 (b) 695 (c) −349

4 Estimate the value of $49 \times 39 - 81 \div 19$ and check the accuracy by evaluating the exact value.

5 Evaluate: (a) $5.21 - 4.02 \times 1.64 + 3.28 \div 4.59$
(b) $(62.16 - 5.23) \times 1.89 - 72.13 \div (41.72 - 4.39)$
(c) $(-13.45) \times (17.26) - (-4.01) \div (-3.26)$

6 Write each of the following numbers to (i) three significant figures and (ii) two decimal places:
(a) 53.354 (b) −9.045 (c) 29.949

7 Evaluate: (a) $(4^{-3} \times 3^{-2} \div 16^{-1}) \times (3^3 \div 2^{-3})$ without a calculator
(b) $5.243^{-9.061}$ with a calculator

8 Multiply and divide each of the following numbers by 10, 100 and 1000:
(a) 1.01 (b) 0.0283 (c) -172.1012

9 Write each of the following numbers in scientific notation:
(a) 325.67 (b) 0.005 24

10 Assuming that the following contains numbers obtained by measurement, estimate a value and then use a calculator to find the value to the correct level of accuracy:

$$\frac{8.01^3}{(9.3214)(7.528)}$$

11 Reduce each of the following fractions to their lowest terms:
(a) $\frac{75}{175}$ (b) $\frac{51}{68}$ (c) $\frac{55}{45}$

12 Evaluate: (a) $\frac{1}{8}+\frac{3}{4}$ (b) $\frac{5}{8}-\frac{2}{3}$ (c) $\frac{4}{3}\times\frac{9}{5}$ (d) $\frac{6}{7}\div\frac{3}{14}$ (e) $\frac{5}{9}/\frac{2}{3}$ (f) $\frac{1}{9}$ of $\frac{27}{35}$
(g) $27^{-1/3}$ (h) $25^{-3/2}$

13 Convert to decimals: (a) $-\frac{3}{7}$ (b) $\frac{5}{11}$ (c) $\frac{22}{9}$

14 A fast-food outlet sells hamburgers, fish and chips in the following proportions. Hamburgers account for $\frac{1}{4}$ of all sales, fish account for $\frac{1}{5}$ of all sales and chips account for the remainder of sales. Give the component sales in ratio form.

15 What is: (a) 37.5% of 32 (b) 9 as a percentage of 45

16 How are irrational numbers treated within an arithmetical evaluation?

17 Which real numbers can we never manipulate in their decimal form?

18 Simplify into the form aj where a is a real number and $\mathrm{j}^2 = -1$;

(a) $\sqrt{-25}$ (b) $(-625)^{1/2}$ (c) $(-4)^{-3/2}$ (d) j^5 (e) $\left(\frac{-1}{\mathrm{j}^3}\right)^2$

Module 2: Data Handling

Aim: To demonstrate how to manipulate numerical data from its collection, through its organization to its analysis.

Objectives: When you have read this module you will be able to:

- ▶▶ Understand that the process of measurement is subject to error.
- ▶▶ Estimate the likely error in a measurement.
- ▶▶ Estimate the likely error when evaluating an expression using measured values.
- ▶▶ Organize data prior to its being statistically analysed.
- ▶▶ Evaluate averages and measures of the spread of the data about the mean value of the data.
- ▶▶ Contrast two sets of data by comparing their means and spreads about the mean.

Unit 1 Measurement

Test yourself with the following:

1 A rectangular block has three sides of measured length 15.01 m, 52.32 m and 0.82 m. Give the volume as accurately as possible.

2 Find both the fractional and percentage errors in the following. The first number is the actual value of a parameter and the second is its measured value.

(a) 9, 9.09 (b) 16.4, 16.0 (c) −3.05, −2.95

3 If two parameters a and b with actual lengths 121 cm and 314 cm are measured as 119.8 cm and 315.7 cm respectively find both the fractional and percentage errors in the calculation of:

(a) ab (b) $\frac{a}{b}$ (c) $a+b$ (d) $a-b$

4 If the radius r of a sphere is measured to an accuracy of 4.5% what is the accuracy of the calculated

(a) volume $= \frac{4}{3}\pi r^3$ (b) surface area $= 4\pi r^2$

5 Three sides of a rectangular block are measured to be 101 ± 0.4 cm, 53 ± 0.2 cm, 12.5 ± 0.1 cm. What is

(a) the error (b) the fractional error (c) the percentage error

in the calculated volume of the rectangular block?

6 Speed is measured by dividing the distance travelled by the time taken to cover that distance. If a car travels 35.6 km in 25 minutes what is the speed in km per hr? If the measured distance is subject to an error of 10% and the measured time taken is subject to an error of 10 seconds per measured hour what is the error in the calculated value of the speed?

Measuring Apparatus

All the measurements that we make are limited in their accuracy. No piece of measuring apparatus can be more accurate than the finest division on its scale. For example, a metre rule graded with divisions of a millimetre can only measure lengths to an accuracy of a millimetre.

Sources of Error

There are two sources of error in any measurement that we need to be aware of: **accuracy** and **natural error**.

Accuracy

The accuracy of a measured parameter depends upon the fineness of the measuring scale. For example, if a kilogram scale is graduated in grams it can only be used for measurements as fine as a gram. This sort of error we can estimate.

Natural Error

Natural error arises from incorrect reading of a measuring scale. This can be due to any one of a number of causes ranging from poor experimental method to poor eyesight. We cannot make a reasonable estimate for this sort of error so it will not be considered further.

Compounding Errors

When performing a calculation using numbers obtained from measurements, the individual errors in each number will accumulate to produce a result that is inaccurate. By defining the **fractional error** in a given measurement we can estimate how these errors will accumulate arithmetically. (In what follows we shall assume that the actual error in a measurement is known.)

Fractional Errors

If a quantity x is measured to be x' then the error is given as

$$e = x' - x$$

so that

$$x' = x + e$$

If we define the fractional error f where:

$$f = \frac{e}{x} \text{ so that } e = fx$$

then:

$$x' = x + fx$$
$$= x(1 + f)$$

The percentage error is then $100f$. For example, if a length of 25 cm is measured as 25.1 cm then the error is 0.1 cm, the fractional error is $0.1/25 = 0.004$ and the percentage error is $(0.1/25) \times 100\% = 0.4\%$.

Error in a Product

If a quantity Q is given as $Q = ab$, where a and b are measured with fractional errors f_a and f_b respectively, the measured value of Q is Q_m where:

$$Q_m = a(1 + f_a)b(1 + f_b)$$
$$= ab(1 + f_a + f_b + f_a f_b)$$

Assuming that products of fractional errors are small enough to neglect, the measured value of Q is taken to be

$$Q_m = ab(1 + f_a + f_b) = ab + ab(f_a + f_b)$$

The error in the measured value of Q is then

$$Q_m - Q = ab(f_a + f_b)$$

The fractional error in the measured value of Q is equal to the sum of the fractional errors of a and b.

For example, if the two sides of a rectangle are measured to 5% accuracy then the resultant area – the product of the two side lengths – will be subject to 5% + 5% = 10% inaccuracy.

Error in a Quotient

If a quantity Q is given by $Q = a/b$ where a and b are measured, the error in the measured value of Q can be shown to be obtained by subtracting the fractional errors of a and b:

$$Q_m - Q = \frac{a}{b}(f_a - f_b)$$

For example, if the mass of a body is measured to 6% accuracy and its volume is measured to 2% accuracy, its resultant calculated density – mass divided by volume – is accurate to 6% – 2% = 4%.

Error in a Sum or Difference

If a quantity Q is given by $Q = a + b$ where a and b are measured, the measured value of Q is

$$Q_m = (a + e_a) + (b + e_b)$$
$$= (a + b) + e_a + e_b$$

The error in the measured value of Q is then

$$Q_m - Q = e_a + e_b$$

the sum of the individual errors in a and b respectively.

For example, if an area of 12 square units is measured as 12.2 square units and a second area of 18 square units is measured as 17.5 square units, the error obtained in adding the two areas is simply

$$0.2 + (-0.5) = -0.3$$

This gives a fractional error of

$$\frac{-0.3}{12 + 18} = \frac{-0.3}{30} = -0.01$$

The percentage error is then

$$(-0.01)(100) = -1\%$$

Estimating Likely Errors in a Measurement

There are a number of different methods of estimating errors as well as a number of different experimental practices designed to reduce errors. For our purposes, however, we shall define the possible error in any given measured value as ± 1 unit of the finest graduation on the scale of the measuring apparatus being used. For example, if a stopwatch is used to time events and the stopwatch scale is marked in subdivisions of 0.1 second then any measurement will be deemed to be accurate to ± 0.1 second. Because of this more than one value will be obtained for the error. For example, we have seen that for the quotient $Q = a/b$ the error in the measured value Q_m is given as:

$$Q_m - Q = (a/b)(f_a - f_b)$$

As a result, if the mass of a body is measured to $\pm 6\%$ accuracy and its volume to $\pm 2\%$ accuracy, there are four possible accumulated errors in the resultant calculated density – mass divided by volume:

f_a	f_b	total error $(f_a - f_b)$
+6%	+2%	4%
+6%	−2%	8%
−6%	−2%	−4%
−6%	+2%	−8%

The accuracy of the calculated density is then taken to be the worst possible case, namely $\pm 8\%$.

EXAMPLES

1 A rectangular block has three sides of measured length 10.15 m, 25.23 m and 1.1 m. Give the volume as accurately as is allowed.

The volume is obtained by multiplying the lengths of the three sides together. So that

$$\text{Volume} = 10.15 \times 25.23 \times 1.1$$

$$= 281.692\,95 \text{ cubic metres (by calculator)}$$

The least number of significant figures used in the components of the calculation is two in the number 1.1. As a consequence the greatest number of significant figures allowed in the answer is three. This gives the most accurate estimation of the volume as 282 m^3 (obtained by rounding).

2 Find both the fractional and percentage errors in the following. The first number is the actual value of a parameter and the second is its measured value.

(a) 20, 20.05 **(b)** 8, 7.84 **(c)** -0.01, $+0.01$

(a) The error is 0.05, the fractional error is $0.05/20 = 0.0025$ and the percentage error is $0.0025 \times 100\% = 0.25\%$.
(b) The error is -0.16, the fractional error is $-0.16/8 = -0.02$ and the percentage error is $-0.02 \times 100\% = -2\%$.
(c) The error is $+0.02$, the fractional error is $0.02/-0.01 = -2$ and the percentage error is -200%.

3 If two parameters a and b, with actual lengths 12 cm and 34 cm, are measured as 11.8 cm and 34.7 cm respectively, find both the fractional and percentage errors in the calculation of:

(a) ab **(b)** $\dfrac{a}{b}$ **(c)** $a + b$ **(d)** $a - b$

The error in a is -0.2 with fractional error $-0.2/12 = -0.0167$. The error in b is 0.7 with fractional error $0.7/34 = 0.0206$.

(a) The fractional error in ab is $(-0.0167) + (0.0206) = 0.0039$. The percentage error is 0.39%.
(b) The fractional error in a/b is $(-0.0167) - (0.0206) = -0.0373$. The percentage error is -3.73%.
(c) The fractional error in $a + b$ is

$$\frac{-0.2 + 0.7}{12 + 34} = 0.011$$

The percentage error is 1.1%.
(d) The fractional error in $a - b$ is

$$\frac{-0.2 - 0.7}{12 + 34} = -0.0196$$

The percentage error is -1.96%.

4 The pressure P, the volume V and the temperature T of a gas are said to obey the formula

$$\frac{PV}{T} = K$$

If P is measured to an accuracy of 1%, V is measured to an accuracy of 5% and T is measured to an accuracy of 3%, find the resultant accuracy in the computed value of K in the formula.

If p, v and t represent the fractional errors in the measurements of P, V and T respectively then the measured value of K is

$$K_m = \left(\frac{PV}{T}\right)(1 + p + v - t)$$

because p and v are added to reflect the product PV and t is subtracted to reflect the division by T. The fractional error in K is then $p + v - t$. This means that the percentage error in K is

$$(100p + 100v - 100t)\% = (1 + 5 - 3)\% = 3\%$$

EXERCISES

1 A rectangular block has three sides of measured length 20.75 m, 35.62 m and 20 cm. Replace each $*$ by the appropriate number in the calculation of the volume.

The volume is obtained by multiplying the lengths of the three sides together. So that

$$\begin{aligned} \text{Volume} &= * \times * \times 0.2 \\ &= 1{*}7.{*}2{*} \text{ cubic metres (by calculator)} \end{aligned}$$

The least number of significant figures used in the components of the calculation is $*$ in the number 0.2 m. As a consequence, the greatest number of significant figures allowed in the answer is $*$. This gives the most accurate estimation of the volume as $1{*}{*}\ \text{m}^3$ (obtained by rounding).

2 Fractional and percentage errors are to be found in each of the following. The first number is the actual value of a parameter and the second is its measured value. Replace each $*$ by the appropriate number.

(a) 7, 7.35 **(b)** 26, 25.4 **(c)** -12.8, -13.2

(a)
$$\begin{aligned} \text{Error} &= * \\ \text{Fractional error} &= */7 \\ &= *.* \\ \text{Percentage error} &= *.* \times 100\% \\ &= * \end{aligned}$$

(b)
$$\begin{aligned} \text{Error} &= -* \\ \text{Fractional error} &= -*/* \\ &= -*.* \\ \text{Percentage error} &= -*.* \times 100\% \\ &= *\% \end{aligned}$$

(c)
$$\begin{aligned} \text{Error} &= * \\ \text{Fractional error} &= */* \\ &= *.* \\ \text{Percentage error} &= *.* \times 100\% \\ &= *\% \end{aligned}$$

3 Two parameters a and b with actual lengths 21 cm and 19 cm are measured as 20.8 cm and 19.4 cm respectively. Replace each $*$ by the appropriate number in the following calculations to find both the fractional and percentage errors:

(a) ab **(b)** $\dfrac{a}{b}$ **(c)** $a + b$ **(d)** $a - b$

Error in a is $-0.*$ cm and error in b is $0.*$ cm. Fractional error in $a = -0.*/21 = -0.0095$ and fractional error in $b = 0.*/19 = 0.021$.

(a) Fractional error in $ab = -* + 0.021 = *$
Percentage error in $ab = * \times 100\% = *.{*}5\%$

(b) Fractional error in $a/b = -0.0095 - * = *$
Percentage error in $a/b = *.*\%$

(c) Fractional error in $a + b$ is

$$\frac{* + *}{* + *} = 0.00***$$

Percentage error is $*\%$.
(d) Fractional error in $a - b$ is

$$\frac{* - *}{*} = -*$$

Percentage error is $-*.*\%$.

4 In a certain formula the quantities X, Y and Z are related by the equation

$$\frac{XY^2}{Z} = \text{constant}$$

If X is measured to an accuracy of 2%, Y is measured to an accuracy of 4% and Z is measured to an accuracy of 1%, find the resultant accuracy in the computed value of the constant in the formula.

If x, y and z represent the fractional errors in the measurements of X, Y and Z respectively then the measured value of the constant is

$$\text{constant} = \left(\frac{XY^2}{Z}\right)(1 + * + 2y - *)$$

The fractional error in the constant is then $* + 2y - *$. This means that the percentage error in the constant is

$$(100* + 200y - 100*)\% = (2 + * - 1)\% = *\%$$

Unit 2 Calculation

Test yourself with the following:

1 Perform the following calculations using a calculator:

(a) $92.03 + 8.95$ (d) $\dfrac{123.45}{9.23}$

(b) $5.67 - 0.93$ (e) $\dfrac{6.46 - 1.484}{9.3}$

(c) 18.65×7.60 (f) $\dfrac{(5.753 \times 24.1) - 6.77}{8.03}$

2 Use the function buttons on your calculator to perform the following:

(a) $\dfrac{1}{15.76}$ (b) 4.02^2 (c) $1.111^{1/2}$ (d) $9.09^{3.92}$

3 Use the calculator memory to perform the following calculations:

(a) $\dfrac{26.43 \times 6 - 18 \times 5.7}{26.7}$ (b) $\dfrac{(4.325 \times 7.439) - (2.104 \times 6.398)}{39.5}$

4 Find the value of e^{π} to three decimal places.

5 Use your calculator to evaluate:

(a) The number of seconds in a century (1 year = 365 days).

(b) The speed in inches per second of a person walking at 4 miles per hour. There are 36 inches in a yard and 1760 yards in a mile.

(c) The cost of 53 sheets of plywood each measuring 3.125 m^2 if the price is £0.35 per square foot (1 square foot = 929 cm^2).

6 A radioactive isotope loses mass according to the formula

$$m = m_0 2^{-t/T}$$

where m_0 g is the mass at time $t = 0$. m g is the mass at time t s and T s is the 'half-life' – the time taken for a given mass to decay to half its original amount.

If the half-life of an isotope is 1.25×10^6 s and the original mass is 34.56 g how much is left after 2 weeks?

Calculators

It is assumed that most calculations will be performed using a calculator. The typical electronic hand calculator contains three types of key. The **number keys**, the **arithmetic operations keys** and the **function keys**. In addition the calculator has a **memory** that is capable of storing a single number.

Number Keys

The number keys range from 0 to 9 and by pressing them in sequence natural numbers can be displayed on the calculator screen. There is also a decimal point key for decimal numbers. Some calculators will even display and manipulate fractions.

Operations Keys

The operations keys permit addition, subtraction, multiplication, division and raising to a power. Using these, arithmetic operations can be performed on the numbers entered into the calculator. For example, the key marked:

x^y

is used to raise the displayed number x to the power y. Try it:

enter the number 4

press the x^y key

enter the number 2 and press =

The result is 16 which is 4 raised to the power 2.

Function Keys

The function keys permit a variety of complicated operations to be performed on the numbers entered into the calculator. For example, the $\sqrt{x}$ function key will produce the positive square root of the entered number. So if the number 9 is entered and the $\sqrt{x}$ function key is pressed, the display changes to 3 which is equal to $\sqrt{9}$ – the positive square root of 9.

Memory

All operations are performed on the number that is displayed on the calculator screen. Sometimes it may be desired to store the end result of a calculation for use at a later time. This can be done by placing the number into the memory by using the Min key. At any later time this number can be recalled from memory by using the MR key. For example, to perform the calculation

$$\frac{6.2 + 2.4}{1.1 - 5.4}$$

calculate the difference $1.1 - 5.4$ to give -4.3. Place this number into the memory by pressing Min. Clear the display and calculate the sum $6.2 + 2.4$ to give 8.6. Now press the ÷ operation key to indicate that the number in the display is about to be divided by something. Now press the memory return key, MR, to display the number -4.3 from memory. Now press = to give the result of the division as -2.

Relevant Figures in a Calculation

Often the result of a calculation on a calculator will be a number with a lot of numerals after the decimal point. This not only produces an ungainly number to handle but also encourages an acceptance that the result is always relevant to so many decimal places. For example, if £100 is invested at 10% per annum compound interest for 3 years the value of the investment will accrue to

$$100(1.1)^3 = 133.10$$

This is the result given by a calculator. If, however, we wished to find out how much we would have to invest today to have £100 in 3 years time at 10% compound interest the amount is given by

$$100/(1.1)^3 = 75.131\,480\,09$$

Again, this is the result using a calculator. Because we are discussing money, such a number is not relevant beyond two decimal places and so we must round the number to read 75.13.

EXAMPLES

1 Perform the following calculations using a calculator:

(a) $12.45 + 6.70$ **(d)** $12.45 \div 6.70$
(b) $12.45 - 6.70$ **(e)** $(2.34 - 0.013) \div 8.2$
(c) 12.45×6.70 **(f)** $[(7.834 \times 13.2) - 4.53] \div 3.66$

(a) Enter 12.45, press the + button and then enter 6.7 followed by =. The answer 19.15 is displayed.
(b) Repeat (a) but use the − button instead of the + button. The answer is 5.75.
(c) and **(d)** Use the × and ÷ buttons respectively to produce the answers 83.415 and 1.858 209 respectively.
(e) Subtracting 0.013 from 2.34 gives 2.327 which divided by 8.2 gives 0.283 780.
(f) $7.834 \times 13.2 = 103.4088$. Subtract 4.53 to give 98.8788 and then divide by 3.66 to give the final answer 27.016 066.

2 Use the function buttons to perform the following calculations:

(a) $\frac{1}{8.93}$ **(b)** 5.33^2 **(c)** $9.012^{1/2}$ **(d)** $3.25^{2.14}$

(a) Enter 8.93 and press the $1/x$ button to give 0.111 982.
(b) Enter 5.33 and press the x^2 button to give 28.4089.
(c) Enter 9.012 and press the $x^{1/2}$ button to give 3.001 999 3.
(d) Enter 3.25 and press the x^y button. Now enter 2.14 and press =. The result displayed is 12.457 486.

3 Use the calculator memory to perform the following calculations:

(a) $\frac{13.12 \times 4 - 12 \times 3.3}{10.1}$ **(b)** $\left(\frac{25.01 \times 3 - 16 \times 4.2}{13.6}\right)^{0.41}$

(a) Enter 12 and press the × button. Now enter 3.3 and press = to give the result 39.6.

Press Min to enter the displayed number into memory and then clear the screen by pressing AC.

Perform the product 13.12 × 4 to give the result 52.48. Now press the − sign followed by MR to recover the number in memory. Press = to give the result 12.88. Finally divide by 10.1 to give the final answer as 1.275 247 5.

or Many calculators are capable of performing this sort of sequence of operations without resorting to the use of the memory keys. If your calculator is of this type then follow this sequence of operations:

Enter each numeral and operation of the numerator as they appear: 13.12 × 4 − 12 × 3.3. Now press = to give the result as 12.88. Finally divide by 10.1 to give the final answer as 1.275 247 5.

(b) Enter 16, press the × button, enter 4.2 and press = to give 67.2.

In a similar manner to the previous calculation, evaluate the number within the brackets as 0.575 735 2. This number is to be raised to the power 0.41 so press the function button labelled x^y. The number 0.575 735 2 is the x value. Now enter .41 and press = to produce the result 0.797 427 8.

4 Find the values of π and the exponential number e accurate to five decimal places.

To find π just press the π button to produce the display

3.141 592 7

This is the value of the irrational number π as stored by the calculator. To five places of decimals this number is

3.141 59

Another function button is labelled e^x where e is the exponential number. This is the exponential function button. Press 1 to display the value of x and then press the exponential function button to produce the display

2.718 281 8

This is the value of e^1. To five places of decimals this is

2.718 28

EXERCISES

1 Replace each $*$ by an appropriate number in each of the following:

(a) $7.56 + 16.34 = 2*.9$

(b) $13.77 - 2.89 = 1*.8*$

(c) $35.65 \times 7.66 = 2*3.**9$

(d) $\dfrac{46.53}{7.89} = 5.8**\,33*\,4$

(e) $\dfrac{3.22 - 2.213}{7.4} = \dfrac{*.*07}{7.4} = 0.1**\,*81$

(f) $\dfrac{(9.111 \times 35.5) - 49.46}{18.34} = \dfrac{3*3.**05 - 4*.*6}{18.34} = \dfrac{*7*.98**}{18.34} = 14.*3*\,9*9$

2 Replace each $*$ by an appropriate number in each of the following:

(a) $\dfrac{1}{5.55} = 0.**0\,**0\,1$ **(c)** $25.482^{1/2} = 5.*47\,*6*\,9$

(b) $2.98^2 = *.8*04$ **(d)** $6.771^{1.56} = **.7**\,878$

3 Replace each $*$ by an appropriate number in each of the following:

(a) $\dfrac{78.32 \times 5 - 92 \times 3.6}{16.7} = \dfrac{3*1.* - *3*.2}{16.7} = \dfrac{6*.*}{16.7} = 3.*1*\,7**\,5$

(b) $\dfrac{(9.256 \times 7.295) + (3.546 \times 1.487)}{63.9} = \dfrac{6*.5*2*2 + *.2*2*02}{63.9} = \dfrac{*2.*9*\,4*2}{63.9} = 1.***\,*08\,5$

(c) $\left(\dfrac{93.22 \times 5 + 56 \times 9.3}{28.4}\right)^{3.2} = \left(\dfrac{4**.1 + *20.*}{28.4}\right)^{3.2} = \left(\dfrac{9*6.*}{28.4}\right)^{3.2}$

$= *4.*5^{3.2} = 85\,***.8*2$

4 Replace the $*$ by an appropriate number in the following:

$\pi^2 = 9.*7$ to two decimal places

Unit 3 Collecting and Tabulating Data

Test yourself with the following:

1 A farmer recorded the number of eggs laid by a flock of 60 chickens on one particular day. The results were listed on a sheet of paper as the eggs were gathered. The numbers were:

2	3	2	1	0	4	3	2	2	3
2	1	0	1	2	1	3	0	1	0
3	4	2	4	1	3	1	2	2	3
0	1	2	1	3	2	0	1	0	0
1	2	2	2	4	3	2	1	3	2
4	4	3	3	3	2	1	0	3	2

Construct a frequency table of the eggs laid.

2 The heights of a group of 25 people were measured in metres to the nearest centimetre with the following results:

1.98	1.76	1.87	1.55	1.48
1.65	1.69	1.72	1.75	1.85
1.91	1.40	1.56	1.82	1.72
1.71	1.89	1.78	1.72	1.64
1.57	1.53	1.66	1.68	1.75

Construct a frequency table with suitable groups for this data.

3 Draw a pie chart to represent the following data:

XYZ Wholesalers
Annual Proportionate Sales of Liquor

Wine	25%
Beer	65%
Spirits	10%

4 Draw a suitable pictogram to represent the following data:

Upville College Student Enrolment

Year	Men	Women
1	105	45
2	175	85
3	250	110
4	400	200

Data and Information

A **datum** is a sequence of alphanumeric characters that can exist without reference to context or meaning. When that datum is placed within a context so as to give it meaning the datum contains **information**.

For example, the sequence of characters B858BHD is a datum. As soon as it is placed on a car number-plate as B 858 BHD it is given context and has meaning. In this latter form it contains information.

Collection of Data

Data are collected whenever an experiment is performed and measurements made. For example, recording the temperature of the air every hour gives rise to data in the form of related pairs of numbers connecting time and temperature. Before data can be manipulated it must be organized.

Tabulation of Data

Data can be organized into tables of suitably labelled, regular rows and columns. Once the table is constructed the data are then entered into it. For example, the following tabulated data give the number of lunches served in a canteen during a particular week:

Day	Number of lunches
Mon	25
Tue	35
Wed	15
Thu	23
Fri	27

Discrete and Continuous Data

When the data that we gather come in the form of a collection of numbers then we refer to the data as being **discrete**. For example, if we were to count the number of drinks served by 20 different hot drinks machines on a college campus then the numbers collected would form discrete data.

Sometimes, however, despite the fact that the data are discrete we record the data as falling within continuous ranges of values. For example, if we performed a survey to record the take-home pay of all the people employed by a particular company and recorded the value of every single pay-packet it would be entirely possible for every recorded number to be different. This is because take-home pay is a particular value on a continuous scale of values. Continuous, that is, to the nearest penny. In such a case it is much more convenient to record an individual take-home pay as being within a continuous range of money values. The following table illustrates this:

Take-home pay (£) (to nearest £)	Number of employees
1–10 000	123
10 001–20 000	257
20 001–30 000	96
30 001–	74

Here we have recorded the number of employees whose take-home pay falls within defined ranges of money values. This is known as **grouping** the data. Notice that in this table the groups represent values that are continuous to the nearest £ so that rounding has to be applied. For example, an employee whose take-home pay is £10 000.49 is recorded in the first group whereas an employee with a take-home pay of £10 000.50 is recorded in the second group.

Frequency

With ungrouped data the **frequency** of a datum is the number of times that it appears amongst the data. With grouped data the number of data items falling into a particular group is called the **group frequency** of that particular group.

Pictorial Display

Data can often be displayed in the form of pictures such as **pie charts** and **pictograms**.

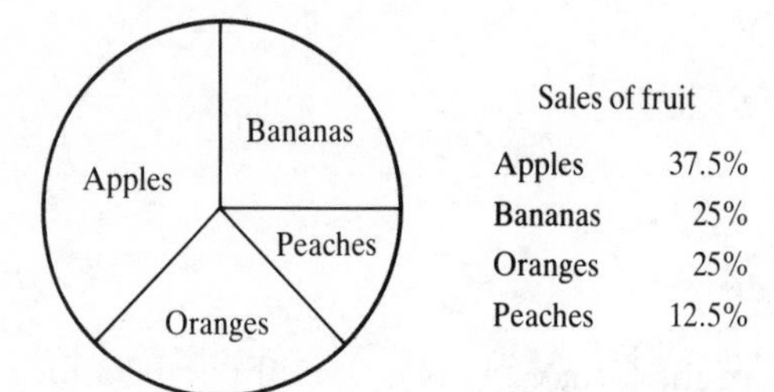

Sales of fruit

Apples	37.5%
Bananas	25%
Oranges	25%
Peaches	12.5%

Figure 2.1

A pie chart is used to represent data that exists in proportions. The pie, a circle, is taken to represent the whole, and relative proportions of the circle represent the appropriate proportions of the data. A pictogram produces a display that is designed to attract attention. These sorts of display are useful for giving an overall impression of the data rather than the numerical detail.

EXAMPLES

1 A yoghurt farm sells yoghurt in cartons that are supposed to weigh 50 g. A sample of 30 cartons from a single day's production yielded the following weights, measured to the nearest 0.1 g:

49.9	50.1	49.9	50.2	50.0
50.4	50.2	50.0	49.9	50.3
50.1	50.1	49.9	50.0	50.4
50.0	50.2	50.4	49.9	50.3
50.4	50.3	50.1	49.9	50.0
50.0	50.0	50.1	50.0	49.9

Tabulate this data in a frequency table.

Weight (g)	49.9	50.0	50.1	50.2	50.3	50.4	Total
Frequency	7	8	5	3	3	4	30

The frequency is the number of times a particular datum appears. The total of all the frequencies is included as a check that the counting has been done correctly. The total of all the frequencies equals the number of data items.

2 To open a till when no sale transaction is made the No Sale button is pressed. In a chain of 40 franchised newsagents the number of times the No Sale button was pressed during a particular week was recorded as follows:

20	35	14	1	19	23	32	22
17	21	15	19	20	4	11	35
6	0	18	12	44	28	25	38
14	27	35	40	3	9	10	12
0	13	17	15	27	6	14	23

Tabulate this data into groups 0–9, 10–19, 20–29, 30–39, 40, 49.

No-sale till openings:

	Groups					
	0–9	10–19	20–29	30–39	40–49	Total
Frequency	8	15	10	5	2	40

Here the total of all the frequencies is given as a check that the counting has been done correctly.

3 Draw a pie chart to represent the following data:

Faddy Fast Foods
Proportion of Sales of Fast Food

Hamburgers	30%
Fish and chips	40%
Pizza	20%
Other	10%

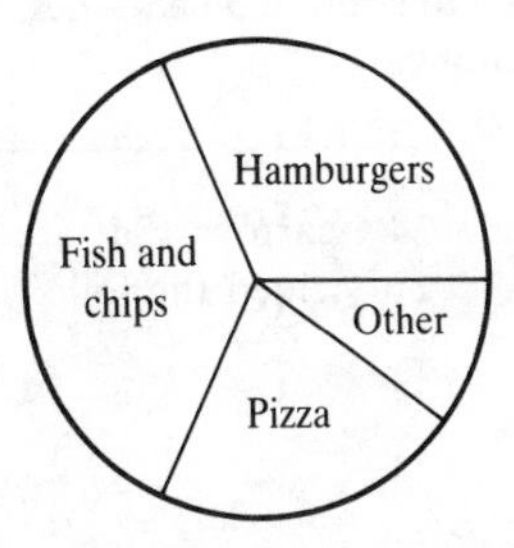

Figure 2.2

4 Draw a pictogram to represent the following data:

People in household	Average number of pets owned
1	1.1
2	1.8
3	2.6
4	2.9
5	3.0

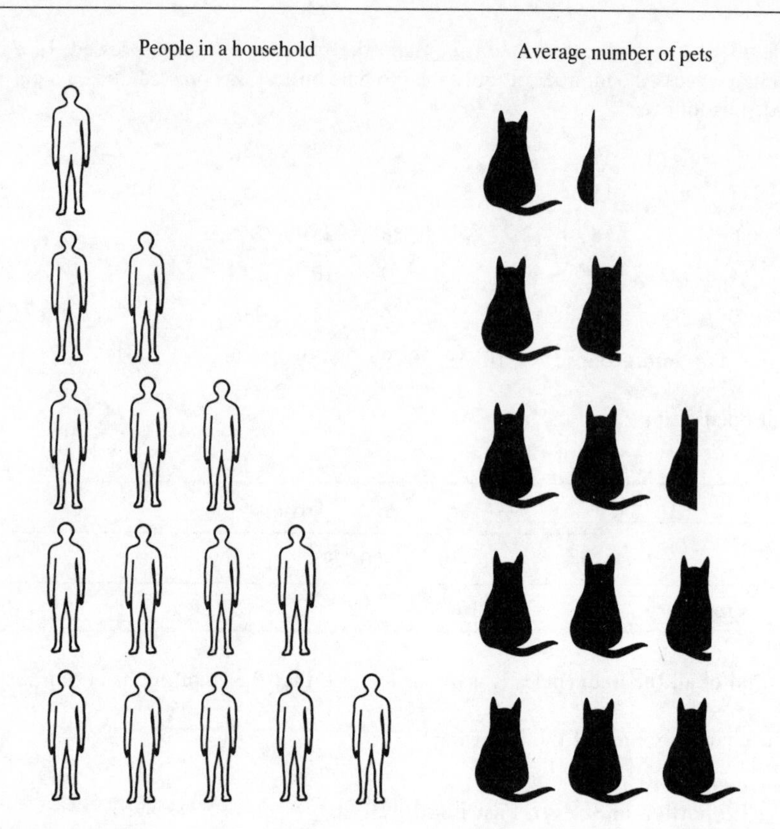

Figure 2.3

EXERCISES

1 A matchbox is supposed to contain 35 matches. A sample of 30 boxes from a single day's production yielded the following numbers:

35	34	33	35	33
33	34	34	36	34
33	35	35	35	35
34	36	34	33	34
32	33	33	34	35
35	37	36	36	34

Replace each * by the appropriate number in the following tabulation of this data in a 'frequency' table.

Matches per box	32	33	34	35	36	37
Frequency	*	*	*	*	*	*

2 In a darts competition the following scores were obtained from the successive throws of three darts:

17	60	23	12	5	95	87	33
28	88	93	73	67	3	46	77
19	57	48	38	29	94	83	38
57	67	36	70	31	45	11	22
40	53	37	95	67	13	44	73

Replace each * by the appropriate number in the following tabulation of data into the groups 0–19, 20–39, 40–59, 60–79, 80–99:

Score group	0–19	20–39	40–59	60–79	80–99
Frequency	*	*	*	*	*

3 Draw a pie chart to represent the following data:

Middleboro Local Authority
Annual Proportions of Heating Fuels Used

Electricity	18%
Coal	16%
Gas	39%
Oil	27%

4 Draw a pictogram to represent the following data:

Brick production (1000s)

Jan	Feb	Mar	Apr	May	Jun	Jul	Aug	Sep	Oct	Nov	Dec
25	32	38	45	49	61	58	39	24	19	58	65

Unit 4 Calculating Means, Medians and Modes

Test yourself with the following:

1 The following data represents annual tonnage of coal sold by a coal merchant for the last 12 years.

850	970	1010	1280	1399	2004
1950	1370	1604	1850	1923	1888

Calculate the average tonnage sold per year for the last 12 years.

2 On 36 rolls of two dice the following scores were recorded:

2	12	10	4	5	9
11	8	9	11	3	8
9	2	8	2	11	9
10	4	12	7	12	7
6	10	4	2	9	11
6	8	9	3	5	12

Calculate the mean, the median and the mode score.

3 The weights of 24 people were measured to the nearest lb and recorded as follows:

141	195	169	155	171	155
182	146	178	184	152	166
163	197	157	143	173	167
194	158	186	172	164	188

By grouping the data into the groups 140–149, 150–159, ..., 190–199 construct a frequency table from which the mean weight may be calculated.

4 Calculate the median and the modal group of the data in Question 2.

Statistics

A **statistic** is a number or fact that summarizes a collection of data and is derived from the data by various arithmetic methods.

Averages

The simplest single statistics that are used to summarize data are **averages**. There are many different averages including the **arithmetic mean**, the **median** and the **mode**.

Arithmetic Mean

The arithmetic mean, or simply the mean, is obtained by adding a number of data items together and dividing by the number of data items added. For example, the mean of 14, 12 and 40 is given as:

$$\frac{14 + 12 + 40}{3} = \frac{66}{3} = 22$$

Notice that it is not necessary for the mean to be one of the data items. When data values are recorded in a frequency table, the mean is found by multiplying each data item by its frequency and adding up all these products. This is equal to the sum of all the data values. The mean is then given by this sum divided by the sum of all the frequencies. For example, in the following frequency table:

Datum value	Frequency	Frequency × Datum value
2	3	6
4	2	8
6	5	30
8	3	24
10	2	20
Totals	15	88

$$\text{Mean} = \frac{88}{15}$$

$$= 5.9 \text{ to one dec. pl.}$$

Grouped Data

For data recorded in a grouped frequency table the datum value corresponding to a particular group is taken to be the midpoint value of that group. For example, in the grouping of employee's take-home pay:

Take-home pay (£) (to nearest £)	Number of employees
1–10 000	123
10 001–20 000	257
20 001–30 000	96
30 001–	74

The average take-home pay is obtained by adding the products of each frequency with the midpoint value of each group and dividing by the total number of employees. That is:

Group (£)	Midpoint	Frequency	Frequency × Midpoint
1–10 000	5 000.5	123	615 061.5
10 001–20 000	15 000.5	257	3 855 128.5
20 001–30 000	25 000.5	96	2 400 048.0
30 001–	35 000.5	74	2 590 037.0
	Totals	550	9 460 275.0

$$\text{Mean} = \frac{9\,460\,275}{550}$$

$$= 17\,201 \text{ to the nearest £}$$

The Median

When the data are arranged in descending or ascending order the median is the middle datum. There are as many data values less than the median as there are data values greater than the median. For example, the median of 1, 3, 5, 7, 9 is 5 – the middle datum.

If there are an even number of data there is no middle datum. In this case the median is the arithmetic mean of the two middle data. For example, the data 1, 3, 5, 7, 9, 11 has no middle datum. The median is, therefore, the mean of the middle two data, namely, $(5 + 7)/2 = 6$. Just like the mean, the median does not need to be one of the data.

The Mode

The mode is that datum that occurs the most in a collection of data. For example, the mode of the following data: 1, 1, 1, 2, 2, 3 is 1 as it occurs the most.

For grouped data, the modal group is the group with the largest frequency.

Comparative Merits

The mode is the easiest average to compute but it has very little worth beyond that. The median is unaffected by extremes as it only indicates the centre point when data are arranged according to size. The mean is the most important average because it is both sensitive to the actual numerical values of the data and can be dealt with mathematically. For example, the following seven numbers:

1, 1, 2, 2, 2, 3, 3

have mode, median and mean of 2. If the numbers 1 and 2 are added to this list the mode and the median are unaffected but the mean is changed to 1.89 to two decimal places.

EXAMPLES

1 Calculate the arithmetic mean, median and mode of the following:

10, 12, 11, 17, 18, 29, 34, 25, 47, 64, 76, 10, 12, 10

The sum of the numbers = 375
The number of numbers = 14
The average number = 375/14 = 26.8 to one dec. pl.

Arranged in ascending order the numbers are:

10, 10, 10, 11, 12, 12, 17, 18, 25, 29, 34, 47, 64, 76

The middle numbers are 17 and 18. Therefore

$$\text{Median} = \frac{17 + 18}{2}$$

$$= 17.5$$

The mode is 10 as that is the most frequently occurring number.

2 Calculate the arithmetic means of both the grouped and the ungrouped data in Example 2 of Unit 3. Example 2 reads:

To open a till when no sale transaction is made the No Sale button is pressed. In a chain of 40 franchised newsagents the number of times the No Sale button was pressed during a particular week was recorded as follows:

20	35	14	1	19	23	32	22
17	21	15	19	20	4	11	35
6	0	18	12	44	28	25	38
14	27	35	40	3	9	10	12
0	13	17	15	27	6	14	23

Ungrouped

The sum of the numbers = 744
The number of numbers = 40

$$\text{The average number} = \frac{744}{40} = 18.6$$

Grouped

The data are tabulated into groups 0–9, 10–19, 20–29, 30–39, 40–49 in the following table:

	Groups					
	0–9	10–19	20–29	30–39	40–49	Total
Frequency	8	15	10	5	2	40

Here the total of all the frequencies is given as a check that the counting has been done correctly.

The average can now be calculated using the following table where the Average column records the midpoint value of the corresponding group:

Group	Average	Frequency	Frequency × Average
0–9	4.5	8	36.0
10–19	14.5	15	217.5
20–29	24.5	10	245.0
30–39	34.5	5	172.5
40–49	44.5	2	89.0
Total		40	760.0
		Average	19.0

Notice the difference. The ungrouped data has an average of 18.6 and the grouped data has an average of 19.0. The average of the ungrouped data is accurate whereas the average of the grouped data contains an inaccuracy due to the assumption that all the data within a group possess the midpoint value.

3 Calculate the median and the mode of the data in Question 2.

Median

If the data are ranked starting from the lowest value we obtain the following listing:

0	0	1	3	4	6	6	9	10	11	12	12	13	14	14	14
15	15	17	17	18	19	19	20	20	21	22	23	23	25	27	27
28	32	35	35	35	38	40	44								

As there is an even number of data there is no middle datum. Instead we select the middle two data which are 17 and 18. The average of these two, 17.5, is the median.

Mode

The mode of the ungrouped data is not unique. Both 14 and 35 appear three times. The collection of data is said to be **bimodal** – it has two modes. The modal group in this specifically chosen grouping, however, is unique – it is the 10–19 group.

EXERCISES

1 Replace each * by the appropriate number in the calculation of the arithmetic mean, median and mode of the following data:

21, 34, 21, 45, 21, 33, 34, 65, 87, 45, 12, 19, 32, 21

The sum of the numbers = *
The number of numbers = *
The average number = */* = *

Arranged in ascending order the numbers are:

*, *, *, . . ., *

The middle numbers are * and *. Therefore the median = *.
The mode is * as that is the most frequently occurring number.

2 Replace each * by the appropriate number in the calculation of the arithmetic mean of the data in Example 1, Unit 4. Example 1 reads:

A yoghurt farm sells yoghurt in cartons that are supposed to weigh 50 g. A sample of 30 cartons from a single day's production yielded the following weights measured to the nearest 0.1 g:

49.9	50.1	49.9	50.2	50.0
50.4	50.2	50.0	49.9	50.3
50.1	50.1	49.9	50.0	50.4
50.0	50.2	50.4	49.9	50.3
50.4	50.3	50.1	49.9	50.0
50.0	50.0	50.1	50.0	49.9

These data are then tabulated as follows:

Weight	Frequency	Weight × Frequency
49.9	*	*
50.0	*	*
50.1	*	*
50.2	*	*
50.3	*	*
50.4	*	*
Total	30	*

Average 50.1 to one dec. pl.

3 Replace each * by the appropriate number in the calculation of the median and the mode of the data in the previous exercise.

If the 30 data items are listed in ascending order the median will be the average of the 15th and the 16th items in the list. From the table it is seen that the 15th item is * and the 16th is *. The median is therefore *.

The most frequently occurring data item is * and this is the mode.

Unit 5 Calculating Mean Deviations and Standard Deviations

Test yourself with the following:

1 Write down the absolute values of the following numbers:

$$-342,\ 0.014,\ -0.002,\ -(0.2)^3,\ 1,\ -1$$

2 Calculate the mean absolute deviation and the standard deviation of the ungrouped data in Test Question 1 of Unit 3 of this module.

3 Calculate the mean absolute deviation and the standard deviation for the grouped data in Test Question 2 of Unit 3 of this module.

4 By comparing means and standard deviations contrast the following examination results of two students:

	Alan	Beth
Maths	66	61
English	64	65
Physics	62	60
History	65	63
French	68	70
Geography	61	65
Computing	65	67
Chemistry	69	69

Spread of Data About the Mean

Any collection of data will have their values spread about the mean value. For example, the two sets of numbers:

10 20 30 40 50

and

28 29 30 31 32

both have the same mean of 30 but their spreads about the mean are quite different.

Mean Deviation

The **deviation** of a datum is obtained by subtracting the value of the datum from the arithmetic mean. Unfortunately, the mean deviation is zero. For example, in the above two sets of numbers the deviations from the mean are respectively:

20 10 0 -10 -20 which has an average of 0

and

2 1 0 -1 -2 which also has an average of 0

If the negative signs are removed then the first set of deviations has a mean of $60/5 = 12$ and the second set has a mean of $6/5 = 1.2$. The relative sizes of 12 and 1.2 indicate the relative spreads of each set of data about the mean. Removing the negative signs from the deviations of the data provides a way of obtaining a sensible measure of spread. There are two ways of removing the minus signs:

(1) taking the **absolute value** of the deviation, and
(2) squaring the deviation.

Absolute Value of a Number

The absolute value of a number is the magnitude of the number regardless of its sign. For example,

the absolute value of -3 is 3

the absolute value of 4 is 4

The absolute value is denoted by the number within parallel lines, thus ||. For example $|-2| = 2$ and $|6| = 6$.

Mean Absolute Deviation

The **Mean Absolute Deviation** (MAD) of the data is the average of the absolute values of all the individual deviations. For example, the MAD of:

10 20 30 40 50

is 12 and the MAD of:

28 29 30 31 32

is 1.2. This demonstrates how the first set of numbers are more widely spread about the mean than the second set of numbers.

Variance

The negative signs can be removed by squaring the deviations. The **variance** is the average of the squares of the deviations from the mean. For example, the variance of:

10 20 30 40 50 (with mean 30)

is given as

$$\frac{(30-10)^2+(30-20)^2+(30-30)^2+(30-40)^2+(30-50)^2}{5}=\frac{1000}{5}=200$$

and the variance of:

28 29 30 31 32 (with mean 30)

is given as:

$$\frac{(30-28)^2+(30-29)^2+(30-30)^2+(30-31)^2+(30-32)^2}{5}=\frac{10}{5}=2$$

The units of variance are the square of the units of the data. As such, it is inappropriate to use the variance as a measure of the spread of the data about the mean. A more appropriate measure, in the same units as the data, is the **standard deviation**.

Standard Deviation

The standard deviation is the square root of the variance. For example, the standard deviation of the first set of numbers is:

$\sqrt{200} = 14.14$ to two dec. pl.

and of the second is:

$\sqrt{2} = 1.41$ to two dec. pl.

Comparative Merits

The mean absolute deviation is a good measure of the spread of the data about the mean as it uses all the data. However, it is very difficult to handle mathematically as opposed to arithmetically. On the other hand, the standard deviation can be analysed mathematically, so for this reason the standard deviation is the more commonly used measure of spread.

Comparing Two Sets of Data

Two sets of data can be compared by comparing their means and spreads. For example, at a Weight Watchers meeting the mean weight of the group and the spread about the mean were measured on two successive weeks. If the two measures of spread were the same but the second mean was less than the first then it could be reasoned that the group as a whole were successfully losing weight. If, however, the two means were the same but the second spread was greater than the first then it could be concluded that some were successful at losing weight while others gained weight, leaving the group's average weight unaltered.

EXAMPLES

1 Find the absolute value of each of the following numbers:

$12, -34, 0, -1.2, 0.34$

The absolute values are:

$|12| = 12, |-34| = 34, |0| = 0, |-1.2| = 1.2, |0.34| = 0.34$

respectively.

2 Calculate the mean absolute deviation and the standard deviation for the data in Example 1 of Unit 3 of this module. Example 1 reads:

A yoghurt farm sells yoghurt in cartons that are supposed to weigh 50 g. A sample of 30 cartons from a single day's production yielded the following weights measured to the nearest 0.1 g:

49.9	50.1	49.9	50.2	50.0
50.4	50.2	50.0	49.9	50.3
50.1	50.1	49.9	50.0	50.4
50.0	50.2	50.4	49.9	50.3
50.4	50.3	50.1	49.9	50.0
50.0	50.0	50.1	50.0	49.9

The two measures of dispersion are now calculated in the following tables where, because of space limitations, single letters have been used to represent the tabulated items. For example, W represents weight, f represents frequency and M represents the mean:

Mean absolute deviation

Weight W	Frequency f	$W \times f$	$M - W$	$\|M - W\|$	$f \times \|M - W\|$
49.9	7	349.3	0.2	0.2	1.4
50.0	8	400.0	0.1	0.1	0.8
50.1	5	250.5	0.0	0.0	0.0
50.2	3	150.6	−0.1	0.1	0.3
50.3	3	150.9	−0.2	0.2	0.6
50.4	4	201.6	−0.3	0.3	1.2
Total	30	1502.9			4.3
	Mean (M)	50.1		MAD	0.14

Variance and standard deviation

Weight W	Frequency f	$W \times f$	$M - W$	$(M - W)^2$	$f \times (M - W)^2$
49.9	7	349.3	0.2	0.04	0.28
50.0	8	400.0	0.1	0.01	0.08
50.1	5	250.5	0.0	0.00	0.00
50.2	3	150.6	−0.1	0.01	0.03
50.3	3	150.9	−0.2	0.04	0.12
50.4	4	201.6	−0.3	0.09	0.36
Total	30	1502.9			0.87
	Mean (M)	50.1		Variance	0.029
				Standard deviation	0.17

3 Calculate the mean absolute deviation and the standard deviation for the grouped data in Example 2 of Unit 3 of this module.

This problem is solved using a tabular method:

Group	Midpt	Dev	Abs	f	$f \times$ Abs	Sq dev	$f \times$ Sq dev
0–9	4.5	14.5	14.5	8	116.0	210.25	1682.00
10–19	14.5	4.5	4.5	15	67.5	20.25	303.75
20–29	24.5	−5.5	5.5	10	55.0	30.25	302.50
30–39	34.5	−15.5	15.5	5	77.5	240.25	1201.25
40–49	44.5	−25.5	25.5	2	51.0	650.25	1300.50
Mean	19.0		Total	40	367.0		4790.00
				MAD	9.175	Variance	119.75
						Standard deviation	10.94

Column 1 just lists the groups. Column 2 lists the midpoints of the groups. Column 3 lists the deviations of the group midpoints from the mean. Column 4 lists the absolute values of the deviations in column 3. Column 5 lists the group frequencies. Column 6 lists the products of the absolute deviations and the frequencies. At the bottom is the total and the average – this is the mean absolute deviation.

Column 7 lists the group deviations squared. Column 9 lists the product of the squared deviations and the frequencies. At the bottom is the total and the average squared deviations – the variance. The standard deviation is the square root of the variance.

The advantage of using this tabular method of calculating the standard deviation of a collection of data is that it instils a sense of order into the handling of numbers and it also clearly displays what you are doing at each step of the calculation. It is, however, a laborious method and one that can be replaced by using the statistical mode on a calculator. Every calculator is different so to use the statistical facilities of your calculator you will have to read the manual provided by the manufacturer. This will not only tell you how to put the calculator into statistical mode but it will also tell you how to enter the data into the calculator's memory. Once the data are entered it will then be a matter of pressing the standard deviation key to display the standard deviation. You might try this in all the appropriate examples and exercises.

4 On an assembly line producing units with a critical weight it was suggested that an improvement be made to the manufacturing process to produce a more consistent product. The following two lists of data represent the weights (in grams) of five sample units taken before and after the adjustment. By comparing means and standard deviations contrast the two sets of data. Is the adjustment producing a more consistent product?

Unit	Wt1	Wt2
1	184	180
2	178	185
3	190	190
4	170	187
5	195	175

The mean of both Wt1 and Wt2 is 183.4. The standard deviation of Wt1 is 8.8 and of Wt2 5.3, both calculated to one decimal place. Consequently, the improvement produced a more consistent product because the standard deviation reduced.

EXERCISES

1 Replace each * by an appropriate number in the following:

$$|-4| = *, \ |0.27| = *, \ |*| = 5.4, \ |-101.2| = 10*.*$$

2 Complete the following calculation to find the mean absolute deviation of the following data:

101	101	107	105	102	103	100
105	100	102	101	106	102	106
104	101	101	108	101	106	106
104	108	104	105	108	103	100

Mean absolute deviation

Value V	Frequency f	$V \times f$	$M - V$	$\|M - V\|$	$f \times \|M - V\|$
100					
101					
102					
103					
104					
105					
106					
107					
108					
Total	28	2900			64.0
	Mean (M)	*		MAD	*

Standard deviation

Value V	Frequency f	$V \times f$	$M - V$	$(M - V)^2$	$f \times (M - V)^2$
100					
101					
102					
103					
104					
105					
106					
107					
108					
Total	28	2900			186.88
	Mean (M)	*		Variance	*
				Standard deviation	*

3 Complete the following calculation to find the standard deviation for the data in Exercise 2 where the data are grouped into the groups 100–103, 104–106, 105–108.

Standard deviation

Group	Mid-point V	Frequency f	$V \times f$	$M - V$	$(M - V)^2$	$f \times (M - V)^2$
100–102	101					
103–105	104					
106–108	107					
Total		28	2900			174.88
		Mean (M)	*		Variance	*
					Standard deviation	*

4 By comparing means and standard deviations contrast the following monthly hours worked by two employees:

Month	A	B
1	165	152
2	155	143
3	160	172
4	162	175

The means of data sets A and B are both the same at *. The standard deviation of set A is * and the standard deviation of set B is 13.43 which is nearly four times as large as that of set A. Consequently, employee A is more/less consistently employed than employee B.

Module 2: Further exercises

1 The distance s travelled in time t seconds by a ball thrown vertically upwards is given by the formula

$$s = ut - \tfrac{1}{2}gt^2$$

where $g = 98\ \text{cm s}^{-2}$ is the acceleration due to gravity and $u\ \text{cm s}^{-1}$ is its initial velocity.

(a) Find s when $t = 1$, 2.5 and 5 seconds if $u = 2.45\ \text{m s}^{-1}$.

(b) Explain the results of part (a).

2 If two electrical resistances p ohms and q ohms are wired in parallel their combined resistance is R where R is given by the formula

$$\frac{1}{R} = \frac{1}{p} + \frac{1}{q}$$

If $p = 5$ ohms what is the value of q that makes their combined resistance $R = 2$ ohms?

3 In each of the following:

(i) estimate a reasonable answer;

(ii) find the accurate answers to the designated number of decimal places.

(a) $\dfrac{0.98 + 8.8}{0.19}$ 1 dec. pl.

(b) $\dfrac{(1.2)(9.01) - (8.3)(7.2)}{16.12}$ 2 dec. pl.

(c) $3.01^{2.91}$

(d) $\dfrac{8.1^{-4.05}}{16.2^{0.51}}$ 1 sig. fig.

4 Perform (a) and (c) below by hand and (b) and (d) using a calculator:

(a) $\dfrac{0.96 + 8.4}{0.12}$ (c) $\left[\dfrac{5(7.2)}{6}\right] \times \left(\dfrac{1.21}{36}\right)^{1/2}$

(b) $\dfrac{0.97 + 8.5}{0.13}$ (d) $\left[\dfrac{(5.2)(7.15)}{5.9}\right] \times \left(\dfrac{1.22}{37}\right)^{1/2}$

5 In measuring the speed of a train a 5% error was made in the measure of the distance it had travelled and a 2% error in the measurement of the time taken to cover the distance. What is the resultant error in the calculated value of the speed?

6 In measuring the area of a triangle a 1% error was made in the length of the base and a 2% error was made in the height. What was the resultant error in the area?

7 The following data represents the daily temperature in degrees Celsius for the month of June:

18	17	19	22	19	17	15	18	20	23
16	16	18	15	14	13	15	15	18	20
21	22	24	24	25	24	21	18	16	14

(a) Tabulate this data and complete the frequency table.
(b) Tabulate this data into groups of 2 degrees and complete the frequency table.

8 Calculate the mean temperature from the data in Question 7:
(a) for the ungrouped data,
(b) for the grouped data in groups 11–15, 16–20, 21–25.

9 Find the median and the mode of the ungrouped data in Question 7.

10 Find the modal group of the grouped data in Question 8.

11 Calculate the mean deviation and the standard deviation for the grouped data in Question 8.

12 The five minutes past noon train runs every weekday. For the past three weeks its time of departure has been recorded in the form of hr:mins:secs as:

0:05:50	0:05:10	0:05:50	0:05:05	0:05:30
0:05:00	0:05:25	0:05:59	0:06:15	0:06:55
0:05:03	0:05:00	0:06:01	0:07:10	0:05:05

What is the average time of departure? (*Hint*: Convert the times to seconds.)

13 Calculate the standard deviation of the departure times from the mean departure time from the data in Question 12.

14 A train is considered as late departing if it leaves one minute after it is due to leave. How many standard deviations is this?

15 If a train is considered late departing if its departure time is greater than 0.5 standard deviations later than its average departure time, how many times was the train late departing during those recorded three weeks?

16 In an archery contest between Colin and Dorothy the number of bull's-eyes scored in successive rounds are given in the table. By comparing means and standard deviations contrast their relative performances:

Round	Colin	Dorothy
1	5	3
2	4	3
3	3	4
4	2	3
5	1	2

Module 3: Relationships

Aims: To exhibit linear relationships between two sets of numbers and to model such relationships in the form of equations.

Objectives: When you have read this module you will be able to:

- ▶▶ Draw parallel scales for two specified sets of numbers and generate an equation that represents their relationship.
- ▶▶ Understand the terms 'independent' and 'dependent' variable and use a specified equation to interpolate and extrapolate further information.
- ▶▶ Construct a cartesian graph of discrete points from a given set of data. Highlight the graph using jagged lines or vertical bars.

Unit 1 Parallel Scales

Test yourself with the following:

1 Plot the following sets of numbers on two parallel scales and from the plot derive an equation that relates km to miles:

Miles	5	10	15	20	25	30
Km	8	16	24	32	40	48

2 From the following sets of numbers derive an equation that relates the amount of a loan outstanding to the number of payments made:

Loan outstanding	5000	4000	3000	2000	1000	0
Payments made	4	8	12	16	20	24

3 In each of the following, indicate:
(a) the subject of the equation
(b) the dependent and the independent variables

(i) $u = 3v - 5$ (ii) $s = ut + 3$ (iii) $T = 2\pi\sqrt{L}$

The Real Line

Running the length of a thermometer is a scale in the form of a line on which are marked a series of numbers that represent temperatures. We saw in Module 1 that any point on a straight line can be distinguished from any other point by writing a number alongside it. When we have given every point a number then we find that not only have we identified every point on the line, but we have also exhausted our supply of numbers. The line is the **real line** and this is our starting point.

A line extends indefinitely in either direction so in practice we only consider a segment of the line of sufficient length to display numbers over a desired range. For example, if we were to draw a line to represent the Celsius temperature scale from 0 °C to 10 °C then the line would just be of sufficient length to cover the range of numbers 0 to 10 as shown in Figure 3.1. Note that the only numbers written are the even numbers, the odd numbers being indicated by short bars across the line. If all the numbers were to be written then the scale could not be so easily read – the effect would be one of clutter. When numbering the points on a line we only put in sufficient numbers to make the scale clear with intermediate numbers being indicated by short bars or enlarged points.

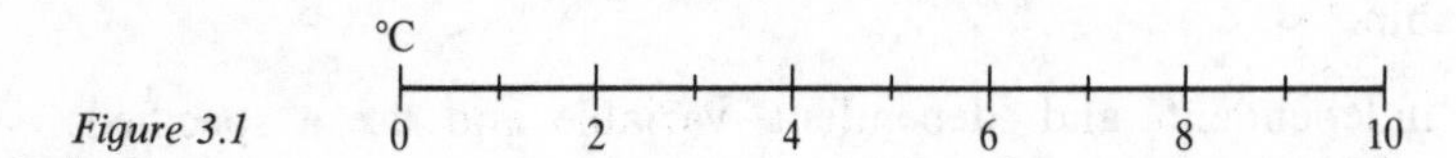

Figure 3.1

Parallel Scales

To draw a straight line requires the use of a ruler. This ruler has two parallel edges in the form of two straight lines. One edge is marked off in inches and tenths of an inch and the other edge is marked off in centimetres and millimetres. On the ruler in the diagram the zero points for inches and centimetres are directly opposite each other and directly opposite the 1 inch marker is the marker indicating 2.5 centimetres. Indeed, directly opposite every marker on the inches side is an appropriate marker on the centimetre side – there is a relationship between inches and centimetres.

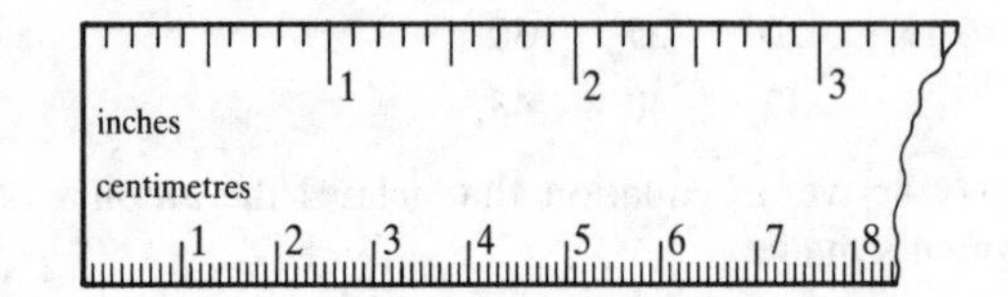

Figure 3.2

This relationship can be seen graphically by using parallel scales. Each side of the ruler can be thought of as a segment of a real line. The lengths of each line are the same but the numberings are different.

Because different ranges of numbers can be marked on the same length of line we can compare one range with another. To illustrate this take the middle of the ruler away to leave a single line. On the top of the line are marked inches and on the bottom of the line are marked centimetres.

From this diagram we can construct a table of inches and corresponding centimetres.

Inches	1.0	2.0	3.0	4.0	5.0	6.0	7.0	8.0	9.0	10.0
Centimetres	2.5	5.0	7.5	10.0	12.5	15.0	17.5	20.0	22.5	25.0

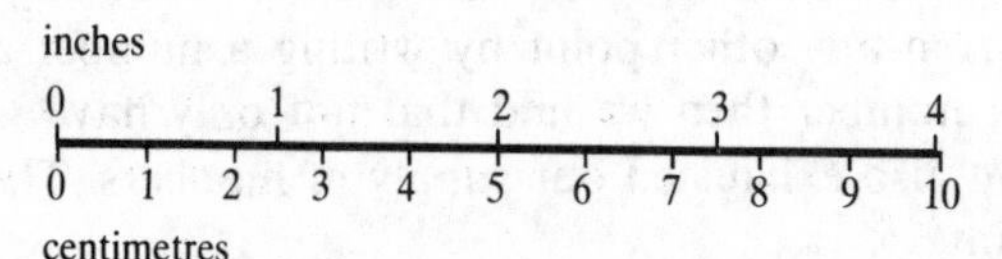

Figure 3.3

Here we see the relationship expressed numerically. Every inch is equal to 2.5 centimetres. We can express this relationship in the form of an equation:

$$1 \text{ inch} = 2.5 \text{ centimetres}$$

If we were to measure the length of a line using inches then we could convert our measurement to a length in terms of centimetres by using this relationship:

$$\text{Length in centimetres} = 2.5 \times \text{Length in inches}$$

or, using abbreviations:

$$\text{cm} = 2.5 \times \text{in}$$

Labelling

The numbers on the edge of the ruler correspond to units of measurement and you will notice that each scale is annotated to indicate the units. The labelling states that the top line of points represents inches and the bottom line represents centimetres. Sometimes you will see rulers where the labelling is an abbreviation such as *cm* for centimetres or *in* for inches. If we are to draw a numbered line where the numbers represent some definite quantity then the labelling should make that clear. Either the full name of the quantity must be used or an abbreviation. If the abbreviation is not a common one then a key must be provided to indicate to what the labelling refers. For example, Figure 3.4 displays a line on one side of which the numbers 0 to 100 are marked off. On the other side the numbers 32 to 212 are marked off so that the 0 and the 32 coincide and the 100 and 212 coincide. By inspection we can see that the points 25, 50 and 75 on the top of the line correspond to the points 77, 122 and 167 on the bottom of the line. Indeed, just like the ruler, for every point on the top of the line there is a corresponding point on the bottom of the line. Again there is a relationship between the two scales.

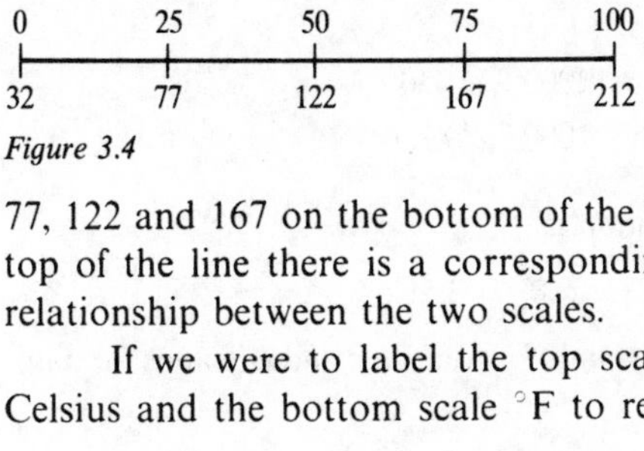

0 25 50 75 100

32 77 122 167 212

Figure 3.4

If we were to label the top scale °C to represent temperature measured in degrees Celsius and the bottom scale °F to represent temperature measured in degrees Fahrenheit then the relationship would express the connection between the two temperature scales. The relationship between degrees Celsius and degrees Fahrenheit is not as straightforward as that between inches and centimetres because the zeros on the two scales no longer coincide. However, we can see that:

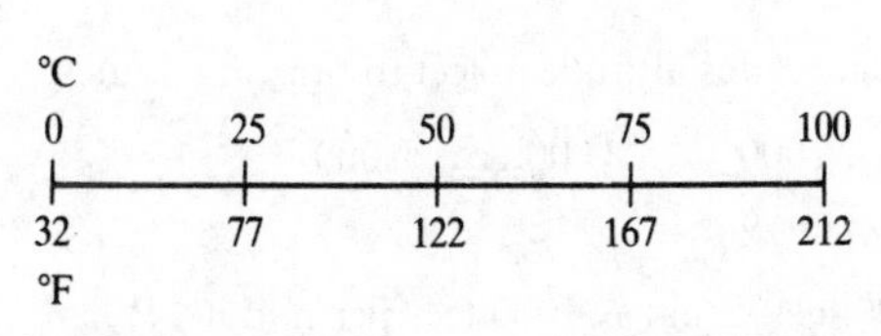

Figure 3.5

$$100 \text{ Celsius units} = 180 \text{ Fahrenheit units}$$

180 being the difference between 212 and 32. If we divide both sides of this equation by 100 then we find that:

$$1 \text{ Celsius unit} = 1.8 \text{ Fahrenheit unit}$$

Also we see that 0 °C corresponds to 32 °F so any temperature measured in degrees Celsius can be converted to a measure in degrees Fahrenheit by using the following equation:

$$°\text{F} = 32° + (1.8) \times °\text{C}$$

Or, put more simply:

$$F = 32 + 1.8C$$

where C stands for the number of degrees Celsius of a measured temperature and F stands for the equivalent measurement of the same temperature in degrees Fahrenheit.

Variables

In the equation:

$$F = 32 + 1.8C$$

the temperature readings represented by F and C can be any value we wish to choose. If we choose C to be a particular value then substituting that value into the equation permits us to compute the corresponding value of F. Because we can assign a variety of values to C and F we call them **variables**. The variable F is the **subject** of the equation; it is F that is to be found after we have substituted a value for the variable C. Because F depends upon the value of C we call F the **dependent variable** and C the **independent variable**.

EXAMPLES

1 Plot the following sets of numbers on two parallel scales and from the plot derive an equation that relates units used with total cost:

Units used	500	1000	1500	2000	2500
Total cost	1750	3500	5250	7000	8750

Units U

500 1000 1500 2000 2500

1750 3500 5250 7000 8750

Cost T_c

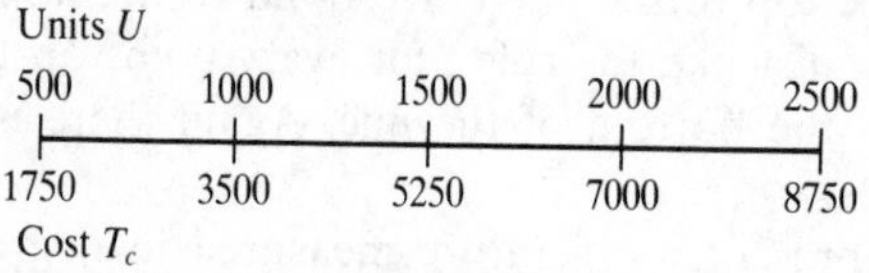

Figure 3.6

The equation is:

$$T_C = 3.5U$$

where T_C stands for total cost and U stands for units used.

2 From the following sets of numbers derive an equation that relates altitude in feet to time in minutes:

Altitude (feet)	32 000	29 000	26 000	23 000	20 000
Time (minutes)	3	6	9	12	15

For each 3 minute interval the altitude decreases by 3000 feet – that is 1000 feet per minute. When $T = 0$ the altitude would have been 35 000 feet. Consequently, the equation is:

$$A = 35\,000 - 1000T$$

where A is the altitude in feet and T is the time in minutes.

3 In each of the following, indicate:

(i) the subject of the equation
(ii) the dependent and the independent variables

(a) $y = 8 - 4x$ **(b)** $p = lm - 2$ **(c)** $t = 2\pi/\omega$

(a) Subject and dependent variable y, independent variable x.
(b) Subject and dependent variable p, independent variables l and m.
(c) Subject and dependent variable t, independent variable ω.

EXERCISES

1 Plot the following sets of numbers on two parallel scales and from the plot derive an equation that relates units sold to profit made:

Units sold	100	300	500	700	900
Profit	2.5	7.5	12.5	17.5	22.5

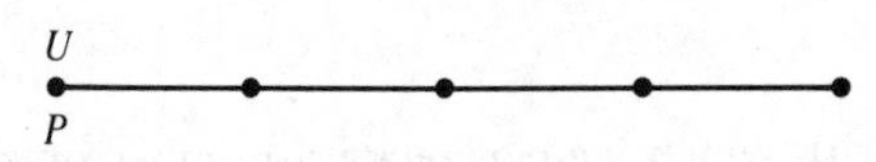

Figure 3.7

The equation is:

$$P = *U$$

where P is the profit and U stands for the number of units sold.

2 From the following sets of numbers derive an equation that relates tasks completed to time taken in minutes:

Tasks completed	2	4	6	8	10
Time taken	35	45	55	65	75

The equation is:

$$T = * + *N$$

where T is the time taken in minutes and N represents the number of tasks completed.

3 In each of the following, indicate:
(i) the subject of the equation
(ii) the dependent and the independent variables

(a) $r = 15 - 3s$ **(b)** $u = 12 - vw$ **(c)** $d = \sqrt{[4 - (x^2 + y^2)]}$

(a) Subject and dependent variable *, independent variable is *.
(b) Subject and dependent variable is *, independent variables are * and *.
(c) Subject and dependent variable is *, independent variables are * and *.

Unit 2 Interpolation and Extrapolation

Test yourself with the following:

1 From the following data derive an equation that relates radians to degrees:

Degrees	30	60	90	120	150	180
Radians	0.52	1.04	1.56	2.08	2.60	3.12

When this is done find the values of the degrees or radians that correspond to the following:
(a) 45 degrees (b) 1.17 radians (c) 200 degrees (d) 0.13 radians

Which conclusions are obtained by interpolation and which by extrapolation?

2 By using the following data derive an equation that relates units made to total manufacturing costs:

Units made	5	10	15	20	25	30
Total cost	390	655	920	1185	1450	1715

When this is done find the number of units made or total cost that correspond to the following:
(a) 7 units (b) Total cost 1291 (c) 1 unit (d) Total cost 2245
Which conclusions are obtained by interpolation and which by extrapolation?

Interpolation

When we construct parallel scales, the accuracy with which we can relate a number on one scale to the corresponding number on the other scale depends upon the quality of the construction. For example, when we related Celsius to Fahrenheit using the scales drawn in Figure 3.5 the accuracy was limited to the finest markers – the 25 °C lines. To relate temperatures within that limit requires the use of the equation, for example, the Fahrenheit equivalent of a Celsius reading of 28.75 °C is given as:

$$F = 32 + (1.8)(28.75)$$

$$= 83.75\,^{\circ}\mathrm{F}$$

Calculating related values between the marked values on the scale is known as **interpolation**.

Extrapolation

To relate temperatures that lie beyond the outer limits of the scale is not possible without extending the scale in the appropriate direction. However, once again the equation comes to our rescue. For example, the Fahrenheit equivalent of a Celsius reading of 250 °C is given as:

$$F = 32 + (1.8)(250)$$

$$= 482\,^{\circ}\mathrm{F}$$

Calculating related values beyond the boundary of the scale is known as **extrapolation**.

The equation that represents the relationship between the two sets of numbers comes from the tabulated numbers. Because of this it is not necessarily true that the relationship holds for numbers outside the tabulated range. For this reason extrapolation must be performed with this caution in mind.

EXAMPLES

1 From the following data derive an equation that relates distance in miles travelled to time taken in hours:

Miles	5	10	15	20
Time (hrs)	11.75	23.5	35.25	47.00

When this is done find the values of hours or miles that correspond to the following:

(a) 17.5 miles **(b)** 44 hours **(c)** 2.5 miles **(d)** 52 hours

Which conclusions are obtained by interpolation and which by extrapolation?

The equation relating miles to time is:

$$T = 2.35M$$

where M represents the number of miles travelled and T represents the number of hours taken.

(a) 17.5 miles in $T = 41.125$ hours — Interpolation
(b) 44 hours to travel $M = 18.7$ miles — Interpolation (to one dec. pl.)
(c) 2.5 miles in $T = 5.875$ hours — Extrapolation
(d) 52 hours to travel $M = 22.1$ miles — Extrapolation (to one dec. pl.)

2 Using the following data, derive an equation that relates the length of a spring in cm to the load applied in kg:

Load applied (kg)	2	4	6	8	10
Length (cm)	180	210	240	270	300

When this is done find the length of the spring or the load applied that corresponds to the following:

(a) 5.5 kg **(b)** 253 cm **(c)** 13 kg **(d)** 160 cm

Which conclusions are obtained by interpolation and which by extrapolation?

The equation relating the length of the spring to the load applied is:

$$L = 150 + 15K$$

where L is the length of the spring and K is the load applied.

(a) $K = 5.5$ kg so $L = 232.5$ cm — Interpolation
(b) $L = 253$ cm so $K = 6.87$ kg — Interpolation
(c) $K = 13$ kg so $L = 345$ cm — Extrapolation
(d) $L = 160$ cm so $K = \frac{2}{3}$ kg — Extrapolation

EXERCISES

1 From the following data derive an equation that relates the volume-to-surface-area ratio of a sphere to its radius in cm:

Radius (cm)	15	21	27	33	39
Ratio	5	7	9	11	13

When this is done find the radius or the ratio that corresponds to the following:

(a) Radius 18 cm **(b)** Ratio 8 **(c)** Radius 12 cm **(d)** Ratio 15

Which conclusions are obtained by interpolation and which by extrapolation?

The equation is

*

(a) Radius = 18 cm so Ratio = * — Interpolation
(b) Ratio = 8 so Radius = * — *polation
(c) Radius = 12 cm so Ratio = * — *polation
(d) Ratio = 15 so Radius = * — *polation

2 Using the following data, derive an equation that relates the number of units sold to the percentage discount given:

Units sold	1060	1120	1180	1240	1300
Discount (%)	3	6	9	12	15

When this is done find the units sold or the discount given that correspond to the following:
(a) Units sold 1150 **(b)** Discount 10% **(c)** Units sold 1500 **(d)** Discount 1%
Which conclusions are obtained by interpolation and which by extrapolation?

The equation relating units sold to discount is

*

(a) $U = 1150$ so $D = *\%$ *polation
(b) $D = 10\%$ so $U = *$ Interpolation
(c) $U = 1500$ so $D = *\%$ *polation
(d) $D = 1\%$ so $U = *$ Extrapolation

Unit 3 Relationships and Graphs

Test yourself with the following:

Plot the following set of data on a cartesian graph and highlight the plotted points in an appropriate manner:

p	5	10	15	20	25
q	6	−4	6	10	−3

Relationships

We have seen how relationships between two sets of numbers can be displayed using parallel scales and how it is possible to create an equation to describe the relation in symbolic form. There are, however, many relationships between two sets of numbers that cannot be displayed in this manner. For example, consider the following record of the time taken for a particle to travel certain distances:

Time (s)	1.0	2.0	3.0	4.0	5.0
Distance (m)	0.5	2.0	4.5	8.0	12.5

If you look carefully at the numbers you will see that it is not possible to record both sets on a pair of parallel scales in the manner that we have done previously. The times 1.0 s to 5.0 s could be plotted on one side of a parallel scale but the corresponding distances could not be plotted on the other side in any meaningful way because they do not increase by regular amounts for each lapsed second. We need an alternative method of displaying the information given in this table. This alternative method makes use of what is called the **cartesian coordinate system**.

Cartesian Coordinates

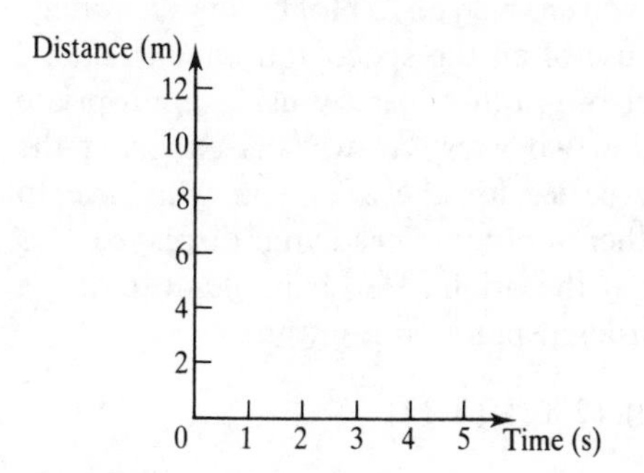

Figure 3.8

Instead of plotting two ranges of numbers on the two sides of one line we shall plot each range on a separate line, the lines being drawn perpendicular to each other. The number ranges that we shall plot are from 0 to 5 on the horizontal line, each point representing a particular second. The number range we shall plot on the vertical line will be from the 0 to the maximum distance 12.5.

To exhibit the relationship between time and distance we proceed as follows:

(1) Draw a vertical line through the point on the horizontal axis denoting time 1 second.
(2) Draw a horizontal line through the number 0.5 that represents the distance travelled at that particular time.
(3) Where these two lines intersect mark a point.
(4) Repeat this with every pair of numbers representing time and the corresponding distance.

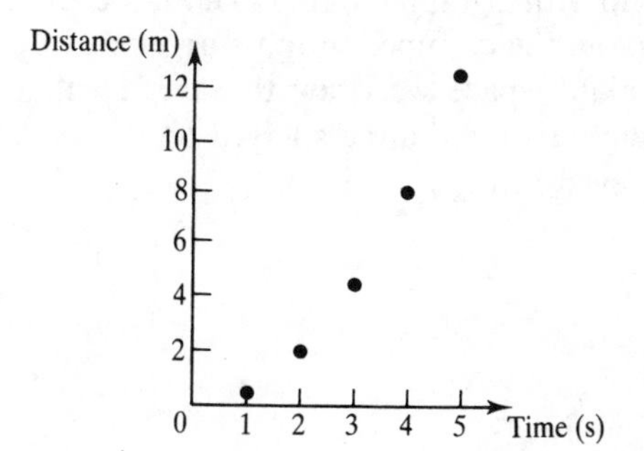

Figure 3.9

The final effect is as shown in Figure 3.9. The diagram is a pictorial display of the relationship that exists between the two pairs of numbers. It is called a **cartesian graph**.

Cartesian Graphs

Any two sets of numbers that are related to each other can be plotted on a cartesian graph. The two lines against which each set of data is plotted are called **axes** and they are taken to intersect at their common zero points – called the **origin**. The pairs of numbers that are used to plot each point of the graph are called the **coordinates** of the point and they appear within brackets in a strict order – they are called ordered pairs of numbers. The first number of the pair is called the horizontal coordinate and the second the vertical coordinate or **ordinate**.

This ensures, for example, that the point represented by the pair (2, 3) is different from the point represented by the pair (3, 2).

Drawing a Graph

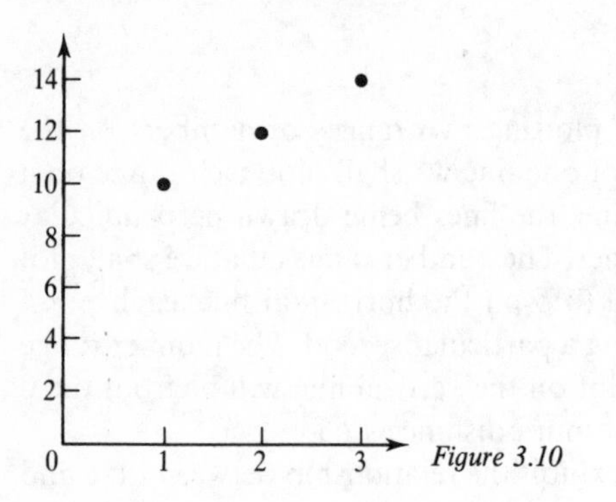

Figure 3.10

Whenever you draw a graph always use graph paper. This will enable you to plot points accurately. Also make use of all the space you have available on the sheet of graph paper by using appropriate scales on the two axes. As well as choosing the appropriate scales for the axes you will have to decide whether or not you need your displayed axes to intersect at the origin. For example, to plot the following ordered pairs on a graph:

$$(1, 10), (2, 12), (3, 14)$$

we note that the horizontal axis must contain the range of numbers 1 to 3 whereas the vertical axis needs to cover the range 10 to 14. If we included the origin of the coordinate system in the graph we would end up with a graph looking like Figure 3.10.

To avoid the cramped graph surrounded by a mass of blank space we draw the axes so that only the ranges required are displayed on the axes. This is achieved in Figure 3.11.

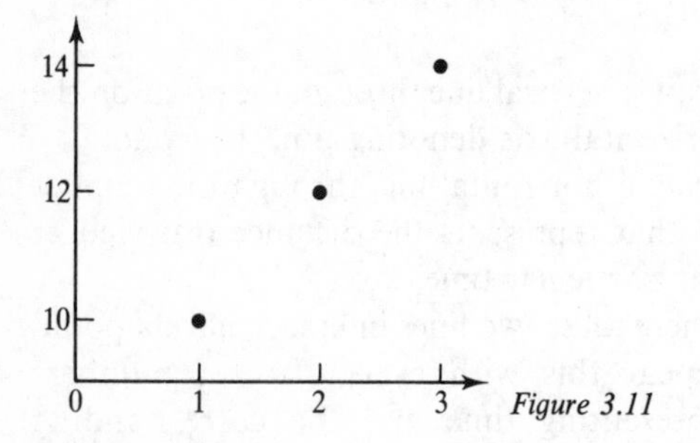

Figure 3.11

Highlighting the Graph

When pairs of points are plotted to form a cartesian graph the isolated points can be highlighted by various means.

Vertical Bars

By drawing in vertical bars from the horizontal axis to the point the graph is made clearer.

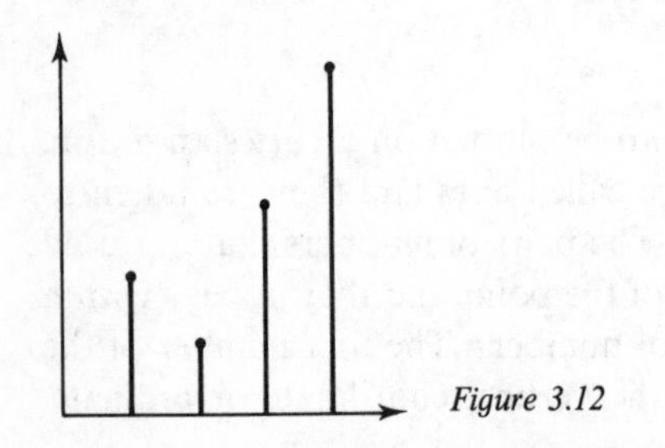
Figure 3.12

Jagged Lines

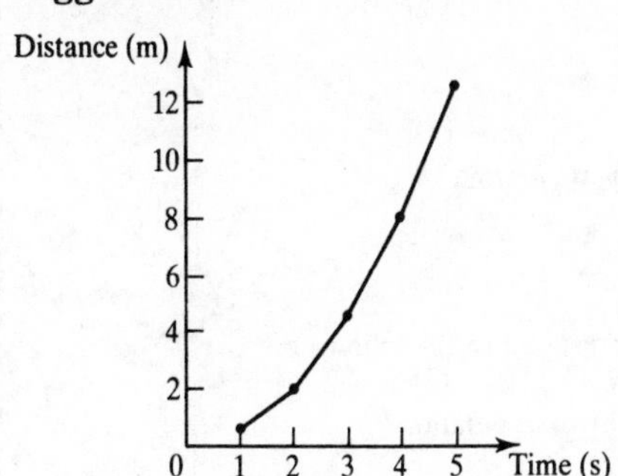

In the case of the time versus distance graph each point can be joined by a straight line to form a jagged line effect. Again, the plotted points are highlighted. In each of the preceding two cases the highlighting effect has nothing to do with the graph itself which simply consists of isolated points. Such graphs are called **discrete graphs** – they are plots of separate and distinct ordered pairs of numbers.

Figure 3.13

EXAMPLE

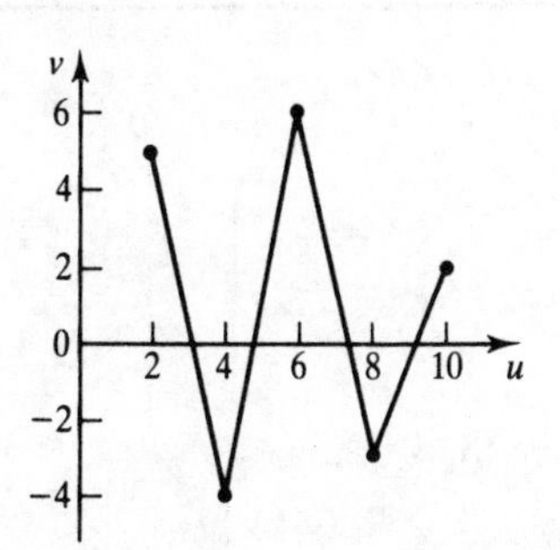

Plot the following set of data on a cartesian graph and highlight the plotted points in an appropriate manner:

u	2	4	6	8	10
v	5	−4	6	−3	2

Figure 3.14

EXERCISE

Plot the following set of data on a cartesian graph and highlight the plotted points in an appropriate manner:

s	1	3	5	7	9
t	4	8	−6	2	−5

Module 3: Further exercises

1 Plot the following sets of numbers on two parallel scales and from the plot derive an equation that relates ounces to grams:

Grams	0	50	100	150	200
Ounces	0	1.75	3.50	5.25	7.00

2 From the following sets of numbers derive an equation that relates the number of overtime hours worked to the overtime payment:

Overtime (hours)	2	4	6	8	10
Overtime pay (£)	14.5	29.0	43.5	58.0	72.5

3 In each of the following, indicate:

(i) the subject of the equation

(ii) the dependent and the independent variables

(a) $p = 4 + 3q$ (b) $z = xy$ (c) $s = ut + \frac{1}{2}t^2$

4 From the following data, derive an equation that relates mass to volume:

Mass (g)	10	20	30	40	50
Volume (cm^3)	14	28	42	56	70

When this is done find the values of mass or volume that correspond to the following:

(a) 15.5 g (b) 35 cm^3 (c) 2.5 g (d) 62 cm^3

Which conclusions are obtained by interpolation and which by extrapolation?

5 Plot the following set of data on a cartesian graph and highlight the plotted points in an appropriate manner:

a	4	8	12	16	20
b	−10	5	15	0	−5

Module 4: Displaying Relationships

Aims: To display linear relationships between two sets of numbers in graphical form and to link linear equations and linear inequalities to their graphs.

Objectives: When you have read this module you will be able to:

- ▶▶ Create a continuous straight-line graph from discrete data.
- ▶▶ Recognize the equation of a straight line and derive information from the equation.
- ▶▶ Model simple physical situations where two variables are linearly related and make predictions from the model.
- ▶▶ Solve the problem of two simultaneous linear equations.
- ▶▶ Solve problems involving linear inequalities and optimize from a feasible region.

Unit 1 The Straight Line

Test yourself with the following:

1 On a sheet of graph paper draw the following straight lines:
(a) $y = -4x - 6$ (b) $y = -2x + 2$ (c) $y = 9x - 7$ (d) $5x + 5y = 5$

2 Determine the gradients and the vertical intercepts of the following straight lines:
(a) $y = 2x - 1$ (b) $y = -4x + 2$ (c) $3x + 6y = 17$ (d) $x - y = 4$

3 Find the equation of the line through $(5, -6)$ that is parallel to:
(a) $y = 2x - 5$ (b) $y = -x + 1$ (c) $x + 4y = 2$ (d) $x - 5y = 10$

4 Write down the equation of the straight line that:
(a) has gradient 8, vertical intercept -2
(b) has gradient -3 and passes through $(1, 0)$
(c) passes through $(2, -3)$ and $(1, 4)$
(d) has vertical intercept -8 and passes through $(4, 2)$

5 In each of the following equations a value is given for one of the variables. Find the corresponding value of the other variable:
(a) $p = 8q + 10$ $p = 18$
(b) $y = 5 - 4x$ $y = -3$
(c) $2u + 3v = 4$ $u = 5$
(d) $5r - 4s = 2$ $r = 2$

Continuity from Discrete Data

The equation that expresses the relationship between degrees Celsius and degrees Fahrenheit is:

$$F = 32 + 1.8C$$

where C is the temperature measured in degrees Celsius and F is the same temperature measured in degrees Fahrenheit. From this equation we can construct a table of corresponding values of C and F from which we can construct ordered pairs:

$$(0, 32), (25, 77), (50, 122), (75, 167), (100, 212)$$

These ordered pairs can then be plotted against cartesian axes to produce the graph shown in Figure 4.1. The graph consists of a collection of isolated points – it is called a **discrete graph**. The graph represents only the actual points we have plotted and as such contains only that amount of information. If we were to plot more points – calculated from the equation – we would put more information into the graph (see Figure 4.2). The maximum amount of information the graph can contain is achieved when all possible points are plotted which will give the straight line shown in Figure 4.3. This is, of course, an ideal because to plot the straight line a point at a time would take for ever. The ideal, however, is what we are trying to achieve and the graph that we end up with in this ideal case is called a **continuous graph**. What we do in reality is to plot a finite number of points and join them together with a straight line. Indeed, for a straight line we need only plot two points.

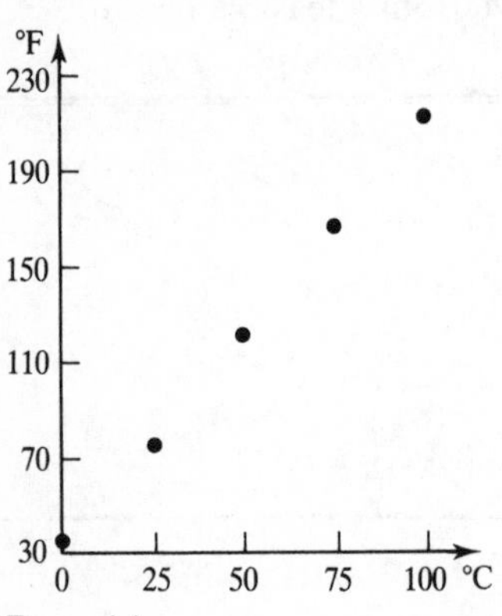

Figure 4.1

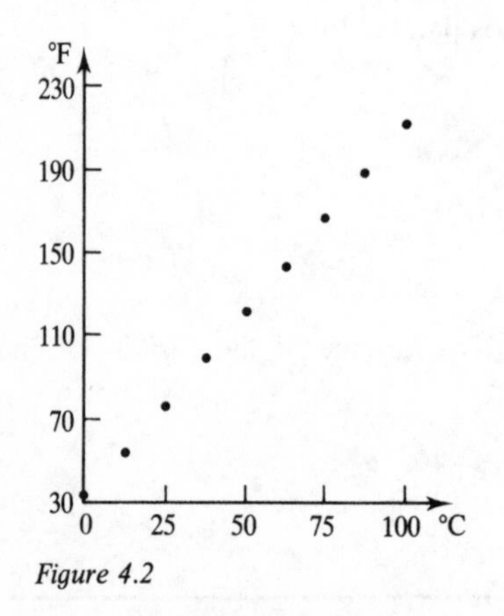

Figure 4.2

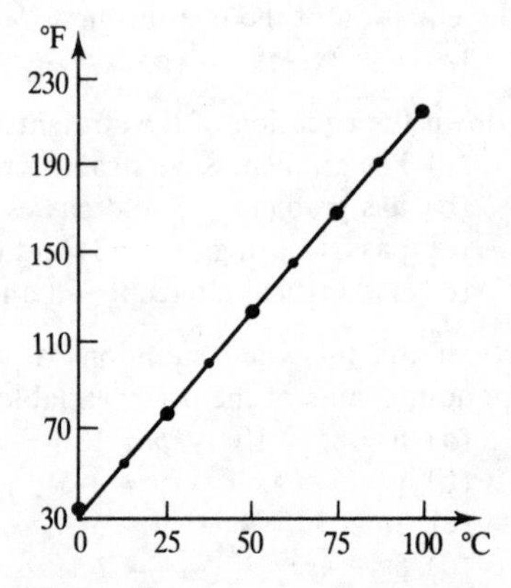

Figure 4.3

Straight Lines

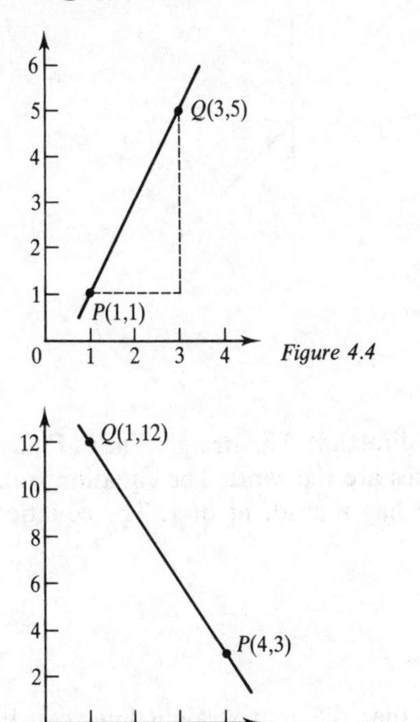

Figure 4.4

Figure 4.5

Every straight line is inclined to the horizontal axis and the **gradient** of the line is a measure of this inclination. If P and Q are two distinct points on the straight line the gradient is the ratio:

$$\frac{\text{Difference in vertical coordinates of } P \text{ and } Q}{\text{Difference in horizontal coordinates of } P \text{ and } Q}$$

which is equal to the increase in the vertical direction per unit increase in the horizontal direction. For example, in Figure 4.4 the vertical coordinates of points on the line increase as the corresponding horizontal coordinates increase. The gradient is 2. In Figure 4.5, the vertical coordinates of points on the line decrease as the corresponding horizontal coordinates increase. We call a decrease a **negative increase** so that the gradient of the line in this case is -3.

Special Straight Lines

In the vertical straight line of Figure 4.6 every point has the same q value, namely 3. The line could, therefore, be plotted from the equation $q = 3$. No matter what the p value is the q value is always 3. Notice that this line does not have a defined gradient. The increase in the vertical direction occurs over a zero increase in the horizontal direction and we have not defined division by zero.

A similar reasoning applies to the horizontal line of Figure 4.7 which can be plotted from the equation $p = -4$. Notice that the gradient of this line is zero. For any change in

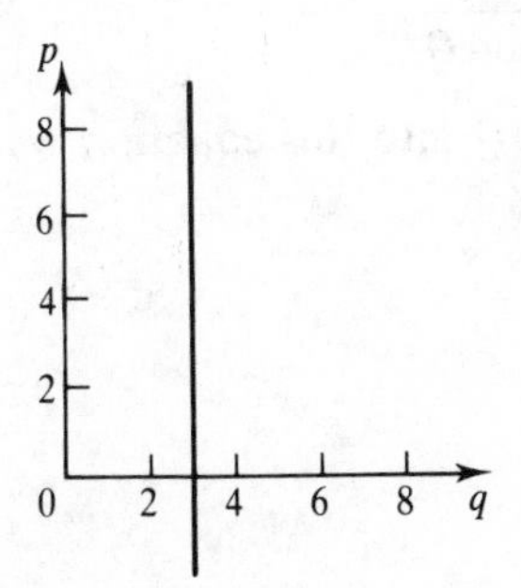

Figure 4.6

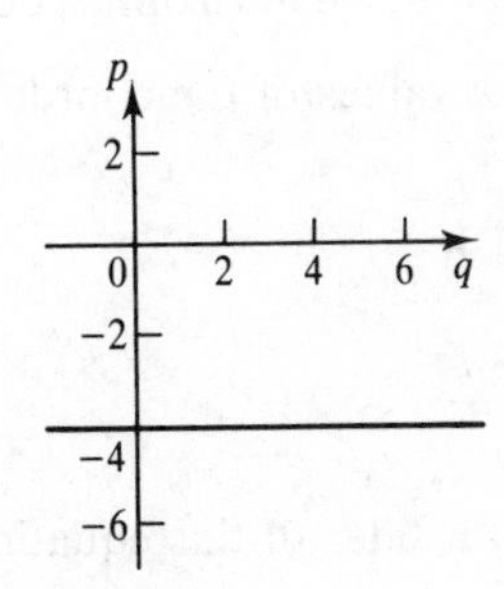

Figure 4.7

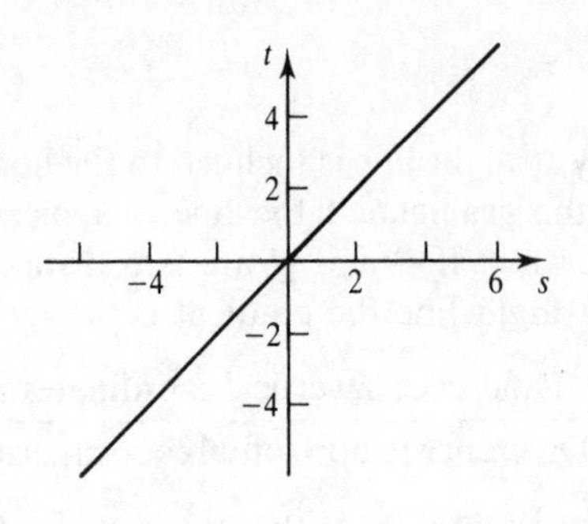

Figure 4.8

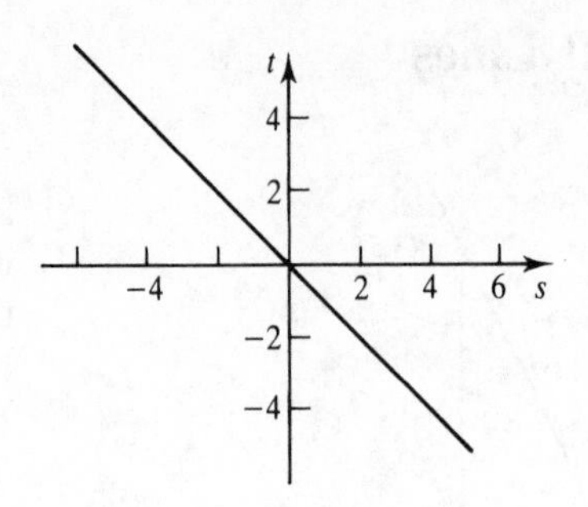

Figure 4.9

the horizontal direction there is no change in the vertical direction. The straight line in Figure 4.8 is such that at any point on the line the s and t values are the same. The equation from which such a line could be plotted is then $s = t$ which has a gradient of 1. The equation $s = -t$ gives the line with gradient -1 of Figure 4.9.

The General Straight Line

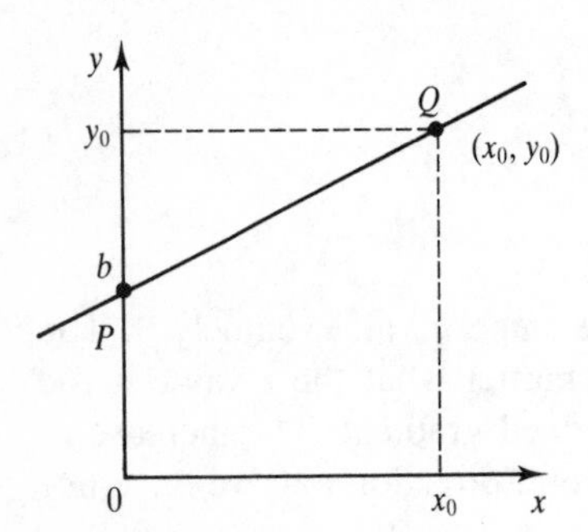

Figure 4.10

We have seen that different straight lines can be plotted from different equations. However, all these equations are variations of a general form. A typical straight line is the one drawn in Figure 4.10.

In Figure 4.10 the straight line intercepts the vertical axis at the point labelled P with coordinate $(0, b)$ and the other point labelled Q on the line has coordinates (x_0, y_0). Here the symbols x_0 and y_0 have the subscripts 0 attached to them to indicate that the point Q on the line is a fixed point and not a variable point with variable coordinates (x, y).

If we let a denote the gradient of the line then between any two distinct points on the line:

$$a = \frac{\text{Difference in vertical coordinates of } P \text{ and } Q}{\text{Difference in horizontal coordinates of } P \text{ and } Q}$$

Substituting the values of the coordinates of P and Q into this equation we find that:

$$a = \frac{y_0 - b}{x_0 - 0}$$

$$= \frac{y_0 - b}{x_0}$$

Multiplying both sides of this equation by x_0 yields:

$$ax_0 = y_0 - b$$

Adding b to both sides of this equation yields:

$$ax_0 + b = y_0$$

Interchanging sides of this equation, to make y_0 the subject, gives:

$$y_0 = ax_0 + b$$

This equation is satisfied by both the coordinates of point P and the coordinates of point Q. Indeed, because point Q is merely representative of any point on the line, we drop the subscript 0 and state that any point on the line will have coordinates that satisfy the equation

$$y = ax + b$$

This equation is referred to as the **equation of a straight line** where a is the gradient and b is the vertical intercept. Notice that this equation is identical in form to

$$F = 32 + 1.8C$$

where the variables are C and F instead of x and y. This equation has the straight-line graph of Figure 4.3 with gradient $a = 1.8$ and vertical intercept $b = 32$.

The Equation of a Straight Line

The equation of a straight line

$$y = ax + b$$

contains four quantities. These are two variables x and y, whose values are the coordinates of a point lying on the line, and two constants. The constants are the gradient a and the vertical intercept b. To find the value of any one of these quantities we require the values of the other three. The value of the fourth is then obtained by balancing the equation. Sometimes balancing the equation is a simple matter of substituting numbers. For example, if the equation of a straight line is given as

$$y = 5x - 2$$

and a value of x is given as 4 then the corresponding value of y is found by substituting numbers into the equation. That is

$$\begin{aligned} y &= (5 \times 4) - 2 \\ &= 18 \end{aligned}$$

At other times the equation may require some manipulation before the result can be found. For example, if, in the above equation, we are given that $y = 33$ how do we find x? Simple substitution on one side of the $=$ sign is insufficient:

$$33 = 5x - 2$$

We need to adjust the equation to isolate x and obtain an equation of the form:

$$x = \text{some number} \qquad \text{or} \qquad \text{some number} = x$$

To perform this particular case we add the number 2 to both sides to give

$$33 + 2 = 5x - 2 + 2$$

that is

$$35 = 5x$$

Finally, dividing both sides by 5 gives the required result:

$$7 = x$$

Balancing the equation means that if an arithmetic operation is performed on one side of the equation we must perform the identical operation on the other side to maintain the balance of the equation. For example, if

$$y = ax + b$$

then

$$y - b = ax + b - b$$

$$= ax$$

and

$$\frac{y - b}{a} = \frac{ax}{a}$$

$$= x$$

so that

$$x = \frac{y - b}{a}$$

Here we have changed the subject of the equation from y to x.

An Alternative Form

An alternative form for the equation of a straight line is:

$$px + qy = r$$

where p, q and r are constants. In this form the gradient is given by $-p/q$ and the vertical intercept by r/q. For example, the equation:

$$3x + 4y = 5$$

represents a straight line. To show that this equation can be rewritten in the standard form of the straight-line equation we proceed as follows:

(1) Subtract $3x$ from both sides of the equation:

$$3x + 4y - 3x = 5 - 3x$$

That is

$$4y = 5 - 3x$$

(2) Divide both sides of this equation by 4:

$$\frac{4y}{4} = \frac{5 - 3x}{4}$$

$$= \frac{5}{4} - \frac{3x}{4}$$

That is

$$y = \left(\frac{-3}{4}\right)x + \frac{5}{4}$$

which is the standard equation for a straight line with gradient $-3/4$ and vertical intercept $5/4$.

Notice that the two lines:

$$px + qy = r$$

and

$$px + qy = s$$

are parallel as they both have the same gradient $-p/q$ but different vertical intercepts r/q and s/q respectively.

EXAMPLES

1 On a sheet of graph paper draw the following straight lines and determine both their gradients and their vertical intercepts:

(a) $y = 2x + 3$ **(b)** $y = -3x + 4$ **(c)** $y = 4x - 2$ **(d)** $3x + 4y = 24$

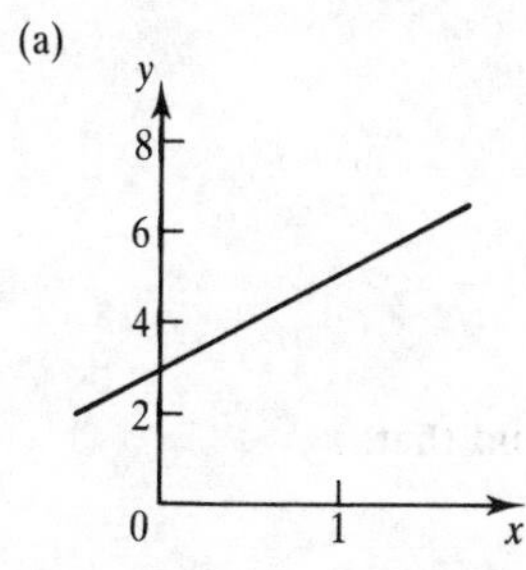

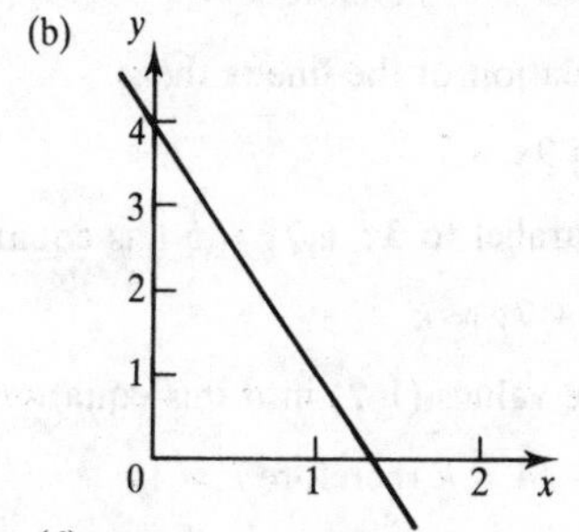

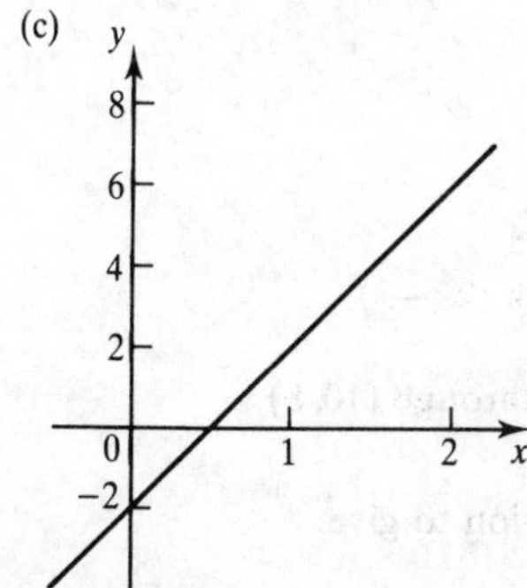

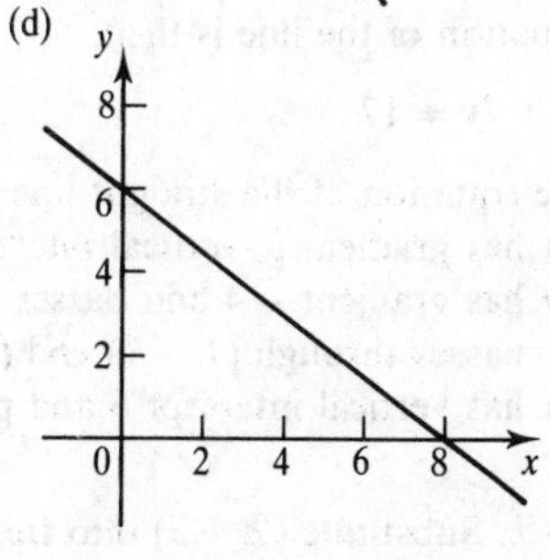

Figure 4.11

(a) Gradient 2, vertical intercept 3.
(b) Gradient -3, vertical intercept 4.
(c) Gradient 4, vertical intercept -2.
(d) Gradient $-3/4$, vertical intercept 6. The standard form of this equation is then $y = (-3/4)x + 6$.

2 Determine the gradients and the vertical intercepts of the following straight lines:
(a) $y = 5x - 4$ **(b)** $y = -7x + 8$ **(c)** $5x + 2y = 10$ **(d)** $2x - 4y = 16$

(a) Gradient 5, vertical intercept -4.
(b) Gradient -7, vertical intercept 8.
(c) The equation $5x + 2y = 10$ can be rewritten as

$$y = -\left(\frac{5}{2}\right)x + 5$$

This is a straight line of gradient $-5/2$ and vertical intercept 5.
(d) The equation $2x - 4y = 16$ can be rewritten as

$$y = \tfrac{1}{2}x - 4$$

This is a straight line of gradient $\frac{1}{2}$ and vertical intercept -4.

3 Find the equation of the line through (1, 2) that is parallel to:
(a) $y = 9x - 4$ **(b)** $3x + 7y = 5$

(a) Any line parallel to $y = 9x - 4$ has the same gradient but a different vertical intercept. Hence, the equation is of the form:

$$y = 9x + k$$

Substituting the values (1, 2) into this equation it is found that:

$$2 = 9 + k \text{ therefore } k = -7$$

The desired equation of the line is then:

$$y = 9x - 7$$

(b) Any line parallel to $3x + 7y = 5$ has equation

$$3x + 7y = k$$

Substituting the values (1, 2) into this equation it is found that:

$$3 + 14 = k \text{ therefore } k = 17$$

The desired equation of the line is then:

$$3x + 7y = 17$$

4 Write down the equation of the straight line that:
(a) has gradient 3, vertical intercept -5
(b) has gradient -4 and passes through $(2, -3)$
(c) passes through $(1, -1)$ and $(2, 6)$
(d) has vertical intercept 6 and passes through (10, 8)

(a) $y = 3x - 5$
(b) $y = -4x + b$. Substitute $(2, -3)$ into this equation to give:

$$-3 = -8 + b \text{ therefore } b = 5$$

The equation of the line is then $y = -4x + 5$.

(c) $y = ax + b$. Because the line passes through the points $(1, -1)$ and $(2, 6)$ the gradient of the line is

$$a = \frac{\text{Increase in vertical}}{\text{Increase in horizontal}}$$

$$= \frac{6 - (-1)}{2 - 1}$$

$$= 7$$

The equation is then

$$y = 7x + b$$

Substituting the coordinates $(1, -1)$ into this equation yields

$$-1 = 7 + b$$

Therefore

$$b = -8$$

The final equation is then

$$y = 7x - 8$$

(d) Vertical intercept 6 and passes through $(10, 8)$. Here the equation is of the form:

$$y = ax + 6$$

Substituting the coordinates $(10, 8)$ into the equation yields

$$8 = 10a + 6$$

That is

$$2 = 10a$$

So that

$$a = \tfrac{1}{5}$$

So the equation is

$$y = \tfrac{1}{5}x + 6$$

5 In each of the following equations a value is given for one of the variables. Find the corresponding value of the other variable:

(a) $y = 3x - 4$ $\quad y = 2$
(b) $l = -4 - 2m$ $\quad l = 0$
(c) $3a + 4b = -5$ $\quad b = 1$
(d) $6p - 5q = -13$ $\quad p = 2$

(a) $y = 3x - 4$ $\quad y = 2$ Therefore

$$2 = 3x - 4$$

Adding 4 to both sides of this equation gives

$$2 + 4 = 3x - 4 + 4$$

that is

$$6 = 3x$$

Dividing both sides by 2 gives the result:

$$2 = x$$

(b) $l = -4 - 2m \qquad l = 0$ Therefore

$$0 = -4 - 2m$$

Adding 4 to both sides of this equation gives

$$0 + 4 = -4 - 2m + 4$$

that is

$$4 = -2m$$

Dividing both sides by -2 gives the result:

$$-2 = m$$

(c) $3a + 4b = -5 \qquad b = 1$ Therefore

$$3a + 4 = -5$$

Subtracting 4 from both sides of this equation gives

$$3a + 4 - 4 = -5 - 4$$

that is

$$3a = -9$$

Dividing both sides by 3 gives the result:

$$a = -3$$

(d) $6p - 5q = -13 \qquad p = 2$ Therefore

$$12 - 5q = -13$$

Subtracting 12 from both sides of this equation gives

$$12 - 5q - 12 = -13 - 12$$

that is

$$-5q = -25$$

Dividing both sides by -5 gives the result:

$$q = 5$$

EXERCISES

1 On a sheet of graph paper draw the following straight lines and determine their gradients and vertical intercepts:

(a) $y = 5x - 9$ **(b)** $8x + 9y = 72$

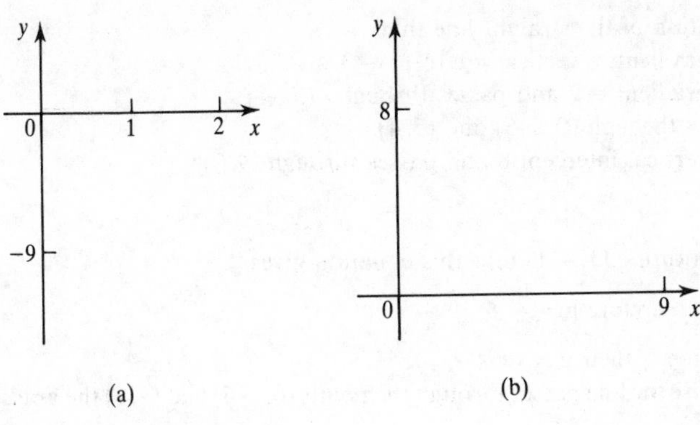

Figure 4.12

(a) Gradient ∗, vertical intercept −∗.
(b) Gradient 8/∗, vertical intercept ∗.

2 Determine the gradients and the vertical intercepts of the following straight lines:
(a) $y = 2x + 3$ **(b)** $y = -6x - 10$ **(c)** $7x + 4y = 11$ **(d)** $5x - 3y = 12$

(a) Gradient ∗, vertical intercept 3.
(b) Gradient −6, vertical intercept ∗.
(c) The equation $7x + 4y = 11$ can be rewritten as

$$y = \frac{-*}{4} + \frac{*}{4}$$

This is a straight line with slope $-*/4$ and vertical intercept $*/4$.

(d) The equation $5x - 3y = 12$ can be rewritten as

$$y = *x - *$$

This is a straight line with slope ∗ and vertical intercept ∗.

3 Find the equation of the line through (2, 3) that is parallel to:
(a) $y = 8x - 3$ **(b)** $7x - 2y = 6$

(a) Any line parallel to $y = 8x - 3$ has the equation

$$y = *x + k$$

Substituting (2, 3) into this equation gives

$$3 = * + k \text{ therefore } k = -*$$

The equation for the line is then $y = *x - *$.

(b) Any line parallel to $7x - 2y = 6$ has the equation

$$*x - *y = k$$

Substituting (2, 3) into this equation gives

$$* - * = k \text{ therefore } k = *$$

The equation for the line is then $*x - *y = *$.

4 Write down the equation of the straight line that:

(a) has gradient 5, vertical intercept -3
(b) has gradient -2 and passes through $(1, -1)$
(c) passes through $(0, -5)$ and $(3, 4)$
(d) has vertical intercept 8 and passes through $(9, 0)$

(a) $y = *x + *$
(b) $y = *x + b$. Substituting $(1, -1)$ into this equation gives

$$* = * + b \text{ therefore } b = *$$

The equation of the line is then $y = *x + *$.
(c) $y = ax + b$. Because the line passes through the points $(0, -5)$ and $(*, *)$ the gradient of the line is:

$$a = \frac{\text{Increase in vertical}}{\text{Increase in horizontal}}$$

$$= \frac{* - *}{* - *}$$

$$= *$$

The equation is then

$$y = *x + b$$

Substituting the coordinates $(*, *)$ into this equation yields

$$* = * + b$$

Therefore

$$b = *$$

The final equation is then

$$y = *x + *$$

(d) $y = mx + *$. Substituting $(9, 0)$ into this equation gives

$$* = *m + * \text{ therefore } m = *$$

The equation of the line is then $y = *x + *$.

5 In each of the following equations a value is given for one of the variables. Find the corresponding value of the other variable:

(a) $u = 4v - 2$ $\quad u = 10$ $\qquad$ **(c)** $2y - 8x = 2$ $\quad x = 1$
(b) $s = -9 - 2t$ $\quad s = 3$ $\qquad$ **(d)** $3l + 4m = -6$ $\quad l = -2$

(a) $u = 4v - 2$ $\quad u = 10$ therefore

$$* = 4v - 2$$

Adding $*$ to both sides of this equation gives

$$* = 4v$$

Dividing both sides of this equation by $*$ gives the result:

$$* = v$$

(b) $s = -9 - 2t$ $s = 3$ therefore

$$* = -9 - 2t$$

Adding * to both sides of this equation gives

$$* = -2t$$

Dividing both sides of this equation by * gives the result:

$$* = t$$

(c) $2y - 8x = 2$ $x = 1$ therefore

$$2y - * = 2$$

Adding * to both sides of this equation gives

$$2y = *$$

Dividing both sides of this equation by * gives the result:

$$y = *$$

(d) $3l + 4m = -6$ $l = -2$ therefore

$$* + 4m = -6$$

Adding * to both sides of this equation gives:

$$4m = *$$

Dividing both sides of this equation by * gives the result:

$$m = *$$

Unit 2 Simultaneous Linear Equations

Test yourself with the following:

1 Draw a graph to determine at what points the following pairs of lines cross:

(a) $y = x - 1$, $y = -2x + 5$ (b) $2x + 2y = 2$, $-2x - 4y = 8$

2 A triangle is formed by the straight lines:

$$2x - 5y = 5, \quad 4x + 6y = -6 \quad \text{and} \quad y = 2x + 3$$

What are the coordinates of the vertices?

3 A straight line given by $y = x + 1$ is intersected by the two lines:

$$x + y = 1 \quad \text{and} \quad y = -2x - 5$$

What are the coordinates of the point midway between the two points of intersection?

Two Lines

Two straight lines are either parallel or they intersect at a point. If two lines are parallel then they have the same gradient and their equations will differ only in the value of the vertical intercept. For example, the lines with equations

$$y = 5x - 2$$

$$y = 5x + 6$$

are parallel, as are

$$2x + 3y = 4$$

$$2x + 3y = -1$$

If the lines are not parallel then their equations will differ in the values of their gradients. For example, the lines with equations

$$y = 2x + 4$$

$$y = 7x - 1$$

intersect at a single point and the coordinates of that point will satisfy both equations. We describe such a pair of equations as **simultaneous equations** because they possess a common, simultaneous solution in the form of the coordinates of their point of intersection. It is our object to devise a method of finding that point.

The Point of Intersection Found Graphically

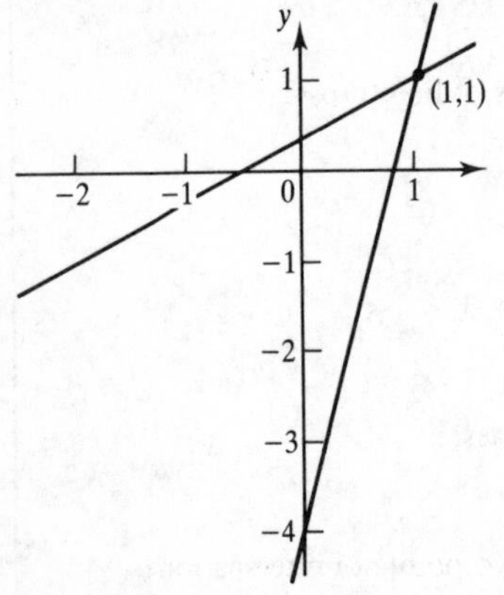

Figure 4.13

If the two equations are plotted on a sheet of graph paper then their point of intersection can be found graphically, as shown in Figure 4.13.

On the graph the two straight lines given by the equations

$$y = 5x - 4$$

$$2x - 3y = -1$$

are seen to intersect at the point with coordinates $x = 1$ and $y = 1$. These values for the two variables can be seen to satisfy both equations because the same point lies on both lines.

EXAMPLES

At what point do the following pairs of lines cross?

(a) $y = 3x - 4$ **(b)** $3x + 3y = 11$
$y = -2x + 6$ $2x - 6y = 18$

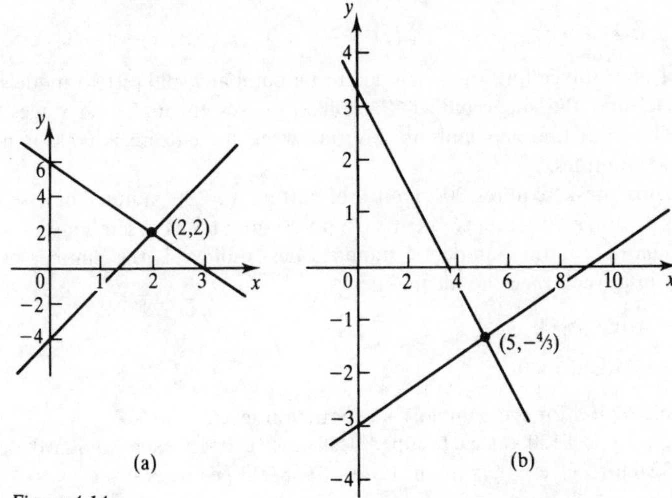

Figure 4.14

(a) From the plot of the two lines it can be seen that they intersect at $x = 2$, $y = 2$.
(b) From the plot of the two lines it can be seen that they intersect at $x = 5$, $y = -4/3$.

EXERCISES

At what point do the following pairs of lines cross?

(a) $y = x - 4$ **(b)** $x + y = 2$
$y = -4x + 6$ $3x - 5y = 14$

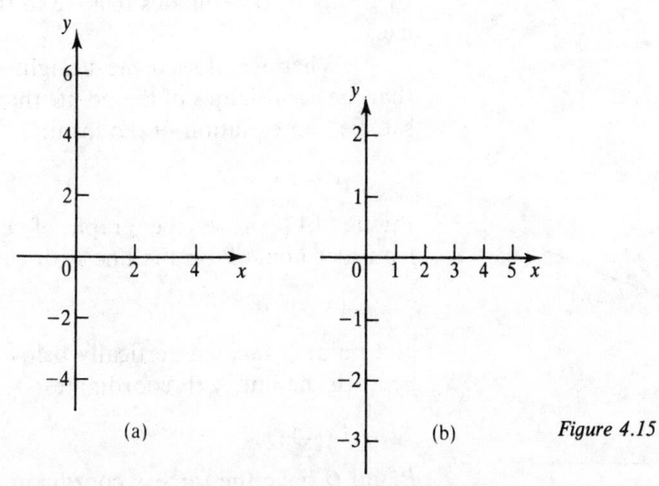

Figure 4.15

(a) From the plot of the two lines it can be seen that they intersect at $x = *$, $y = *$.
(b) From the plot of the two lines it can be seen that they intersect at $x = *$, $y = *$.

Unit 3 Linear Inequalities

Test yourself with the following:

1 On a cartesian graph with x- and y-axes, shade in the regions defined by the following inequalities:

(a) $x < 4$ (b) $y \geqslant -x$ (c) $x + y \leqslant 1$ (d) $2x - y > 3$

2 A department of an office furniture manufacturing company makes two models of desk, the standard model and the super model. Each item passes through two stages, cutting and assembling. The total time available in any one week for cutting is 6000 minutes and for assembly 10 000 minutes.

Each super desk requires 20 minutes of cutting and 30 minutes of assembling. Each standard desk requires 10 minutes of cutting and 20 minutes of assembling.

Show that if x is the number of standard desks and y is the number of super desks manufactured in a week then the inequalities:

$$10x + 20y \leqslant 6000$$

$$20x + 30y \leqslant 10\,000$$

define a feasible region for the company's production levels.

If the profit is £120 on each super desk and £70 on each standard desk, find the production of each to give a maximum profit of £38 000 per week.

The Plane and Linear Inequalities

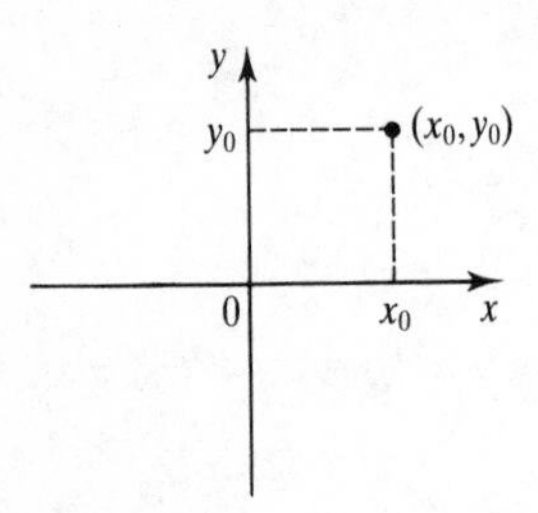

Figure 4.16

Cartesian axes are drawn on a flat area called a **plane** and any point in that plane can be located by giving its coordinates relative to the coordinate axes.

When we plotted the straight line we found that the coordinates of the points that lay on a line satisfied an equation of the form:

$$y = ax + b$$

Figure 4.17 shows the graph of a typical line. Consider point P on the line with coordinates:

$$(x_0, y_0)$$

and point Q located vertically below point P and beneath the line with coordinates:

$$(x_0, y_1)$$

P and Q have the same x coordinate because Q is vertically below P and, for the same reason, the y coordinate of Q is less than the y coordinate of P.

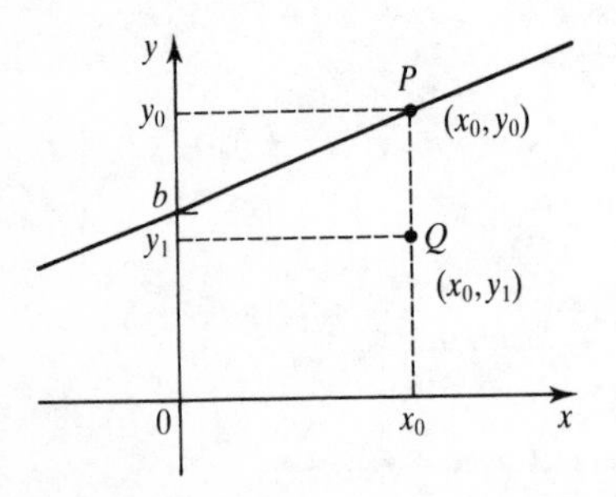

Figure 4.17

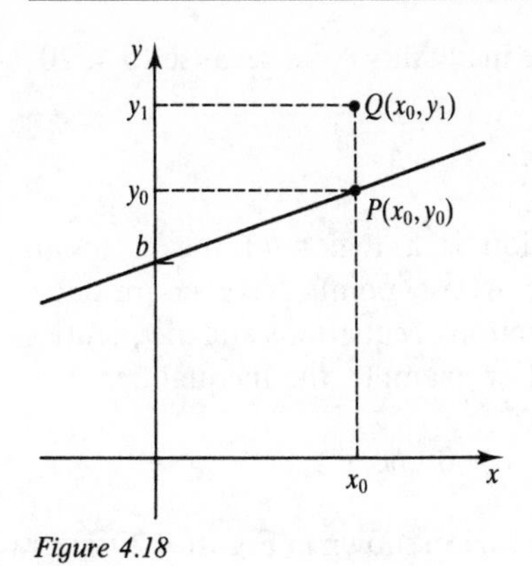

Figure 4.18

This is expressed as

$$y_1 < y_0$$

and

$$y_0 = ax_0 + b$$

so that

$$y_1 < ax_0 + b$$

Indeed, any point that is below the line will have coordinates that satisfy the inequality

$$y < ax + b$$

Similarly, the inequality

$$y > ax + b$$

is satisfied by the coordinates of any point in the plane that lies above the line. Because these inequalities represent regions of the plane bounded by a straight line they are called **linear inequalities**.

Modelling with Linear Inequalities

Many problems can be modelled by using a set of linear equations and linear inequalities. For example, a company manufactures two types of machine, machine S and machine T. Each month it manufactures more machines of type S than it does of type T and it manufactures at least 10 of type T to meet a constant demand. We shall now plot on a graph all the possible options for the numbers of each type of machine that the company manufactures. First we translate the information given into inequalities.

Let s represent the number of machines of type S made per month and t represent the number of type T. From the information given we know that:

- The number of machines type S is greater than the number of machines type T. This is expressed by the inequality

  $$t < s$$

- There are at least 10 machines of type T. This is expressed by the inequality

  $$t \geqslant 10$$

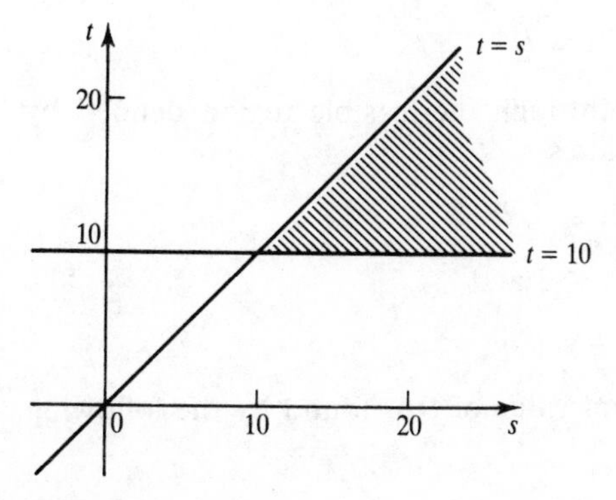

Figure 4.19

These two inequalities can then be represented graphically as in Figure 4.19. Any point above the horizontal line has coordinates that satisfy the inequality $t \geqslant 10$. Any point below the sloping line has coordinates that satisfy the inequality $t < s$. Any point in the shaded region is both on or above the horizontal line and below the sloping line, so it has coordinates that satisfy both inequalities simultaneously.

The coordinates of any point in the shaded region represent the permissible number of machines of type S and type T. The shaded region is called the **feasible region**. For example, the point (20, 15) satisfies the inequality $t < s$ because $15 < 20$. It also satisfies the inequality $t \geqslant 10$ because $15 \geqslant 10$.

The Feasible Region

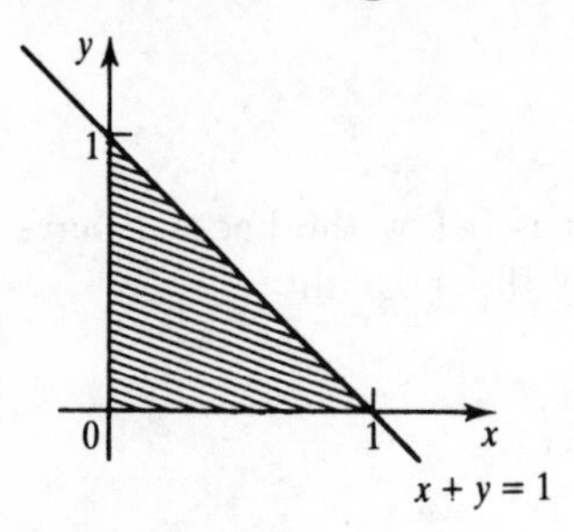

Figure 4.20

The feasible region is a region of the cartesian coordinate plane whose points have coordinates that satisfy a collection of equations and inequalities simultaneously. For example, the inequalities

$$x + y \leqslant 1, \; x \geqslant 0 \text{ and } y \geqslant 0$$

define the feasible region shown in Figure 4.20. Here $x \geqslant 0$ signifies all points on and to the right of the vertical axis, $y \geqslant 0$ signifies all points on and above the horizontal axis and $x + y \leqslant 1$ signifies all points on and below the line $x + y = 1$. The only points that satisfy all three inequalities simultaneously are those in the shaded region – the feasible region.

Optimization

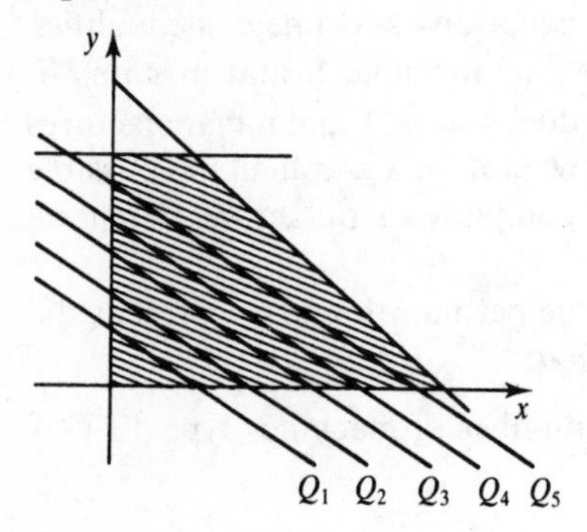

Figure 4.21

We have seen that the equation

$$px + qy = Q$$

is the equation of a straight line with gradient $-p/q$ and vertical intercept Q/q. Keeping p and q as fixed constants and plotting lines with different Q-values will produce a family of parallel straight lines. If the only members of this family of straight lines to be considered are those that pass through the feasible region, then for one of those lines Q will have a minimum value and for another line Q will have a maximum value. The process of finding the maximum or minimum value of Q is called **optimization**. For example, if the family of straight lines given by

$$x + 2y = Q$$

is to pass through the feasible region defined by the inequalities

$$0 \leqslant x \leqslant 5,$$

$$0 \leqslant y \leqslant 3 \text{ and}$$

$$y \leqslant -x + 6$$

the optimum value of Q is found by the following procedure:

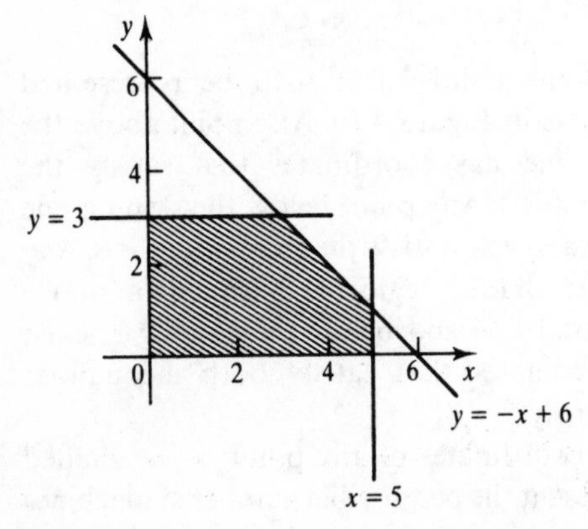

Figure 4.22

(1) Draw the feasible region. (See Figure 4.22.)

(2) Choose a value of Q and draw a single line:

$$y + 2x = Q$$

where the value of Q is chosen to make the line pass through the feasible region – here Q is chosen to be 2. (See Figure 4.23.)

(3) Optimize the value of Q by noting that all lines parallel to $y + 2x = Q$ have the same form for the equation but with different Q-values. From the diagram it can be seen that the line with the greatest Q-value is the one passing through $x = 5$ and $y = 1$. The line with the smallest Q-value is the line passing through the origin. The minimum value of Q is zero and the maximum value of Q is 11.

Figure 4.23

EXAMPLES

1 On a cartesian graph with x- and y-axes, shade in the regions defined by the following inequalities:

(a) $y \geqslant 5$ **(b)** $x > -y$ **(c)** $x - y \geqslant -2$ **(d)** $3x + 2y < 6$

(a)

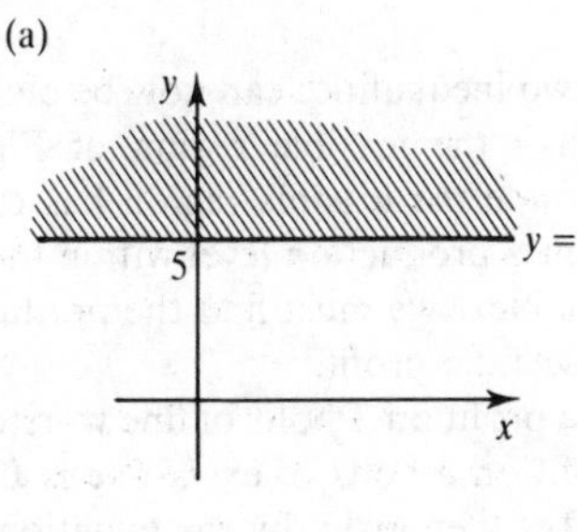

(b)

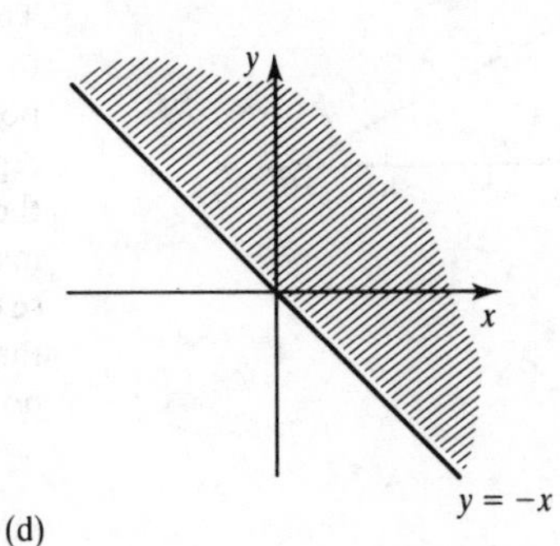

(c)

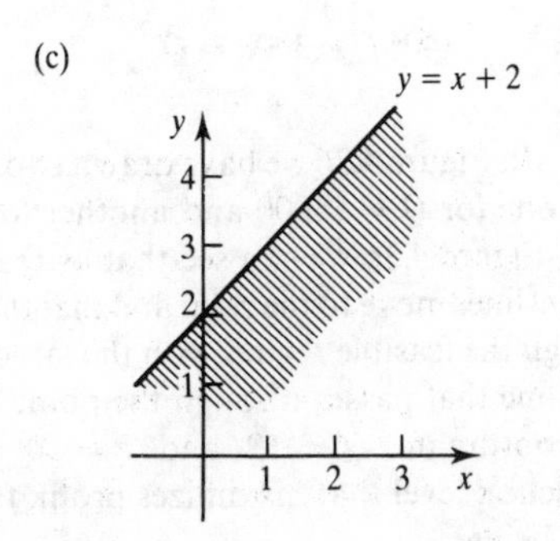

(d)

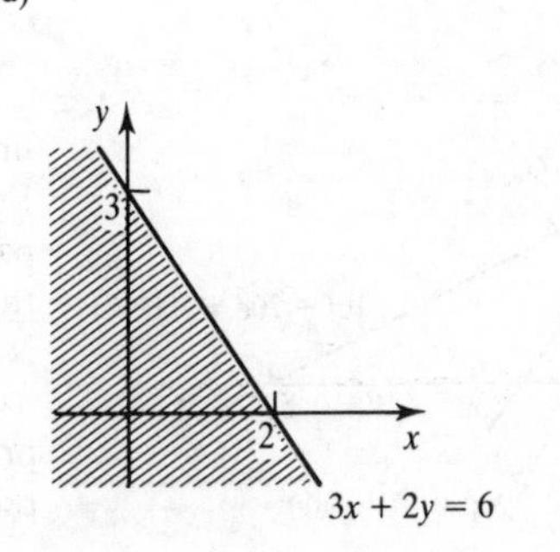

Figure 4.24

2 A woollen mill manufactures two qualities of worsted fabric, fine worsted and extra-fine worsted. Each fabric passes through two stages of manufacture, weaving and finishing. Production of worsted is performed according to the following criteria:

- **(a)** Each bolt of fine worsted requires 20 hours of weaving and each bolt of extra-fine worsted requires 10 hours of weaving. In any one week no more than 400 hours of weaving time is available.
- **(b)** Each bolt of fine worsted requires 10 hours of finishing and each bolt of extra-fine worsted requires 20 hours of finishing. In any one week no more than 500 hours of finishing time is available.
- **(c)** Profit on a bolt of fine worsted is £400 and profit on a bolt of extra-fine worsted is £300.

Find the weekly production of each type of worsted to give a maximum profit.

To solve this problem we must first model the criteria in the form of equations and inequalities. Let f be the number of bolts of fine worsted manufactured in a week and e the number of bolts of extra-fine worsted.

(a) The time taken to weave f bolts of fine is $20f$ hours and the time taken to weave e bolts of extra-fine is $10e$ hours. The maximum weaving time available is 400 hours so that:

$$20f + 10e \leqslant 400$$

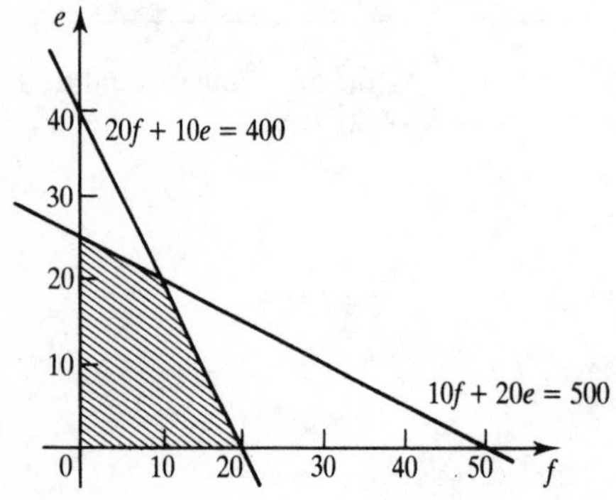

Figure 4.25

(b) The time taken to finish f bolts of fine is $10f$ hours and the time taken to finish e bolts of extra-fine is $20e$ hours. The maximum finishing time available is 500 hours so that:

$$10f + 20e \leqslant 500$$

These two inequalities can now be plotted on a graph to display the feasible region of Figure 4.25. Any point inside the feasible region has coordinates that represent a production level within the capabilities of the mill. Next we must find the production level that maximizes the profit.

(c) The profit on f bolts of fine worsted is £$400f$ and the profit on e bolts of extra-fine is £$300e$. The total profit Q is then given by the equation:

$$400f + 300e = Q$$

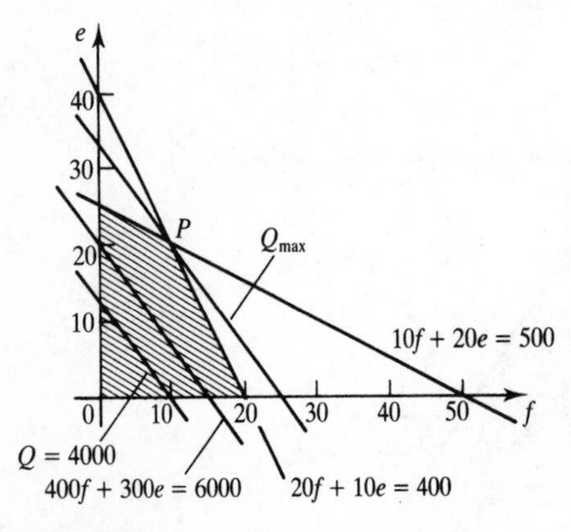

Figure 4.26

In Figure 4.26 we have drawn two typical profit lines, one for $Q = £4000$ and another for $Q = £6000$. From Figure 4.26 we can see that as Q increases, the parallel lines move to the right and that the line passing through the feasible region with the largest value of Q is the line that passes through the point P. This point has coordinates $f = 10$ and $e = 20$. This is the production level that maximizes profit, the maximum profit being:

$$400 \times 10 + 300 \times 20 = £10\,000$$

EXERCISES

1 On a cartesian graph with x- and y-axes, shade in the regions defined by the following inequalities:

(a) $x \geqslant -3$ **(b)** $y < -2x$ **(c)** $x < 1 - y$ **(d)** $5x - 4y \geqslant 20$

(a)

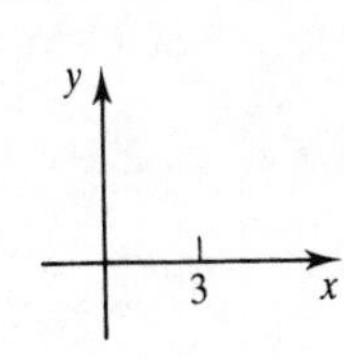

(b)

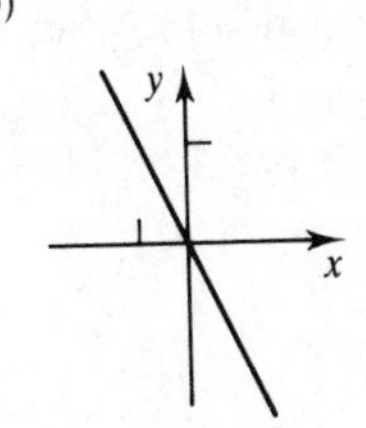

(c)

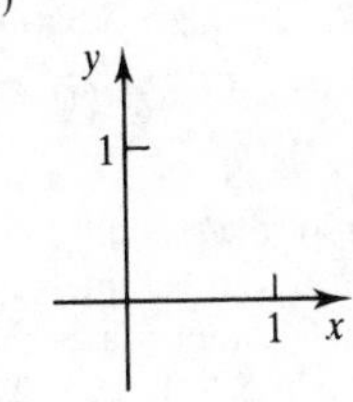

(d)

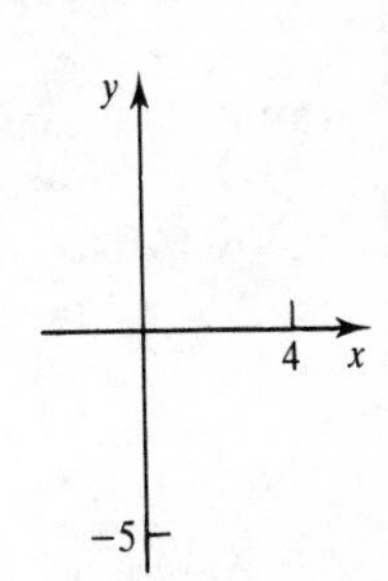

Figure 4.27

2 A factory has two processing divisions RM (Raw Materials) and FP (Finished Parts). Two of the products manufactured, namely X101 and Y304 have the following time requirements per unit:

	RM	FP
X101	10	25
Y304	20	10

The total time available in each division is:

RM	FP
1000	1000

If each finished unit yields a stock value of £1000 find the production which will maximize the stock value.

If we let x represent the number of X101s and y the number of Y304s the inequalities derived from the two tables are:

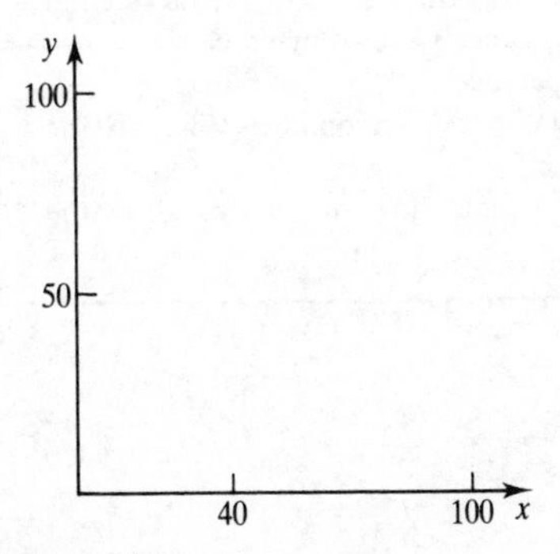

Figure 4.28

- Time used in RM is given by:

  $$10x + 20y \leqslant *$$

- Time used in FP is given by:

  $$25x + *y \leqslant 1000$$

- The equation for stock value V is given as:

  $$V = 1000x + *$$

Plotting these inequalities on a sheet of graph paper and a typical stock value line for $V = *$ demonstrates that the maximum value of V within the feasible region is given by $x = *$ and $y = *$.

Module 4: Further exercises

1 On a sheet of graph paper draw the following straight line:
(a) $y = -3x - 3$ (b) $y = -3x + 1$ (c) $y = 8x + 5$ (d) $2x + 2y = 4$

2 Determine the gradients and the vertical intercepts of the following straight lines:
(a) $y = 5x - 3$ (b) $y = -2x + 2$ (c) $2x + 4y = 16$ (d) $x + y = 8$

3 Find the equation of the line through $(1, -1)$ that is parallel to:
(a) $y = 3x + 4$ (b) $y = -2x - 1$ (c) $2x + 3y = 6$ (d) $4x - 2y = 12$

4 Write down the equation of the straight line that:
(a) has gradient 6, vertical intercept 4
(b) has gradient -5 and passes through $(4, -3)$
(c) passes through $(5, -1)$ and $(2, 6)$
(d) has vertical intercept -4 and passes through $(3, 0)$

5 At what point do the following pairs of lines cross?
(a) $y = 2x + 4$, $y = -4x + 22$
(b) $3x + 4y = -5$, $-4x + 12y = 24$

6 A triangle is formed by the straight lines:

$$x - y = 2,\ x + y = 3 \text{ and } y = 2x - 1$$

What are the coordinates of the vertices?

7 A straight line given by $y = -3x + 5$ is intersected by the two lines:

$$x - y = 3 \text{ and } y = 2x$$

What are the coordinates of the point midway between the two points of intersection?

8 In each of the following equations a value is given for one of the variables. Find the corresponding value of the other variable:
(a) $s = 3r + 4$ $s = 12$ (c) $u + 2v = 3$ $u = -5$
(b) $y = 5 - 6x$ $y = -7$ (d) $5m - 4n = -3$ $m = -3$

9 Solve the following pair of simultaneous equations graphically:

$$6x - 2y = 2 \qquad y = 2x - 4$$

10 On a cartesian graph with x- and y-axes, shade in the regions defined by the following inequalities:
(a) $y < -2$ (b) $3x < -6y$ (c) $x - 2y \leqslant -4$ (d) $2x + 8y \geqslant -16$

11 A chemical plant produces two compounds, compound A and compound B. The plant has a contract to supply at least 5 tonnes of compound B per week but cannot obtain sufficient supplies to manufacture more than 10 tonnes of each compound per week. Because of the continuous nature of the production process the plant manufactures twice as much compound A as compound B and to avoid closing down any part of the plant the total tonnage must be at least 20 tonnes per week.

It takes four times as long to produce one tonne of compound A as compound B. What are the production quantities that will minimize production time?

If a revolutionary new process reversed the manufacturing times how would this affect the production quantities for minimum production time?

Chapter 1: Miscellaneous exercises

1 Simplify: (a) $\left(\dfrac{-1}{125}\right)^{-2/3}$ (b) $81^{-2^{-2}}$

2 A car travels $x/10$ metres in 10 seconds. If this rate is maintained for 10 minutes, how many metres does it travel in 10 minutes?

3 Let r be the result of doubling both the base and the index of 3^2. If $r = (3^2) \times (z^2)$ find z.

4 In the base 10 the number 123 means

$$1 \times 10^2 + 2 \times 10 + 3$$

On the planet Betzelgar, however, numbers are written in the base b. Fred Perfect purchases a car for 550 Betzelgar dollars. He gives the salesmen a note for 1000 dollars and receives in change 120 dollars. What is the base b of the number system?

5 The annual changes in the population of a beehive for four successive years are 30% increase, 25% increase, 25% decrease and 30% decrease. What is the net percentage change over the four years to the nearest percent?

6 Simplify:

(a) $\dfrac{1^{-3/4}}{5^{-1} + 3^{-1}}$ (b) $\sqrt{\tfrac{4}{3}} - \sqrt{\tfrac{3}{4}}$

7 Given the collection of n numbers, $n > 1$, of which one of them is $1 + 1/n$, and all the others are 1, find the arithmetic mean and the standard deviation.

8 Find the collection of x-values that satisfy the inequalities

$$2 \leqslant |x - 1| \leqslant 5$$

9 The arithmetic mean of a collection of 50 numbers is 38. If two of the numbers, namely 45 and 55, are discarded what is the arithmetic mean of the remainder?

10 A merchant bought 832 widgets. He sold 800 of them for the price paid for the 832. The remaining widgets were sold at the same price per widget as the other 800. Based on the cost, find the percentage gain on the entire transaction.

11 Write the decimal $0.\dot{3}\dot{6}$ as a fraction in its lowest terms.

12 Which of the following are correct statements:

(a) $[\sqrt{(-9)}][\sqrt{(-81)}] = \sqrt{[(-9)(-81)]}$ (c) $\sqrt{729} = 27$

(b) $\sqrt{[(-9)(-81)]} = \sqrt{729}$

13 The symbol 31_b represents a two-digit number in the base b. If the number 102_b is double the number 31_b find b.

2

■ ■ ■ Algebra

The aims of this chapter are to:

1 Review the arithmetic work done with numbers in a symbolic form.

2 Demonstrate how to factorize polynomials and how to expand brackets into polynomials.

3 Demonstrate how to solve polynomial equations and how to separate algebraic fractions into partial fractions.

4 Display the use of matrices as stores of information and introduce the concepts of vectors and geometric vectors.

5 Review the elements of trigonometry and thereby lay the groundwork for an extended description of the trigonometric functions in Module 13 of Chapter 3.

6 Develop an awareness of the existence of complex numbers and demonstrate their elementary properties in cartesian form.

This chapter contains six modules:

Module 5: Algebraic Expressions
The foundations of algebra are laid in this module. Symbols are defined, as are the rules for combining them with the arithmetic operations.

Module 6: Polynomials and the Binomial Theorem
By considering the graphical description of polynomials, this module demonstrates how polynomials can be factorized. The module ends with a discussion of binomial expansions for positive integer powers.

Module 7: Solving Algebraic Equations
The algebraic manipulations covered in the previous two modules are called into play to demonstrate how the solutions to algebraic equations are obtained. The module ends with a discussion of partial fractions.

Module 8: Matrices and Vectors
Matrices are considered as stores of information. Their arithmetic and the inverses of 2×2 matrices are considered. Vectors and geometric vectors are introduced to lay the foundation for a further course on vector analysis.

Module 9: The Right-angled Triangle
The various geometric features of the right-angled triangle are discussed and the elements of trigonometry introduced. Pythagoras' theorem is demonstrated and special right-angled triangles are considered.

Module 10: Complex Numbers
Complex numbers were briefly introduced in Module 7 when solutions to quadratic equations were considered. In this module they are more formally defined and their elementary properties and algebra are considered within the cartesian representation.

Module 5: Algebraic Expressions

Aim: To review the arithmetic work done with numbers in a symbolic form.

Objectives: When you have read this module you will be able to:

- ▶▶ Use symbols in place of numbers and combine symbols using all the operations of arithmetic.
- ▶▶ Simplify algebraic expressions by rearrangement and by abstracting common factors.
- ▶▶ Evaluate expressions by substituting numbers or alternative symbols.

Unit 1 Symbols and Their Algebra

Test yourself with the following:

1 Remove the brackets from the following:

(a) $5\{x - 4[y + 3(z - 6)]\}$ (c) $8\{7[x + 2(y - 4)] - 5(z + 3)\}$

(b) $[2(x - 4) - 5(y + 3)]$

2 Is subtraction a commutative operation?

3 Is division an associative operation?

4 Simplify each of the following into a single algebraic fraction:

(a) $\frac{a}{b} + \frac{b}{a}$ (c) $\frac{a}{b} - \frac{c}{b^2} + \frac{d}{a}$

(b) $\frac{a^2}{b^3} \div \frac{a}{b^2}$

Symbols

In the previous chapter we considered the equation

$$F = 32 + 1.8C$$

which related a temperature measured in degrees Fahrenheit (F) to the same temperature measured in degrees Celsius (C). The symbols F and C refer to the number of degrees Fahrenheit and the number of degrees Celsius, respectively. To obtain the equivalent Fahrenheit temperature for a given Celsius temperature we substitute the Celsius value in the equation and work out the Fahrenheit equivalent using the rules of arithmetic. When we substitute numbers for symbols in this way we are said to assign values to them. There are two types of symbol that concern us here: **variables** and **constants**.

Variables

Symbols that can be assigned more than one value are called variables. For example, in the equation

$$x + y = 1$$

the letters x and y can assume many different values and still satisfy the equation. For instance $x = 1$, $y = 0$ and $x = -36$, $y = 37$ are two such pairs of values. The letters x and y in this equation are accordingly referred to as variables.

Constants

Symbols that can be assigned just one value are called constants. For example, for a given straight line the equation is given as

$$y = ax + b$$

Here the symbols x and y are variables because they can be assigned the coordinates of any point that lies on the line. The symbols a and b, however, can only be assigned a single value each for a specific line. The symbol a is the gradient and the symbol b is the vertical intercept and for a specific line these are constant.

Algebra

The word 'algebra' is used to describe a mathematical system of objects and operations between objects. Arithmetic has as objects the real numbers and as operations addition, subtraction, and so on. We could call arithmetic the 'algebra of numbers'. Used as a single word, algebra is traditionally taken to mean arithmetic when we include alphabetic symbols along with the numerals.

Operations

The operations of algebra are the same as those of arithmetic, namely addition, subtraction, multiplication, division and raising to a power. When combining symbols together using these operations, all the usual arithmetic conventions apply. For example, the sum of p and q is written as $p + q$, and the product as $p \times q$ or, more usually, pq. Division is denoted by $p \div q$ or, more usually, p/q and symbols can be bracketed together.

Brackets

As we saw in the previous chapter, brackets are used to group symbols together. To remove the brackets from an algebraic expression the precedence rules dictate that those operations within brackets are performed first. To remove brackets that are nested within each other the innermost brackets are removed first. For example,

$$\begin{aligned} 3\{x - [5 + 2(x - y)]\} &= 3[x - (5 + 2x - 2y)] \\ &= 3(x - 5 - 2x + 2y) \\ &= 3x - 15 - 6x + 6y \\ &= 6y - 3x - 15 \end{aligned}$$

Notice that

$$
\begin{aligned}
-(x-y) &= -1(x-y) \\
&= (-1)x - (-1)y \\
&= -x + y
\end{aligned}
$$

The Rules of Algebra

To regulate the operations of algebra we make use of a number of rules that govern the way we manipulate symbols and numbers. There are a number of such rules but only three will be mentioned here.

Commutativity

The order in which we add or multiply two numbers together does not matter.

$$x + y = y + x$$

$$xy = yx$$

Notice that if

$$xy = 0$$

then either $x = 0$ or $y = 0$, or both $x = 0$ and $y = 0$.

Associativity

The order in which we group numbers when adding them or when multiplying them does not matter.

$$x + (y + z) = (x + y) + z$$

$$x(yz) = (xy)z$$

Distributivity

This rule dictates how the operations of multiplication and division distribute themselves over the operation of addition and subtraction. Here we see that multiplication is both left and right distributive but that division is only right distributive.

$$x(y + z) = xy + xz$$

$$x(y - z) = xy - xz$$

$$(x + y)z = xz + yz$$

$$(x - y)z = xz - yz$$

$$(x \pm y) \div z = x \div z \pm y \div z$$

Algebraic Fractions

Numerical fractions are obtained by dividing one integer by another. Division of symbols follows the same rules to create algebraic fractions. For example,

$$5 \div 3 \text{ can be written as } \frac{5}{3} \text{ so } a \div b \text{ can be written as } \frac{a}{b}$$

Addition and Subtraction

The addition and subtraction of an algebraic fraction follow the same rules as addition and subtraction of numerical fractions. For example,

$$\frac{a}{b} + \frac{c}{d} = \frac{ad}{bd} + \frac{cd}{bd} = \frac{ad + cb}{bd}$$

where bd is the LCM of b and d.

Multiplication

Algebraic fractions are multiplied by multiplying their numerators and denominators. For example,

$$\frac{a}{b} \times \frac{c}{d} = \frac{ac}{bd}$$

Division

To divide by an algebraic expression we multiply by its reciprocal. For example,

$$\frac{a}{b} \div \frac{c}{d} = \frac{a}{b} \times \frac{d}{c}$$

$$= \frac{ad}{bc}$$

Powers

All the previously noted rules for powers apply:

$$a^1 = a$$

$$a^0 = 1$$

$$a^3 \times a^2 = a^{3+2} = a^5$$

$$a^5 \div a^2 = a^{5-2} = a^3$$

$$(a^3)^4 = a^{3 \times 4} = a^{12}$$

$$a^{-2} = \frac{1}{a^2}$$

and fractional powers denote roots.

EXAMPLES

1 Remove the brackets from the following:

(a) $9\{x + 2[y - (z + 4)]\}$ **(c)** $6\{4[x - (y + 3)] + 2(z - 1)\}$
(b) $[7(3 + x) - (y - 2)4]$

(a)
$$\begin{aligned} 9\{x + 2[y - (z + 4)]\} &= 9[x + 2(y - z - 4)] \\ &= 9(x + 2y - 2z - 8) \\ &= 9x + 18y - 18z - 72 \end{aligned}$$

(b)
$$\begin{aligned} [7(3 + x) - (y - 2)4] &= (21 + 7x - 4y + 8) \\ &= 7x - 4y + 29 \end{aligned}$$

(c)
$$\begin{aligned} 6\{4[x - (y + 3)] + 2(z - 1)\} &= 6[4(x - y - 3) + 2z - 2] \\ &= 6(4x - 4y - 12 + 2z - 2) \\ &= 6(4x - 4y + 2z - 14) \\ &= 24x - 24y + 12z - 84 \end{aligned}$$

2 Is division a commutative operation?

If division were a commutative operation then for any pair of numbers represented by x and y it would be true that:

$$x \div y = y \div x$$

This is not the case as can be seen by the simple substitution of 1 for x and 2 for y:

$$\begin{aligned} \text{left-hand side} &= 1 \div 2 \\ &= \tfrac{1}{2} \\ \text{right-hand side} &= 2 \div 1 \\ &= 2 \end{aligned}$$

so that the left-hand side does not equal the right-hand side.

3 Is division distributive over addition from the left?

If division were distributive over addition from the left then for any three numbers represented by x, y and z it would be true that:

$$x \div (y + z) = (x \div y) + (x \div z)$$

This is not the case as can be seen by the simple substitution of 1 for x, 2 for y and 3 for z:

$$\begin{aligned} \text{left-hand side} &= 1 \div (2 + 3) \\ &= \tfrac{1}{5} \\ \text{right-hand side} &= (1 \div 2) + (1 \div 3) \\ &= \tfrac{5}{6} \end{aligned}$$

so that the left-hand side does not equal the right-hand side. In particular $1/(x + y)$ does not equal $1/x + 1/y$.

4 Simplify each of the following into a single algebraic fraction:

(a) $\dfrac{a}{b} - \dfrac{b}{a}$ **(b)** $\dfrac{a}{b^2} \div \dfrac{b}{a^2}$ **(c)** $\dfrac{a}{c} + \dfrac{c}{b} + \dfrac{a^2}{bc}$

$$\textbf{(a)}\ \frac{a}{b} - \frac{b}{a} = \frac{aa - bb}{ab} \qquad ab \text{ is the LCM of the denominators}$$

$$= \frac{a^2 - b^2}{ab}$$

$$\textbf{(b)}\ \frac{a}{b^2} \div \frac{b}{a^2} = \frac{a}{b^2} \times \frac{a^2}{b}$$

$$= \frac{aa^2}{b^2 b}$$

$$= \frac{a^3}{b^3}$$

$$= \left(\frac{a}{b}\right)^3$$

$$\textbf{(c)}\ \frac{a}{c} + \frac{c}{b} + \frac{a^2}{bc} = \frac{ab + cc + a^2}{bc}$$

$$= \frac{ab + c^2 + a^2}{bc}$$

EXERCISES

1 Remove the brackets from the following:

(a) $-\{2x + 3[4y - 6(z - 1)]\}$ **(c)** $8\{[x + 2(5 - y)] + 4(z + 2)\}$

(b) $[(x - 3)4 - 2(y + 3)]$

$$\textbf{(a)}\ -\{2x + 3[4y - 6(z - 1)]\} = -[2x + 3(4y - * + *)]$$

$$= -(2x + * - * + *)$$

$$= -2x * 12y * 18z * 18$$

$$\textbf{(b)}\ [(x - 3)4 - 2(y + 3)] = (* - * - * - *)$$

$$= 4x - 2y - 18$$

$$\textbf{(c)}\ 8\{[x + 2(5 - y)] + 4(z + 2)\} = 8[(x + 10 - 2y) + *]$$

$$= 8(* + *)$$

$$= *$$

$$= 8x - 16y + 32z + 144$$

2 Is subtraction an associative operation?

If subtraction were an associative operation then for any three numbers represented by x, y and z it would be true that:

$$(x * y) * x = x * (y * z)$$

This is not the case as can be seen by the simple substitution of 1 for x, 2 for y and 3 for z:

$$\text{left-hand side} = *$$

$$= -4$$

$$\text{right-hand side} = *$$

$$= 2$$

so that the left-hand side does not equal the right-hand side.

3 Is division distributive over subtraction from the left?

If division were distributive over subtraction from the left then for any three numbers represented by x, y and z it would be true that:

$$x * (y * z) = (x * y) * (x * z)$$

This is not the case as can be seen by the simple substitution of 2 for x, 3 for y and 4 for z:

$$\begin{aligned} \text{left-hand side} &= * \\ &= -2 \\ \text{right-hand side} &= * \\ &= \tfrac{1}{6} \end{aligned}$$

so that the left-hand side does not equal the right-hand side.
In particular

$$\frac{1}{x - y} \text{ does not equal } \frac{1}{x} - \frac{1}{y}$$

4 Simplify each of the following into a single algebraic fraction:

(a) $\frac{a}{b} + \frac{c}{a}$ **(b)** $\frac{a}{b^2} \cdot \frac{b}{a^2}$ **(c)** $\frac{a}{bc} + \frac{c}{ab} + \frac{b}{ac}$

(a) $\frac{a}{b} + \frac{c}{a} = \frac{a* + c*}{*}$ $\quad *$ is the LCM of the denominators

$$= \frac{* + *}{*}$$

(b) $\frac{a}{b^2} \cdot \frac{b}{a^2} = \frac{*}{*}$

$$= \frac{*}{*^2}$$

$$= \frac{1}{*}$$

$$= (ab)^{-1}$$

(c) $\frac{a}{bc} + \frac{c}{ab} + \frac{b}{ac} = \frac{a* + c* + b*}{*}$

$$= \frac{*^2 + *^2 + *^2}{abc}$$

Unit 2 Expressions: Their Simplification and Evaluation

Test yourself with the following:

1 Simplify each of the following expressions:

(a) $2x + 3yz - x + 4yz$

(b) $st - 3tu - ps + 3pu$

(c) $abc + ab + a$

(d) $2\{[6(x - 2) + 4(3 - y)] - 7[(x + 2) - 5(y - 4)]\}$

2 A simple pendulum consists of a compact mass suspended by a cord and performing small oscillations about the vertical. If the length of the cord is L cm and the time taken to perform a single complete swing is T s, the periodic time, the equation linking L and T is

$$T = 2\pi\sqrt{\frac{L}{g}}$$

where $g = 979.05\ \mathrm{cm\,s^{-1}}$ is the acceleration due to gravity.
Calculate the periodic time T for pendula with lengths 50 cm, 100 cm and 200 cm.

3 If a coil of wire is tightly wound into n circular coils each of radius r cm and a current of I amperes is passed through it, the magnetic field H gauss a distance x from the centre of the coils is given by:

$$H = 2\pi n r^2 I(r^2 + x^2)^{-3/2}$$

If $n = 100$, $r = 5$ cm, $I = 5$ amperes and $x = 35$ cm find the value of H.

4 Rewrite the following expressions using the equations given:
(a) $x^2 + y^2$ where $x = at$ and $y = a\sqrt{(1 - t^2)}$
(b) $u^2 + 2as$ where $s = ut + \frac{1}{2}at^2$

Algebraic Expressions

An **algebraic expression** is a collection of variables, constants and numbers connected together by the arithmetic operations. For example,

$$[3 - a(x^2 - y^2)]^{1/2}$$

is an algebraic expression involving the variables x and y, the constant a and the numbers 3, 2 and $\frac{1}{2}$.

Terms and Coefficients

An algebraic expression consists of **terms** and **coefficients.** The terms are symbols or groups of symbols and the coefficients are the numbers that accompany the terms. For example, in the expression

$$15s^2 - 3t$$

the number 15 is the coefficient of the s^2 term and -3 is the coefficient of the t term.

Simplifying Expressions

Expressions can be simplified by combining **like terms** and by factorizing **similar terms.**

Like Terms

Like terms can be added together. For example, the expression

$$3r - 5pq + 2r + 7pq$$

can be rearranged as

$$3r + 2r - 5pq + 7pq$$

which is simplified to

$$5r + 2pq$$

Similar Terms and Factors
Similar terms are terms with a symbol in common and common symbols can be factorized. For example, the expression

$$2xz - 6xy$$

can be simplified by recognizing that each term is a multiple of $2x$. This can be factored out to give

$$2x(z - 3y)$$

This procedure is called **factorizing the expression.**

Expanding Brackets

Sometimes it may be desired to reverse the process of factorizing an expression by removing the brackets. For example, the brackets in the expression

$$3a(b + c) - 2a(b - c)$$

can be removed to give

$$3ab + 3ac - 2ab + 2ac$$

This process is known as **expanding the brackets**. Having expanded the brackets we can further simplify this expression to

$$ab + 5ac$$

Finally we can factorize the a to obtain

$$a(b + 5c)$$

Evaluating Expressions

Expressions can be evaluated by substituting numbers for symbols or even by substituting alternative symbols.

Substituting Numbers
When numerical values are assigned to variables and constants in an expression, the expression itself assumes a numerical value. For example, if $L = 20$ and $g = 980$ then the expression

$$2\pi\sqrt{\frac{L}{g}}$$

is evaluated as

$$2\pi\sqrt{(20/980)} = 0.898 \text{ to three decimal places}$$

Because an expression can be numerically evaluated we can assign its values to another variable and so form an equation. For example, the equation

$$p = 3u^2 - 4v$$

states that the variable p can be assigned a value obtained by evaluating $3u^2 - 4v$. The variable p is the dependent variable whose value depends on the values of the independent variables u and v.

Substituting Symbols

Sometimes a variable in an expression is related to other variables. In such a case the other variables can be substituted. For example, if p is related to v by the equation

$$p = v^2$$

the expression

$$5pq - p^2$$

can be written in terms of v as

$$5(v^2)q - (v^2)^2$$

which is

$$5v^2q - v^4$$

EXAMPLES

1 Simplify each of the following expressions:

(a) $3p - 4q + 5p - 2q$
(b) $2r - 4pr - r + 6pr$
(c) $4\{[s(t-1) - t(7+s)] - [2(st+3) - 5(1-st)]\}$
(d) $\sqrt{a} + \dfrac{b}{\sqrt{a}}$
(e) $(x+y)^{1/2} - x(x+y)^{-1/2}$

(a) $$\begin{aligned} 3p - 4q + 5p - 2q &= 3p + 5p - 4q - 2q \\ &= 8p - 6q \end{aligned}$$

(b) $$\begin{aligned} 2r - 4pr - r + 6pr &= r + 2pr \\ &= r(1+2p) \end{aligned}$$

(c) $$\begin{aligned} &4\{[s(t-1) - t(7+s)] - [2(st+3) - 5(1-st)]\} \\ &= 4[(st - s - t7 - ts) - (2st + 6 - 5 + 5st)] \\ &= 4(st - s - 7t - st - 2st - 6 + 5 - 5st) \\ &= 4(st - st - 2st - 5st - s - 7t - 6 + 5) \\ &= 4(-7st - s - 7t - 1) \\ &= 4[-s(7t+1) - (7t+1)] \\ &= 4[(-s-1)(7t+1)] \\ &= -4(s+1)(7t+1) \end{aligned}$$

(d) $\sqrt{a} + b/\sqrt{a}$. To simplify this expression we make note of the fact that

$$\frac{1}{\sqrt{a}} = \frac{\sqrt{a}}{a}$$

so that

$$\begin{aligned}\sqrt{a} + \frac{b}{\sqrt{a}} &= \sqrt{a} + \sqrt{a}\left(\frac{b}{a}\right)\\ &= \sqrt{a}\left(1 + \frac{b}{a}\right)\\ &= \frac{\sqrt{a}(a+b)}{a}\\ &= \frac{(a+b)}{\sqrt{a}}\end{aligned}$$

(e) $$\begin{aligned}(x+y)^{1/2} - x(x+y)^{-1/2} &= (x+y)^{-1/2}[(x+y) - x]\\ &= \frac{y}{(x+y)^{1/2}}\end{aligned}$$

2 In an electrical circuit a resistance of R ohms is placed in series with n batteries each of internal resistance of r ohms. Each battery produces an electromotive force (e.m.f.) of E volts. The formula for this arrangement is given by

$$I = \frac{nE}{R + nr}$$

where I is the current flowing in the circuit measured in amperes.

If $R = 22.35$ ohms, $n = 5$, $E = 2.01$ volts and $r = 1.02$ ohms, find the value of I, the current flowing in the circuit.

Substituting the numerical values for the parameters in the formula produces

$$\begin{aligned}I &= \frac{5 \times 2.01}{22.35 + 5 \times 1.02}\\ &= \frac{10.05}{27.45}\\ &= 0.3661\ldots \text{ amperes}\end{aligned}$$

The dots indicate that the number extends to many more decimal places. However, since the resistances and the e.m.f. were only given to two places of decimals the value obtained for the current should be rounded to two places also. This gives the more appropriate result:

$$I = 0.37 \text{ amperes (approx.)}$$

3 If an account of money £P is invested in an account that earns r% per annum compound interest, the amount £A in the account after n years is given by the formula

$$A = P(1 + r/100)^n$$

If £1000 is invested in this account, what is the value of the investment after 5 years if the interest rate is 11% per annum?

Here we have that $P = 1000$, $n = 5$ and $r = 11$. We are required to find A. Substituting the numbers for the parameters in the formula we find that

$$\begin{aligned} A &= 1000(1 + 11/100)^5 \\ &= 1000(1.11)^5 \\ &= 1685.0581\ldots \end{aligned}$$

This result is only valid to the nearest penny so we round to the second decimal place:

$$A = £1685.06$$

4 Rewrite the following expressions using the equations given:

(a) $(p^2 + q^2)/(p^2 - q^2)$ where $p = \sqrt{(q^2 - 1)}$
(b) $a^2 + 2a^2b + a^2b^2$ where $b = c - 1$

(a)
$$\begin{aligned} \frac{p^2 + q^2}{p^2 - q^2} &= \frac{[\sqrt{(q^2 - 1)}]^2 + q^2}{[\sqrt{(q^2 - 1)}]^2 - q^2} \\ &= \frac{q^2 - 1 + q^2}{q^2 - 1 - q^2} \\ &= \frac{2q^2 - 1}{-1} \\ &= 1 - 2q^2 \end{aligned}$$

(b)
$$\begin{aligned} a^2 + 2a^2b + a^2b^2 &= a^2 + 2a^2(c - 1) + a^2(c - 1)^2 \\ &= a^2 + 2a^2c - 2a^2 + a^2(c - 1)(c - 1) \\ &= 2a^2c - a^2 + a^2[c(c - 1) - 1(c - 1)] \\ &= 2a^2c - a^2 + a^2[c^2 - c - c + 1] \\ &= 2a^2c - a^2 + a^2[c^2 - 2c + 1] \\ &= 2a^2c - a^2 + a^2c^2 - 2a^2c + a^2 \\ &= a^2c^2 \\ &= (ac)^2 \end{aligned}$$

EXERCISES

1 Simplify each of the following expressions:

(a) $-10x + 16ab + x - 4ab$ **(c)** $ab + bc + ac + b^2$

(b) $(l - m)l - m(l + m)$ **(d)** $\sqrt{\frac{x}{y}} + \sqrt{\frac{y}{x}}$

(a)
$$\begin{aligned} -10x + 16ab + x - 4ab &= -10x + x + 16ab - 4ab \\ &= *x + *ab \\ &= *(*ab - 3x) \end{aligned}$$

(b)
$$\begin{aligned} (l - m)l - m(l + m) &= l^2 - ml - ml - m^2 \\ &= *l^2 + *ml + *m^2 \end{aligned}$$

(c)
$$\begin{aligned} ab + bc + ac + b^2 &= ab + ac + b^2 + bc \\ &= *(b + c) + b(*) \\ &= (*)(*) \end{aligned}$$

(d)
$$\begin{aligned} \sqrt{\frac{x}{y}} + \sqrt{\frac{y}{x}} &= \sqrt{(*)}\left[1 + \sqrt{\frac{y}{x}}\sqrt{\frac{y}{x}}\right] \\ &= \sqrt{(*)}(1 + *) \\ &= \frac{* + *}{\sqrt{(*)}} \end{aligned}$$

2 Given that $v = u + at$ where $u = 10$ and $a = 5$, find v for $t = 5, 10, 15$ and 20.

When $t = 5, \quad v = 3*$
$t = 10, \quad v = 6*$
$t = 15, \quad v = *5$
$t = 25, \quad v = *$

3 Fill in each $*$ in the following: given $T = 2\pi[(K^2 + h^2)/gh]^{1/2}$ where $\pi = 3.17$, $K = 5$, $h = 1$ and $g = 980$ then

$$
\begin{aligned}
T &= 2 \times 3.17\left(\frac{*^2 + 1^2}{(980 \times *)*}\right) \\
&= *\left(\frac{*}{*}\right)^{1/2} \\
&= *(0.026\,530\,6)^{1/2} \\
&= * \times * \\
&= 1.032\,673\,2 \\
&= * \text{ to four significant figures}
\end{aligned}
$$

4 Rewrite the following expressions using the equations given:

(a) $x^2 + y^2 \quad$ where $x = \dfrac{at}{\sqrt{(1 + t^2)}}$ and $y = \dfrac{a}{\sqrt{(1 + t^2)}}$

(b) $\dfrac{s^2 - 1}{s^2 - t^2} \quad$ where $s = \sqrt{(1 - x^2)}$ and $t = \sqrt{(1 + x^2)}$

(a)
$$
\begin{aligned}
x^2 + y^2 &= \left[\frac{at}{\sqrt{(1 + t^2)}}\right]^2 + \left[\frac{a}{\sqrt{(1 + t^2)}}\right]^2 \\
&= * + * \\
&= */(1 + t^2) \\
&= *
\end{aligned}
$$

(b)
$$
\begin{aligned}
\frac{s^2 - 1}{s^2 - t^2} &= \frac{[\sqrt{(1 - x^2)}]^2 - 1}{[\sqrt{(1 - x^2)}]^2 - *} \\
&= \frac{-x^2}{*} \\
&= *
\end{aligned}
$$

Unit 3 Multiplication and Division of Expressions

Test yourself with the following:

1 Perform the following multiplications and simplify your results:

(a) $(3x - 4)(2x + 6)$ (b) $(2a + 5)(a^2 - 2a + 3)$

2 Perform the following divisions:

(a) $\dfrac{x^2 - 9x + 20}{x - 4}$ (c) $\dfrac{a^2 - b^2}{a + b}$

(b) $\dfrac{x^2 + x + 1}{x + 1}$ (d) $\dfrac{p^3 + 2p + 1}{p^2 + p + 1}$

Multiplying Expressions

When multiplying expressions within brackets the distributive laws are used. For example,

$$\begin{aligned}(a+b)(x-y) &= a(x-y)+b(x-y)\\ &= ax-ay+bx-by\end{aligned}$$

Notice, again, that

$$\begin{aligned}(a+1)(x-y) &= a(x-y)+1(x-y)\\ &= a(x-y)+(x-y)\end{aligned}$$

the 1 multiplying the second bracket on the right is usually suppressed.

When multiplying algebraic expressions that involve powers of variables, all similar powers are accumulated together. The final expression is then written down in ascending or descending order of the powers. For example,

$$\begin{aligned}(x-4)(x^2-x+3) &= x(x^2-x+3)-4(x^2-x+3)\\ &= x^3-x^2+3x-4x^2+4x-12\\ &= x^3-5x^2+7x-12\end{aligned}$$

Dividing Expressions

Long division of algebraic expressions is a straightforward process when mastered but mastering it takes a little effort.

To divide x^2+5x+6 by $x+2$ we note that division is repetitive subtraction so we are required to find out how many times $x+2$ can be subtracted from x^2+5x+6.

We consider the leading term x^2 in the numerator and the leading term x in the denominator. Because

$$x+x+\cdots+x \text{ (} x \text{ times)} = x.x$$

we see that x can be subtracted x times from x^2. Recognizing that we are not subtracting just x but $x+2$ we proceed as follows:

$$\begin{aligned}x^2+5x+6-x(x+2) &= x^2+5x+6-x^2-2x\\ &= 3x+6\end{aligned}$$

Subtracting $x+2$ from x^2+5x+6 x times leaves a remainder of $3x+6$. Now we subtract $x+2$ from the remainder noting that x can be subtracted 3 times from $3x$:

$$\begin{aligned}3x+6-3(x+2) &= 3x+6-3x-6\\ &= 0\end{aligned}$$

Hence

$$x^2+5x+6-x(x+2)-3(x+2)=0$$

that is

$$x^2+5x+6-(x+3)(x+2)=0$$

that is

$$x^2 + 5x + 6 = (x + 3)(x + 2)$$

so that

$$\frac{x^2 + 5x + 6}{x + 2} = x + 3$$

There is an alternative procedure for performing this repetitive subtraction:

$$\begin{array}{rrl}
 & x + 3 & \\
x + 2 & \overline{)\, x^2 + 5x + 6} & \text{Divide the leading } x^2 \text{ by the leading } x \text{ to give } x \\
 & x^2 + 2x & \text{Multiply } x + 2 \text{ by } x \text{ and subtract from } x^2 + 5x + 6 \\
 & \overline{\quad 3x + 6} & \text{Divide the leading } 3x \text{ by the leading } x \text{ to give } 3 \\
 & 3x + 6 & \text{Multiply } (x + 2) \text{ by } 3 \text{ and subtract from } 3x + 6 \\
 & \overline{\quad 0} & \text{Remainder zero}
\end{array}$$

For more involved divisions we use this latter procedure. For example, to divide $x^3 - y^3$ by $x - y$ proceed as follows:

$$\begin{array}{rrl}
 & x^2 + xy \qquad + y^2 & \\
x - y & \overline{)\, x^3 \qquad\qquad - y^3} & \text{Divide leading } x^3 \text{ by leading } x \text{ to give } x^2 \\
 & x^3 - x^2y \qquad\quad & \text{Multiply } x - y \text{ by } x^2 \\
 & & \text{Subtract } x^2(x - y) \text{ from } x^3 - y^3 \\
 & \overline{+ x^2y \qquad - y^3} & \text{Divide leading } x^2y \text{ by leading } x \text{ to give } xy \\
 & x^2y - xy^2 \quad & \text{Multiply } x - y \text{ by } xy \\
 & & \text{Subtract } xy(x - y) \text{ from } x^2y - y^3 \\
 & \overline{+ xy^2 - y^3} & \text{Divide leading } xy^2 \text{ by leading } x \text{ to give } y^2 \\
 & xy^2 - y^3 & \text{Multiply } x - y \text{ by } y^2 \\
 & & \text{Subtract } y^2(x - y) \text{ from } xy^2 - y^3 \\
 & \overline{\quad 0} & \text{Remainder } 0
\end{array}$$

Hence

$$(x^3 - y^3) \div (x - y) = x^2 + xy + y^2$$

EXAMPLES

1 Perform the following multiplications and simplify your result:
(a) $(5b + 2)(3b - 8)$ **(b)** $(3x - 2)(2x^2 + x - 5)$

(a) $(5b + 2)(3b - 8) = 5b(3b - 8) + 2(3b - 8)$
$= 15b^2 - 40b + 6b - 16$
$= 15b^2 - 34b - 16$

(b) $(3x-2)(2x^2+x-5) = 3x(2x^2+x-5) - 2(2x^2+x-5)$
$$= 6x^3 + 3x^2 - 15x - 4x^2 - 2x + 10$$
$$= 6x^3 - x^2 - 17x + 10$$

2 Perform the following divisions:

(a) $\dfrac{x - 2x - 8}{x + 2}$ **(c)** $\dfrac{r^3 - 8s^3}{r - 2s}$

(b) $\dfrac{x^2 + 2x + 3}{x + 4}$ **(d)** $\dfrac{2x^3 - 3x + 4}{x^2 - x - 3}$

(a)

$$\begin{array}{r|l}
 & x \quad -4 \\
\hline
x+2 & x^2 - 2x - 8 \\
 & x^2 + 2x \\
\hline
 & \quad - 4x - 8 \\
 & \quad - 4x - 8 \\
\hline
 & \qquad\quad 0
\end{array}$$

so that

$$\frac{x^2 - 2x - 8}{x + 2} = x - 4$$

(b)

$$\begin{array}{r|l}
 & x \quad -2 \\
\hline
x+4 & x^2 + 2x + 3 \\
 & x^2 + 4x \\
\hline
 & \quad - 2x + 3 \\
 & \quad - 2x - 8 \\
\hline
 & \qquad\quad 11 \quad \text{Remainder}
\end{array}$$

Here we have a remainder of 11 and we cannot subtract x from this in any meaningful way. The consequence is that

$$x^2 + 2x + 3 - (x - 2)(x + 4) = 11$$

so that

$$x^2 + 2x + 3 = (x - 2)(x + 4) + 11$$

and hence:

$$\frac{x^2 + 2x + 3}{x + 4} = (x - 2) + \frac{11}{x + 4}$$

(c)

$$
\begin{array}{r|l}
 & r^2 + 2rs \;\; + 4s^2 \\
\hline
r - 2s & r^3 \qquad\qquad\qquad -8s^3 \\
 & r^3 - 2r^2s \\
\hline
 & \quad 2r^2s \qquad\quad - 8s^3 \\
 & \quad 2r^2s - 4rs^2 \\
\hline
 & \qquad\qquad 4rs^2 - 8s^3 \\
 & \qquad\qquad 4rs^2 - 8s^3 \\
\hline
 & \qquad\qquad\qquad 0 \quad \text{Remainder}
\end{array}
$$

Hence:

$$\frac{r^3 - 8s^3}{r - 2s} = r^2 + 2rs + 4s^2$$

(d)

$$
\begin{array}{r|l}
 & 2x \;\; + 2 \\
\hline
x^2 - x - 3 & 2x^3 \qquad\qquad - 3x + \;4 \\
 & 2x^3 - 2x^2 - 6x \\
\hline
 & \qquad 2x^2 + 3x + \;4 \\
 & \qquad 2x^2 - 2x - \;6 \\
\hline
 & \qquad\qquad 5x + 10 \quad \text{Remainder}
\end{array}
$$

Hence

$$\frac{2x^3 - 3x + 4}{x^2 - x - 3} = (2x + 2) + \frac{5x + 10}{x^2 - x - 3}$$

EXERCISES

1 Perform the following multiplications and simplify your results:

(a) $(2p - 9)(4 + 6p)$ **(b)** $(5 + 2x)(6x^2 - 7x - 8)$

(a) $(2p - 9)(4 + 6p) = *(4 + 6p) - 9(*)$
$= *p + *p^2 - * - *p$
$= *p^2 - 46p - *$
$= *(*p^2 - *p - 18)$

(b) $(5 + 2x)(6x^2 - 7x - 8) = 5(*) + *(6x^2 - 7x - 8)$
$= *x^3 + *x^2 + *x - 40$

2 Perform the following divisions:

(a) $\dfrac{x^2 + 2x - 3}{x - 1}$ **(c)** $\dfrac{a^3 + b^3}{a + b}$

(b) $\dfrac{2x^2 - x - 1}{x + 2}$ **(d)** $\dfrac{3x^3 + 2x^2 - 1}{3x^2 + 2x + 1}$

(a)

$$
\begin{array}{r|l}
 & x + * \\
\hline
x - 1 & x^2 + 2x - 3 \\
 & * - * \\
\hline
 & \quad\;\; 3x - 3 \\
 & \quad\;\; * - * \\
\hline
 & \qquad\;\; *
\end{array}
$$

Therefore

$$\frac{x^2 + 2x + 3}{x - 1} = x + *$$

(b)

$$
\begin{array}{r|l}
 & * - * \\
\hline
x + 2 & 2x^2 - x - 1 \\
 & * + *x \\
\hline
 & \quad - *x - 1 \\
 & \quad - *x - * \\
\hline
 & \qquad\quad *
\end{array}
$$

Therefore

$$\frac{2x^2 - x - 1}{x + 2} = (* - *) + \frac{*}{x + 2}$$

(c)

$$
\begin{array}{r|l}
 & *^2 - *b + b^2 \\
\hline
a + b & a^3 \qquad\qquad + b^3 \\
 & a^3 + *^2 b \\
\hline
 & \quad - *^2 b \qquad + b^3 \\
 & \quad - *^2 b - *b^2 \\
\hline
 & \qquad\qquad *b^2 \\
 & \qquad\qquad *b^2 + b^3 \\
\hline
 & \qquad\qquad\quad * \quad \text{Remainder}
\end{array}
$$

Hence

$$\frac{a^3 + b^3}{a + b} = *^2 - *b + b^2$$

(d)

$$\begin{array}{r|l} & \quad * \\ \hline 3x^2 + 2x + 1 & 3x^3 + 2x^2 \qquad - 1 \\ & 3x^3 + *x^2 + *x \\ \hline & \qquad\qquad *x - 1 \quad \text{Remainder} \end{array}$$

Hence:

$$\frac{3x^3 + 2x - 1}{3x^2 + 2x + 1} = * + \frac{*x - 1}{3x^2 + 2x + 1}$$

Unit 4 Equations and Graphs

Test yourself with the following:

1 Which of the following equations are identities?

(a) $y = \sqrt{(1 - x^2)}$ (c) $s = ut + \frac{1}{2}at^2$

(b) $(a - b) \div c = \frac{a}{c} - \frac{b}{c}$

2 Given that $y = x^2 - 1$, plot x against y for $x = 0, \pm 1, \pm 2, \pm 3, \pm 4$. How would you convert this discrete graph into a continuous graph?

Equations

An equation is a statement of the equality of two expressions. For example,

$$T = 2\pi\left(\frac{l}{g}\right)^{1/2}$$

is an equation that equates the periodic time T of a simple pendulum to the length l of the pendulum and the acceleration due to gravity g. The symbol T on the left-hand side is called the **subject** of the equation.

There are different types of equation.

Conditional Equation

A conditional equation, usually just called an equation, is true only for certain values of the symbols involved. For example, $v^2 = 1$ is an equation that is only true for each of the two values $v = +1$ and $v = -1$.

Identity

An identity is a statement of equality that is true for all the values of the symbols. For example,

$$3(A + 2) \equiv 3A + 6$$

This equation is true for any value of A – it is an identity. The expression on the left of the equation is not just equal to the expression on the right, it is **equivalent** to it – one expression is an alternative form of the other. Hence the connective $\equiv$ which stands for **equivalence** and reads 'is equivalent to'.

Defining Equation

A defining equation is a statement of equality that defines an expression. For example:

$$a^2 \triangleq a \times a$$

defines the meaning of a^2.

Assigning Equation

An assigning equation is a statement of equality that assigns a specific value to a variable. For example:

$$z := 3$$

This assigns the value 3 to the variable z.

The uses of $\equiv$, $\triangleq$ and $:=$ as connectives are often substituted by the $=$ sign. While it is not strictly correct to do so, it is acceptable.

Graphs of Equations

An expression involving a single independent variable x can be evaluated and have its value assigned to a dependent variable y. This procedure generates ordered pairs of numbers which can be plotted against a pair of cartesian axes to form the graph of the equation. For example, the equation

$$y = 2x + 1$$

can be used to generate ordered pairs of numbers that, when plotted against cartesian axes, lie on a straight line of gradient 2 and vertical intercept 1, as shown in Figure 5.1. In this way certain types of expression can be associated with particular shapes of graph as we shall see in the next unit.

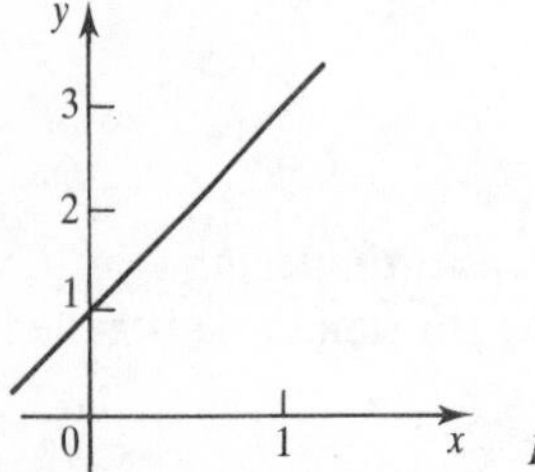

Figure 5.1

EXAMPLES

1 Which of the following equations are identities?

(a) $x + y = y + x$

(b) $v^2 = u^2 + 2as$

(c) $\dfrac{a^2+b^2-c^2}{abc} = \dfrac{a}{bc} + \dfrac{b}{ac} - \dfrac{c}{ab}$

(a) $x + y = y + x$ This equation is the commutative rule of addition and is true for all values of x and y. As a consequence it is an identity.

(b) $v^2 = u^2 + 2as$ This equation is not true for all possible choices of values for v, u, a and s. For example, equality is not preserved for $v = 1$ and $u = a = s = 0$. As a consequence this is a conditional equation.

(c) $\dfrac{a^2+b^2-c^2}{abc} = \dfrac{a}{bc} + \dfrac{b}{ac} - \dfrac{c}{ab}$

The left-hand side of this equation can be converted into the right-hand side by applying the rules of algebra. In other words, the equality holds true regardless of the numerical values of a, b and c. As a consequence this equation is an identity.

2 Given $y = x^2$, plot x against y for $x = 0, \pm 1, \pm 2, \pm 3, \pm 4$. How would you create a continuous graph from this?

We create a table of x versus y values:

x	-4	-3	-2	-1	0	$+1$	$+2$	$+3$	$+4$
y	16	9	4	1	0	1	4	9	16

These values are then plotted to produce the graph of Figure 5.2. It appears that these points all lie on a curve. This curve would be more clearly defined if many more points were plotted. The curve is, indeed, the ideal – the end result of plotting all possible pairs of x–y values. The ideal is impossible to plot as it would require the plotting of an infinite number of points. Instead, the continuous curve is inferred from the equation after plotting sufficient points to give a clear indication of the shape of the curve. This curve is given in Figure 5.3.

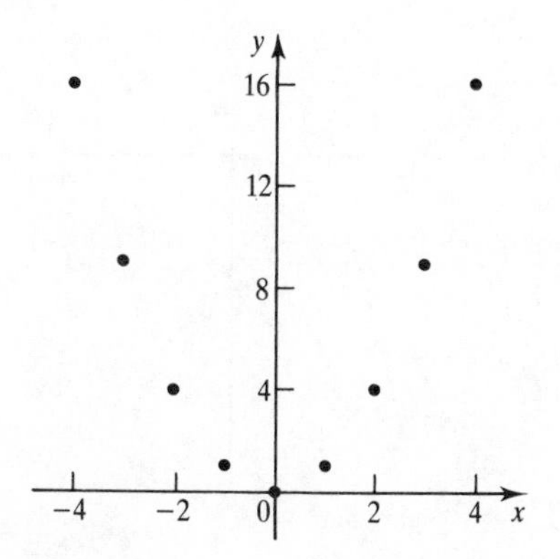

Figure 5.2

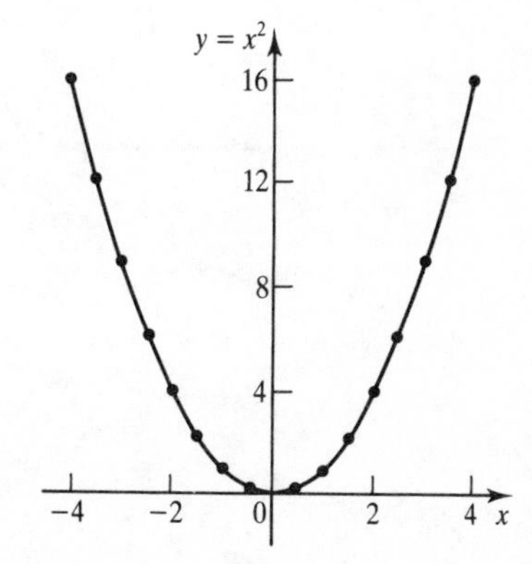

Figure 5.3

EXERCISES

1 Which of the following equations are identities:

(a) $l = \dfrac{T^2 g}{4\pi^2}$

(b) $\dfrac{r^3 - s^3}{r - s} = r^2 + rs + s^2$

(c) $\dfrac{ab^2c^{-3}}{a^{-1}bc^2} = \dfrac{a^2b}{c^5}$

(a) This equality is not true for all possible values of the variables. For example, if $T = g = l = 1$ the equality is not upheld. Consequently, this equation is only true for certain values of the variables and as such is $*$.
(b) The right-hand side of this equation is obtained from the left-hand side by performing the division and so the equation is true regardless of the values of the variables, provided that $r \neq s$. Consequently, this equation is $*$.
(c) The left-hand side of this equation can be converted into the right-hand side by applying the rules of algebra. In other words, the equality holds true regardless of the permissible numerical values of a, b or c. As a consequence this equation is $*$.

2 Given $y = x^3$, plot x against y for $x = 0, \pm 1, \pm 2, \pm 3, \pm 4$. How would you create a continuous graph from this?

We create a table of x versus y values:

x	-4	-3	-2	-1	0	$+1$	$+2$	$+3$	$+4$
y	*	*	*	*	*	*	*	*	*

These values are then plotted to produce the graph of Figure 5.4. It appears that these points all lie on a $*$. This $*$ would be more clearly defined if many more points were plotted. The $*$ is, indeed, the ideal – the end result of plotting all possible pairs of x–y values. The ideal is impossible to plot as it would require the plotting of an infinite number of points. Instead, the continuous curve is inferred from the $*$ after $*$. This curve is given in Figure 5.5.

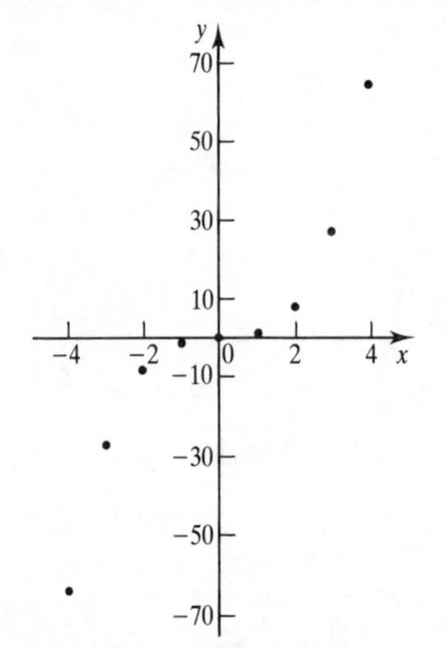

Figure 5.4

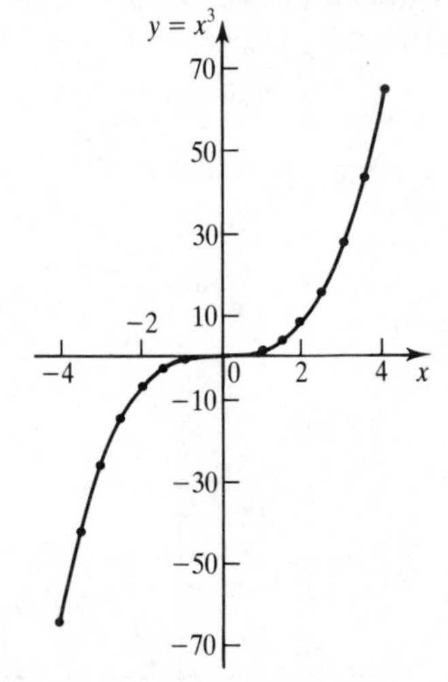

Figure 5.5

Module 5: Further exercises

1 Remove the brackets from the following:

(a) $-3\{p - 7[q - 8(3 - r)]\}$ (c) $-2\{9[2x - 3(4y - 6)] - 9(3z - 5)\}$
(b) $[-10(s - 4) - (5 - t)3]$

2 Simplify each of the following to a single algebraic fraction:

(a) $\frac{ab}{c} - 1$ (b) $\frac{ab}{c} \div \frac{cb}{a}$ (c) $\left(\frac{ab}{c} + \frac{ac}{b}\right) \div \frac{bc}{a}$

3 Simplify each of the following expressions:

(a) $9u + 3v - 2u - v$
(b) $-2\{3[u(5 - w) - 2w(4 - v)] - 8[3(1 - vw) - (9 - 3vw)]\}$
(c) $\sqrt{\left[1 - \frac{2xy + y^2}{(x + y)^2}\right]}$

4 In an electrical circuit a resistance of R ohms is placed in parallel with n series-connected batteries, each of internal resistance r ohms. Each battery produces an electromotive force (e.m.f.) of E volts. The formula for this arrangement is given by

$$I = \frac{nE(R + nr)}{Rnr}$$

where I is the current flowing in the circuit measured in amperes. If $R = 54.72$ ohms, $n = 10$, $E = 2.01$ volts and $r = 1.02$ ohms find the value of I, the current flowing in the circuit. What is the value of I if the resistance R is increased by a factor of 2?

5 A plank of length $2l$ and thickness $2t$ is balanced across a rough cylinder of radius r. If the plank is slightly disturbed it will oscillate without slipping with a periodic time T where

$$T = 2\pi\sqrt{\frac{l^2 + 4t^2}{3g(r - t)}}$$

If $l = 110$ cm, $t = 1$ cm, $r = 5$ cm and $g = 980$ cm s^{-2} find the periodic time T. What is the effect of doubling the thickness? What is the effect of doubling the length?

6 Rewrite the following expressions using the equations given:

(a) $a^2 - 2a^2b + a^2b^2$ where $b = c + 1$
(b) $r^2 - s^2$ where $r + s = 1$

7 Perform the following multiplications and simplify your results:

(a) $(4 - 3p)(2p + 6)$ (b) $(1 - 5x)(2 - 7x - 3x^2)$

8 Perform the following divisions:

(a) $\frac{x^2 + 5x - 6}{x - 1}$ (c) $\frac{a^3 - (8b)^3}{a + 2b}$

(b) $\frac{x^2 - x + 1}{x - 1}$ (d) $\frac{-3x^3 + 9x - 1}{3x^2 + 2}$

9 Which of the following are identities?

(a) $x^2 + y^2 = r^2$ (b) $x^2 - y^2 = (x - y)(x + y)$ (c) $x = 0$

10 Given $y = x^3 + 1$ plot x against y for $x = 0, \pm 1, \pm 2, \pm 3, \pm 4$. How would you create a continuous graph from this?

Module 6: Polynomials and the Binomial Theorem

Aims: To demonstrate how to factorize polynomials and how to expand brackets into polynomials.

Objectives: When you have read this module you will be able to:

▶▶ Recognize quadratic and cubic expressions and their associated graphs.

▶▶ Find the roots of the factors of simple polynomials.

▶▶ List the factors of standard quadratic and cubic expressions.

▶▶ Expand brackets raised to a natural number power and use the binomial theorem.

Unit 1 Polynomials

Test yourself with the following:

1 Give the order and list the coefficients of each of the following polynomials:

(a) $8x^4 - 3x^3 + 7x - 9$ (b) $9x^6 + 2x^2 - 4x + 8$

2 Find the values of the unknown coefficients in each of the following identities in the variable x:

(a) $-3x^2 - 2x + 11 \equiv ax^2 + bx + c$

(b) $12x^4 + 13 \equiv ax^4 + bx^3 + cx^2 + dx + e$

(c) $(a + 7)x^3 - (b - 3)x + 2c - 4 \equiv 0$

3 Express each of the following as a quadratic in the form $ax^2 + bx + c$:

(a) $(x - 3)(x + 4)$ (c) $(5 - 4x)^2$

(b) $(3x - 4)(2 + x)$

4 Express each of the following as a cubic in the form $ax^3 + bx^2 + cx + d$:

(a) $(x - 3)(2x + 5)(3x + 1)$ (c) $(3x - 2)^3$

(b) $(2 - 3x)^2(x + 7)$

5 Factorize each of the following:

(a) $x^2 + 5x + 6$ (d) $27x^3 + 8$

(b) $x^2 - 6x + 9$ (e) $64x^3 - 27$

(c) $4x^2 - 25$

Polynomial Expressions

A **polynomial** is an expression involving sums of powers of a variable. For example:

$$8x^5 - 7x^4 - 6x^3 + 4x^2 - 3x + 2$$

is a polynomial in the variable x. The numbers, 8, -7, -6, 4, -3 and 2 are called the

polynomial coefficients. The **order** of the polynomial is given by the highest power of x – this polynomial is of order 5. Some types of polynomial have special names. For example:

- $3x - 4$ is a polynomial of order 1. It is also referred to as a **linear expression**. Here the coefficient of x is 3 and -4 is the coefficient of the constant.
- $2x^2 - 4x + 5$ is a polynomial of order 2. It is also referred to as a **quadratic expression**. Here the coefficient of x^2 is 2, the coefficient of x is -4 and the coefficient of the constant is 5.
- $x^3 - x^2 + x - 1$ is a polynomial of order 3. It is also referred to as a **cubic expression**. Here the coefficient of x^3 is 1, the coefficient of x^2 is -1, the coefficient of x is 1 and the coefficient of the constant is -1.

The Equality of Polynomials

Two polynomials are equal for all values of the variable if and only if their respective coefficients are equal. For example, if, in some calculation it is found that

$$2x^2 - 3x + 4 = ax^3 + bx^2 + cx + d$$

it can be concluded that

$$a = 0,\ b = 2,\ c = -3 \text{ and } d = 4$$

Graphs and Factors of Simple Polynomial Expressions

By selecting a value for x and substituting this into the polynomial expression, the expression can be evaluated. If this value is plotted against a vertical cartesian axis and the corresponding value of x plotted on the horizontal axis then a graph results. The graph is said to be the graph associated with the given expression.

Linear Expressions

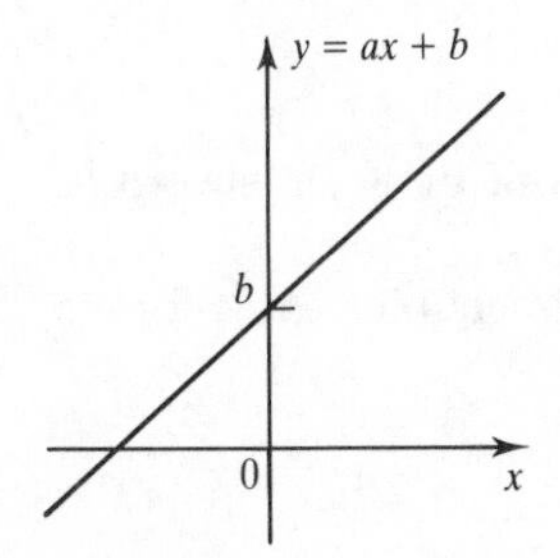

Figure 6.1

As we saw in the previous chapter, the linear expression has an associated graph in the form of a straight line.

Quadratic Expressions

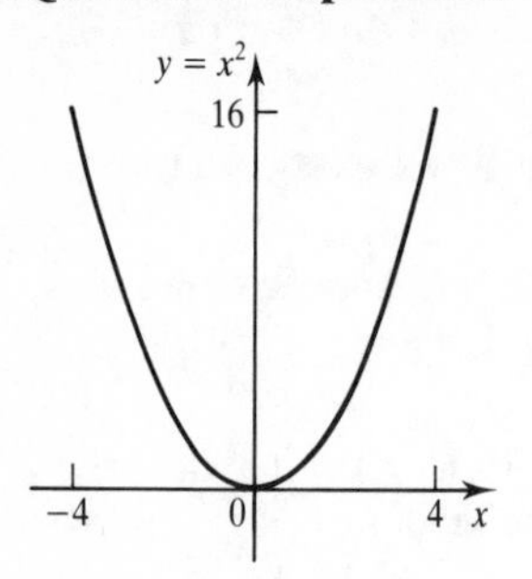

Figure 6.2

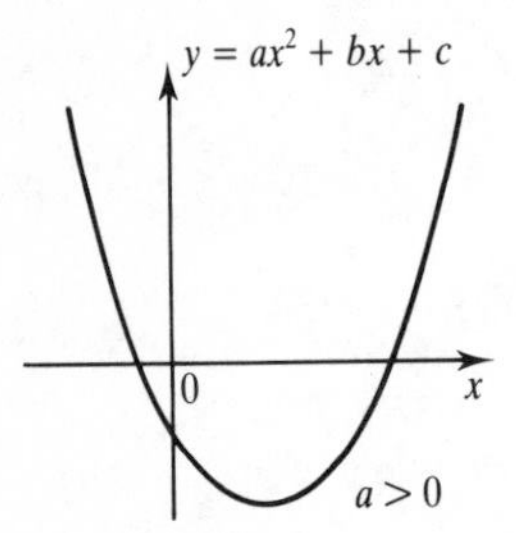

Figure 6.3

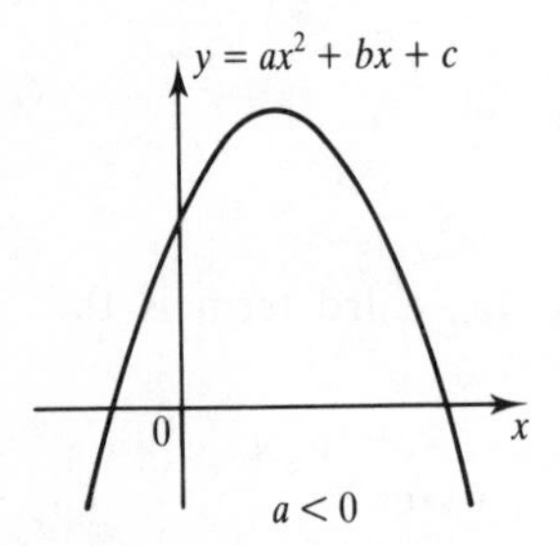

Figure 6.4

The simplest quadratic expression is given as x^2 and plotting this against x gives the graph of Figure 6.2. The shape of this graph is called a **parabola**. The general quadratic expression is of the form

$$ax^2 + bx + c$$

which also has an associated graph in the form of a parabola. If $a > 0$ then the parabola has a minimum (Figure 6.3), and if $a < 0$ then the parabola has a maximum (Figure 6.4).

The product of two linear expressions gives a quadratic expression. For example,

$$\begin{aligned}(x + 1)(x + 2) &\equiv x(x + 2) + 1(x + 2)\\ &\equiv x^2 + 2x + x + 2\\ &\equiv x^2 + 3x + 2\end{aligned}$$

The linear expressions $x + 1$ and $x + 2$ are called the **factors** of the quadratic expression $x^2 + 3x + 2$. Other quadratic expressions can be expressed as the square of a single linear expression. For example,

$$\begin{aligned}(x + 3)^2 &\equiv (x + 3)(x + 3)\\ &\equiv x(x + 3) + 3(x + 3)\\ &\equiv x^2 + 3x + 3x + 9\\ &\equiv x^2 + 6x + 9\end{aligned}$$

Here, the quadratic is said to have two coincident factors.

Cubic Expressions

The simplest cubic expression is x^3 and this has the graph of Figure 6.5.

The general cubic expression is of the form

$$ax^3 + bx^2 + cx + d$$

If $a > 0$ then the graph looks like Figure 6.6 and if $a < 0$ the graph is of the form shown in Figure 6.7.

The product of a linear expression and a quadratic expression gives a cubic expression, as does the product of three linear expressions. For example,

$$\begin{aligned}(2x - 4)(x^2 + x + 1) &\equiv 2x(x^2 + x + 1) - 4(x^2 + x + 1)\\ &\equiv 2x^3 + 2x^2 + 2x - 4x^2 - 4x - 4\\ &\equiv 2x^3 - 2x^2 - 2x - 4\end{aligned}$$

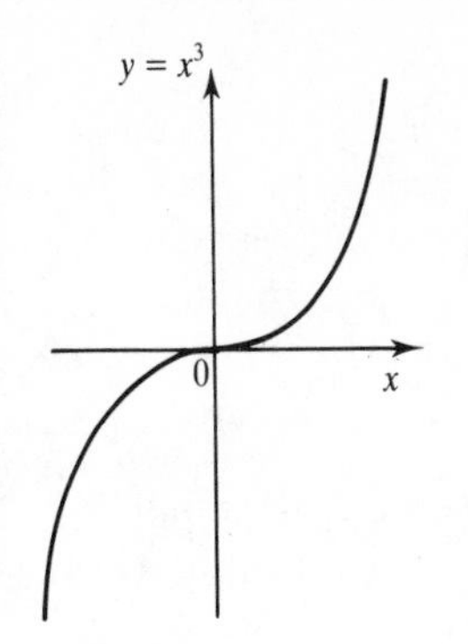

Figure 6.5

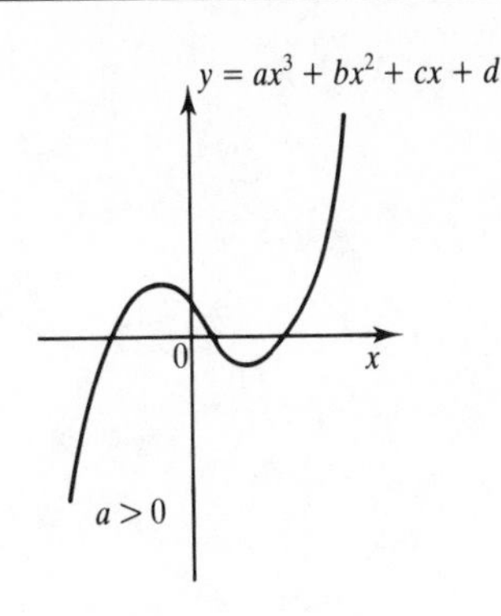

Figure 6.6

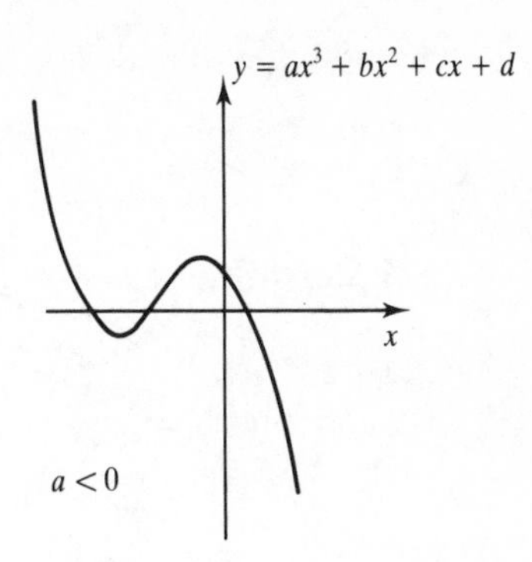

Figure 6.7

and

$$(x + 1)(x + 2)(x - 3) = x^3 - 7x - 6$$

Standard Factorizations

We have seen quadratic and cubic expressions written as products of linear factors. Here we list five standard factorizations for polynomial expressions that should be committed to memory as they occur frequently.

(1) $$(x + a)(x + b) \equiv x^2 + (a + b)x + ab$$

Note that the coefficient of the x-term is the sum $a + b$ and the third term is the product ab. Using these facts enables simple quadratics to be factorized by inspection. For example, to factorize

$$x^2 + 5x + 6 = (x + a)(x + b)$$

we note that $2 + 3 = 5$ and $2 \times 3 = 6$ so that $a = 2$ and $b = 3$. Thus

$$x^2 + 5x + 6 = (x + 2)(x + 3)$$

(2) $$(x + a)^2 \equiv x^2 + 2ax + a^2$$

Here we have two identical factors. Notice the form – twice a for the second coefficient and a squared for the third. For example, to factorize

$$x^2 - 6x + 9 = (x + a)(x + b)$$

we note that $2 \times (-3) = 6$ and $(-3)(-3) = 9$ so that $a = b = -3$. Thus

$$x^2 - 6x + 9 = (x - 3)^2$$

(3) $$(x + a)(x - a) \equiv x^2 - a^2$$

This factorization is referred to as the **difference of two squares**. For example, to factorize

$$x^2 - 4$$

we note that $4 = 2^2$ so that

$$x^2 - 4 = (x + 2)(x - 2)$$

(4) $$(x + a)(x^2 - ax + a^2) \equiv x^3 + a^3$$

The **sum of two cubes**. For example, to factorize

$$x^3 + 8$$

we note that $8 = 2^3$ so that

$$x^3 + 8 = (x + 2)(x^2 - 2x + 4)$$

(5) $$(x - a)(x^2 + ax + a^2) \equiv x^3 - a^3$$

The **difference of two cubes**. For example,

$$x^3 - 8 = (x - 2)(x^2 + 2x + 4)$$

EXAMPLES

1 Give the order and list the coefficients of each of the following polynomials:
(a) $5x^3 + 4x^2 - 3x + 2$ **(b)** $10x^8 - 5x^6 + 6x^4 - 3$

(a) This is a third-order polynomial – a cubic – and the coefficients are

5, 4, -3 and 2 in descending order of powers

(b) This is an eighth-order polynomial and the coefficients are

10, 0, -5, 0, 6, 0, 0, 0 and -3 in descending order of powers

2 Find the values of the unknown coefficients in each of the following identities in the variable x:
(a) $6x^2 + 3x - 9 \equiv ax^2 + bx + c$
(b) $5x^3 - 9 \equiv ax^3 + bx^2 + cx + d$
(c) $(8 - a)x^2 + (9 + b)x + c \equiv 0$

An identity is an equation that is true for all values of the variable. Here the variable is x and in each of these three identities the polynomial on the left is identical to the polynomial on the right for all values of x. As a result, corresponding coefficients must be equal:
(a) $a = 6$, $b = 3$ and $c = -9$
(b) $5x^3 - 9$ can be rewritten as

$$5x^3 + 0x^2 + 0x - 9$$

Hence $a = 5$, $b = 0$, $c = 0$ and $d = -9$.
(c) This identity can be rewritten as

$$(8 - a)x^2 + (9 + b)x + c \equiv 0x^2 + 0x + 0$$

Hence $(8 - a) = 0$, $(9 + b) = 0$ and $c = 0$. This gives

$$a = 8, b = -9 \text{ and } c = 0$$

3 Factorize each of the following:
(a) $x^2 + 3x + 2$ **(d)** $8x^3 - 27$
(b) $x^2 + 4x + 4$ **(e)** $2\sqrt{2} + x^3$
(c) $9x^2 - 16$

(a) $x^2 + 3x + 2 = (x + a)(x + b)$ where $a + b = 3$ and $ab = 2$. By trial and error it is found that $a = 1$ and $b = 2$. Hence

$$x^2 + 3x + 2 = (x + 1)(x + 2)$$

(b) $x^2 + 4x + 4 = (x + a)(x + b)$ where $a + b = 4$ and $ab = 4$. By trial and error it is found that $a = b = 2$. Hence

$$x^2 + 4x + 4 = (x + 2)^2$$

(c) $9x^2 - 16$ This is the difference of two squares:

$$9x^2 - 16 = (3x)^2 - 4^2$$

and as such has the standard factorization:

$$(3x)^2 - 4^2 = (3x - 4)(3x + 4)$$

(d) $8x^3 - 27$ This is the difference of two cubes:

$$8x^3 - 27 = (2x)^3 - 3^3$$

and as such has the standard factorization:

$$\begin{aligned}(2x)^3 - 3^3 &= (2x - 3)((2x)^2 + (3)(2x) + 3^2)\\ &= (2x - 3)(4x^2 + 6x + 9)\end{aligned}$$

(e) $2\sqrt{2} + x^3$ This is the sum of two cubes:

$$2\sqrt{2} + x^3 = (\sqrt{2})^3 + x^3$$

and as such has the standard factorization:

$$\begin{aligned}(\sqrt{2})^3 + x^3 &= (\sqrt{2} + x)((\sqrt{2})^2 - \sqrt{2}x + x^2)\\ &= (\sqrt{2} + x)(2 - \sqrt{2}x + x^2)\end{aligned}$$

EXERCISES

1 Give the order and list the coefficients of the following polynomials (replace each $*$ by the appropriate number):

(a) $9x^4 + 8x^3 + 2x^2 - 3x + 5$ **(b)** $100x^8 - 13$

(a) The order is $*$ and the coefficients in descending order of powers are: $*, 8, *, *, *$.
(b) The order is $*$ and the coefficients in descending order of powers are: $*, 0, *, 0, *, 0, *, 0, *$.

2 Replace each $*$ by the appropriate number or symbol for each of the following:

(a) $12x^2 - 10x + 8 \equiv ax^2 + bx + c$
(b) $8x^3 - 16 \equiv ax^3 + bx^2 + cx + d$
(c) $(13 - a)x^3 + (b - 5)x + (2 - c) \equiv 0$

(a) $a = *, * = -10, c = *$
(b) $a = *, b = 0, c = *, d = *$
(c) The 0 on the right-hand side of the identity can be rewritten as

$$*x^3 + *x^2 + *x + *$$

so that

$$\begin{aligned}13 - a &= *, \text{ so } a = *\\ b - 5 &= *, \text{ so } b = *\\ 2 - c &= *, \text{ so } c = 2\end{aligned}$$

3 Factorize each of the following:

(a) $x^2 + 7x + 12$ **(d)** $x^3 - 8a^3$
(b) $x^2 - 10x + 25$ **(e)** $x^3 + \frac{1}{8}$
(c) $16x^2 - 36$

(a) $x^2 + 7x + 12 = (x + a)(x + b)$ where $a + b = *$ and $ab = *$. By trial and error it is found that $a = *$ and $b = *$. Hence

$$x^2 + 7x + 12 = (x + *)(x * 4)$$

(b) $x^2 - 10x + 25 = (x + a)(x + b)$ where $a + b = *$ and $ab = 25$. By trial and error it is found that $a = b = *$. Hence

$$x^2 - 10x + 25 = (x + *)^2$$

(c) $16x^2 - 36$ This is the * of two *:

$$16x^2 - 36 = *^2 - *^2$$

and as such has the standard factorization:

$$*^2 - *^2 = (* - *)(* + *)$$

(d) $x^3 - 8a^3$ This is the * of two *:

$$x^3 - 8a^3 = *^3 - *^3$$

and as such has the standard factorization:

$$\begin{aligned} *^3 - *^3 &= (* - *)(*^3 + * + *^2) \\ &= (x - *)(x^2 + 2* + *) \end{aligned}$$

(e) $x^3 + \frac{1}{8}$ This is the * of two *:

$$x^3 + \tfrac{1}{8} = x^3 + *^3$$

and as such has the standard factorization:

$$\begin{aligned} x^3 + *^3 &= (x + *)(x^2 - *x + *^2) \\ &= (x + *)(x^2 - * + \tfrac{1}{*}) \end{aligned}$$

Unit 2 Expansions

Test yourself with the following:

1 Obtain the expansions of each of the following by performing the necessary multiplications:

(a) $(x + 3)^2$ (c) $(6x + 1)^3$
(b) $(x - 5)^3$ (d) $(8x - 3)^4$

2 Use Pascal's triangle to expand the following:

(a) $(x + 1)^8$ (c) $(5x + 1)^3$
(b) $(x - 4)^4$ (d) $(9x - 6)^4$

3 Evaluate: (a) $7!$ (c) $\dfrac{16!}{4!4!4!}$

(b) $\dfrac{12!}{10!}$ (d) $\dfrac{(n-1)!}{n!}$

4 Write in factorial form:

(a) $21 \times 22 \times 23$ (c) $(n + 1) \times (n + 2) \times (n + 3) \times (n + 4)$

(b) $85 \times 86 \times 87$

5 Factorize: (a) $8! - 6!$ (b) $n! + (n + 1)! + (n + 2)!$

6 Expand each of the following using the binomial theorem:

(a) $(1 + x)^8$ (d) $(8 - x)^4$

(b) $(1 + 9x)^7$ (e) $\left(7 + \frac{x}{14}\right)^6$

(c) $\left(1 - \frac{x}{7}\right)^5$

7 Find: (a) the seventh term of the expansion of $(1 - x/3)^{16}$

(b) the eighth term of the expansion of $(3 - 6x)^{12}$

8 Find the coefficient of:

(a) x^8 in the expansion of $(1 + 5x)^{14}$

(b) x^5 in the expansion of $(4 + 5x)^{10}$

Products of Factors and Their Expansions

If the expression:

$$(x + 1)^2$$

is written in the form

$$x^2 + 2x + 1$$

the second form is referred to as the **expansion** of the first form. The numbers that multiply the various powers of x are referred to as the **coefficients of the expansion**. In the above expansion the coefficients are: 1, 2 and 1.

Pascal's Triangle

The French mathematician Pascal developed a neat way of obtaining the coefficients of the expansion of

$$(a + b)^n$$

where n is a natural number. This is demonstrated in **Pascal's triangle** as follows:

1	Coefficients of $(a + b)^0$
1 1	Coefficients of $(a + b)^1$
1 2 1	Coefficients of $(a + b)^2 = a^2 + 2ab + b^2$
1 3 3 1	Coefficients of $(a + b)^3 = a^3 + 3a^2b + 3ab^2 + b^3$
1 4 6 4 1	Coefficients of $(a + b)^4 = a^4 + 4a^3b + 6a^2b^2 + 4ab^3 + b^4$

Each of these can be verified by performing the appropriate expansion. Notice that the numbers in a row of the triangle are formed by starting and ending with 1. The intervening numbers are obtained by adding the two numbers above immediately to the left and immediately to the right.

This triangle can be extended indefinitely but for large n this becomes unwieldy. Fortunately, there is an alternative method of arriving at the expansion.

Pascal's triangle can be re-drawn in regular, numbered rows and columns with directional arrows as follows:

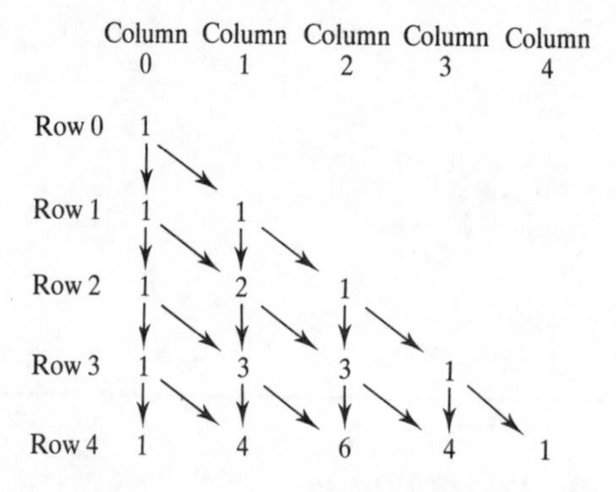

Every number in the triangle can now be seen to be located at the intersection of a given row and a given column.

By following the arrows from the top of the triangle you will find that the number of different routes that can be taken to a given location is the same as the number at that location. Try it. The number of different routes from the top of the triangle to the intersection of row 3, column 1 is 3. We shall denote this number by the symbol:

$$\binom{3}{1}$$

Pascal's triangle can be constructed by counting such routes, noting that it is only possible to arrive at a given location either from directly above or from above and immediately to the left. For example, the number of routes to the intersection of the 4th row with the 2nd column is $\binom{4}{2} = 4$, which is equal to the number of routes to the intersection of the 3rd row and the second column – $\binom{3}{2} = 3$ – plus the number of routes to the intersection of the 3rd row with the 1st column – $\binom{3}{1} = 1$. We can, therefore, write:

$$\binom{3}{1} + \binom{3}{2} = \binom{4}{2}$$

Following this reasoning we can, in general, say that to arrive at the intersection of the $(n + 1)$th row and the $(r + 1)$th column one can only do so from immediately above – the intersection of the nth row and $r + 1$th column or from above and immediately to the left – the intersection of the nth row and rth column:

$$\binom{n}{r} + \binom{n}{r+1} = \binom{n+1}{r+1}$$

Our aim is to be able to give a value to $\binom{n}{r}$ for any n and any r, but before we can do this we need to know about **factorials** and **combinations**.

Factorials

The number $3 \times 2 \times 1$ is referred to as '3 factorial' and is written as 3!. The number 5! (5 factorial) is given as $5 \times 4 \times 3 \times 2 \times 1 = 120$. In general, the number n factorial, $n!$, is given as

$$n! = n(n-1)(n-2)(n-3)(\ldots)3.2.1$$

where n is a positive integer. In addition to this is the proviso that:

$$0! = 1$$

For large values of n it is a long process to find the value of $n!$. On many calculators the factorial is a feature of the function keys. For example, using such a facility on a calculator we can easily show that $10! = 3\,628\,800$.

Combinations

If at the beginning of the week you had available in your freezer two frozen fish dinners and three frozen meat dinners, you could decide on Sunday night to have fish on Monday and Tuesday, giving a weekly meal arrangement of:

FFMMM

where F stands for a fish dinner and M stands for a meat dinner. Alternatively, you could select to have fish on Wednesday and Friday, giving the arrangement:

MMFMF

How many different weekly arrangements are there? To answer this we need only consider when to select to have the fish dinners because the meat dinners automatically fill the gaps when we do not select to have fish.

The first fish dinner can be selected to be on any one of 5 days, leaving 4 remaining days from which to select the second fish dinner. So there are:

$$5 \times 4 = 20$$

way of selecting when to have the two fish dinners. However, not all of these selections are different. For example, if your first choice for fish was to be Monday and your second choice was to be Friday, this would be the same selection as a first choice of Friday and a second choice of Monday. Every selection is repeated twice so there are only:

$$(5 \times 4)/2 = 10$$

different arrangements of the two fish dinners over the 5 days.

Now,

$$\begin{aligned}(5 \times 4)/2 &= (5 \times 4 \times 3 \times 2 \times 1)/[(3 \times 2 \times 1)(2 \times 1)] \\ &= 5!/[3!2!] \\ &= 5!/[(5-2)!2!]\end{aligned}$$

This result is quite general. The number of arrangements or combinations of two identical items among five locations is given by:

$$5!/[(5-2)!2!]$$

and the number of combinations of r identical items among n locations is:

$$n!/[(n-r)!r!]$$

Combinations have a particular property that permits us to link them directly to Pascal's triangle. For example:

$$\begin{aligned} 3!/[(3-1)!1!] + 3!/[(3-2)!2!] &= 3!/[2!1!] + 3!/[1! \\ &= (3!/[2!2!])\{2+2\} \\ &= (3! \times 4)/[2!2!] \\ &= 4!/[2!2!] \\ &= 4!/[(4-2)!2!] \end{aligned}$$

Recalling from Pascal's triangle:

$$\binom{3}{1} + \binom{3}{2} = \binom{4}{2}$$

We can correlate these two equations by saying that:

$$\binom{3}{1} = 3!/[(3-1)!1!], \quad \binom{3}{2} = 3!/[(3-2)!2!] \quad \text{and} \quad \binom{4}{2} = 4!/[(4-2)!2!]$$

Indeed, we can say, quite generally, that:

$$\binom{n}{r} = n!/[(n-r)!r!]$$

which now gives us a way to evaluate the numbers in Pascal's triangle without constructing the triangle first. For example:

$$\begin{aligned} \binom{5}{2} &= 5!/[(5-2)!2!] \\ &= 5!/[3!2!] \\ &= (5 \times 4)/2 \\ &= 10 \end{aligned}$$

So we now know that the number at the intersection of row 5 and column 2 is 10 without needing to know the numbers on row 4.

The Binomial Theorem

The **binomial theorem** duplicates the information contained within Pascal's triangle. For example, the expansion of $(x + 2)^4$ can now be found to be:

$$(x + 2)^4 = \binom{4}{0}x^4 2^0 + \binom{4}{1}x^{4-1}2^1 + \binom{4}{2}x^{4-2}2^2 + \binom{4}{3}x^{4-3}2^3 + \binom{4}{4}x^{4-4}2^4$$

$$= x^4 2^0 + 4x^{4-1}2^1 + 6x^{4-2}2^2 + 4x^{4-3}2^3 + x^{4-4}2^4$$

$$= x^4 + 4x^3 2 + 6x^2 4 + 4x8 + 16$$

$$= x^4 + 8x^3 + 24x^2 + 32x + 16$$

This pattern can be extended to provide the binomial expansion of $(a + b)^n$ as:

$$(a + b)^n = \binom{n}{0}a^n b^0 + \binom{n}{1}a^{n-1}b^1 + \binom{n}{2}a^{n-2}b^2 + \ldots + \binom{n}{r}a^{n-r}b^r$$

$$+ \ldots + \binom{n}{n}a^{n-n}b^n$$

where $\binom{n}{r} = n!/[(n - r)!r!]$ so that:

$$a^n + na^{n-1}b + \frac{n(n-1)a^{n-2}b^2}{2!} + \frac{n(n-1)(n-2)a^{n-3}b^3}{3!} + \ldots + b^n$$

For example, the binomial expansion of $(x + 2)^4$ can again be found to be:

$$x^4 + 4x^3 2 + \frac{(4)(3)x^2 2^2}{2!} + \frac{(4)(3)(2)x2^3}{3!} + \frac{(4)(3)(2)(1)2^4}{4!}$$

$$= x^4 + 8x^3 + \frac{(12)4x^2}{2} + \frac{(24)8x}{6} + 2^4$$

$$= x^4 + 8x^3 + 24x^2 + 32x + 16$$

This can be easily verified either by multiplying $x + 2$ by itself four times or by referring to Pascal's triangle. Although the binomial expansion can be extended to all real number powers, we shall only be concerned with natural number powers at the moment.

Notice that in the expansion of $(a + b)^n$ the powers of a and b in any term add up to n.

EXAMPLES

1 Obtain the expansions of each of the following by performing the necessary multiplications:
(a) $(x + 2)^2$ **(b)** $(x - 3)^3$ **(c)** $(2x + 1)^2$ **(d)** $(3x - 4)^3$

(a) $(x + 2)^2 = (x + 2)(x + 2)$
$= x(x + 2) + 2(x + 2)$
$= x^2 + 2x + 2x + 4$
$= x^2 + 4x + 4$

(b) $(x - 3)^3 = (x - 3)(x - 3)(x - 3)$
$= (x - 3)(x^2 - 6x + 9)$
$= x(x^2 - 6x + 9) - 3(x^2 - 6x + 9)$
$= x^3 - 6x^2 + 9x - 3x^2 + 18x - 27$
$= x^3 - 9x^2 + 27x - 27$

(**c**) $(2x+1)^2 = (2x+1)(2x+1)$
$= 2x(2x+1) + (2x+1)$
$= 4x^2 + 4x + 1$

(**d**) $(3x-4)^3 = (3x-4)(9x^2 - 24x + 16)$
$= 27x^3 - 108x^2 + 144x - 64$

2 Use Pascal's triangle to expand the following:

(**a**) $(x+1)^4$ (**b**) $(x-3)^5$ (**c**) $(2x+1)^4$ (**d**) $(3x-4)^5$

(**a**) The coefficients of $(a+b)^4$ are, from Pascal's triangle:

1, 4, 6, 4, 1

Hence

$$(x+1)^4 = x^4 + 4x^3.1 + 6x^2.1^2 + 4x.1^3 + 1$$
$$= x^4 + 4x^3 + 6x^2 + 4x + 1$$

(**b**) The coefficients of $(a+b)^5$ are:

1, 5, 10, 10, 5, 1

To expand $(x-3)^5$ consider first $(x-3)$. This can be written as:

$$(x-3) = [x + (-3)]$$

Hence

$$(x-3)^5 = x^5 + 5x^4(-3) + 10x^3(-3)^2 + 10x^2(-3)^3 + 5x(-3)^4 + (-3)^5$$
$$= x^5 - 15x^4 + 90x^3 - 270x^2 + 405x - 243$$

(**c**) $(2x+1)^4 = (2x)^4 + 4(2x)^3 1 + 6(2x)^2 1^2 + 4(2x)1^3 + 1^4$
$= 16x^4 + 32x^3 + 24x^2 + 8x + 1$

(**d**) $(3x-4)^5 = [3x + (-4)]^5$
$= (3x)^5 + 5(3x)^4(-4) + 10(3x)^3(-4)^2 + 10(3x)^2(-4)^3 + 5(3x)(-4)^4 + (-4)^5$
$= 243x^5 - 1620x^4 + 4320x^3 - 5760x^2 + 3840x - 1024$

3 Evaluate: (**a**) $3!$ (**b**) $\dfrac{6!}{4!}$ (**c**) $\dfrac{8!}{2!4!6!}$ (**d**) $\dfrac{n!}{(n-1)!}$

(**a**) $3! = 3 \times 2 \times 1 = 6$

(**b**) $\dfrac{6!}{4!} = \dfrac{6 \times 5 \times 4 \times 3 \times 2 \times 1}{4 \times 3 \times 2 \times 1} = 6 \times 5 = 30$

(**c**) $\dfrac{8!}{2!4!6!} = \dfrac{8 \times 7 \times 6 \times 5 \times 4 \times 3 \times 2 \times 1}{(2 \times 1)(4 \times 3 \times 2 \times 1)(6 \times 5 \times 4 \times 3 \times 2 \times 1)}$
$= \dfrac{8 \times 7}{(2 \times 1)(4 \times 3 \times 2 \times 1)}$
$= \dfrac{7}{6}$

Check each of these using the $n!$ function on a calculator.

(**d**) $\dfrac{n!}{n-1!} = \dfrac{n(n-1)(n-2)(\ldots) \times 2 \times 1}{(n-1)(n-2)(\ldots) \times 2 \times 1}$
$= n$

4 Write in factorial form:

(a) $6 \times 5 \times 4$ **(b)** $28 \times 27 \times 26$ **(c)** $n(n-1)(n-2)$

(a) $6 \times 5 \times 4 = \dfrac{6 \times 5 \times 4 \times 3 \times 2 \times 1}{3 \times 2 \times 1} = \dfrac{6!}{3!}$

(b) $28 \times 27 \times 26 = \dfrac{28!}{25!}$

(c) $n(n-1)(n-2) = \dfrac{n!}{(n-3)!}$

5 Factorize: **(a)** $11! - 10!$ **(b)** $n! - (n-1)!$

(a)
$$\begin{aligned} 11! - 10! &= 11 \times 10! - 10! \\ &= 10!(11-1) \\ &= 10 \times 10! \end{aligned}$$

(b)
$$\begin{aligned} n! - (n-1)! &= n(n-1)! - (n-1)! \\ &= (n-1)!(n-1) \end{aligned}$$

6 Expand each of the following using the binomial theorem:

(a) $(1+x)^7$ **(b)** $\left(1 - \dfrac{x}{2}\right)^6$ **(c)** $(2+x)^4$ **(d)** $(4-3x)^3$

(a)
$$\begin{aligned} (1+x)^7 &= (x+1)^7 \\ &= x^7 + 7.1x^6 + \frac{7.6.1^2x^5}{2!} + \frac{7.6.5.1^3x^4}{3!} + \frac{7.6.5.4.1^4x^3}{4!} \\ &\quad + \frac{7.6.5.4.3.1^5x^2}{5!} + \frac{7.6.5.4.3.2.1^6x}{6!} + \frac{7.6.5.4.3.2.1.1^7}{7!} \\ &= x^7 + 7x^6 + 21x^5 + 35x^4 + 35x^3 + 21x^2 + 7x + 1 \end{aligned}$$

in agreement with Pascal's triangle.

(b)
$$\begin{aligned} (1 - x/2)^6 &= [(-x/2) + 1]^6 \\ &= (-x/2)^6 + 6(-x/2)^5 + \frac{6.5(-x/2)^4}{2!} + \frac{6.5.4(-x/2)^3}{3!} + \frac{6.5.4.3(-x/2)^2}{4!} \\ &\quad + \frac{6.5.4.3.2(-x/2)}{5!} + \frac{6.5.4.3.2.1}{6!} \\ &= \frac{x^6}{64} - \frac{3x^5}{16} + \frac{15x^4}{16} - \frac{5x^3}{2} + \frac{15x^2}{4} - 3x + 1 \end{aligned}$$

(c)
$$\begin{aligned} (2+x)^4 &= (x+2)^4 \\ &= x^4 + 4x^3.2 + \frac{4.3x^2 2^2}{2!} + \frac{4.3.2x2^3}{3!} + \frac{4.3.2.1.2^4}{4!} \\ &= x^4 + 8x^3 + 24x^2 + 32x + 16 \end{aligned}$$

(d)
$$\begin{aligned} (4 - 3x)^3 &= [(-3x) + 4]^3 \\ &= (-3x)^3 + 3(-3x)^2 4 + \frac{3.2(-3x)4^2}{2!} + \frac{3.2.1.4^3}{3!} \\ &= -27x^3 + 108x^2 - 144x + 64 \end{aligned}$$

7 (a) Find the third term of the expansion of $(1 - 5x)^{10}$.
(b) Find the fifth term of the expansion of $(3 + 2x)^{11}$.

(a) Expanding $(1 - 5x)^{10}$ as $[(-5x) + 1]^{10}$ the third term is the term involving x^8. This is

$$\frac{10.9(-5x)^8}{2!} = 17\,578\,125x^8$$

(b) Expanding $(3 + 2x)^{11}$ as $(2x + 3)^{11}$ we find that the fifth term is the term involving x^7. This is

$$\frac{11.10.9.8(2x)^7 3^4}{4!} = 3\,421\,440x^7$$

8 (a) Find the coefficient of x^4 in the expansion of $(1 + 4x)^{20}$.
(b) Find the coefficient of x^5 in the expansion of $(12 - 5x)^{18}$.

(a) Expanding $(1 + 4x)^{20}$ as $(4x + 1)^{20}$ the term involving x^4 is

$$\frac{20.19.18.17.16.15.14.13.12.11.10.9.8.7.6.5(4x)^4}{16!}$$
$$= \frac{20.19.18.17(4x)^4}{4!}$$
$$= 1\,240\,320x^4$$

The coefficient of the x^4 term is therefore 1 240 320.
(b) Expanding $(12 - 5x)^{18}$ as $[(-5x) + 12]^{18}$ we find that the term involving x^5 is

$$\frac{18.17.16.15.14.13.12.11.10.9.8.7.6(-5x)^5 12^{13}}{13!}$$

which has the value

$$-(8568)(3125)(1.069\,93 \times 10^{14})x^5$$

which to a reasonable accuracy is:

$$-2.86 \times 10^{21} x^5$$

The coefficient is then -2.86×10^{21}.

EXERCISES

1 Obtain the expansions of each of the following by performing the necessary multiplications and replacing the $*$ by the appropriate number or symbol:
(a) $(x - 1)^3$ (b) $(2x + 4)^2$ (c) $(3x - 4)^3$

(a)
$$\begin{aligned}(x - 1)^3 &= (x - 1)(x - 1)(* - 1)\\ &= [*(x - 1) - (x - 1)](x - 1)\\ &= (x^2 - x - x + 1)(x - 1)\\ &= (x^2 - *x + 1)(x - 1)\\ &= x(x^2 - *x + 1) - (x^2 - *x + 1)\\ &= x^3 - *x^2 + x - x^2 + *x - 1\\ &= x^3 - *x^2 + *x - 1\end{aligned}$$

(b) $(2x + 4)^2 = (2x + 4)(2x + 4)$
$= *(2x + 4) + *(2x + 4)$
$= *x^2 + *x + *x + *$
$= *x^2 + *x + *$

(c) $(3x - 4)^3 = (3x - 4)^2(3x - 4)$
$= (*x^2 - *x + *)(3x - 4)$
$= *x^3 - *x^2 - *x^2 + *x + *x - *$
$= *x^3 - 108x^2 + *x - 64$

2 Use Pascal's triangle to expand the following:
(a) $(x + 1)^7$ **(b)** $(x - 2)^4$ **(c)** $(5x - 6)^3$

(a) From Pascal's triangle the coefficients of $(x + 1)^7$ are:

1, *, 21, *, 35, *, 7, *

Accordingly,

$$(x + 1)^7 = x^7 + *x^6 + 21x^5 + *x^4 + 35x^3 + *x^2 + 7x + *$$

(b) $(x - 2)^4 = [x + (-2)]^4$
$= x^4 + *x^3(-2) + 6x^*(-2)^2 + *x(-2)^3 + (-2)^*$
$= x^4 - 8x^3 + *x^2 - 32x + *$

(c) $(5x - 6)^3 = [5x + (-6)]^3$
$= (*)^3 + *(*)^2(*) + *(*)(*)^2 + (*)^3$
$= 125x^3 - *x^2 + 540x - *$

3 Evaluate: **(a)** $5!$ **(b)** $\frac{8!}{5!}$ **(c)** $\frac{9!}{3!3!}$ **(d)** $\frac{(n + 2)!}{n!}$

(a) $5! = 5 \times * \times 3 \times * \times 1 = *$

(b) $\frac{8!}{5!} = \frac{(8 \times * \times 6 \times * \times 4 \times * \times 2 \times *)}{(5 \times * \times 3 \times * \times 1)} = * \times * \times * = *$

(c) $\frac{9!}{3!3!} = \frac{(* \times 8 \times * \times 6 \times * \times 4 \times * \times 2 \times *)}{(*)(*)}$
$= * \times 8 \times * \times * \times *$
$= *$

(d) $\frac{(n + 2)!}{n!} = \frac{(n + 2)(*)(n)(*)(\ldots)3 \times 2 \times 1}{n(*)(\ldots)3 \times 2 \times 1}$
$= (*)(* + 1)$
$= *^2 + *n + *$

4 Write in factorial form:
(a) $16 \times 15 \times 14$ **(b)** $30 \times 31 \times 32$ **(c)** $(n + 1)n(n - 1)$

(a) $16 \times 15 \times 14 = \frac{16 \times 15 \times 14 \times * \times * \times \ldots \times 3 \times 2 \times 1}{* \times * \times \ldots \times 3 \times 2 \times 1}$
$= \frac{16!}{*!}$

(b) $$30 \times 31 \times 32 = \frac{1 \times 2 \times 3 \times \ldots \times * \times 30 \times 31 \times *}{1 \times 2 \times 3 \times \ldots \times *}$$
$$= \frac{*!}{*!}$$

(c) $$(n+1)n(n-1) = \frac{(n+1)(n)(n-1)(*)(\ldots) \times 3 \times 2 \times 1}{(*)(\ldots) \times 3 \times 2 \times 1}$$
$$= \frac{(n+1)!}{(*)!}$$

5 Factorize: **(a)** $6! - 4!$ **(b)** $(n+1)! - (n-1)!$

(a) $$6! - 4! = * \times * \times 4! - 4!$$
$$= 4!(* - 1)$$
$$= * \times 4!$$

(b) $$(n+1)! - (n-1)! = (*)(*)(n-1)! - (n-1)$$
$$= (n-1)!(* - 1)$$
$$= (n^* + * - *)(n-1)!$$

6 Expand each of the following using the binomial theorem:

(a) $(1-x)^6$ **(b)** $\left(1 + \frac{x}{4}\right)^8$ **(c)** $(3-x)^6$ **(d)** $(6+2x)^4$

(a) $$(1-x)^6 = [(-x)+1]^6$$
$$= (-x)^* + 6(-x)^* + \frac{6.5(-x)^*}{*!} + \frac{6.5.4(-x)^*}{*!} + \frac{6.5.4.3(-x)^*}{*!}$$
$$+ \frac{6.5.4.3.2(-x)}{*!} + \frac{6.5.4.3.2.1}{*!}$$
$$= x^6 - 6x^5 + 15x^4 - 20x^3 + 15x^2 - 6x + 1$$

(b) $$(1 + x/4)^8 = (x/4 + 1)^8$$
$$= (x/4)^8 + *(x/4)^7 + \frac{*(x/4)^6}{*!} + \frac{*(x/4)^5}{*!} + \frac{*(x/4)^4}{*!}$$
$$+ \frac{*(x/4)^3}{*!} + \frac{*(x/4)^2}{*!} + \frac{*(x/4)}{*!} + \frac{*}{*!}$$
$$= *x^8 + *x^7 + *x^6 + *x^5 + \ldots + *x + *$$

(c) $$(3-x)^6 = [(-x)+3]^6$$
$$= (-x)^6 + *(-x)^5 3 + \frac{6.5(-x)^4 *^2}{2!} + \frac{*(-x)^3 3^3}{3!} + \frac{6.5.4.3(-x)^2 *^4}{4!}$$
$$+ \frac{*(-x)3^5}{5!} + \frac{6.5.4.3.2.1.*^6}{6!}$$
$$= x^6 + *x^5 + *x^4 + \ldots + *x + 729$$

(d) $$(6+2x)^4 = (2x+6)^4$$
$$= (2x)^4 + *(2x)^3 * + \frac{*(2x)^2 *^2}{*!} + \frac{*(2x)*^3}{*!} + \frac{*.*^4}{*!}$$
$$= *x^4 + 192x^3 + *x^2 + 1728x + *$$

7 **(a)** Find the fourth term of the expansion of $(1 + 7x)^{11}$.
(b) Find the sixth term of the expansion of $(5 - 3x)^9$.

(a) Extending Pascal's triangle and evaluating $(1 + 7x)^{11}$ as $(7x + 1)^{11}$, the fourth term is

$$\frac{*.*.*(7x)^8}{*!} = *x^8$$

(b) $(5 - 3x)^9 = [(-3x) + 5]^9$ The sixth term is the term involving x^4. This is

$$\frac{9.8.7.6.5(-3x)^4 **}{*!} = *x^4$$

8 **(a)** Find the coefficient of x^3 in the expansion of $(1 - x/5)^{15}$.
(b) Find the coefficient of x^7 in the expansion of $(11 + 10x)^{13}$.

(a) From Pascal's triangle, evaluating $(1 - x/5)^{15}$ as $[(-x/5) + 1]^{15}$ we find that the coefficient of x^3 is $*$.

$$\frac{15.14.*.*.\ldots.*(-1/5)^*}{*!} = *$$

(b) $(11 + 10x)^{13} = (10x + 11)^{13}$ The coefficient of the x^7 term is

$$\frac{13.12.11.10.9.8(*)^7(*)^6}{*!} = *$$

Module 6: Further exercises

1 Obtain the expansions of each of the following by performing the necessary multiplications:
(a) $(x + 4)^2$ (c) $(5x + 3)^3$
(b) $(x - 6)^3$ (d) $(6x - 2)^4$

2 Use Pascal's triangle to expand the following:
(a) $(x + 1)^5$ (c) $(3x + 2)^5$
(b) $(x - 6)^6$ (d) $(7x - 4)^6$

3 Evaluate: (a) $9!$ (b) $\frac{15!}{16!}$ (c) $\frac{12!}{3!2!1!}$ (d) $\frac{(n+2)!}{(n-2)!}$

4 Write in factorial form:
(a) $47 \times 48 \times 49$ (b) $15 \times 16 \times 17$ (c) $(n + 3) \times (n + 2) \times (n + 1) \times (n)$

5 Factorize: (a) $12! - 8!$ (b) $n! + (n + 2)! - (n + 1)!$

6 Expand each of the following using the binomial theorem:
(a) $(1 + x)^5$ (c) $(6 + x)^7$
(b) $\left(1 - \frac{x}{3}\right)^7$ (d) $\left(8 - \frac{x}{4}\right)^5$

7 (a) Find the sixth term of the expansion of $(1 - 2x/5)^{12}$.
(b) Find the third term of the expansion of $(9 + 3x/2)^8$.

8 (a) Find the coefficient of x^8 in the expansion of $(1 + 8x)^{11}$.
(b) Find the coefficient of x^4 in the expansion of $(7 - 3x)^9$.

Module 7: Solving Algebraic Equations

Aims: To demonstrate how to solve polynomial equations and how to separate algebraic fractions into partial fractions.

Objectives: When you have read this module you will be able to:

- Recognize quadratic and cubic equations and the graphical interpretation of their solutions.
- Solve linear, quadratic and cubic equations algebraically.
- Use the factor theorem to factorize general polynomials.
- Separate algebraic fractions into partial fractions.

Unit 1 Solving Equations Graphically

Test yourself with the following:

1 By constructing a table of values within the stated ranges and plotting them on a cartesian graph, solve the following equations:

(a) $2x^2 + 7x - 15 = 0 \qquad -6 \leqslant x \leqslant 3$
(b) $25x^2 - 30x + 9 = 0 \qquad 0 \leqslant x \leqslant 1$
(c) $x^3 - 3x^2 - 13x + 15 = 0 \qquad -5 \leqslant x \leqslant 5$

2 Each of the following sets of data represents corresponding values of two variables that are in proportion to one another. Determine the nature of the proportionality and proportionality constant in each (each top variable is to be plotted against the horizontal axis):

(a)

Time t	0	10	20	30	40
Distance s	0	5	10	15	20

(b)

Time t	0	18	36	54	72	90
Price p	0	−3	−6	−9	−12	−15

(c)

Object u	1	2	4	5
Image v	5.00	2.50	1.25	1.00

(d)

x	15	12	6	3
y	4	5	10	20

3 Two variables p and q are assumed to be in direct proportion. Two values for these variables are $p = 64$ when $q = 4$. Determine the proportionality constant.

4 Two variables s and t are assumed to be in inverse proportion. Two values for these variables are $s = 81$ when $t = 3$. Determine the proportionality constant.

Polynomial Equations

A polynomial expression equated to zero is called a **polynomial equation**. There are a number of specific polynomial equations of interest.

Linear Equations

A **linear equation** is any equation of the form

$$ax + b = 0$$

The graph associated with a linear expression is a straight line. The point on the line that corresponds to the value of the expression being zero is where the line crosses the x-axis. The value of x at the point where the line crosses the x-axis is called the **solution** of the linear equation.

For example, the graph of the expression $x + 2$ is shown in Figure 7.2. The line crosses the x-axis when $x = -2$ and this is the solution to the equation

$$x + 2 = 0$$

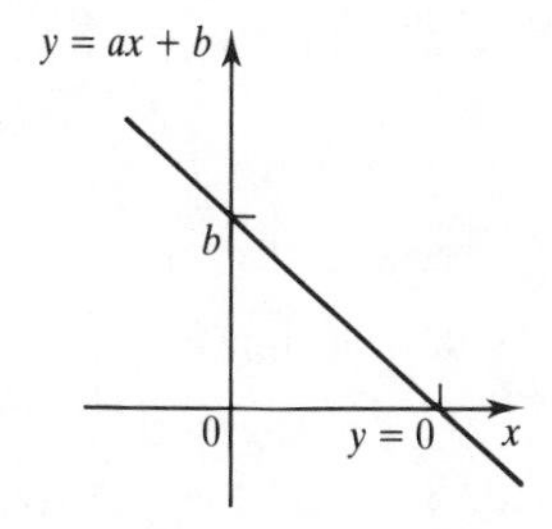

Figure 7.1

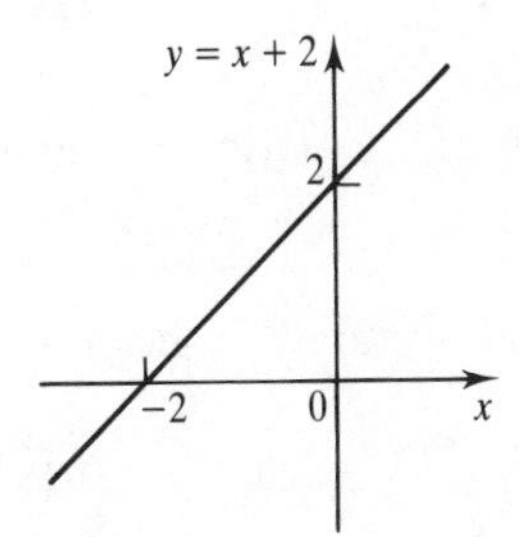

Figure 7.2

Quadratic Equations

A **quadratic equation** is any equation of the form

$$ax^2 + bx + c = 0$$

A quadratic expression has an associated graph in the form of a parabola. Just as with the linear equation, the quadratic expression has the value 0 when the graph – the parabola – crosses the x-axis. This means that the solution to the quadratic equation can consist of 2, 1 or no real number values of x.

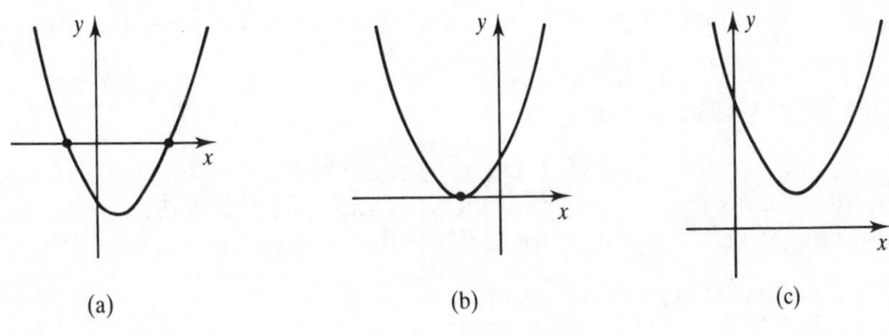

Figure 7.3

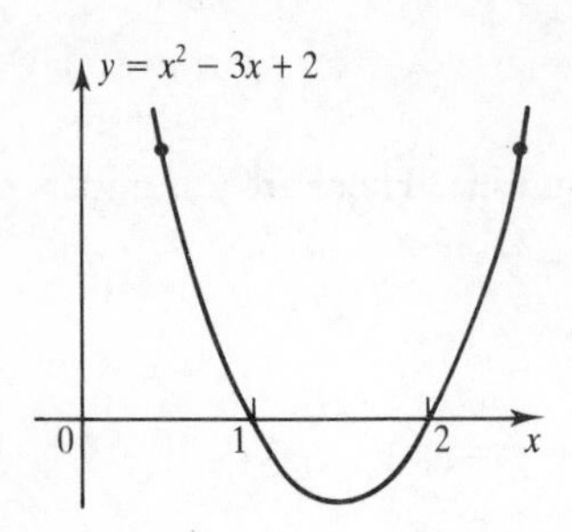

Figure 7.4

For example, the graph of the quadratic expression

$$x^2 - 3x + 2$$

is shown in Figure 7.4. The parabola crosses the x-axis at the points $x = 1$ and $x = 2$. These two values for x are each the solutions of the two linear equations

$$x - 1 = 0 \text{ and } x - 2 = 0$$

respectively. Furthermore, these two values of x are also solutions to the equation

$$(x - 1)(x - 2) = 0$$

which expands to the quadratic equation

$$x^2 - 3x + 2 = 0$$

The equations $x = 1$ and $x = 2$ are called the **roots** of the quadratic equation because, as we see, the quadratic equation can be generated from them.

Cubic Equations

A **cubic equation** is any equation of the form

$$ax^3 + bx^2 + cx + d = 0$$

As can be seen from Figure 7.5, the cubic equation has a solution that may consist of 1, 2 or 3 x-values.

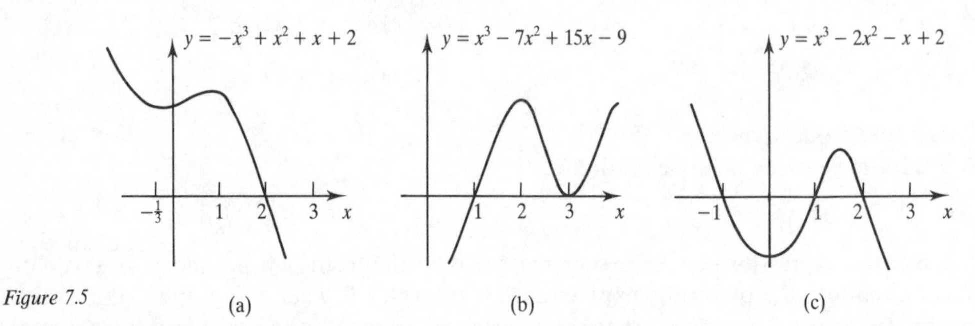

Figure 7.5

Note that the cubic always has at least one real value of x as a solution because the cubic expression has values ranging from negative to positive or from positive to negative. Consequently, there must be at least one value of x where the curve crosses the x-axis.

Direct Proportion

If a change in one variable x causes a corresponding change in another variable y such that the ratio of their values remains constant then the variables are said to be in **direct proportion** to one another. The relationship between the two variables can then be written as

$$y/x = k \text{ (constant)} \qquad \text{or} \qquad y = kx$$

where k is called the **proportionality constant**.

For example, in a perfect gas of constant volume the pressure P is directly proportional to the temperature T so that we can write

$$P = kT$$

Using this equation the proportionality constant can be deduced from a single temperature reading and a single pressure reading.

Clearly, the graph of two variables in direct proportion is a straight line passing through the origin where the constant of proportionality is the gradient of the line.

Squared Proportionality

If a change in one variable x causes a corresponding change in another variable y such that the ratio of y to the square of x remains constant then the variables are said to be in **squared proportion** to one another. The relationship between the two variables can then be written as

$$\frac{y}{x^2} = k \text{ (constant)} \qquad \text{or} \qquad y = kx^2$$

where k is the proportionality constant. The graph of two variables in squared proportionality to one another is a parabola.

For example, the power P dissipated in a resistor R by an electrical current I is proportional to the square of the current flowing. The constant of proportionality is the resistance of the resistor:

$$P = RI^2$$

Inverse Proportion

If a change in one variable x causes a corresponding change in another variable y such that the value of their product remains constant then the variables are said to be in **inverse proportion** to one another. The relationship between the two variables can then be written as

$$yx = k \text{ (constant)} \qquad \text{or} \qquad y = \frac{k}{x}$$

where k is the proportionality constant. For example, in a perfect gas at constant temperature the pressure P is inversely proportional to the volume V so that we can write

$$P = \frac{k}{V}$$

Inverse Square Proportionality

The repulsive force F between two electrons is inversely proportional to the square of their separation r:

$$F = \frac{k}{r^2}$$

The constant of proportionality depends on the electronic charge and the physical properties of the medium in which they reside.

Simultaneous Proportion

The gravitational force between two masses is directly proportional to each mass and inversely proportional to the square of their separations. This means that we can write:

Force F proportional to mass m_1
Force F proportional to mass m_2
Force F inversely proportional to inverse square of separation r

These simultaneous proportionalities can be combined together to give the following relationships:

$$F = \frac{k(m_1 m_2)}{r^2}$$

where k is the combined proportionality constant. According to Newton's theory of gravitation $k = G$ – the gravitational constant.

EXAMPLES

1 By constructing a table of values between the stated ranges and plotting those values on a cartesian graph, solve each of the following equations:

(a) $2 + 2x - 4x^2 = 0 \qquad -2 \leqslant x \leqslant 2$
(b) $49x^2 + 56x + 16 = 0 \qquad -1.5 \leqslant x \leqslant 0$
(c) $-3x^3 - 28x^2 - 52x + 48 = 0 \qquad -7 \leqslant x \leqslant 2$

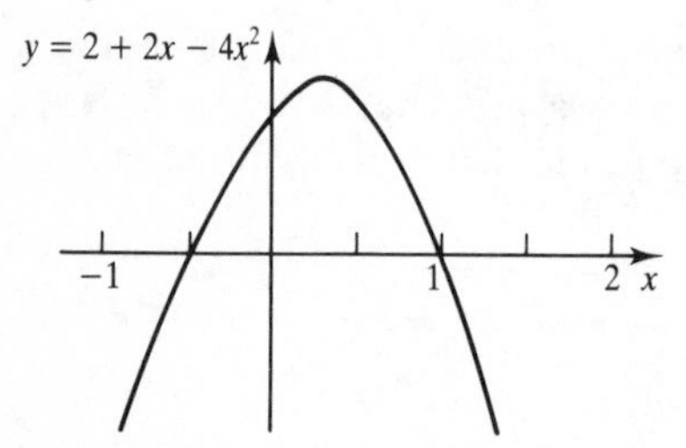

Figure 7.6

(a) The graph of this quadratic crosses the x-axis at $x = 1$ and $x = -\frac{1}{2}$ (see Figure 7.6). These two points are the solution of this equation.
(b) This graph touches the x-axis at $x = -4/7$ (see Figure 7.7). There are two coincident roots to this quadratic at this point.
(c) The graph of this cubic crosses the x-axis at $x = 2/3$, $x = -4$ and $x = -6$ (see Figure 7.8). These three values are the solution of the cubic equation.

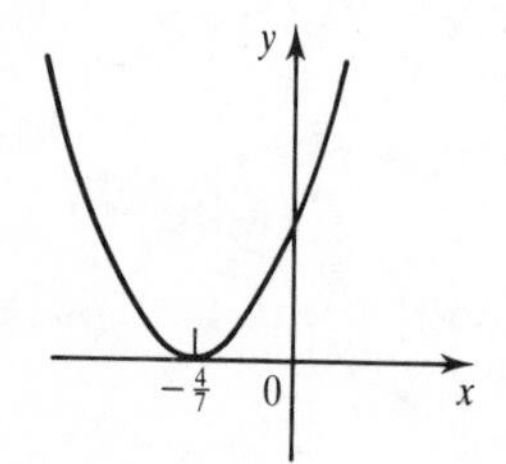

Figure 7.7

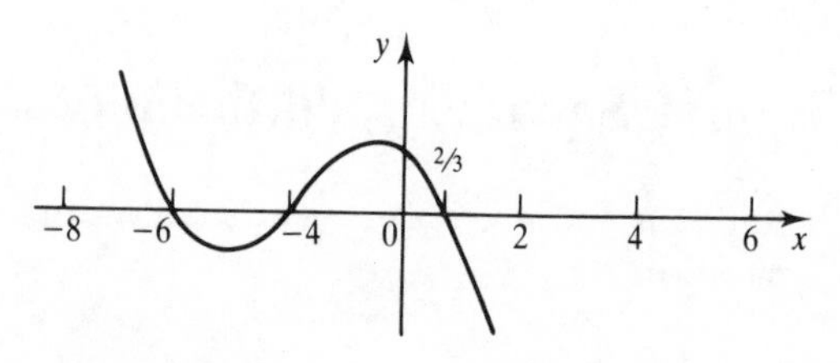

Figure 7.8

2 Each of the following sets of data represent corresponding values of two variables that are in proportion to one another. Determine the nature of the proportionality and proportionality constant in each case (the top variable is to be plotted against the horizontal axis):

(a)

Time t	0	3	6	9	12
Distance s	0	2	4	6	8

(b)

Time t	0	1	2	3	4	5
Price fall p	0	-2	-4	-6	-8	-10

(c)

Object u	10	20	30	40	50
Image v	1	0.5	0.3	0.25	0.2

(d)

x	1	2	3	4	5
y	1.00	0.50	0.33	0.25	0.20

(a) $s = 2t/3$; $k = 2/3$ **(c)** $u = 10/v$; $k = 10$
(b) $p = -2t$; $k = -2$ **(d)** $y = 1/x$; $k = 1$

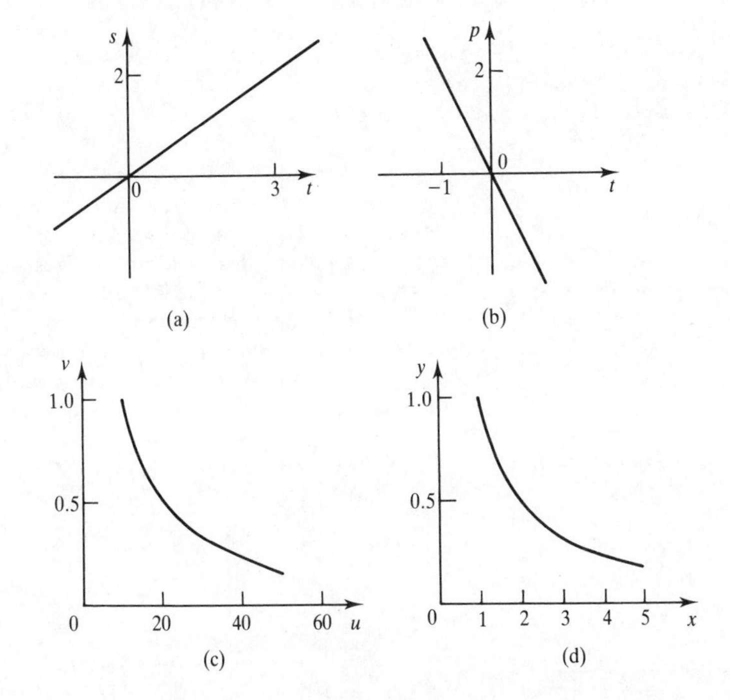

Figure 7.9

3 Two variables p and q are assumed to be in direct proportion. Two values for these variables are $p = 35$ when $q = 7$. Determine the proportionality constant.

Since p and q are in direct proportion they are related as follows:

$$p = kq$$

Substituting the values for p and q it is found that

$$35 = k7$$

therefore $k = 5$ and the relationship is $p = 5q$.

4 Two variables s and t are assumed to be in inverse proportion. Two values for these variables are $s = 22$ when $t = 3$. Determine the proportionality constant.

Since s and t are in inverse proportion they are related as follows:

$$s = \frac{k}{t}$$

Substituting the values for s and t it is found that

$$22 = \frac{k}{3}$$

therefore $k = 66$ and the relationship is $s = 66/t$.

EXERCISES

1 By constructing a table of values between the stated ranges and then plotting them, solve each of the following equations:

(a) $27x^3 + 108x^2 + 144x + 64 = 0 \qquad -6 \leqslant x \leqslant 4$
(b) $4 + 6x - 4x^2 = 0 \qquad -1 \leqslant x \leqslant 3$
(c) $1 - 6x + 11x^2 - 6x^3 = 0 \qquad -4 \leqslant x \leqslant 6$

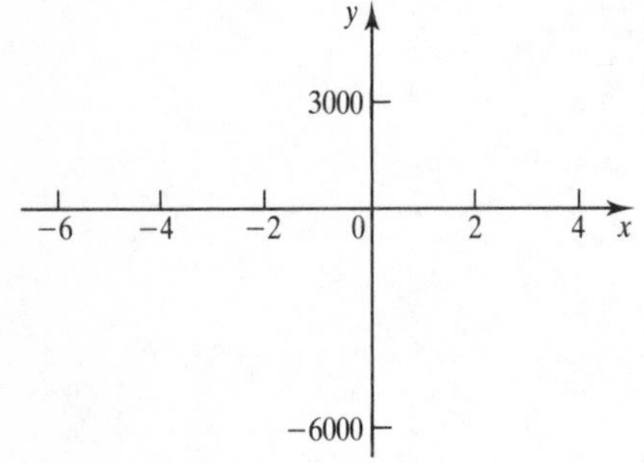

Figure 7.10

(a) The graph of this cubic crosses the x-axis at $x = *$ (see Figure 7.10). There are $*$ coincident roots at this point.
(b) The graph of this quadratic crosses the x-axis at $x = *$ and $x = *$ (see Figure 7.11). These two points form the $*$ to the quadratic equation.
(c) The graph of this cubic crosses the x-axis at $x = *$, $x = *$ and $x = *$ (see Figure 7.12). These three points are the $*$ of the cubic.

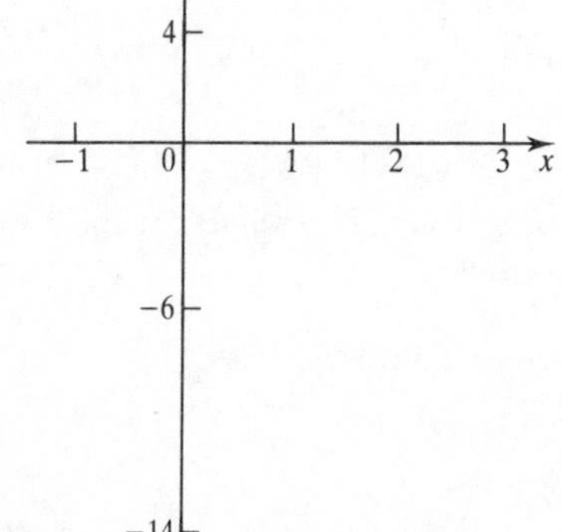
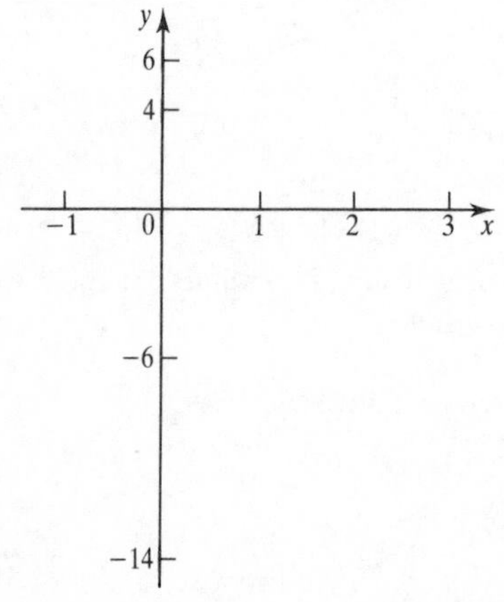

Figure 7.11

Figure 7.12

2 Each of the following sets of data represent corresponding values of two variables that are in proportion to one another. Determine the nature of the proportionality and proportionality constant in each case (the top variable is to be plotted against the horizontal axis):

(a)

Time t	0	12	24	36	48
Distance s	0	4	8	12	16

(b)

Time t	0	4	8	12	16	20
Price p	0	-1	-2	-3	-4	-5

(c)

Object u	100	200	400	500
Image v	10	5.0	2.5	2.0

(d)

x	16	8	4	2
y	-32	-16	-8	-4

(a) $s = t/*$; $k = *$ **(c)** $u = */v$; $k = *$
(b) $p = *t$; $k = *$ **(d)** $y = *x$; $k = *$

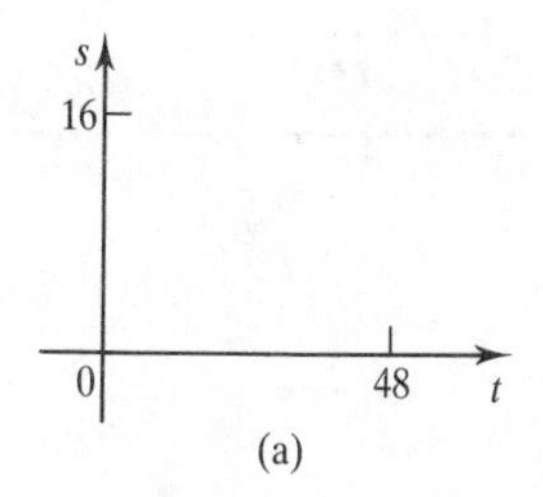

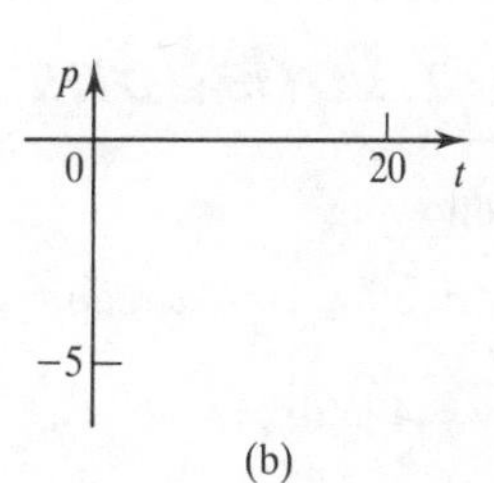

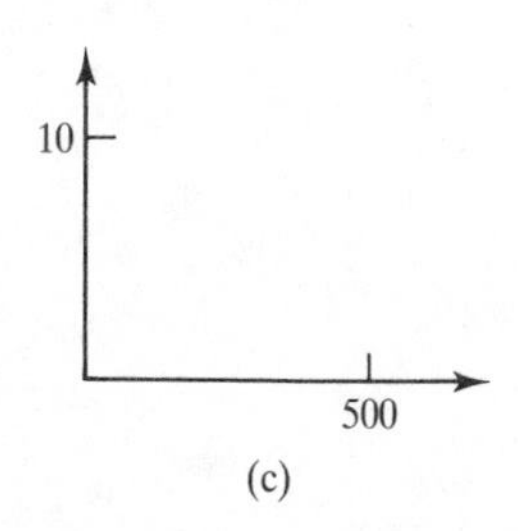

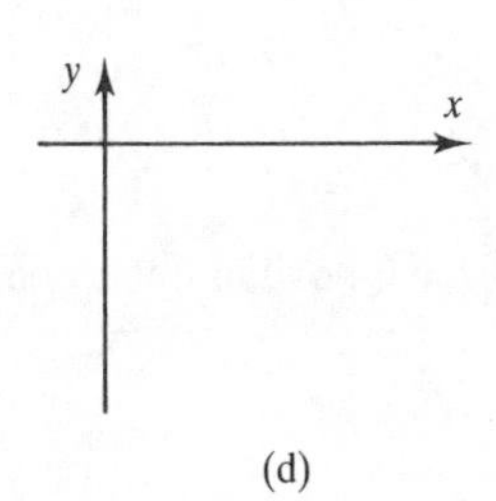

Figure 7.13

3 Two variables p and q are assumed to be in direct proportion. Two values for these variables are $p = 108$ when $q = 18$. Determine the proportionality constant.

Since p and q are in direct proportion they are related as follows:

$$p = kq$$

Substituting the values for p and q it is found that

$$* = k*$$

therefore $k = *$ and the relationship is $p = *q$.

4 Two variables s and t are assumed to be in inverse proportion. Two values for these variables are $s = 1.3$ when $t = 5$. Determine the proportionality constant.

Since s and t are in inverse proportion they are related as follows:

$$s = \frac{k}{t}$$

Substituting the values for s and t it is found that

$$* = \frac{k}{*}$$

therefore $k = *$ and the relationship is $s = */t$.

Unit 2 Solving Equations Algebraically

Test yourself with the following:

1 Solve the following simultaneous equations and determine how many times their graphs intersect each other:

(a) $2x + 2y = 4$ and $x^2 - y^2 = 16$
(b) $3x + 5y = 11$ and $3x^2 + 5y^2 = 17$

2 Solve each of the following quadratic equations:

(a) $x^2 - 3x + 2 = 0$
(b) $8x^2 - 32x + 32 = 0$
(c) $x^2 - 8x - 3 = 0$
(d) $x^2 + 14x + 39 = 0$
(e) $3x^2 - 9x = 0$
(f) $3x^2 - 1 = 5x^2 - 3x$
(g) $5x^2 + 35x - 55 = 0$
(h) $13x^2 - 26x + 13 = 0$

3 Find the solutions to the following quadratic equations:

(a) $x^2 + 5x + 7 = 0$ (b) $8x^2 - 12x + 5 = 0$

4 Solve the following:

(a) $\dfrac{1}{x+1} + \dfrac{1}{x+2} = 1$ (b) $\dfrac{1}{x^2} - 1 = \dfrac{1}{x} - 1$

5 If p and q are the solutions to

$$ax^2 + bx + c = 0$$

show that

$$acx^2 - (b^2 - 2ac)x + ac = 0$$

has roots p/q and q/p.

6 Show that real solutions of

$$kx^2 + 2x - (k - 2) = 0$$

can be found for any value of k.

Transposing Equations

An algebraic equation represents a balancing of numerical values on either side of the equation expressed in symbolic form. As a consequence, any arithmetic operation performed on one side of an equation must be duplicated on the other side so as to maintain the balance of the equation.

The typical equation will consist of the subject of the equation as the dependent variable on the left-hand side by itself and the independent variable embedded within some expression on the right-hand side. For example,

$$p = (3q^2 - 5)^{1/3}$$

Here, p is the dependent variable and subject of the equation, and q is the independent variable. By performing a sequence of arithmetical operations simultaneously on both sides of this equation, the variables p and q can be **transposed** to form a different equation where q is the dependent variable and subject of the new equation and p is the independent variable.

We have already transposed variables in simple equations, but before we proceed to more complicated equations we must adopt a strategy.

The expression containing the independent variable on the right-hand side will have been constructed by performing a sequence of arithmetic operations. To effect the transposition we must undo each one of these arithmetic operations in turn, starting with the last of the sequence and working through to the first of the sequence. We undo an arithmetic operation by performing its inverse operation. For example, consider the construction of the expression on the right-hand side of the equation

$$p = (3q^2 - 5)^{1/3}$$

Imagine you are using a calculator and, starting with a value for q, you are going to find the corresponding value for p.

Enter a value for q

Operation	*Inverse operation*
raise to power 2	raise to power 1/2
multiply by 3	divide by 3
subtract 5	add 5
raise to power 1/3	raise to power 3

This gives the value of p.

The sequence of operations for transposition is then given by the list of inverse operations in reverse order, namely:

raise to power 3
add 5
divide by 3
raise to power 1/2

Let us try it:

$$p = (3q^2 - 5)^{1/3}$$

Raise both sides of the equation to power 3:

$$\begin{aligned} p^3 &= [(3q^2 - 5)^{1/3}]^3 \\ &= (3q^2 - 5)^1 \\ &= 3q^2 - 5 \end{aligned}$$

Note the identical operation performed on the left-hand side so as to maintain the balance of the equation.

Add 5 to both sides of the equation:

$$\begin{aligned} p^3 + 5 &= 3q^2 - 5 + 5 \\ &= 3q^2 \end{aligned}$$

Divide both sides of the equation by 3:

$$\begin{aligned} \frac{p^3 + 5}{3} &= \frac{3q^2}{3} \\ &= q^2 \end{aligned}$$

Raise both sides of the equation to the power 1/2:

$$\begin{aligned} \left(\frac{p^3 + 5}{3}\right)^{1/2} &= (q^2)^{1/2} \\ &= q \end{aligned}$$

Finally, to put the equation into the normal form of subject on the left and expression on the right, we interchange sides of the equation to produce the final, transposed result:

$$q = \left(\frac{p^3 + 5}{3}\right)^{1/2}$$

The secret of effecting such a strategy is, of course, to know the sequence in which the arithmetic operations were performed to build up the original expression in the first place.

Solving Equations

If numbers are substituted for the variables in a conditional equation and the equation still holds true then those numbers are called **solutions** of the equation. The process of finding all the numerical values of the variables that satisfy the equation is called **solving** the equation. If the variable whose value we are trying to find is embedded within an arithmetic expression then the equation will probably have to be transposed prior to the solution being found. For example, if

$$y = 5\sqrt{\frac{x}{3}}$$

and we are given that $y = 10$ then we must transpose x to find its value. That is

$$10 = 5\sqrt{\frac{x}{3}}$$

so that

$$2 = \sqrt{\frac{x}{3}}$$

and

$$4 = \frac{x}{3}$$

giving

$$x = 12$$

Hence, $x = 12$, $y = 10$ is a solution of $y = 5\sqrt{(x/3)}$.

Linear Equations

In the general linear equation $ax + b = 0$ it is assumed that the values of a and b are known. To find the value of x we must transpose the equation as follows:

$ax + b = 0$	The original equation
$ax + b - b = 0 - b$	Subtract b from both sides
$ax = -b$	
$\frac{ax}{a} = \frac{-b}{a}$	Divide both sides by a
$x = \frac{-b}{a}$	

The value of x is then obtained by simple substitution of the values of a and b in this latter, transposed equation.

Simultaneous Equations

Simultaneous equations are equations that have solutions in common. Many equations have an infinite number of solutions. For example, the equation of the straight line

$$x + 2y = 8$$

is satisfied by the coordinates of any point lying on the line. Similarly, the equation of the straight line

$$2x - 4y = -8$$

also has an infinite number of solutions. Because these two lines are not parallel they intersect at one point. The coordinates of that point will then satisfy both equations – the two equations are simultaneously satisfied by this common solution.

In Unit 2 of Module 4 we saw how to solve simultaneous linear equations using a graphical method. Here we obtain the solution by algebraic methods and we proceed by example.

To solve the pair of simultaneous linear equations:

$$x + 2y = 8$$
$$2x - 4y = -8$$

the first step is to eliminate one of the variables to produce an equation in a single variable. To do this, we transpose the first equation to obtain an equation giving y in terms of an expression involving x:

$$\begin{aligned} x + 2y &= 8 \\ x + 2y - x &= 8 - x \\ 2y &= 8 - x \\ y &= \frac{8 - x}{2} \\ &= 4 - \frac{x}{2} \end{aligned}$$

In the second of the simultaneous equations we now substitute for y this expression involving x:

$$\begin{aligned} 2x - 4y &= -8 \\ 2x - 4\left(4 - \frac{x}{2}\right) &= -8 \\ 2x - 16 + 2x &= -8 \\ 4x - 16 &= -8 \\ 4x - 16 + 16 &= -8 + 16 \\ 4x &= 8 \\ x &= 2 \end{aligned}$$

This is the value of x that satisfies both simultaneous equations. To find the corresponding value of y we substitute into $y = 4 - x/2$ to find

$$\begin{aligned} y &= 4 - \tfrac{2}{2} \\ &= 3 \end{aligned}$$

Finally, as a check that no mistakes have been made, substitute the two values we have found into each of the simultaneous equations in turn:

$$x + 2y = 2 + 6 = 8 \text{ and } 2x - 4y = 4 - 12 = -8$$

Because they agree with the simultaneous equations our work is correct and their solution is, therefore

$$x = 2,\ y = 3$$

Any pair of equations involving two variables, not necessarily linear, could be candidates for simultaneous solution. For example, to solve the equations

$$x^2 = 8y^3 \text{ and } 2x^{1/3} = y$$

simultaneously we note that from the second equation

$$y = 2x^{1/3}$$

We now substitute for y in the first equation to give

$$\begin{aligned} x^2 &= 8(2x^{1/3})^3 \\ &= 8(2^3 x) \\ &= 64x \end{aligned}$$

This equation can be written as

$$x^2 - 64x = 0$$

Factorizing the x on the left-hand side yields

$$x(x - 64) = 0$$

This equation has solutions $x = 0$ and $x = 64$. We now substitute these two values for x to find the corresponding values for y: for $x = 0$ we have

$$y = 2x^{1/3} = 0$$

for $x = 64$ we have

$$y = 2x^{1/3} = 2(64)^{1/3} = 8$$

Again, checking by substitution into the second of the simultaneous equations:

$$x^2 = 8y^3$$

shows our working to be correct. The solutions of the simultaneous equation are, therefore

$$x = 0,\ y = 0 \text{ and } x = 64,\ y = 8$$

If we were to draw the graphs of these two equations against the same pair of axes we would find that they intersect at $(0, 0)$ and $(64, 8)$.

Quadratic Equations and Completing the Squares

The general quadratic equation is given in the form

$$ax^2 + bx + c = 0$$

and often its solution may be found by straightforward factorizing. For example, the quadratic equation

$$x^2 + 5x + 6 = 0$$

can be solved by noticing that the quadratic expression on the left-hand side of the equation can be factorized to give

$$x^2 + 5x + 6 = (x + 2)(x + 3)$$

As a consequence, the quadratic equation can be rewritten in the form

$$(x + 2)(x + 3) = 0$$

This equation means that $x + 2 = 0$ or $x + 3 = 0$, which further means that $x = -2$ or $x = -3$. These two values for x represent the solution of the quadratic equation.

Perfect Squares

Many quadratic equations can be solved by inspecting and recognizing their factors. For example,

$$x^2 + 2ax + a^2 = 0$$

is solved by recognizing that this quadratic factorizes into a perfect square:

$$x^2 + 2ax + a^2 = (x + a)^2 = 0$$

Therefore $x + a = 0$ so that $x = -a$ is the solution to the quadratic equation.

Completing the Squares

Many quadratic equations do not have an immediately obvious factorization. For example,

$$x^2 + 4x + 1 = 0$$

cannot be factorized into simple integer factors. To solve this problem we make use of a method known as **completing the squares**. The first two terms of the quadratic on the left are the same as the first two terms of a quadratic which is a perfect square:

$x^2 + 4x$ are the first two terms of $(x + 2)^2$

Indeed, because

$$(x + 2)^2 \equiv x^2 + 4x + 4$$

we can write

$$x^2 + 4x = (x + 2)^2 - 4$$

Therefore

$$\begin{aligned} x^2 + 4x + 1 &= (x + 2)^2 - 4 + 1 \\ &= (x + 2)^2 - 3 \end{aligned}$$

The quadratic equation:

$$x^2 + 4x + 1 = 0$$

can now be written as

$$(x + 2)^2 - 3 = 0$$

so that

$$\begin{aligned} (x + 2)^2 &= 3 \\ x + 2 &= \pm\sqrt{3} \\ x &= -2 \pm \sqrt{3} \end{aligned}$$

These are then the two solutions to the quadratic, namely,

$$x = -2 + \sqrt{3} \text{ and } x = -2 - \sqrt{3}$$

The General Quadratic Equation

The general quadratic equation is of the form

$$ax^2 + bx + c = 0$$

The solutions to the general quadratic equation can be found by employing the method of completing the squares.

Dividing both sides of this equation by a gives

$$x^2 + \frac{bx}{a} + \frac{c}{a} = 0$$

To find the solutions to this form of the general quadratic consider first the perfect square:

$$\left(x + \frac{b}{2a}\right)^2 \equiv x^2 + \frac{bx}{a} + \left(\frac{b}{2a}\right)^2$$

Subtracting $(b/2a)^2$ from both sides of this equation yields

$$\left(x + \frac{b}{2a}\right)^2 - \left(\frac{b}{2a}\right)^2 \equiv x^2 + \frac{bx}{a}$$

The expression on the right-hand side of this equation consists of the first two terms of the general quadratic, so if we add c/a to both sides of this equation to complete the quadratic on the right-hand side, we find that

$$\left(x + \frac{b}{2a}\right)^2 - \left(\frac{b}{2a}\right)^2 + \frac{c}{a} \equiv x^2 + \frac{bx}{a} + \frac{c}{a}$$

Here the expression on the left-hand side of the equation is equivalent to the general quadratic expression. This means that the quadratic equation can be rewritten as

$$\left(x + \frac{b}{2a}\right)^2 - \left(\frac{b}{2a}\right)^2 + \frac{c}{a} = 0$$

This means that

$$\begin{aligned}\left(x + \frac{b}{2a}\right)^2 &= \left(\frac{b}{2a}\right)^2 - \frac{c}{a} \\ &= \frac{b^2}{4a^2} - \frac{c}{a} \\ &= \frac{b^2 - 4ac}{4a^2}\end{aligned}$$

Taking square roots of both sides of this equation gives

$$\begin{aligned}x + \frac{b}{2a} &= \pm\sqrt{\frac{b^2 - 4ac}{4a^2}} \\ &= \pm\frac{\sqrt{(b^2 - 4ac)}}{2a}\end{aligned}$$

Finally, subtracting $b/2a$ from both sides of this equation, we find

$$\begin{aligned}x &= \frac{-b}{2a} \pm \frac{\sqrt{(b^2 - 4ac)}}{2a} \\ &= \frac{-b \pm \sqrt{(b^2 - 4ac)}}{2a}\end{aligned}$$

These are the two solutions to the general quadratic equation

$$ax^2 + bx + c = 0$$

For example, in the equation

$$3x^2 + 2x - 5 = 0$$

the coefficients are $a = 3$, $b = 2$ and $c = -5$. Accordingly, the solution is given as

$$\begin{aligned} x &= \frac{-2 \pm \sqrt{2^2 - 4 \times 3 \times (-5)}}{2 \times 3} \\ &= \frac{-2 \pm \sqrt{4 + 60}}{6} \\ &= \frac{-2 \pm \sqrt{64}}{6} \\ &= \frac{-2 \pm 8}{6} \\ &= \frac{-2 - 8}{6} \quad \text{or} \quad \frac{-2 + 8}{6} \\ &= \frac{-5}{3} \quad \text{or} \quad 1 \end{aligned}$$

These are the two solutions to the quadratic.

The Discriminant

For the general quadratic equation

$$ax^2 + bx + c = 0$$

the quantity

$$b^2 - 4ac$$

in the solution is called the **discriminant**, and the nature of the solution to the quadratic depends upon whether this discriminant is positive, zero or negative.

Positive Discriminant

If the discriminant is a positive number then there are two distinct solutions to the quadratic equation corresponding to the positive and negative values of the square root of the discriminant. Graphically, this corresponds to the case where the graph of $y = ax^2 + bx + c$ crosses the x-axis at two points.

Zero Discriminant

If the discriminant is zero there is one real solution to the quadratic equation. To maintain the pattern of description we say that there are two coincident solutions. Graphically, this corresponds to the case where the graph of $y = ax^2 + bx + c$ touches the x-axis at one point.

Negative Discriminant

If the discriminant is negative the square root of the discriminant is an imaginary number. Graphically, this corresponds to the case where the graph of $y = ax^2 + bx + c$ neither crosses nor touches the x-axis. In this case we have two solutions but they are not real numbers. Each solution is a mixture of real and imaginary numbers.

For example, consider the quadratic equation

$$x^2 - 2x + 2 = 0$$

Here, $a = 1$, $b = -2$ and $c = 2$. Applying the quadratic formula gives

$$\begin{aligned} x &= \frac{2 \pm \sqrt{(4-8)}}{2} \\ &= 1 \pm \frac{\sqrt{-4}}{2} \\ &= 1 \pm \mathrm{j} \end{aligned}$$

The numbers $1 + \mathrm{j}$ and $1 - \mathrm{j}$ are neither real nor imaginary numbers but a mixture or complex of real and imaginary numbers. We call them **complex numbers** and we shall consider them further in Module 10.

EXAMPLES

1 Solve the following simultaneous equations and determine how many times their graphs intersect:

(a) $x + y = 1$ and $x^2 - y^2 = 9$
(b) $2x - 4y = 6$ and $3x^2 + 5y^2 = 8$

(a) From the first equation $y = 1 - x$. Substitute this into the second equation to give

$$x^2 - (1 - x)^2 = 9$$

that is

$$x^2 - (1 - 2x + x^2) = 9$$

that is

$$2x - 1 = 9 \text{ hence } x = 5 \text{ and } y = -4$$

The graphs of these two equations intersect in a single point at $(5, -4)$.

(b) From the first equation $x = 3 + 2y$. Substitution gives

$$3(3 + 2y)^2 + 5y^2 = 8$$

that is

$$3(9 + 12y + 4y^2) + 5y^2 = 8$$

that is

$$17y^2 + 36y + 19 = 0$$

that is

$$(17y + 19)(y + 1) = 0$$

So that $y = -1$ or $-19/17$. When $y = -1$ then $x = 3 - 2 = 1$ and when $y = -19/17$ then $x = 3 - 38/17 = 13/17$. The solutions are then

$$(1, -1) \text{ and } (13/17, -19/17)$$

The graphs of these two equations intersect at two points, $(1, -1)$ and $(13/17, -19/17)$ respectively.

2 Solve each of the following quadratic equations:

(a) $x^2 + 3x + 2 = 0$	**(e)** $x^2 + 5x - 2 = 0$
(b) $5x^2 - 25x + 30 = 0$	**(f)** $x^2 - 12x + 36 = 0$
(c) $x^2 - 2x = 0$	**(g)** $3x^2 - 6x - 3 = 0$
(d) $x^2 + 2 = 2x^2 - x$	**(h)** $5x^2 - 30x + 45 = 0$

(a) $x^2 + 3x + 2 = (x + a)(x + b)$ where $a + b = 3$ and $ab = 2$. Therefore $a = 2$ and $b = 1$. Hence

$$x^2 + 3x + 2 = (x + 2)(x + 1) = 0$$

Hence:

$$x + 2 = 0 \text{ so } x = -2$$

or

$$x + 1 = 0 \text{ so } x = -1$$

The solution is $x = -1$ or $x = -2$.

(b) $5x^2 - 25x + 30 = 0$ Factorize to give

$$5(x^2 - 5x + 6) = 0$$

That is

$$5(x - 3)(x - 2) = 0$$

Hence

$$x = 3 \text{ or } x = 2$$

(c) $x^2 - 2x = 0$ Factorize to give

$$x(x - 2)$$

Hence

$$x = 0 \text{ or } x = 2$$

(d) $x^2 + 2 = 2x^2 - x$ Transpose to give

$$x^2 - x - 2 = 0$$

that is

$$(x - 2)(x + 1) = 0$$

Hence

$$x = 2 \text{ or } x = -1$$

(e) $x^2 + 5x - 2 = 0$ Here, $a = 1$, $b = 5$, $c = -2$ and the discriminant is equal to 33. As this does not have an integer root we use the quadratic formula.

$$\textit{Solution} \quad x = \frac{-5 \pm \sqrt{25 + 8}}{2}$$

$$= \frac{-5 \pm \sqrt{33}}{2} \qquad \text{Two distinct solutions}$$

(f) $x^2 - 12x + 36 = 0$ Here, $a = 1$, $b = -12$, $c = 36$ and the discriminant is equal to 0. As this has a root of zero we do not need to use the quadratic formula.

Solution $x^2 - 12x + 36 = (x + a)(x + b)$

where $a + b = -12$ and $ab = 36$. Thus $a = b = -6$ so that this quadratic has the single solution $x = 6$.

(g) $3x^2 - 6x - 3 = 0$ Here, $a = 3$, $b = -6$, $c = -3$ and the discriminant is equal to 72. As this does not have an integer root we use the quadratic formula.

Solution
$$\begin{aligned} x &= \frac{6 \pm \sqrt{36 + 36}}{6} \\ &= 1 \pm \frac{\sqrt{72}}{6} \\ &= 1 \pm \sqrt{2} \quad \text{Two distinct solutions} \end{aligned}$$

(h) $5x^2 - 30x + 45 = 0$ Here, $a = 5$, $b = -30$, $c = 45$ and the discriminant is equal to 0. As this has a root of zero we do not need to use the quadratic formula.

Solution
$$\begin{aligned} 5x^2 - 30x + 45 &= 5(x^2 - 6x + 9) \\ &= 5(x - 3)(x - 3) \end{aligned}$$

So that this quadratic has the single solution $x = 3$.

3 Find the solutions to the following quadratic equations:

(a) $x^2 + x + 1 = 0$ **(b)** $4x^2 - 3x + 2 = 0$

(a) $a = 1$, $b = 1$, $c = 1$

Solution
$$\begin{aligned} x &= \frac{-1 \pm \sqrt{1 - 4}}{2} \\ &= -\frac{1}{2} \pm \frac{1}{2}\sqrt{-3} \\ &= -\frac{1}{2} \pm \frac{1}{2}\sqrt{3}j \quad \text{Two complex solutions} \end{aligned}$$

(b) $a = 4$, $b = -3$, $c = 2$

Solution
$$\begin{aligned} x &= \frac{3 \pm \sqrt{9 - 32}}{8} \\ &= \frac{3}{8} \pm \frac{\sqrt{-23}}{8} \\ &= \frac{3}{8} \pm \frac{\sqrt{23}j}{8} \quad \text{Two complex solutions} \end{aligned}$$

4 Solve the following:

(a) $\dfrac{(1 - x)}{x} = \dfrac{1}{(1 - x)}$ **(b)** $\dfrac{1}{(x + 2)} - \dfrac{2}{x} = \dfrac{3}{(x - 1)}$

(a) Multiply both sides by the product of the denominators $x(1 - x)$ to give

$$(1 - x)^2 = x$$

This expands and transposes to

$$x^2 - 3x + 1 = 0$$

with solution

$$x = \frac{3 \pm \sqrt{5}}{2}$$

(b) Multiply both sides of the equation by $x(x - 1)(x + 2)$ to give

$$x(x - 1) - 2(x - 1)(x + 2) = 3x(x + 2)$$

This expands and transposes to

$$4x^2 + 9x - 4 = 0$$

with solution

$$x = \frac{-9 \pm \sqrt{81 + 64}}{8}$$

$$= \frac{-9}{8} \pm \frac{\sqrt{145}}{8}$$

5 If p and q are the solutions to

$$ax^2 + bx + c = 0$$

Show that $p + q = -b/a$ and $pq = c/a$.

The two solutions to the quadratic are given as

$$p = \frac{-b + \sqrt{(b^2 - 4ac)}}{2\text{a}}$$

and

$$q = \frac{-b - \sqrt{(b^2 - 4ac)}}{2a}$$

Consequently

$$p + q = \frac{-b + \sqrt{(b^2 - 4ac)}}{2a} + \frac{-b - \sqrt{(b^2 - 4ac)}}{2a}$$

$$= \frac{-b}{a}$$

and

$$pq = \frac{(-b + \sqrt{(b^2 - 4ac)})(-b - \sqrt{(b^2 - 4ac)})}{4a^2}$$

$$= \frac{b^2 - (b^2 - 4ac)}{4a^2} \qquad \text{difference of two squares}$$

$$= \frac{4ac}{4a^2}$$

$$= \frac{c}{a}$$

6 For what values of k can real solutions to the following quadratic equation be found?

$$x^2 + 3x + k = 0$$

The solution is given as

$$x = \frac{-3 \pm \sqrt{9 - 4k}}{2}$$

Real solutions can only be found if

$$9 - 4k \geqslant 0$$

that is

$$k \leqslant \tfrac{9}{4}$$

If $k = 9/4$ one real solution can be found and if $k < 9/4$ two distinct real solutions can be found. If $k > 9/4$ the solutions are complex.

EXERCISES

1 Replace each * by the appropriate number in the solutions to the following simultaneous equations:

(a) $x - y = 3$ and $x^2 + y^2 = 5$
(b) $x - y = 4$ and $x^2 - y^2 = 8$

(a) From the first equation $x = y + 3$. Substituting this into the second equation gives

$$(y + 3)^2 + y^2 = 5$$

that is

$$2y^2 + *y + * = 2(y^2 + *y + *) = 0$$

This has solutions

$$y = * \text{ or } y = *$$

When $y = *$ then $x = *$ and when $y = *$ then $x = *$. The solutions are then

$$(*, *) \text{ and } (*, *)$$

The graphs of these two equations intersect in *.

(b) Recognizing that $x^2 - y^2 = (x - y)(x + y)$, the second equation can be written as

$$*(x + y) = *$$

This means that the two equations reduce to a pair of linear equations:

$$x - y = 4$$
$$x + y = *$$

The solution to these equations is $x = *$ and $y = *$. The graphs of these two equations intersect in *.

2 Replace each * by the appropriate number in the solutions of the following quadratic equations:

(a) $x^2 - 3x + 2 = 0$
(b) $4x^2 + 8x - 12 = 0$
(c) $16x^2 - 32x = 0$
(d) $2x^2 - 6 = x^2 + x$
(e) $x^2 - 3x + 1 = 0$
(f) $x^2 - 6x + 9 = 0$
(g) $2x^2 - 14x - 20 = 0$
(h) $3x^2 - 24x + 48 = 0$

(a) $x^2 - 3x + 2 = (x + a)(x + b)$ where $a + b = -*$ and $ab = *$. Hence $a = -*$ and $b = -*$. The quadratic equation can then be written as

$$(x - *)(x - *) = 0$$

so that $x = *$ or $x = *$.

(b) $4x^2 + 8x - 12 = *(x^2 + 2x - 3) = *(x + a)(x + b)$ where $a + b = *$ and $ab = *$. Hence $a = *$ and $b = *$. The quadratic equation can then be rewritten as

$$*(x - *)(x + *) = 0$$

so that $x = *$ or $x = *$.

(c) $16x^2 - 32x = *x(x - *)$ so that $x = *$ or $x = *$.

(d) $2x^2 - 6 = x^2 + x$ so that $x^2 - *x - * = 0$. That is

$$(x + a)(x + b) = 0$$

where $a + b = *$ or $x = *$. Hence $a = *$ and $b = *$. Consequently, $x = *$ or $x = *$.

(e) $x^2 - 3x + 1$ Here $a = *$, $b = -3$, $c = *$ and the discriminant is equal to $*$ which does not have an integer square root. As a consequence we use the quadratic formula.

$$\textit{Solution} \quad x = \frac{3 \pm \sqrt{9 - *}}{*} = \frac{3}{*} \pm \frac{\sqrt{*}}{*}$$

(f) $x^2 - 6x + 9$ Here $a = *$, $b = *$, $c = 9$ and the discriminant is equal to $*$ which has a square root of $*$. Accordingly we do not need to use the $*$.

$$\textit{Solution} \quad x^2 - 6x + 9 = (x - *)^2$$

So that the solution to the quadratic equation is $x = *$.

(g) $2x^2 - 14x - 20$ Here $a = *$, $b = -14$, $c = *$ and the discriminant is equal to $*$ which does not have an integer square root. As a consequence we use the quadratic formula.

$$\textit{Solution} \quad x = \frac{* \pm \sqrt{* - 160}}{4} = * \pm *$$

(h) $3x^2 - 24x + 48$ Here $a = *$, $b = *$, $c = *$ and the discriminant is equal to $*$ which does not have an integer square root. As a consequence we use the quadratic formula.

$$\textit{Solution} \quad x = \frac{* \pm \sqrt{* - *}}{*} = * \pm *$$

3 Find the solutions to the following quadratic equations:

(a) $x^2 - x + 1 = 0$ **(b)** $5x^2 - 2x + 1 = 0$

(a) $a = *$, $b = -1$, $c = *$ Discriminant $= 1 - * = -* < 0$ so real solutions cannot be found. The solutions are the two complex numbers

$$x = \frac{-1 \pm *\mathrm{j}}{*}$$

(b) $a = 5$, $b = *$, $c = 1$ Discriminant $= * - 20 = -* < 0$ so real solutions cannot be found. The solutions are the two complex numbers

$$x = * \pm *$$

4 Solve the following:

$$\textbf{(a)}\ \frac{1}{x^2+2x+1}+\frac{2}{x+1}=\frac{3}{(x+1)^2} \qquad \textbf{(b)}\ \frac{x}{x+2}+3=\frac{4}{x-1}$$

(a) $x^2 + 2x + 1 = (*)^2$ so multiply both sides of the equation by the square of $*$:

$$1 + 2(*) = 3$$

Transposing this equation gives the solution

$$x = *$$

(b) Multiply through by $(*)(*)$ to give

$$x(*) + 3(*)(*) = 4(*)$$

Expanding and transposing this equation gives

$$*x^2 - *x - * = 0$$

which has solution

$$x = * \pm *$$

5 If p and q are the solutions to

$$ax^2 + bx + c = 0$$

find $1/p + 1/q$ and $1/pq$.

Since

$$\frac{1}{p}+\frac{1}{q}=\frac{*}{*}$$

and $p + q = *$ and $pq = *$, then

$$\frac{1}{p}+\frac{1}{q}=*$$

and

$$\frac{1}{pq}=*$$

6 For what values of k can two equal solutions to the following quadratic equation be found?

$$2x^2 + (k-4)x + 2 = 0$$

The solutions to the quadratic are equal if

$$(*)^2 - (4)(*)(*) = *$$

that is, if

$$*k^2 - *k + * = 0$$

that is, if

$$k = * \text{ or } *$$

Unit 3 The Factor Theorem

Test yourself with the following:

Use the factor theorem to factorize:

(a) $x^3 + 6x^2 + 11x + 6$ (c) $x^3 - x^2 - 4x + 4$
(b) $x^3 + 2x^2 - 15x$ (d) $x^3 - 3x^2 + 3x - 1$

Roots of a Polynomial

If a polynomial expression is written as a product of its factors then the values of the variable for which the factors are zero are called the roots of the polynomial. For example, the polynomial equation

$$x^3 - 6x^2 + 11x - 6 = 0$$

can be written as

$$(x - 1)(x - 2)(x - 3) = 0$$

Consequently, the values of x which satisfy this equation are

$$x = 1, x = 2 \text{ and } x = 3$$

as these are the values at which the factors are respectively zero. These solutions of the polynomial equation are the roots of the polynomial.

Notation

The general nth-order polynomial, which we shall denote by P is

$$P = a_n x^n + a_{n-1} x^{n-1} + \ldots + a_1 x + a_0$$

The value of this polynomial for $x = 2$, say, is then denoted by $P_{x=2}$ where

$$P_{x=2} = a_n 2^n + a_{n-1} 2^{n-1} + \ldots + a_1 2 + a_0$$

The polynomial equation

$$a_n x^n + a_{n-1} x^{n-1} + \ldots + a_1 x + a_0 = 0$$

we shall denote by

$$P = 0$$

The Factor Theorem

If a polynomial P is such that

$$P_{x=a} = 0$$

then $x - a$ is a factor of the polynomial and $x = a$ is a solution of the polynomial equation $P = 0$. For example, to factorize the polynomial

$$P = x^3 + 4x^2 - x - 4$$

it is noted that

$$P_{x=1} = 1^3 + 4 \times 1^2 - 1 - 4 = 0$$

so that $(x - 1)$ is a factor of the polynomial.

By long division it can be shown that

$$\frac{P}{x-1} \equiv x^2 + 5x + 4$$

so that

$$P = (x - 1)(x^2 + 5x + 4)$$

The quadratic factor of this polynomial factorizes as

$$x^2 + 5x + 4 \equiv (x + 1)(x + 4)$$

Hence

$$P \equiv (x - 1)(x + 1)(x + 4)$$

Notice also that

$$P_{x=1} = 0, \qquad P_{x=-1} = 0 \qquad \text{and} \qquad P_{x=-4} = 0$$

This means that the roots $x = 1$, $x = -1$ and $x = -4$ are the solution of the polynomial equation $P = 0$.

EXAMPLES

1 Use the factor theorem to factorize the following:

(a) $x^3 - x^2 - 4x + 4$ **(c)** $x^3 + 3x^2 - x - 3$
(b) $x^3 - x^2 - 2x$ **(d)** $x^3 + x^2 - 7x + 5$

(a) $P = x^3 - x^2 - 4x + 4$ By trial and error it is found that $x = 1, 2$ and -2 satisfy this equation. That is

$$P_{x=1} = 1 - 1 - 4 + 4 = 0 \quad \text{so } (x - 1) \text{ is a factor}$$
$$P_{x=2} = 8 - 4 - 8 + 4 = 0 \quad \text{so } (x - 2) \text{ is a factor}$$
$$P_{x=-2} = -8 - 4 + 8 + 4 = 0 \quad \text{so } (x + 2) \text{ is a factor}$$

There are three factors to a cubic therefore

$$P = x^3 - x^2 - 4x + 4 = (x - 1)(x - 2)(x + 2)$$

(b) $P = x^3 - x^2 - 2x$ and

$$P_{x=0} = 0 \quad \text{so } x \text{ is a factor}$$
$$P_{x=-1} = -1 - 1 + 2 = 0 \quad \text{so } (x + 1) \text{ is a factor}$$

Thus

$$x^3 - x^2 - 2x = x(x + 1)(ax + b)$$

Since the coefficient of x^3 on the left is 1 the coefficient a on the right must also be 1. Further, the coefficient of x on the left is -2 so the coefficient b must also be -2. Thus

$$P = x^3 - x^2 - 2x = x(x + 1)(x - 2)$$

(c) $P = x^3 + 3x^2 - x - 3$ and

$$P_{x=1} = 1 + 3 - 1 - 3 = 0 \qquad \text{so } (x - 1) \text{ is a factor}$$

Dividing P by $(x - 1)$ gives

$$\frac{x^3 + 3x^2 - x - 3}{x - 1} = x^2 + 4x + 3$$

and $x^2 + 4x + 3$ factorizes into $(x + 1)(x + 3)$. As a result

$$P = (x - 1)(x + 1)(x + 3)$$

(d) $P = x^3 + x^2 - 7x + 5$ and

$$P_{x=1} = 1 + 1 - 7 + 5 = 0 \qquad \text{so } (x - 1) \text{ is a factor}$$

Dividing

$$\frac{x^3 + x^2 - 7x + 5}{x - 1} = x^2 + 2x - 5$$

This quadratic expression does not factorize into simple roots so it is left as it is, giving

$$P = x^3 + x^2 - 7x + 5 = (x - 1)(x^2 + 2x - 5)$$

EXERCISES

1 Use the factor theorem to factorize the following:

(a) $x^3 - 6x^2 + 11x - 6$ **(c)** $x^3 - 5x^2 - x + 5$
(b) $x^3 + 6x^2 + 8x$ **(d)** $x^3 + 2x^2 + 2x + 1$

(a) $P = x^3 - 6x^2 + 11x - 6$ and

$$P_{x=*} = 1 - 6 + 11 - 6 = 0 \qquad \text{so } (x - *) \text{ is a factor}$$
$$P_{x=2} = * - * + * - * = 0 \qquad \text{so } (x - 2) \text{ is a factor}$$
$$P_{x=*} = 27 - 54 + 33 - 6 = 0 \qquad \text{so } (*) \text{ is a factor}$$

Hence

$$P = (x - *)(* - 2)(* - *)$$

(b) $P = x^3 + 6x^2 + 8x$ and

$$P_{x=*} = 0 \qquad \text{so } x \text{ is a factor}$$
$$P_{x=-2} = * + * - * = 0 \qquad \text{so } (*) \text{ is a factor}$$

Consequently

$$x^3 + 6x^2 + 8x = x(*)(ax + b)$$

The coefficient of x^3 on the left is $*$ so $* = 1$ and the coefficient of x on the left is $*$ so $b = *$. Hence

$$P = x(*)(*x + *)$$

(c) $P = x^3 - 5x^2 - x + 5$ and

$$P_{x=*} = 1 - 5 - 1 + 5 = 0 \qquad \text{so } (*) \text{ is a factor}$$

Dividing

$$\frac{x^3 - 5x^2 - x + 5}{*} = *x^2 - *x - *$$

This quadratic expression factorizes as

$$x^2 - *x - * = (*)(x - 5)$$

Hence

$$P = (*)(*)(x - 5)$$

(d) $P = x^3 + 2x^2 + 2x + 1$ and

$$P_{x=*} = -1 + 2 - 2 + 1 = 0 \qquad \text{so } (*) \text{ is a factor}$$

Dividing

$$\frac{x^3 + 2x^2 + 2x + 1}{*} = *x^2 + *x + *$$

The discriminant of the latter quadratic is negative so this expression is left as it is:

$$P = (*)(*x^2 + *x + *)$$

Unit 4 Partial Fractions

Test yourself with the following:

Break the following into partial fractions:

(a) $\dfrac{5x + 1}{(x - 1)(x + 2)}$ (c) $\dfrac{2x}{(x - 5)^2(x + 1)}$

(b) $\dfrac{4x - 3}{(x + 1)(x^2 + x + 1)}$ (d) $\dfrac{x^3 - 1}{(x - 2)(x - 3)}$

Algebraic Fractions

Algebraic fractions are expressions in the form of a ratio of two polynomials. For example,

$$\frac{x^2(x + 2)}{x^2 - 4}$$

is an algebraic fraction.

Many algebraic fractions can be simplified by factorizing numerator and denominator and cancelling common factors. For example,

$$\frac{x^2(x + 2)}{x^2 - 4} \equiv \frac{x^2(x + 2)}{(x + 2)(x - 2)}$$

$$\equiv \frac{x^2}{x - 2} \qquad \text{provided } x \neq -2$$

Notice the proviso $x \neq -2$. The process of cancelling common factors in the numerator and the denominator of a fraction is division in disguise. The one operation we cannot perform is division by zero. Consequently, cancelling the common factor $(x + 2)$ cannot be performed if x takes on the value $x = -2$ that makes this common factor equal to zero.

Partial Fractions

An algebraic fraction with a polynomial denominator can be separated into a number of simpler fractions, called **partial fractions**. The denominators of the partial fractions are the factors of the original denominator polynomial. For example,

$$\frac{1}{x^3 + 5x + 6}$$

can be written as

$$\frac{1}{(x + 2)(x + 3)}$$

which can be written as

$$\frac{1}{x + 2} - \frac{1}{x + 3}$$

This can be readily confirmed by adding the two fractions together. These last two fractions are referred to as the partial fractions of the original fraction.

Procedure

The following example serves to illustrate the procedure to be followed when separating a reciprocal polynomial expression into partial fractions. The expression to consider is

$$\frac{1}{x^2 + 5x + 6}$$

- *Step 1* Factorize the denominator:

$$\frac{1}{x^2 + 5x + 6} \equiv \frac{1}{(x + 3)(x + 2)}$$

- *Step 2* Assume that the expression can be separated into partial fractions of the form

$$\frac{1}{(x + 3)(x + 2)} \equiv \frac{A}{x + 3} + \frac{B}{x + 2}$$

where A and B are two constants to be determined.

- *Step 3* Add the two partial fractions on the right-hand side to give

$$\frac{1}{(x + 3)(x + 2)} \equiv \frac{A}{x + 3} + \frac{B}{x + 2} \equiv \frac{A(x + 2) + B(x + 3)}{(x + 3)(x + 2)}$$

- *Step 4* Here two fractions with equal denominators are equivalent so equating numerators shows that

$$1 \equiv A(x + 2) + B(x + 3)$$

- *Step 5* Substitute values for x to find the values of A and B. Here we see that if we let $x = -2$ the above equation becomes

$$1 = A(-2 + 2) + B(-2 + 3)$$

that is

$$1 = B$$

This procedure is referred to as the 'cover-up' rule because it eliminates – covers up – the term involving A.

Now substitute $x = -3$ in the above equation – cover-up the term involving B – and we find that

$$1 = A(-3 + 2) + B(-3 + 3)$$

that is

$$1 = -A \text{ so } A = -1$$

The original algebraic fraction can now be written in terms of partial fractions as

$$\frac{1}{x^2 + 5x + 6} \equiv \frac{1}{x + 2} - \frac{1}{x + 3}$$

Be aware The values of $x = -2$ and $x = -3$ that we substituted into the identity that equates numerators in Step 4 are *not allowed values* for either the original algebraic fraction or the partial fractions. This is because when $x = -2$ or -3 their denominators are zero and *we have not defined division by zero.* We can, however, justify the substitution of these values into the identity that equates numerators because this identity is true in its own right independently of its relationship to the denominators.

The procedure for separating an algebraic fraction into partial fractions is subject to a set of rules.

Rules of Partial Fractions

The rules correspond to three specific forms of the original algebraic fractions. These are where:

(1) *There are no repeated factors in the denominator and the numerator polynomial of a partial fraction is of one degree less than its denominator polynomial.*

If one of the denominator polynomial factors is a quadratic then the numerator of the appropriate partial fraction must be a linear polynomial. For example,

$$\frac{1}{(x + 1)(x^2 + x + 1)} \equiv \frac{A}{x + 1} + \frac{Bx + C}{x^2 + x + 1}$$

(2) *Repeated factors exist in the denominator polynomial.*

If factors are repeated in the denominator polynomial then they must be repeated in the partial fraction separation. For example,

$$\frac{1}{(x+1)(x+2)^2} \equiv \frac{A}{x+1} + \frac{B}{(x+2)^2} + \frac{C}{x+2}$$

(3) *The numerator polynomial is of a higher degree than the denominator polynomial.*

If, in the expression to be separated into partial fractions, the numerator is a polynomial of a greater degree than the denominator polynomial then the denominator must be divided into the numerator before the partial fraction separation can be undertaken. For example,

$$\frac{x^3}{x^2+3x+2}$$

Here we must first perform the division to yield

$$\frac{x^3}{x^2+3x+2} \equiv x - 3 + \frac{7x+6}{x^2+3x+2}$$

The latter algebraic fraction can now be separated into partial fractions.

General Partial Fractions

To avoid much repetitious work the following list of general partial fractions can be employed:

- $$\frac{ax+b}{(x-c)(x-d)} = \frac{A}{x-c} + \frac{B}{x-d}$$

  where

  $$A = \frac{ac+b}{c-d}$$

  $$B = \frac{ad+b}{d-c}$$

- $$\frac{ax^2+bx+c}{(x-d)(x^2+ex+f)} = \frac{A}{x-d} + \frac{Bx+C}{x^2+ex+f}$$

  where

  $$A = \frac{ad^2+bd+c}{d^2+ed+f}$$

  $$B = \frac{d(a+b) - A(d+de+f)+c}{d(1-d)}$$

  $$C = \frac{Af-c}{d}$$

- $$\frac{ax^2+bx+c}{(x-d)^2(x-e)} = \frac{A}{(x-d)^2} + \frac{B}{x-d} + \frac{C}{x-e}$$

where

$$A = \frac{ad^2+bd+c}{d-e}$$
$$B = \frac{c+Ae-Cd^2}{ed}$$
$$C = \frac{ae^2+be+c}{(e-d)^2}$$

EXAMPLES

1 Separate the following into partial fractions:

(a) $\dfrac{6}{(x+1)(x-1)}$ **(c)** $\dfrac{3x+7}{(x+2)^2(x+3)}$

(b) $\dfrac{6x+12}{x(x-2)(x+3)}$ **(d)** $\dfrac{x^2+3x+4}{(x+5)(x-2)}$

(a) Assume

$$\frac{6}{(x+1)(x-1)} \equiv \frac{A}{x+1} + \frac{B}{x-1}$$
$$\equiv \frac{A(x-1)+B(x+1)}{(x+1)(x-1)}$$

Equating numerators:

$$6 = A(x-1) + B(x+1)$$

Cover-up A by letting $x = 1$, giving the equation

$$6 = B(1+1) \qquad \text{so that } B = 3$$

Cover-up B by letting $x = -1$, giving the equation

$$6 = A(-1-1) \qquad \text{so that } A = -3$$

The final form of the partial fractions is then

$$\frac{6}{(x+1)(x-1)} \equiv \frac{-3}{x+1} + \frac{3}{x-1}$$

(b) Assume

$$\frac{6x+12}{x(x-2)(x+3)} \equiv \frac{A}{x} + \frac{B}{x-2} + \frac{C}{x+3}$$
$$\equiv \frac{A(x-2)(x+3)+Bx(x+3)+Cx(x-2)}{x(x-2)(x+3)}$$

Equating numerators:

$$6x+12 = A(x-2)(x+3) + Bx(x+3) + Cx(x-2)$$

Cover-up B and C by choosing $x = 0$, giving the equation

$$0 + 12 = A(0 - 2)(0 + 3)$$

that is

$$12 = -6A \qquad \text{so that } A = -2$$

Cover-up A and C by choosing $x = 2$, giving the equation

$$12 + 12 = 2B(2 + 3)$$

that is

$$24 = 10B \qquad \text{so that } B = \frac{12}{5}$$

Cover-up A and B by choosing $x = -3$, giving the equation

$$-18 + 12 = -3C(-3 - 2)$$

that is

$$-6 = 15C \qquad \text{so that } C = \frac{-2}{5}$$

The final form of the partial fractions is then

$$\frac{6x + 12}{x(x - 2)(x + 3)} \equiv \frac{-2}{x} + \frac{12}{5(x - 2)} - \frac{2}{5(x + 3)}$$

(c) Assume

$$\frac{3x + 7}{(x + 2)^2(x + 3)} \equiv \frac{A}{(x + 2)^2} + \frac{B}{x + 2} + \frac{C}{x + 3}$$

$$\equiv \frac{A(x + 3) + B(x + 2)(x + 3) + C(x + 2)^2}{(x + 2)^2(x + 3)}$$

Equating numerators:

$$3x + 7 = A(x + 3) + B(x + 2)(x + 3) + C(x + 2)^2$$

Cover-up A and B by choosing $x = -3$, giving the equation

$$-9 + 7 = C(-3 + 2)^2$$

that is

$$-2 = C \qquad \text{so that } C = -2$$

Cover-up B and C by choosing $x = -2$, giving the equation

$$-6 + 7 = A(-2 + 3)$$

that is

$$1 = A$$

To find B we cannot use the cover-up rule as we cannot simultaneously cover-up A and C. Instead we substitute the values found for A and C and choose x to be any value other than -2 or -3: we choose $x = 1$. This gives the equation

$$3 + 7 = 1(1 + 3) + B(1 + 2)(1 + 3) - 2(1 + 2)^2$$

that is

$$10 = 4 + 12B - 18 \qquad \text{so that } B = 2$$

The final form of the partial fractions is then

$$\frac{3x + 7}{(x + 2)^2(x + 3)} \equiv \frac{1}{(x + 2)^2} + \frac{2}{x + 2} - \frac{2}{x + 3}$$

(d) In this fraction the degree of the numerator is equal to the degree of the denominator. Before a break into partial fractions can be achieved the fraction must be divided out.

$$\frac{x^2 + 3x + 4}{(x + 5)(x - 2)} \equiv \frac{x^2 + 3x + 4}{x^2 + 3x - 10}$$

$$\equiv \frac{x^2 + 3x - 10 + 14}{x^2 + 3x - 10}$$

$$\equiv 1 + \frac{14}{x^2 + 3x - 10}$$

$$\equiv 1 + \frac{14}{(x + 5)(x - 2)}$$

By separating into partial fractions it can be shown that

$$\frac{14}{(x + 5)(x - 2)} \equiv \frac{-2}{x + 5} + \frac{2}{x - 2}$$

The final form of the partial fractions is then

$$\frac{x^2 + 3x + 4}{(x + 5)(x - 2)} \equiv 1 - \frac{2}{x + 5} + \frac{2}{x - 2}$$

EXERCISES

1 Replace each * by the appropriate number or symbol in the following separations into partial fractions:

(a) $\dfrac{3x}{(x - 5)(x + 1)}$ **(c)** $\dfrac{5}{(x + 1)^2(x - 1)}$

(b) $\dfrac{3}{x(x - 1)(x + 1)}$ **(d)** $\dfrac{x^2 + x + 1}{(x + 1)(x + 2)}$

(a) Assume

$$\frac{3x}{(x - 5)(x + 1)} \equiv \frac{A}{x - 5} + \frac{B}{x + 1}$$

$$\equiv \frac{*(x + 1) + *(x - 5)}{(x - 5)(x + 1)}$$

Equating numerators:

$$3x = *(x + 1) + *(x - 5)$$

Cover-up $*$ by choosing $x = -1$, giving the equation

$$-3 = *(-1 - 5)$$

that is

$$* = *$$

Cover-up $*$ by choosing $x = 5$, giving the equation

$$15 = *(5 + 1)$$

that is

$$* = *$$

The final form is then

$$\frac{3x}{(x-5)(x+1)} \equiv \frac{*}{x-5} + \frac{*}{x+1}$$

(b) Assume

$$\frac{3}{x(x-1)(x+1)} \equiv \frac{A}{x} + \frac{B}{x-1} + \frac{C}{x+1}$$

$$\equiv \frac{*(x-1)(x+1) + *x(x+1) + *x(x-1)}{x(x-1)(x+1)}$$

Equating numerators:

$$3 = *(x-1)(x+1) + *x(x+1) + *x(x-1)$$

Cover-up B and C by choosing $x = *$, giving the equation

$$3 = *$$

Cover-up A and C by choosing $x = *$, giving the equation

$$3 = *$$

Cover-up A and B by choosing $x = *$, giving the equation

$$3 = *$$

The final form is then

$$\frac{3}{x(x-1)(x+1)} \equiv \frac{*}{x} + \frac{*}{*(x-1)} + \frac{*}{*(x+1)}$$

(c) Assume

$$\frac{5}{(x+1)^2(x-1)} \equiv \frac{A}{(x+1)^2} + \frac{B}{x+1} + \frac{C}{x-1}$$

$$\equiv \frac{*(x-1) + *(x+1)(x-1) + *(x+1)^2}{(x+1)^2(x-1)}$$

Equating numerators:

$$5 = *(x - *) + *(x+1)(x-1) + *(x + *)^2$$

Cover-up $*$ by choosing $x = *$, giving the equation

$$* = * \qquad \text{so that } * = *$$

Cover-up * by choosing $x = *$, giving the equation

$$* = * \qquad \text{so that } * = *$$

To find * we cannot use the cover-up rule as we cannot simultaneously cover-up * and *. Instead we substitute the values found for * and * and choose x to be any value other * or *: we choose $x = *$. This gives the equation

$$* = * \qquad \text{so that } * = *$$

The final form is then

$$\frac{5}{(x+1)^2(x-1)} \equiv \frac{*}{(x+1)^2} - \frac{*}{(x+1)} + \frac{*}{(x-1)}$$

(d) In the fraction

$$\frac{x^2+x+1}{(x+1)(x+2)}$$

the numerator is of the same degree as the denominator so division is required.

$$\begin{aligned}\frac{x^2+x+1}{(x+1)(x+2)} &\equiv \frac{x^2+x+1}{x^2+3x+2} \\ &\equiv \frac{x^2+3x+2-*x-*}{x^2+3x+2} \\ &\equiv 1 - \frac{*x+*}{x^2+3x+2} \\ &\equiv 1 - \frac{*x+*}{(x+1)(x+2)}\end{aligned}$$

Assume

$$\begin{aligned}\frac{*x+*}{(x+1)(x+2)} &\equiv \frac{A}{x+1} + \frac{B}{x+2} \\ &\equiv \frac{A(*+*)+B(*+*)}{(x+1)(x+2)}\end{aligned}$$

Equating numerators:

$$(*x+*) = A(*+*) + B(*+*)$$

Cover-up A by choosing $x = *$, giving the equation

$$* = * \qquad \text{so that } A = *$$

Cover-up B by choosing $x = *$, giving the equation

$$* = * \qquad \text{so that } B = *$$

which gives the final form as

$$\frac{x^2+x+1}{(x+1)(x+2)} \equiv 1 + \frac{*}{x+1} + \frac{*}{x+2}$$

Module 7: Further exercises

1 By constructing a table of values between the stated ranges and plotting those values on a cartesian graph, solve each of the following equations:

(a) $4x^2 + 4x - 15 = 0 \qquad -3 \leqslant x \leqslant 2$
(b) $4x^2 + 12x + 9 = 0 \qquad -2 \leqslant x \leqslant 0$
(c) $x^3 - x^2 - x + 1 = 0 \qquad -2 \leqslant x \leqslant 2$

2 Each of the following sets of data represents corresponding values of two variables that are in proportion to one another. Determine the nature of the proportionality and proportionality constant in each case (the top variable is to be plotted against the horizontal axis):

(a)

Time t	0	9	18	27	36
Distance s	0	3	6	9	12

(b)

Time t	0	10	20	40	50
Price p	0	−2	−4	−8	−10

(c)

Object u	−1	−2	−5	−10
Image v	4.0	2.0	0.8	0.4

(d)

x	20	15	5	2
y	0.25	0.33	1.00	2.50

3 Two variables p and q are assumed to be in direct proportion. Two values for these variables are $p = 12$ when $q = 84$. Determine the proportionality constant.

4 Two variables s and t are assumed to be in inverse proportion. Two values for these variables are $s = 9$ when $t = 36$. Determine the proportionality constant.

5 Solve the following simultaneous equations and determine how many times their graphs intersect:

(a) $3x + 3y = 9$ and $2x^2 - 2y^2 = 21$
(b) $x + y = 2$ and $x^2 + y^2 = 4$

6 Solve each of the following quadratic equations:

(a) $x^2 - 6x - 16 = 0$
(b) $5x^2 + 10x + 5 = 0$
(c) $4x^2 - 16x = 0$
(d) $2x^2 - 24x = 6x^2 - 20x$
(e) $x^2 + 3x - 10 = 0$
(f) $x^2 + 10x + 12 = 0$
(g) $6x^2 - 8x + 2 = 0$
(h) $11x^2 + 33x + 22 = 0$

7 Find the solutions to the following quadratic equations:

(a) $2x^2 + 8x + 9 = 0$ (b) $5x^2 - 10x + 6 = 0$

8 Solve the following:

(a) $\dfrac{2}{x+2} - \dfrac{3}{x+3} = 1$ (b) $\dfrac{1}{x^2 - 1} = \dfrac{1}{x - 1}$

9 If p and q are the solutions to

$$ax^2 + bx + c = 0$$

show that

$$x^2 - \frac{(a+c)x}{b} + \frac{ac}{b^2} = 0$$

has roots $-1/(p+q)$ and $-pq/(p+q)$.

10 Show that real solutions of

$$x^2 - (2-k)x + 1 = 0$$

can be found for any value of k provided $k \geqslant 4$ or $k \leqslant 0$.

11 Use the factor theorem to factorize the following:

(a) $x^3 - 14x^2 + 56x - 64$ (c) $x^4 - 3x^2 + 2$

(b) $2x^3 - 2x^2 - 112x$ (d) $x^3 + 6x^2 + 12x + 8$

12 Separate the following into partial fractions:

(a) $\dfrac{3x - 2}{(x + 1)(x - 1)}$ (c) $\dfrac{3x}{(x + 2)^2(x - 1)}$

(b) $\dfrac{5x + 1}{(x - 1)(x - 2)(x - 3)}$ (d) $\dfrac{8x^3 - 8}{(x - 2)(x + 3)}$

Module 8: Matrices and Vectors

Aims: To display the use of matrices as stores of information and to introduce the concepts of vectors and geometric vectors.

Objectives: When you have read this module you will be able to:

- ▶▶ Use matrices as stores of information and manipulate such matrices by adding, multiplying by a scalar and multiplying two matrices together.
- ▶▶ Compute the inverse of a 2×2 matrix and use inverse matrices to solve simultaneous linear equations.
- ▶▶ Define and manipulate vectors using addition and multiplication by a scalar.
- ▶▶ Relate vectors to a cartesian coordinate system and manipulate geometric vectors written in terms of the coordinate unit vectors, including computing both the scalar and vector product of two vectors.

Unit 1 Matrices and Their Arithmetic

Test yourself with the following:

1 Describe the type of each of the following matrices:

$$\text{(a) } \begin{bmatrix} 2 & 0 & 3 \end{bmatrix} \quad \text{(b) } \begin{bmatrix} 2 & 3 \\ 0 & 1 \\ 5 & 7 \end{bmatrix} \quad \text{(c) } \begin{bmatrix} 1 & 0 \\ 0 & 1 \end{bmatrix} \quad \text{(d) } \begin{bmatrix} 2 & 4 & 7 \\ 1 & 3 & 3 \\ 2 & 3 & 5 \end{bmatrix}$$

2 Given matrices

$$A = \begin{bmatrix} 0 & 5 \\ 6 & -7 \end{bmatrix} \quad B = \begin{bmatrix} 8 & 9 \\ -1 & 3 \end{bmatrix} \quad C = \begin{bmatrix} 0 & 1 \end{bmatrix} \quad D = \begin{bmatrix} 2 \\ 0 \end{bmatrix}$$

calculate each of the following (where possible):

(a) $A + B$ (d) AB
(b) $B - A$ (e) AD
(c) $6A$ (f) CB

3 Is $AB = BA$ for the matrices A and B in Question 2?

4 Find AA^T where:

(a) $A = \begin{bmatrix} 1 & 4 & -3 \\ 2 & 0 & 6 \end{bmatrix}$

(b) $A = \begin{bmatrix} 1 \\ 0 \\ -7 \\ -1 \end{bmatrix}$

(c) $A = \begin{bmatrix} 0 & 2 & -5 \\ 1 & 0 & 3 \\ -8 & 0 & 9 \end{bmatrix}$

Matrices

A **matrix** is an array of numbers arranged in regular rows and columns. For example,

$$\begin{bmatrix} 1 & 2 \\ 3 & 6 \end{bmatrix} \quad \begin{bmatrix} 4 & -3 & 2 \\ 5 & 0 & 1 \end{bmatrix}$$

The first matrix is referred to as a 2×2 matrix and the second as a 2×3 matrix. If a matrix has i rows and j columns it is called an i by j ($i \times j$) matrix.

Types of Matrix

Matrices are useful methods of storing information, for example if three salespersons Andrew, Belinda and Charles wish to record their monthly sales of two specific items they can do so in matrix form:

	Month 1	
	Item 1	Item 2
Andrew	£1246	£3427
Belinda	£2205	£2439
Charles	£5377	£1026

These sales can then be stored in matrix form as:

$$\begin{bmatrix} 1246 & 3427 \\ 2205 & 2439 \\ 5377 & 1026 \end{bmatrix}$$

There are many types of matrix, some of which are listed below.

Square Matrix

A **square matrix** is a matrix with the same number of rows as columns. For example,

$$\begin{bmatrix} 1 & 2 & 3 \\ 4 & 5 & 6 \\ 7 & 8 & 9 \end{bmatrix}$$

is a 3×3 square matrix.

Diagonal Matrix

A **diagonal matrix** is a square matrix with zeros everywhere except down the leading diagonal. For example,

$$\begin{bmatrix} 1 & 0 & 0 \\ 0 & 5 & 0 \\ 0 & 0 & 9 \end{bmatrix}$$

is a 3×3 diagonal matrix.

Symmetric Matrix

A **symmetric matrix** is a matrix whose (i, j) element is the same as the (j, i) element. The (i, j) element is the element in the ith row and jth column. For example, the matrix

$$\begin{bmatrix} 1 & 0 & 4 \\ 0 & 2 & 5 \\ 4 & 5 & -3 \end{bmatrix}$$

is symmetric – the (1, 3) element is the same as the (3, 1) element, as are the (1, 2) and the (2, 1) and the (2, 3) and the (3, 2) elements.

Unit Matrix

A **unit matrix** is a diagonal matrix with unity down the leading diagonal. For example,

$$\begin{bmatrix} 1 & 0 & 0 \\ 0 & 1 & 0 \\ 0 & 0 & 1 \end{bmatrix}$$

is the 3×3 unit matrix.

Row Vector

A matrix that consists of a single row is called a **row vector**. For example, the matrix

$$\begin{bmatrix} 0 & 2 & 4 & 7 \end{bmatrix}$$

is a row vector.

Column Vector

A matrix that consists of a single column is called a **column vector**. For example, the matrix

$$\begin{bmatrix} -3 \\ 6 \\ 0 \\ 1 \end{bmatrix}$$

is a column vector.

Notation

Matrices can be written in terms of their elements as has just been done. Alternatively matrices can be described using capital letters. For example,

$$A = \begin{bmatrix} 4 & -9 \\ 2 & 7 \end{bmatrix}$$

The Transpose of a Matrix

The transpose A^T of matrix A is the matrix formed from A by interchanging rows for columns. For example, if

$$A = \begin{bmatrix} 1 & 2 & 3 \\ 4 & 5 & 6 \end{bmatrix}$$

then

$$A^T = \begin{bmatrix} 1 & 4 \\ 2 & 5 \\ 3 & 6 \end{bmatrix}$$

where the 1st and 2nd rows of A become the 1st and 2nd columns of A^T.

The Arithmetic of Matrices

Matrices can be added, subtracted, multiplied by a scalar and multiplied together provided they satisfy certain criteria.

Addition and Subtraction

To add two matrices together they must both have the same number of rows and the same number of columns. Their sum is then obtained by adding corresponding elements to form a matrix with the same number of rows and the same number of columns. For example, if

$$A = \begin{bmatrix} 1 & 2 \\ 3 & 4 \\ 0 & 9 \end{bmatrix} \quad \text{and} \quad B = \begin{bmatrix} -2 & 4 \\ 0 & 7 \\ -4 & 5 \end{bmatrix}$$

then

$$A + B = \begin{bmatrix} 1 & 2 \\ 3 & 4 \\ 0 & 9 \end{bmatrix} + \begin{bmatrix} -2 & 4 \\ 0 & 7 \\ -4 & 5 \end{bmatrix}$$

$$= \begin{bmatrix} 1+(-2) & 2+4 \\ 3+0 & 4+7 \\ 0+(-4) & 9+5 \end{bmatrix} = \begin{bmatrix} -1 & 6 \\ 3 & 11 \\ -4 & 14 \end{bmatrix}$$

Multiplication by a Scalar

To multiply a matrix by a single number each element of the matrix is multiplied by that number. For example, if

$$A = \begin{bmatrix} 4 & 8 \\ 2 & 0 \end{bmatrix}$$

then

$$3A = 3 \times \begin{bmatrix} 4 & 8 \\ 2 & 0 \end{bmatrix} = \begin{bmatrix} 3 \times 4 & 3 \times 8 \\ 3 \times 2 & 3 \times 0 \end{bmatrix} = \begin{bmatrix} 12 & 24 \\ 6 & 0 \end{bmatrix}$$

Andrew, Belinda and Charles can add each month's sales figures to find their quarterly sales:

	Month 1		Month 2		Month 3	
	Item 1	Item 2	Item 1	Item 2	Item 1	Item 2
Andrew	£1246	£3427	£2885	£2856	£3766	£2971
Belinda	£2205	£2439	£1994	£3005	£1957	£4301
Charles	£5377	£1026	£3045	£4184	£2886	£3911

By adding each of the three monthly sales matrices they can obtain their quarterly sales figures:

$$\begin{bmatrix} 1246 & 3427 \\ 2205 & 2439 \\ 5377 & 1026 \end{bmatrix} + \begin{bmatrix} 2885 & 2856 \\ 1994 & 3005 \\ 3045 & 4184 \end{bmatrix} + \begin{bmatrix} 3766 & 2971 \\ 1957 & 4301 \\ 2886 & 3911 \end{bmatrix} = \begin{bmatrix} 7897 & 9254 \\ 6156 & 9745 \\ 11308 & 9121 \end{bmatrix}$$

to give

	Quarter 1	
	Item 1	Item 2
Andrew	£7897	£9254
Belinda	£6156	£9745
Charles	£11308	£9121

Multiplication of Matrices

Two matrices can be multiplied together provided that the matrix on the left has as many columns as the matrix on the right has rows. For example, the product of

$$A = \begin{bmatrix} 1 & 2 \\ 3 & 4 \\ 5 & 6 \end{bmatrix} \quad \text{and} \quad B = \begin{bmatrix} 1 & 2 & 3 \\ 4 & 5 & 6 \end{bmatrix}$$

is defined to be

$$AB = \begin{bmatrix} (1 \times 1 + 4 \times 2) & (2 \times 1 + 5 \times 2) & (3 \times 1 + 6 \times 2) \\ (1 \times 3 + 4 \times 4) & (2 \times 3 + 5 \times 4) & (3 \times 3 + 6 \times 4) \\ (1 \times 5 + 4 \times 6) & (2 \times 5 + 5 \times 6) & (3 \times 5 + 6 \times 6) \end{bmatrix} = \begin{bmatrix} 9 & 12 & 15 \\ 19 & 26 & 33 \\ 29 & 40 & 51 \end{bmatrix}$$

The principle is that each number in the nth row of the left-hand matrix is multiplied by each corresponding term of the mth column of the right-hand matrix. These products are then added to form the element in the nth row, mth column of the product matrix. As a consequence, the product of matrices PQ where P is an $n \times r$ matrix and Q is an $r \times m$ matrix is an $n \times m$ matrix. As a further example, suppose our three salespersons receive a quarterly bonus on their sales in the form:

Item 1	1%
Item 2	2%

They can write their bonus matrix as

$$\begin{bmatrix} 0.01 \\ 0.02 \end{bmatrix}$$

If they multiply their quarterly sales matrix by this bonus matrix they can obtain their total bonuses:

$$\begin{bmatrix} 7897 & 9254 \\ 6156 & 9745 \\ 11308 & 9121 \end{bmatrix} \begin{bmatrix} 0.01 \\ 0.02 \end{bmatrix} = \begin{bmatrix} (7897)(0.01) + (9254)(0.02) \\ (6156)(0.01) + (9745)(0.02) \\ (11308)(0.01) + (9121)(0.02) \end{bmatrix} = \begin{bmatrix} 264.05 \\ 256.46 \\ 295.50 \end{bmatrix}$$

Hence their bonuses are

	Bonuses
Andrew	£264.05
Belinda	£256.46
Charles	£295.50

Multiplication by a Unit Matrix

Let I be the 3×3 unit matrix

$$\begin{bmatrix} 1 & 0 & 0 \\ 0 & 1 & 0 \\ 0 & 0 & 1 \end{bmatrix} \quad \text{and} \quad A = \begin{bmatrix} 1 & 2 & 3 \\ 4 & 5 & 6 \end{bmatrix}$$

then:

$$AI = \begin{bmatrix} 1 & 2 & 3 \\ 4 & 5 & 6 \end{bmatrix} \begin{bmatrix} 1 & 0 & 0 \\ 0 & 1 & 0 \\ 0 & 0 & 1 \end{bmatrix}$$

$$= \begin{bmatrix} 1 & 2 & 3 \\ 4 & 5 & 6 \end{bmatrix}$$

$$= A$$

If A is an $i \times j$ matrix and I is the $j \times j$ unit matrix then:

$$AI = A$$

EXAMPLES

1 Describe the type of each of the following matrices:

(a) $[1 \quad 0 \quad 0]$ **(c)** $\begin{bmatrix} 1 & 0 \\ 0 & 1 \end{bmatrix}$

(b) $\begin{bmatrix} 1 \\ 2 \end{bmatrix}$ **(d)** $\begin{bmatrix} 1 & 0 & 2 \\ 0 & 0 & 3 \\ 2 & 3 & 5 \end{bmatrix}$

(a) 1×3 row matrix or row vector.
(b) 2×1 column matrix or column vector.
(c) 2×2 unit matrix.
(d) 3×3 symmetric matrix.

2 Given matrices

$$A = \begin{bmatrix} 1 & 2 \\ -3 & 8 \end{bmatrix} \quad B = \begin{bmatrix} 5 & 0 \\ -2 & 4 \end{bmatrix} \quad C = [5 \quad 3] \quad D = \begin{bmatrix} 1 \\ 6 \end{bmatrix}$$

Calculate each of the following (where possible):

(a) $A + B$ **(d)** AB
(b) $B - A$ **(e)** AD
(c) $2A$ **(f)** CB

(a) $A + B = \begin{bmatrix} 6 & 2 \\ -5 & 12 \end{bmatrix}$ **(d)** $AB = \begin{bmatrix} 1 & 8 \\ -31 & 32 \end{bmatrix}$

(b) $B - A = \begin{bmatrix} 4 & -2 \\ 1 & -4 \end{bmatrix}$ **(e)** $AD = \begin{bmatrix} 13 \\ 45 \end{bmatrix}$

(c) $2A = \begin{bmatrix} 2 & 4 \\ -6 & 16 \end{bmatrix}$ **(f)** $\mathrm{CB} = [19 \quad 12]$

3 Is $AB = BA$ for the matrices A and B in Example 2?

$$BA = \begin{bmatrix} 5 & 10 \\ -14 & 28 \end{bmatrix}$$

No, AB does not equal BA. This is a notable point; matrix multiplication is not necessarily commutative.

4 Find AA^T where:

(a) $A = \begin{bmatrix} 2 & -7 & 0 \\ 3 & 4 & -1 \end{bmatrix}$

(b) $A = [2 \quad -3 \quad 5 \quad 8]$

(c) $A = \begin{bmatrix} 1 & 0 & 2 \\ -4 & 5 & 0 \\ 2 & -1 & 3 \end{bmatrix}$

(a) $A^T = \begin{bmatrix} 2 & 3 \\ -7 & 4 \\ 0 & -1 \end{bmatrix}$ so $AA^T = \begin{bmatrix} 2 & -7 & 0 \\ 3 & 4 & -1 \end{bmatrix} \times \begin{bmatrix} 2 & 3 \\ -7 & 4 \\ 0 & -1 \end{bmatrix}$

$$= \begin{bmatrix} (2)(2) + (-7)(-7) + (0)(0) & (2)(2) + (4)(-7) + (-1)(0) \\ (2)(3) + (-7)(4) + (0)(-1) & (3)(3) + (4)(4) + (-1)(-1) \end{bmatrix}$$

$$= \begin{bmatrix} 53 & -22 \\ -22 & 26 \end{bmatrix}$$

(b) $A^T = \begin{bmatrix} 2 \\ -3 \\ 5 \\ 8 \end{bmatrix}$ so $AA^T = [2 \quad -3 \quad 5 \quad 8] \times \begin{bmatrix} 2 \\ -3 \\ 5 \\ 8 \end{bmatrix}$

$$= [(2)(2) + (-3)(-3) + (5)(5) + (8)(8)]$$

$$= [4 + 9 + 25 + 64]$$

$$= [102]$$

(c) $A^T = \begin{bmatrix} 1 & -4 & 2 \\ 0 & 5 & -1 \\ 2 & 0 & 3 \end{bmatrix}$ so $AA^T = \begin{bmatrix} 1 & 0 & 2 \\ -4 & 5 & 0 \\ 2 & -1 & 3 \end{bmatrix} \times \begin{bmatrix} 1 & -4 & 2 \\ 0 & 5 & -1 \\ 2 & 0 & 3 \end{bmatrix}$

$$= \begin{bmatrix} 5 & -4 & 8 \\ -4 & 41 & -13 \\ 8 & -13 & 14 \end{bmatrix}$$

Notice that AA^T is always a symmetric matrix.

EXERCISES

1 Describe the type of each of the following matrices:

(a) $\begin{bmatrix} 5 \\ 2 \\ 3 \end{bmatrix}$ **(c)** $\begin{bmatrix} 1 & 5 \\ 0 & 1 \end{bmatrix}$

(b) $\begin{bmatrix} 1 & 0 \\ 0 & 3 \end{bmatrix}$ **(d)** $\begin{bmatrix} 1 & 0 & 0 \\ 0 & 1 & 0 \\ 0 & 0 & 1 \end{bmatrix}$

(a) A * × * column matrix or * vector.
(b) A 2 × 2 * matrix.
(c) A 2 × 2 * matrix.
(d) A * × * * matrix.

2 Given matrices

$$A = \begin{bmatrix} 6 & -3 \\ 1 & 4 \end{bmatrix} \quad B = \begin{bmatrix} 0 & 2 \\ -3 & -1 \end{bmatrix} \quad C = [6 \quad -2] \quad D = \begin{bmatrix} 6 \\ 4 \end{bmatrix}$$

Calculate each of the following:

(a) $A + B$ **(d)** AB
(b) $B - A$ **(e)** AD
(c) $-6A$ **(f)** CB

(a) $A + B = \begin{bmatrix} * & -1 \\ -2 & * \end{bmatrix}$ **(d)** $AB = \begin{bmatrix} * & 15 \\ * & * \end{bmatrix}$

(b) $B - A = \begin{bmatrix} * & * \\ -4 & * \end{bmatrix}$ **(e)** $AD = \begin{bmatrix} 24 \\ * \end{bmatrix}$

(c) $6A = \begin{bmatrix} -36 & * \\ * & * \end{bmatrix}$ **(f)** $CB = [6 \quad *]$

3 Is $AB = BA$ for the matrices A and B in Exercise 2?

$$AB = \begin{bmatrix} * & 15 \\ * & * \end{bmatrix} \quad BA = \begin{bmatrix} * & * \\ -19 & 5 \end{bmatrix}$$

so AB is/is not equal to BA.

4 Find AA^T where:

(**a**) $A = \begin{bmatrix} 5 & 3 & 10 \\ 0 & -2 & 7 \end{bmatrix}$

(**b**) $A = [8 \quad 2 \quad -6 \quad 1]$

(**c**) $A = \begin{bmatrix} 3 & -3 & -1 \\ 5 & 0 & 4 \\ 0 & 8 & 6 \end{bmatrix}$

(**a**) $A^T = \begin{bmatrix} * & 0 \\ 3 & * \\ * & * \end{bmatrix}$ so $AA^T = \begin{bmatrix} 5 & 3 & 10 \\ 0 & -2 & 7 \end{bmatrix} \times \begin{bmatrix} * & 0 \\ 3 & * \\ * & * \end{bmatrix}$

$$= \begin{bmatrix} (5)(*) + (3)(3) + (10)(*) & (*)(*) + (*)(*) + (*)(*) \\ (0)(*) + (-2)(3) + (7)(*) & (0)(0) + (*)(*) + (*)(*) \end{bmatrix}$$

$$= \begin{bmatrix} 134 & * \\ * & * \end{bmatrix}$$

(**b**) $A^T = \begin{bmatrix} 8 \\ * \\ * \\ * \end{bmatrix}$ so $AA^T = [8 \quad 2 \quad -6 \quad 1] \times \begin{bmatrix} 8 \\ * \\ * \\ * \end{bmatrix}$

$$= [(8)(8) + (2)(*) + (-6)(*) + (1)(*)]$$

$$= [64 + * + * + *]$$

$$= [105]$$

(**c**) $A^T = \begin{bmatrix} 3 & * & * \\ * & * & * \\ * & * & 6 \end{bmatrix}$ so $AA^T = \begin{bmatrix} 3 & -3 & -1 \\ 5 & 0 & 4 \\ 0 & 8 & 6 \end{bmatrix} \times \begin{bmatrix} 3 & * & * \\ * & * & * \\ * & * & 6 \end{bmatrix}$

$$= \begin{bmatrix} 19 & * & * \\ * & * & * \\ -30 & * & * \end{bmatrix}$$

Unit 2 The Inverse of a Matrix

Test yourself with the following:

1 If possible, find the inverse of each of the following matrices:

(a) $A = \begin{bmatrix} 6 & -3 \\ 0 & 1 \end{bmatrix}$ (b) $B = \begin{bmatrix} 8 & -3 \\ 16 & -6 \end{bmatrix}$

2 Write each pair of simultaneous equations in matrix form:

(a) $4x - 5y = 2$
$x + 3y = 9$

(b) $7x + 3y = 14$
$9y - 7x = -2$

(c) $-2x - 3y = 9$
$4x = 6 + 3y$

(d) $7x - 2y = 15$
$2y + 2 = 0$

3 Use matrix inversion to solve each pair of simultaneous equations in Question 2.

Inverting Matrices

Division of matrices cannot be defined but an operation similar in effect to division is that of multiplication of a matrix by its inverse. If A and B are two square matrices such that

$$AB = I$$

where I is a unit matrix, then matrix B is called the **inverse matrix** of A and is written as

$$A^{-1}$$

We could also claim that A was the inverse of matrix B and write A as

$$B^{-1}$$

More correctly B is the right-hand inverse of A and A is the left-hand inverse of B. We shall only be concerned with the inverses of 2×2 matrices, where the left-hand and right-hand inverses are equal so that

$$AA^{-1} = A^{-1}A = I$$

Determinants

Every square matrix A has a number associated with it called its **determinant**, denoted by:

$$\det A$$

For example, if:

$$A = \begin{bmatrix} 2 & 3 \\ 4 & 5 \end{bmatrix}$$

then $\det A$ is obtained by multiplying diagonally opposite elements

$$\begin{matrix} 2 & & 3 \\ & \times & \\ 4 & & 5 \end{matrix}$$

and then subtracting the products. Hence:

$$\begin{aligned} \det A &= (2 \times 5) - (3 \times 4) \\ &= 10 - 12 \\ &= -2 \end{aligned}$$

In general, if:

$$A = \begin{bmatrix} a & b \\ c & d \end{bmatrix}$$

then:

$$detA = ad - bc$$

Inverse of a 2 × 2 Matrix

If A is a 2×2 matrix of the form

$$\begin{bmatrix} a & b \\ c & d \end{bmatrix}$$

then the inverse A^{-1} is the matrix obtained by interchanging the elements a and d, changing the signs of b and c and dividing the whole by the determinant of A. That is,

$$A^{-1} = \frac{1}{detA}\begin{bmatrix} d & -b \\ -c & a \end{bmatrix} = \frac{1}{ad - bc}\begin{bmatrix} d & -b \\ -c & a \end{bmatrix}$$

For example, the inverse of matrix

$$A = \begin{bmatrix} 2 & 3 \\ 5 & 1 \end{bmatrix}$$

is

$$A^{-1} = \frac{1}{-13}\begin{bmatrix} 1 & -3 \\ -5 & 2 \end{bmatrix}$$

It can be readily seen that

$$\begin{aligned} AA^{-1} &= \frac{-1}{13}\begin{bmatrix} 1 & -3 \\ -5 & 2 \end{bmatrix}\begin{bmatrix} 2 & 3 \\ 5 & 1 \end{bmatrix} \\ &= \frac{-1}{13}\begin{bmatrix} -13 & 0 \\ 0 & -13 \end{bmatrix} \\ &= \begin{bmatrix} 1 & 0 \\ 0 & 1 \end{bmatrix} \end{aligned}$$

Notice that

$$\begin{aligned} A^{-1}A &= \frac{-1}{13}\begin{bmatrix} 2 & 3 \\ 5 & 1 \end{bmatrix}\begin{bmatrix} 1 & -3 \\ -5 & 2 \end{bmatrix} \\ &= \frac{-1}{13}\begin{bmatrix} -13 & 0 \\ 0 & -13 \end{bmatrix} \\ &= \begin{bmatrix} 1 & 0 \\ 0 & 1 \end{bmatrix} \end{aligned}$$

Notice that if $detA = 0$ then A^{-1} *does not exist* because division by zero is not a defined operation.

Matrix Solution to Simultaneous Linear Equations

A pair of simultaneous linear equations can be written as a matrix equation. Using inverse matrices the solution to the equations can then be easily found. For example, the equations

$$2x + 3y = 1$$
$$5x + \ \ y = -4$$

can be written as

$$\begin{bmatrix} 2 & 3 \\ 5 & 1 \end{bmatrix} \begin{bmatrix} x \\ y \end{bmatrix} = \begin{bmatrix} 1 \\ -4 \end{bmatrix}$$

Multiplying from the left by the inverse of the 2×2 matrix it is seen that

$$\frac{-1}{13}\begin{bmatrix} 1 & -3 \\ -5 & 2 \end{bmatrix}\begin{bmatrix} 2 & 3 \\ 5 & 1 \end{bmatrix} \begin{bmatrix} x \\ y \end{bmatrix} = \frac{-1}{13}\begin{bmatrix} 1 & -3 \\ -5 & 2 \end{bmatrix}\begin{bmatrix} 1 \\ -4 \end{bmatrix}$$

that is

$$\begin{bmatrix} 1 & 0 \\ 0 & 1 \end{bmatrix} \begin{bmatrix} x \\ y \end{bmatrix} = \frac{-1}{13}\begin{bmatrix} 13 \\ -13 \end{bmatrix}$$

that is

$$\begin{bmatrix} x \\ y \end{bmatrix} = \begin{bmatrix} -1 \\ 1 \end{bmatrix}$$

So that $x = -1$ and $y = 1$ is the desired solution.

The principles involved in finding this solution can be described in general terms as follows. If a pair of simultaneous equations in variables x and y are represented in matrix form as

$$AX = b$$

where A is formed from the coefficients of the two equations, $X = \left[\begin{smallmatrix} x \\ y \end{smallmatrix}\right]$ and b is the 2×1 matrix formed from constants on the right-hand sides of the equations, then multiplying from the left both sides of this matrix equation by the inverse of A gives

$$A^{-1}AX = A^{-1}b$$

Now, $A^{-1}A = I$, the 2×2 unit matrix, so $A^{-1}AX = IX = X$. Hence,

$$X = A^{-1}b$$

and so the solution is found.

EXAMPLES

1 If possible, find the inverse of each of the following matrices:

(a) $A = \begin{bmatrix} 3 & 2 \\ -6 & 4 \end{bmatrix}$ **(b)** $B = \begin{bmatrix} 5 & 6 \\ 10 & 12 \end{bmatrix}$

(a) $A^{-1} = \dfrac{1}{12+12}\begin{bmatrix} 4 & -2 \\ 6 & 3 \end{bmatrix}$

Check $\dfrac{1}{24}\begin{bmatrix} 3 & 2 \\ -6 & 4 \end{bmatrix}\begin{bmatrix} 4 & -2 \\ 6 & 3 \end{bmatrix} = \dfrac{1}{24}\begin{bmatrix} 24 & 0 \\ 0 & 24 \end{bmatrix} = \begin{bmatrix} 1 & 0 \\ 0 & 1 \end{bmatrix}$

(b) Because the factor $ad - bc = (5)(12) - (6)(10) = 0$, and we cannot divide by zero, the inverse does not exist.

2 Write each pair of simultaneous equations in matrix form:

(a) $x + y = 5$
$2x - 3y = 6$

(b) $6x - 5y = 15$
$-2y + 3x = -3$

(c) $5x + 7y = 0$
$x = 2y + 3$

(d) $3x + 4y = 12$
$5x + 5 = 0$

(a) $\begin{bmatrix} 1 & 1 \\ 2 & -3 \end{bmatrix}\begin{bmatrix} x \\ y \end{bmatrix} = \begin{bmatrix} 5 \\ 6 \end{bmatrix}$ **(c)** $\begin{bmatrix} 5 & 7 \\ 1 & -2 \end{bmatrix}\begin{bmatrix} x \\ y \end{bmatrix} = \begin{bmatrix} 0 \\ 3 \end{bmatrix}$

(b) $\begin{bmatrix} 6 & -5 \\ 3 & -2 \end{bmatrix}\begin{bmatrix} x \\ y \end{bmatrix} = \begin{bmatrix} 15 \\ -3 \end{bmatrix}$ **(d)** $\begin{bmatrix} 3 & 4 \\ 5 & 0 \end{bmatrix}\begin{bmatrix} x \\ y \end{bmatrix} = \begin{bmatrix} 12 \\ -5 \end{bmatrix}$

3 Use matrix inversion to solve each pair of simultaneous equations in Example 2.

(a) $A = \begin{bmatrix} 1 & 1 \\ 2 & -3 \end{bmatrix} \quad A^{-1} = \dfrac{1}{-3-2}\begin{bmatrix} -3 & -1 \\ -2 & 1 \end{bmatrix}$

Therefore

$$\frac{-1}{5}\begin{bmatrix} -3 & -1 \\ -2 & 1 \end{bmatrix}\begin{bmatrix} 1 & 1 \\ 2 & -3 \end{bmatrix}\begin{bmatrix} x \\ y \end{bmatrix} = \frac{-1}{5}\begin{bmatrix} -3 & -1 \\ -2 & 1 \end{bmatrix}\begin{bmatrix} 5 \\ 6 \end{bmatrix}$$

that is

$$\begin{bmatrix} 1 & 0 \\ 0 & 1 \end{bmatrix}\begin{bmatrix} x \\ y \end{bmatrix} = \frac{-1}{5}\begin{bmatrix} -21 \\ -4 \end{bmatrix}$$

Hence

$$\begin{bmatrix} x \\ y \end{bmatrix} = \begin{bmatrix} \frac{21}{5} \\ \frac{4}{5} \end{bmatrix}$$

(b) $A = \begin{bmatrix} 6 & -5 \\ 3 & -2 \end{bmatrix} \quad A^{-1} = \dfrac{1}{-12+15}\begin{bmatrix} -2 & 5 \\ -3 & 6 \end{bmatrix}$

Therefore

$$\frac{1}{3}\begin{bmatrix} -2 & 5 \\ -3 & 6 \end{bmatrix}\begin{bmatrix} 6 & -5 \\ 3 & -2 \end{bmatrix}\begin{bmatrix} x \\ y \end{bmatrix} = \frac{1}{3}\begin{bmatrix} -2 & 5 \\ -3 & 6 \end{bmatrix}\begin{bmatrix} 15 \\ -3 \end{bmatrix}$$

that is

$$\begin{bmatrix} 1 & 0 \\ 0 & 1 \end{bmatrix}\begin{bmatrix} x \\ y \end{bmatrix} = \frac{1}{3}\begin{bmatrix} -45 \\ -63 \end{bmatrix}$$

Hence

$$\begin{bmatrix} x \\ y \end{bmatrix} = \begin{bmatrix} -15 \\ -21 \end{bmatrix}$$

(c) $$A = \begin{bmatrix} 5 & 7 \\ 1 & -2 \end{bmatrix} \quad A^{-1} = \frac{1}{-10-7}\begin{bmatrix} -2 & -7 \\ -1 & 5 \end{bmatrix}$$

Therefore

$$\frac{-1}{17}\begin{bmatrix} -2 & -7 \\ -1 & 5 \end{bmatrix}\begin{bmatrix} 5 & 7 \\ 1 & -2 \end{bmatrix}\begin{bmatrix} x \\ y \end{bmatrix} = \frac{-1}{17}\begin{bmatrix} -2 & -7 \\ -1 & 5 \end{bmatrix}\begin{bmatrix} 0 \\ 3 \end{bmatrix}$$

that is

$$\begin{bmatrix} 1 & 0 \\ 0 & 1 \end{bmatrix}\begin{bmatrix} x \\ y \end{bmatrix} = \frac{-1}{17}\begin{bmatrix} -21 \\ 15 \end{bmatrix}$$

Hence

$$\begin{bmatrix} x \\ y \end{bmatrix} = \begin{bmatrix} \frac{21}{17} \\ \frac{-15}{17} \end{bmatrix}$$

(d) $$A = \begin{bmatrix} 3 & 4 \\ 5 & 0 \end{bmatrix} \quad A^{-1} = \frac{1}{0-20}\begin{bmatrix} 0 & -4 \\ -5 & 3 \end{bmatrix}$$

Therefore

$$\frac{-1}{20}\begin{bmatrix} 0 & -4 \\ -5 & 3 \end{bmatrix}\begin{bmatrix} 3 & 4 \\ 5 & 0 \end{bmatrix}\begin{bmatrix} x \\ y \end{bmatrix} = \frac{-1}{20}\begin{bmatrix} 0 & -4 \\ -5 & 3 \end{bmatrix}\begin{bmatrix} 12 \\ -5 \end{bmatrix}$$

that is

$$\begin{bmatrix} 1 & 0 \\ 0 & 1 \end{bmatrix}\begin{bmatrix} x \\ y \end{bmatrix} = \frac{-1}{20}\begin{bmatrix} 20 \\ -75 \end{bmatrix}$$

Hence

$$\begin{bmatrix} x \\ y \end{bmatrix} = \begin{bmatrix} -1 \\ \frac{15}{4} \end{bmatrix}$$

EXERCISES

1 If possible, find the inverse of each of the following matrices:

(a) $A = \begin{bmatrix} 1 & 5 \\ 3 & -2 \end{bmatrix}$ **(b)** $B = \begin{bmatrix} -1 & 2 \\ 6 & -12 \end{bmatrix}$

(a) $A^{-1} = \dfrac{1}{-2-*}\begin{bmatrix} -2 & * \\ * & 1 \end{bmatrix}$ **(b)** $B^{-1} = \dfrac{1}{*}\begin{bmatrix} * & -2 \\ * & * \end{bmatrix}$

2 Write each pair of simultaneous equations in matrix form:

(a) $x - y = 8$, $2x + 5y = 9$ **(c)** $2x + 2y = 4$, $7x + 5y = 8$

(b) $2x - 4y = 12$, $-5x + y = 6$ **(d)** $9x - 2y = 8$, $4x = 2$

(a) $\begin{bmatrix} 1 & * \\ * & 5 \end{bmatrix}\begin{bmatrix} * \\ * \end{bmatrix} = \begin{bmatrix} 8 \\ 9 \end{bmatrix}$ **(c)** $\begin{bmatrix} * & 2 \\ * & 5 \end{bmatrix}\begin{bmatrix} x \\ y \end{bmatrix} = \begin{bmatrix} 4 \\ * \end{bmatrix}$

(b) $\begin{bmatrix} * & * \\ * & 1 \end{bmatrix}\begin{bmatrix} x \\ y \end{bmatrix} = \begin{bmatrix} * \\ * \end{bmatrix}$ **(d)** $\begin{bmatrix} 9 & * \\ * & * \end{bmatrix}\begin{bmatrix} * \\ * \end{bmatrix} = \begin{bmatrix} 8 \\ * \end{bmatrix}$

3 Use matrix inversion to solve each pair of simultaneous equations in Exercise 2.

(a) $A = \begin{bmatrix} 1 & -1 \\ 2 & 5 \end{bmatrix} A^{-1} = \frac{1}{* - *}\begin{bmatrix} 5 & * \\ -* & 1 \end{bmatrix}$

Therefore

$$\frac{1}{*}\begin{bmatrix} 5 & * \\ -* & 1 \end{bmatrix}\begin{bmatrix} 1 & -1 \\ 2 & 5 \end{bmatrix}\begin{bmatrix} x \\ y \end{bmatrix} = \frac{1}{*}\begin{bmatrix} 5 & * \\ -* & 1 \end{bmatrix}\begin{bmatrix} 8 \\ 9 \end{bmatrix}$$

that is

$$\begin{bmatrix} 1 & 0 \\ 0 & 1 \end{bmatrix}\begin{bmatrix} x \\ y \end{bmatrix} = \frac{1}{*}\begin{bmatrix} * \\ * \end{bmatrix}$$

Hence

$$\begin{bmatrix} x \\ y \end{bmatrix} = \begin{bmatrix} * \\ * \end{bmatrix}$$

(b) $A = \begin{bmatrix} 2 & -4 \\ * & * \end{bmatrix} \quad A^{-1} = \frac{1}{*}\begin{bmatrix} * & * \\ * & 2 \end{bmatrix}$

Therefore

$$\frac{-1}{*}\begin{bmatrix} * & * \\ * & 2 \end{bmatrix}\begin{bmatrix} 2 & -4 \\ * & * \end{bmatrix}\begin{bmatrix} x \\ y \end{bmatrix} = \frac{-1}{*}\begin{bmatrix} * & * \\ * & 2 \end{bmatrix}\begin{bmatrix} * \\ * \end{bmatrix}$$

that is

$$\begin{bmatrix} 1 & 0 \\ 0 & 1 \end{bmatrix}\begin{bmatrix} x \\ y \end{bmatrix} = \frac{-1}{*}\begin{bmatrix} * \\ * \end{bmatrix}$$

Hence

$$\begin{bmatrix} x \\ y \end{bmatrix} = \begin{bmatrix} * \\ * \end{bmatrix}$$

(c) $A = \begin{bmatrix} * & 2 \\ * & * \end{bmatrix} \quad A^{-1} = \frac{1}{*}\begin{bmatrix} * & * \\ -7 & * \end{bmatrix}$

Therefore

$$\frac{-1}{*}\begin{bmatrix} * & * \\ -7 & * \end{bmatrix}\begin{bmatrix} * & 2 \\ * & * \end{bmatrix}\begin{bmatrix} x \\ y \end{bmatrix} = \frac{-1}{*}\begin{bmatrix} * & * \\ -7 & * \end{bmatrix}\begin{bmatrix} * \\ * \end{bmatrix}$$

that is

$$\begin{bmatrix} 1 & 0 \\ 0 & 1 \end{bmatrix}\begin{bmatrix} x \\ y \end{bmatrix} = \frac{-1}{*}\begin{bmatrix} * \\ * \end{bmatrix}$$

Hence

$$\begin{bmatrix} x \\ y \end{bmatrix} = \begin{bmatrix} * \\ * \end{bmatrix}$$

(d) $A = \begin{bmatrix} * & * \\ * & * \end{bmatrix} \quad A^{-1} = \frac{1}{*}\begin{bmatrix} 0 & * \\ * & * \end{bmatrix}$

Therefore

$$\frac{1}{*}\begin{bmatrix} 0 & * \\ * & * \end{bmatrix}\begin{bmatrix} * & * \\ * & * \end{bmatrix}\begin{bmatrix} x \\ y \end{bmatrix} = \frac{1}{*}\begin{bmatrix} 0 & * \\ * & * \end{bmatrix}\begin{bmatrix} * \\ * \end{bmatrix}$$

that is

$$\begin{bmatrix} 1 & 0 \\ 0 & 1 \end{bmatrix}\begin{bmatrix} x \\ y \end{bmatrix} = \frac{1}{*}\begin{bmatrix} * \\ * \end{bmatrix}$$

Hence

$$\begin{bmatrix} x \\ y \end{bmatrix} = \begin{bmatrix} * \\ * \end{bmatrix}$$

Unit 3 Vectors and Their Arithmetic

Test yourself with the following:

1 A deep-sea diver descends towards the sea bed at a constant speed of 5 kilometres per hour. He does so with a tidal current flowing away from him at 5 kph due west. He has to descend 1000 m to a scallop bed from a boat moored due west of the bed a distance x m away from a point directly above the scallop bed. Draw a vector diagram to determine the diver's actual velocity. What is the value of x if the diver wishes to avoid having to walk along the sea bed to the scallops?

2 A light ring is suspended above the ground by three springs. One spring is attached to the floor vertically below the ring and has a tension of 2 newtons. A second spring is attached to the ceiling and is inclined at 45° to the right of the vertical. The tension in this spring is $2\sqrt{2}$ newtons. The third spring is horizontal and attached to a wall on the immediate left of the ring. What is the tension in this spring?

Vectors

Velocity is a quantity possessing both a **magnitude** and a **direction** that can be graphically represented by a **directed line segment**. The length of the line segment represents the magnitude of the velocity – the speed – and the direction of the line segment represents the direction of the velocity. For example, if the wind blows into your face from the east with a speed of 20 mph then we could represent the velocity of the wind by an arrow called a directed line segment as shown in Figure 8.1.

Figure 8.1

Figure 8.2

In Figure 8.1 the direction of the wind is indicated by the arrowhead and its speed is represented by the length of the arrow. Every directed line segment has an initial point and a terminal point and if we label these as A and B respectively then we can represent the directed line segment of Figure 8.2 by the notation

$$\overrightarrow{AB}$$

If the arrow on the directed line segment were pointing in the opposite direction we would denote it as $\overrightarrow{BA}$.

In mathematics we call a quantity that possesses a magnitude and a direction a **vector**. Accordingly, every directed line segment is a pictorial representation of a vector.

Equivalent Directed Line Segments

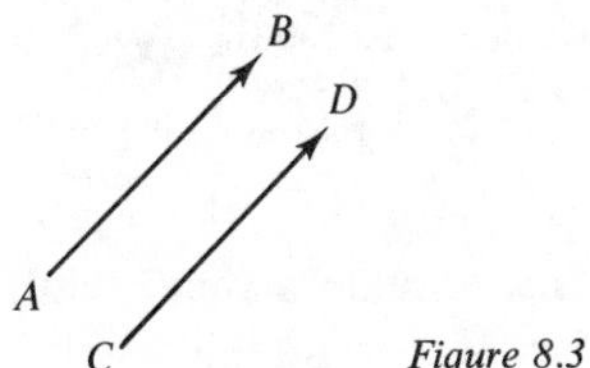

Figure 8.3

Figure 8.3 depicts two directed line segments AB and CD. AB and CD have the same length and the same direction but different locations. They are clearly different directed line segments but they only differ in their location. A vector, however, is only defined in terms of its length and its direction – its location is of no account. As a consequence we say that the directed line segments $\overrightarrow{AB}$ and $\overrightarrow{CD}$ represent the same vector and we express this fact by writing:

$$\overrightarrow{AB} \equiv \overrightarrow{CD}$$

$\overrightarrow{AB}$ is equivalent to $\overrightarrow{CD}$ because they both represent the same vector.

Vector Notation

The notation for a vector, as opposed to a directed line segment, is a lower-case bold letter such as **a** or **f**.

Equality of Vectors

If, in Figure 8.3, the directed line segment AB represents the vector **p** and the directed line segment CD represents the vector **q** then the equivalence of $\overrightarrow{AB}$ and $\overrightarrow{CD}$ is the same as saying that **p** is equal to **q** – vector **p** is the same as vector **q**, that is

$$\mathbf{p} = \mathbf{q}$$

If we are writing by hand we cannot boldface and so we underline:

$$\underline{p} = \underline{q}$$

Vector Arithmetic

The notion of a vector being equivalently represented by two directed line segments with different locations now permits us to define the operation of addition of two vectors.

Addition

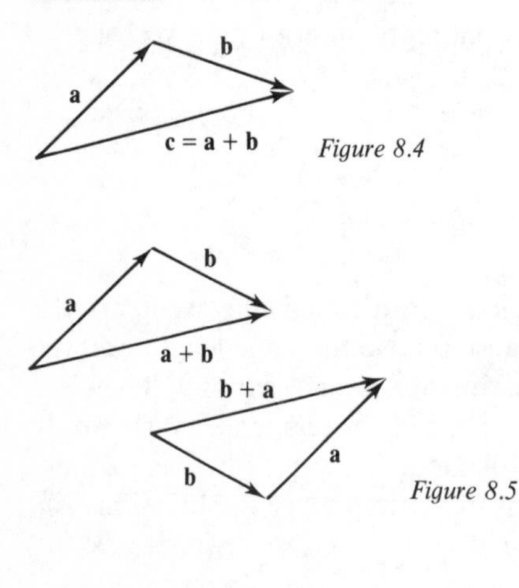

Figure 8.4

Figure 8.5

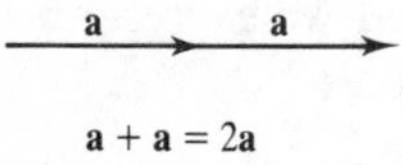

Figure 8.6

When we add vector **a** to vector **b** we select a directed line segment with the same length and direction as **b** but whose initial point coincides with the terminal point of **a**. Their sum is then the vector **c** whose initial point is that of **a** and whose terminal point is that of **b** (see Figure 8.4). Their sum **c** is also referred to as the **resultant** of **a** and **b**.

Addition of vectors is a commutative and associative operation (see Figure 8.5):

$$\mathbf{a} + \mathbf{b} = \mathbf{b} + \mathbf{a}$$

If we add a vector to itself we obtain a vector with the same direction but twice the original length, as Figure 8.6 shows. Clearly we can write $\mathbf{a} + \mathbf{a}$ as $2\mathbf{a}$ which now permits us to define multiplication of a vector by a real number.

Multiplication by a Real Number

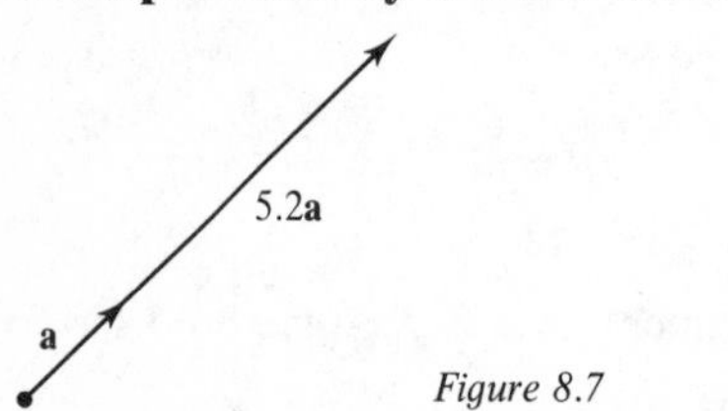

Figure 8.7

If a vector is multiplied by a positive real number β then the result is another vector with the same direction but with a length that is β times the original length. For example, $5.2\mathbf{a}$ is a vector that is in the same direction as **a** but 5.2 times as long.

Unit Vectors

Any vector whose length is unity is called a **unit vector** and is denoted by a boldface (or underlined) letter with a circumflex – or **hat** – above it. For example,

$$\hat{\mathbf{a}}$$

represents a vector of length 1 in the direction of vector **a**.

Because **a** is a vector of length a in the direction of $\hat{\mathbf{a}}$ we can write

$$\mathbf{a} = a\hat{\mathbf{a}}$$

The Zero Vector

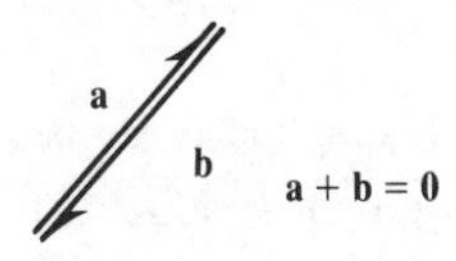

Figure 8.8

If vector **b** is the same length as vector **a** but has the opposite direction then the terminal point of **b** coincides with the initial point of **a**. We say that their sum is a vector with zero length and arbitrary direction – denoted by **0**:

$$\mathbf{a} + \mathbf{b} = \mathbf{0}$$

0 denotes the **zero vector**. This notion of the zero vector permits us to define the operation of subtraction.

Subtraction

If **a** and **b** are two vectors such that

$$\mathbf{a} + \mathbf{b} = \mathbf{0}$$

we say that

$$\mathbf{a} = -\mathbf{b}$$

This means that multiplying a vector by -1 retains the same length but reverses the direction.

EXAMPLES

1 Describe the balance of forces in Figure 8.9 in a vector diagram.

The system is in equilibrium with the 3 and 4 kg weights providing tensions proportional to the weights in the ropes to hold up the 5 kg weight. The resultant of the tensions caused by the 3 and 4 kg weights is an upwards force that balances the downwards force due to the 5 kg weight. The vector diagram representing this balance is as shown in Figure 8.10 where the length of each directed line segment is proportional to the tension it represents.

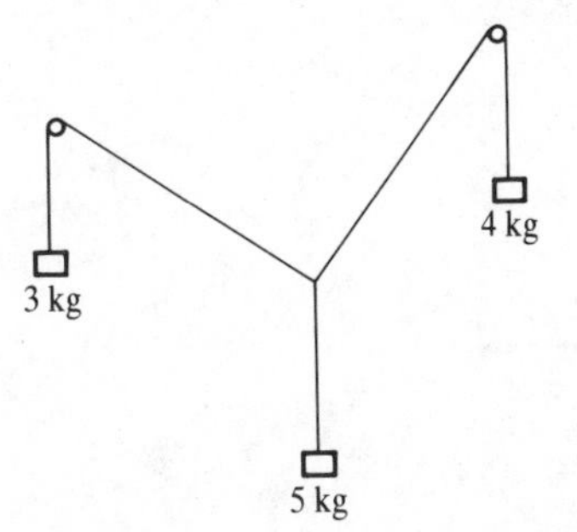

Figure 8.9

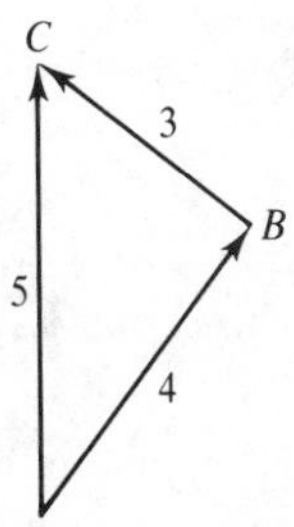

Figure 8.10

2 A rower can row a boat at 2 metres per second. The rower wishes to arrive at point C on the south bank of a river when the river is flowing at 10 metres per second due east. From where on the north bank should the rower start rowing if the river is 200 metres wide and for how long will the boat be travelling?

Let the rower's velocity **a** due south across the river be represented by the directed line segment $\overrightarrow{AB}$ of length 2 units. The river's velocity **b** due east is represented by the directed line segment $\overrightarrow{BC}$ of length 10 units. The resultant velocity of the boat is then represented by the directed line segment $\overrightarrow{AC}$ whose length AC is given by

$$AC^2 = AB^2 + BC^2$$

$$= 104$$

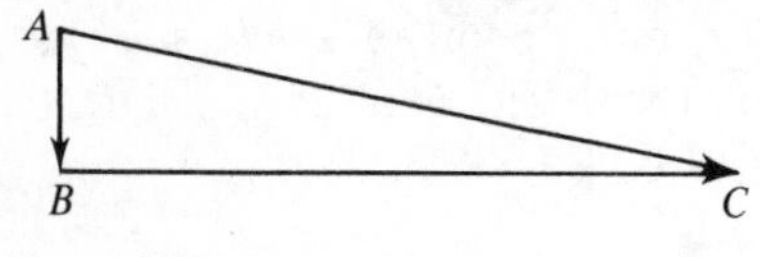

Figure 8.11

therefore $AC = 2\sqrt{26}$ units. From Figure 8.11 it can be seen that since the length of $\overrightarrow{AB}$ is proportional to 200 metres then BC is proportional to 1000 metres and AC is proportional to $200\sqrt{26}$ metres. It can also be seen that the rower must start 1000 metres due west of a point directly opposite C and that the boat will travel for the time it would take to row from A to B in the absence of the current. This is given by

200 metres at 2 metres per second = 100 seconds

= 1 minute 40 seconds

EXERCISES

1 A baseball is struck by the batsman and travels with a velocity of 50 metres per second horizontally and a velocity $(100 - t^2)$ metres per second vertically where t is the time from being struck in seconds. If $\hat{\mathbf{h}}$ and $\hat{\mathbf{v}}$ represent unit vectors in the horizontal and vertical directions respectively, draw diagrams to represent the velocity of the ball at:

(a) $t = 0$ seconds **(b)** $t = 5$ seconds **(c)** $t = 10$ seconds

If **u** is the velocity of the ball then

$$\mathbf{u} = *\hat{\mathbf{h}} + (100 - t^2)\hat{\mathbf{v}}$$

(a) At $t = 0$, $\mathbf{u} = 50\hat{\mathbf{h}} + *\hat{\mathbf{v}}$
(b) At $t = 5$, $\mathbf{u} = *\hat{\mathbf{h}} + *\hat{\mathbf{v}}$
(c) At $t = 10$, $\mathbf{u} = *$

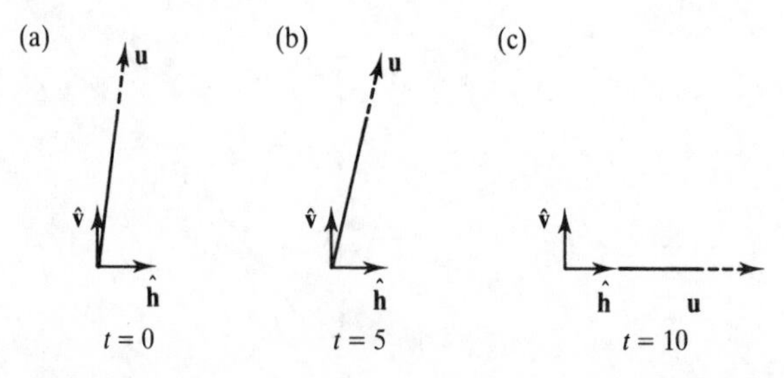

Figure 8.12

2 A light aircraft can fly at 200 kph. If it is to travel due north against a westerly wind of 100 kph in what direction should it head and what will its northerly velocity be?

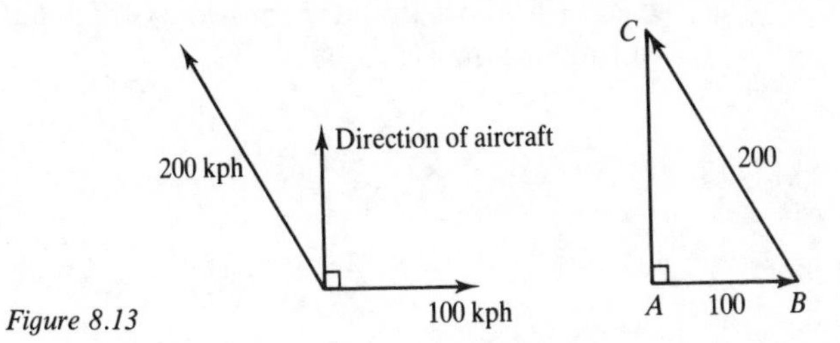

Figure 8.13

The aircraft will have to fly west of north as shown in Figure 8.13. The aircraft will head at an angle of $*^\circ$ west of due north and its northerly velocity will be $*$ kph.

Unit 4 Geometric Vectors

Test yourself with the following:

1 If $\mathbf{a} = 2\mathbf{i} + 3\mathbf{j}$, $\mathbf{b} = -\mathbf{i} + 4\mathbf{j}$, $\mathbf{c} = \mathbf{i} - \mathbf{j} + \mathbf{k}$ and $\mathbf{d} = 3\mathbf{i} + 2\mathbf{j} - 4\mathbf{k}$:

(a) Draw $\mathbf{a}$, $\mathbf{b}$, $\mathbf{a} + \mathbf{b}$ and $4\mathbf{a} - 3\mathbf{b}$

(b) Find: (i) $\mathbf{c} - \mathbf{d}$ (ii) $4\mathbf{c} + 5\mathbf{d}$ (iii) $\mathbf{a}.\mathbf{b}$ (iv) $\mathbf{c}.\mathbf{d}$ (v) $\mathbf{c} \times \mathbf{d}$

2 Find the area of the parallelogram with two adjacent sides given by

$$\mathbf{a} = -\mathbf{i} + \mathbf{j},\ \mathbf{b} = \mathbf{i} - 2\mathbf{j}$$

3 If $\mathbf{a} = \mathbf{i} + \mathbf{j} - \mathbf{k}$, $\mathbf{b} = \mathbf{i} - \mathbf{j} + \mathbf{k}$ and $\mathbf{c} = -\mathbf{i} + \mathbf{j} + \mathbf{k}$ show that

$$\mathbf{a} \times (\mathbf{b} + \mathbf{c}) = (\mathbf{a} \times \mathbf{b}) + (\mathbf{a} \times \mathbf{c})$$

4 Find the volume of the cuboid with three adjacent edges given by

$$\mathbf{a} = \mathbf{i} + \mathbf{j},\ \mathbf{b} = \mathbf{i} - \mathbf{j} \text{ and } \mathbf{c} = -\mathbf{k}$$

Two-dimensional Vectors

All the vectors that we have discussed so far have been drawn without any reference to a specific coordinate system. There has always been an implied coordinate system in the back of our minds, however, because we have talked about length and direction and we can only talk of these with reference to some coordinate system. Now we shall introduce the coordinate system explicitly.

Cartesian Coordinates

If we set up a cartesian coordinate system and draw our vectors with reference to it we call these vectors **geometric vectors**. In particular, in Figure 8.14 we have drawn the geometric

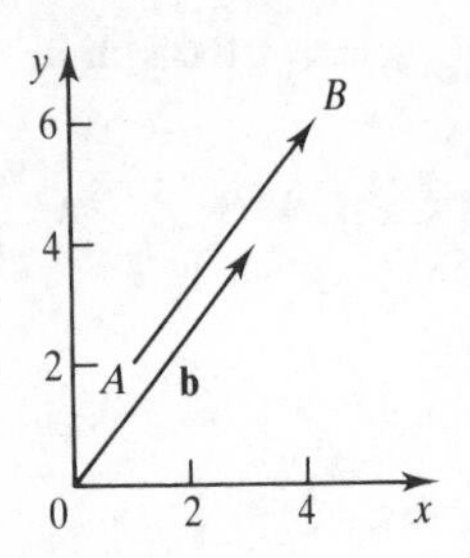

Figure 8.14

vector **a** with initial point A at $(1, 2)$ and terminal point B at $(4, 6)$. Because location of vectors is not important this vector is equal to the geometric vector **b** whose initial point is the origin and whose terminal point is $(3, 4)$.

Unit Vectors i and j

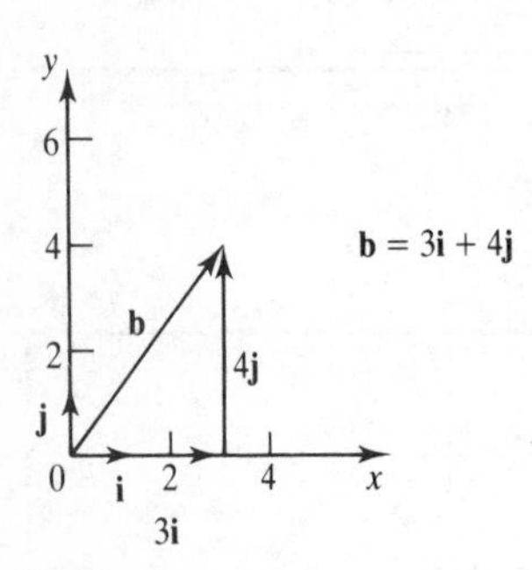

Figure 8.15

A geometric vector connects an initial point to a terminal point and now that we have set up a coordinate system we can regularize this connection. To travel from the origin to the point $(3, 4)$ we first travel to the point $(3, 0)$ along the x-axis and then travel to $(3, 4)$ along a line parallel to the y-axis.

If we set up a unit vector **i** in the direction of the x-axis and a unit vector **j** in the direction of the y-axis then we can represent the geometric vector **b** as

$$\mathbf{b} = 3\mathbf{i} + 4\mathbf{j}$$

The numbers 3 and 4 were called the x- and y-coordinates of **b** respectively and the unit vectors **i** and **j** are called the **coordinate unit vectors**. Notice that they are written by convention without circumflexes.

Addition and Subtraction

Adding or subtracting two geometric vectors is achieved by adding or subtracting their respective coordinates. For example, if

$$\mathbf{a} = 2\mathbf{i} + 3\mathbf{j} \text{ and } \mathbf{b} = -6\mathbf{i} + 4\mathbf{j}$$

then

$$\begin{aligned}\mathbf{a} + \mathbf{b} &= (2\mathbf{i} + 3\mathbf{j}) + (-6\mathbf{i} + 4\mathbf{j}) \\ &= [2 + (-6)]\mathbf{i} + (3 + 4)\mathbf{j} \\ &= -4\mathbf{i} + 7\mathbf{j}\end{aligned}$$

and

$$\begin{aligned}\mathbf{a} - \mathbf{b} &= (2\mathbf{i} + 3\mathbf{j}) - (-6\mathbf{i} + 4\mathbf{j}) \\ &= [2 - (-6)]\mathbf{i} + (3 - 4)\mathbf{j} \\ &= 8\mathbf{i} - \mathbf{j}\end{aligned}$$

Scalar Product

A **scalar** is a quantity that possesses only magnitude and we define the scalar product of two vectors as follows:

$$\mathbf{a}.\mathbf{b} = ab \cos \theta$$

where a and b are the magnitudes of **a** and **b** respectively and θ is the angle between them.

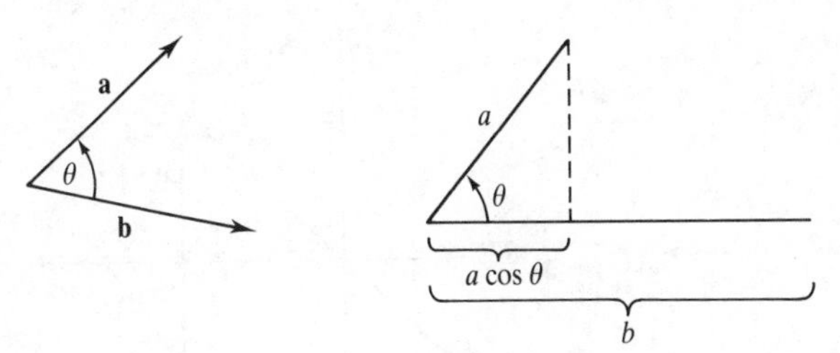

Figure 8.16

The result of the product is a scalar, and when a vector is multiplied by itself in this way the scalar is equal to the square of its length:

$$\begin{aligned}\mathbf{a}.\mathbf{a} &= a^2 \cos 0 \\ &= a^2\end{aligned}$$

The Scalar Product of Geometric Vectors

Because the coordinate base unit vectors are perpendicular to each other their scalar product is zero:

$$\begin{aligned}\mathbf{i}.\mathbf{j} &= 1^2 \cos(\pi/2) \\ &= 0\end{aligned}$$

Because the coordinate unit vectors are of unit length their scalar product with themselves is 1:

$$\mathbf{i}.\mathbf{i} = 1^2 \cos 0 = 1 = \mathbf{j}.\mathbf{j}$$

As a consequence the scalar product of two geometric vectors is equal to the sum of the products of their respective coordinates. For example, the scalar product of

$$\mathbf{a} = 4\mathbf{i} + 3\mathbf{j} \text{ and } \mathbf{b} = 2\mathbf{i} + 5\mathbf{j}$$

is

$$\begin{aligned}\mathbf{a}.\mathbf{b} &= (4\mathbf{i} + 3\mathbf{j}).(2\mathbf{i} + 5\mathbf{j}) \\ &= 4\mathbf{i}.(2\mathbf{i} + 5\mathbf{j}) + 3\mathbf{j}.(2\mathbf{i} + 5\mathbf{j}) \\ &= 8\mathbf{i}.\mathbf{i} + 10\mathbf{i}.\mathbf{j} + 6\mathbf{j}.\mathbf{i} + 15\mathbf{j}.\mathbf{j} \\ &= 8 + 0 + 0 + 15 \\ &= 23\end{aligned}$$

Three-dimensional Vectors

We can extend our coordinate system into three dimensions by including a z-axis as shown in Figure 8.17.

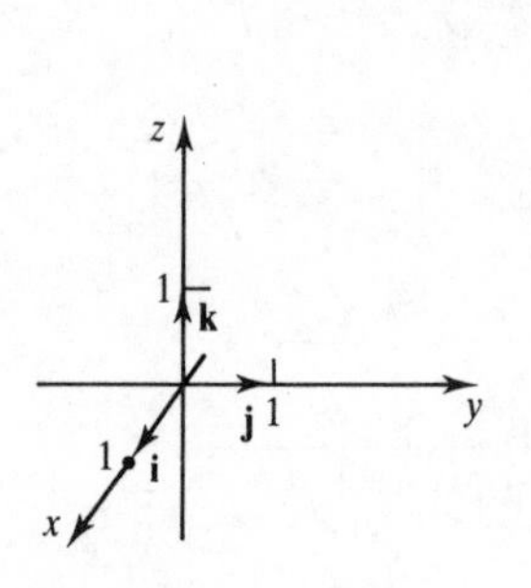

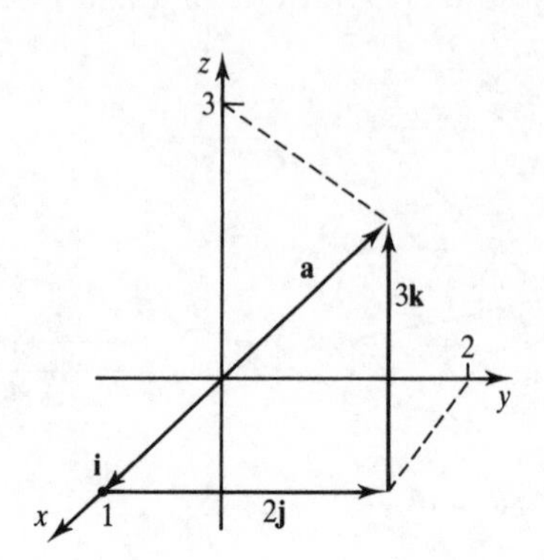

Figure 8.17

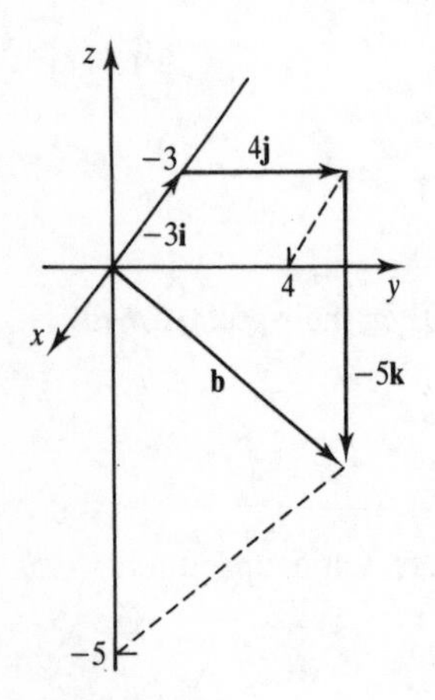

Figure 8.18

Along the z-axis the third coordinate unit vector **k** has been drawn. Now, any vector drawn in three dimensions can be written in terms of **i**, **j** and **k**. For example,

$$\mathbf{a} = \mathbf{i} + 2\mathbf{j} + 3\mathbf{k} \text{ and } \mathbf{b} = -3\mathbf{i} + 4\mathbf{j} - 5\mathbf{k}$$

The arithmetic of such vectors is a simple extension of the arithmetic of two-dimensional vectors. For example,

$$\begin{aligned}\mathbf{a} + \mathbf{b} &= (1 - 3)\mathbf{i} + (2 + 4)\mathbf{j} + (3 - 5)\mathbf{k} \\ &= -2\mathbf{i} + 6\mathbf{j} - 2\mathbf{k}\end{aligned}$$

and

$$\begin{aligned}\mathbf{a} . \mathbf{b} &= (1)(-3) + (2)(4) + (3)(-5) \\ &= -3 + 8 - 15 \\ &= -10\end{aligned}$$

Vector Product

It is possible to define an alternative product between two vectors that produces not a scalar, but another vector. This is done as follows:

$$\mathbf{a} \times \mathbf{b} = ab \sin\theta\, \hat{\mathbf{n}}$$

where $\hat{\mathbf{n}}$ is a unit vector that is perpendicular to both **a** and **b** and whose direction is given by the so-called **corkscrew rule.**

Notice that the vector product of a vector with itself is zero because the angle between a vector and itself is zero and $\sin 0 = 0$. Also, the vector product is not commutative because

$$\mathbf{a} \times \mathbf{b} = -\mathbf{b} \times \mathbf{a}$$

as a consequence of the corkscrew rule.

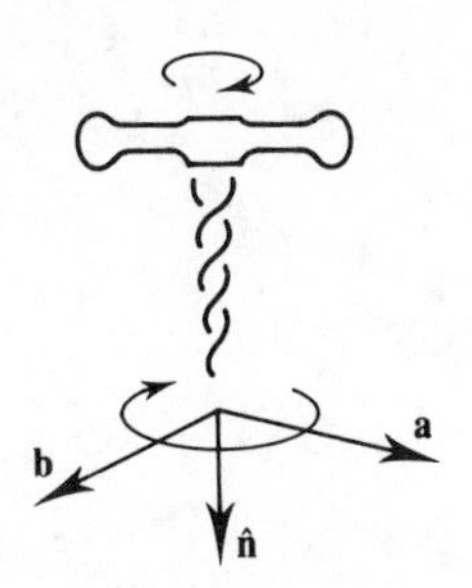

As the corkscrew rotates clockwise from **a** to **b** looking down it travels vertically in the direction $\hat{\mathbf{n}}$

$$\mathbf{a} \times \mathbf{b} = ab \sin \theta \hat{\mathbf{n}}$$

Figure 8.19

Vector Area

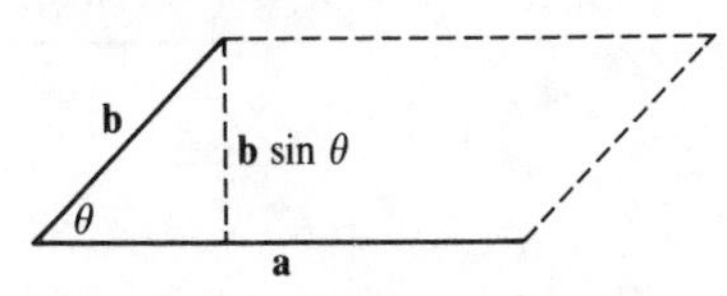

Base × height = $ab \sin \theta$

Figure 8.20

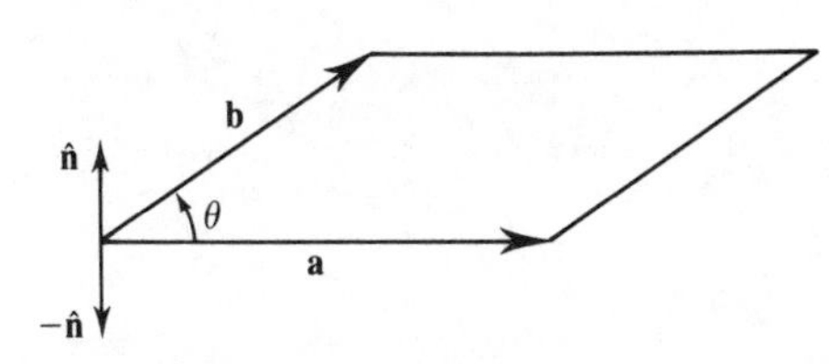

Figure 8.21

The definition of the vector product may appear rather arbitrary at first sight, but it was defined in this way for a reason. If we complete the parallelogram defined by vectors **a** and **b** then we see that the area of this parallelogram is given as

$$ab \sin \theta$$

so that magnitude of the vector product $\mathbf{a} \times \mathbf{b}$ is the area of the parallelogram that **a** and **b** mutually define. So why the unit vector $\hat{\mathbf{n}}$?

A sheet of paper has two surfaces, a top surface and a bottom surface. If we let **a** and **b** represent vectors along two adjacent sides of a flat sheet of paper the top surface can be represented by

$$\mathbf{a} \times \mathbf{b} = ab \sin \theta \hat{\mathbf{n}}$$

where $\hat{\mathbf{n}}$ is in a direction outwards from the surface. In this way we define a **vector area**. The bottom surface is given by

$$\mathbf{b} \times \mathbf{a} = ab \sin \theta(-\hat{\mathbf{n}})$$

which is a vector area in the opposite direction.

The Vector Product of Geometric Vectors

The vector product of two geometric vectors can be obtained in a manner similar to the way we obtained the scalar product, making note of the following:

$$\mathbf{i} \times \mathbf{i} = \mathbf{j} \times \mathbf{j} = \mathbf{k} \times \mathbf{k} = 0$$

and

$$\mathbf{i} \times \mathbf{j} = \mathbf{k},\ \mathbf{i} \times \mathbf{k} = -\mathbf{j} \text{ and } \mathbf{j} \times \mathbf{k} = \mathbf{i}$$

For example, if

$$\mathbf{a} = \mathbf{i} - \mathbf{j} + \mathbf{k} \text{ and } \mathbf{b} = 2\mathbf{i} + 3\mathbf{j} - 4\mathbf{k}$$

then

$$\begin{aligned}\mathbf{a} \times \mathbf{b} &= (\mathbf{i} - \mathbf{j} + \mathbf{k}) \times (2\mathbf{i} + 3\mathbf{j} - 4\mathbf{k}) \\ &= \mathbf{i} \times (2\mathbf{i} + 3\mathbf{j} - 4\mathbf{k}) + (-\mathbf{j}) \times (2\mathbf{i} + 3\mathbf{j} - 4\mathbf{k}) + \mathbf{k} \times (2\mathbf{i} + 3\mathbf{j} - 4\mathbf{k}) \\ &= 2\mathbf{i} \times \mathbf{i} + 3\mathbf{i} \times \mathbf{j} - 4\mathbf{i} \times \mathbf{k} - 2\mathbf{j} \times \mathbf{i} - 3\mathbf{j} \times \mathbf{j} + 4\mathbf{j} \times \mathbf{k} + 2\mathbf{k} \times \mathbf{i} + 3\mathbf{k} \times \mathbf{j} - 4\mathbf{k} \times \mathbf{k} \\ &= \mathbf{0} + 3\mathbf{k} - 4(-\mathbf{j}) - 2(-\mathbf{k}) + \mathbf{0} + 4\mathbf{i} + 2\mathbf{j} + 3(-\mathbf{i}) + \mathbf{0} \\ &= \mathbf{i} + 6\mathbf{j} + 5\mathbf{k}\end{aligned}$$

EXAMPLES

1 If $\mathbf{a} = -3\mathbf{i} + \mathbf{j}$, $\mathbf{b} = 2\mathbf{i} - \mathbf{j}$, $\mathbf{c} = -\mathbf{i} - \mathbf{j} - \mathbf{k}$ and $\mathbf{d} = 4\mathbf{i} - 2\mathbf{j} + 3\mathbf{k}$:

(a) Draw $\mathbf{a}$, $\mathbf{b}$, $\mathbf{a} + \mathbf{b}$ and $2\mathbf{a} + 5\mathbf{b}$.

(b) Find: (i) $2\mathbf{c} + 3\mathbf{d}$ (ii) $\mathbf{a}.\mathbf{b}$ (iii) $\mathbf{c} \times \mathbf{d}$

(a)

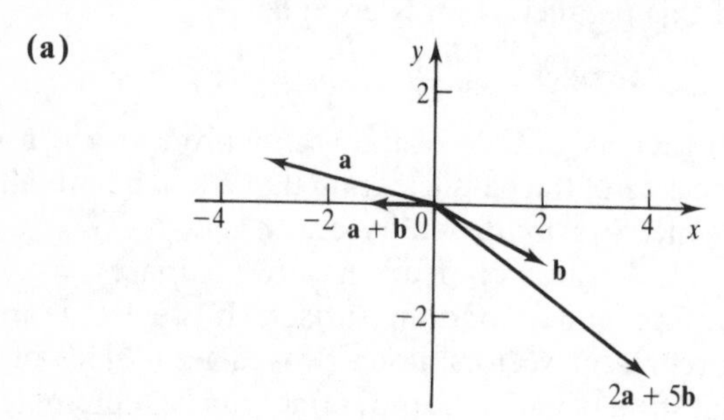

Figure 8.22

(b) (i)
$$\begin{aligned}2\mathbf{c} + 3\mathbf{d} &= 2(-\mathbf{i} - \mathbf{j} - \mathbf{k}) + 3(4\mathbf{i} - 2\mathbf{j} + 3\mathbf{k}) \\ &= (-2 + 12)\mathbf{i} + (-2 - 6)\mathbf{j} + (-2 + 9)\mathbf{k} \\ &= 10\mathbf{i} - 8\mathbf{j} + 7\mathbf{k}\end{aligned}$$

(ii)
$$\begin{aligned}\mathbf{a}.\mathbf{b} &= (-3\mathbf{i} + \mathbf{j}).(2\mathbf{i} - \mathbf{j}) \\ &= -6\mathbf{i}.\mathbf{i} + 3\mathbf{i}.\mathbf{j} + 2\mathbf{j}.\mathbf{i} - \mathbf{j}.\mathbf{j} \\ &= -6 + 0 + 0 - 1 \\ &= -7\end{aligned}$$

(iii)
$$\begin{aligned}\mathbf{c} \times \mathbf{d} &= (-\mathbf{i} - \mathbf{j} - \mathbf{k}) \times (4\mathbf{i} - 2\mathbf{j} + 3\mathbf{k}) \\ &= -4\mathbf{i} \times \mathbf{i} + 2\mathbf{i} \times \mathbf{j} - 3\mathbf{i} \times \mathbf{k} - 4\mathbf{j} \times \mathbf{i} + 2\mathbf{j} \times \mathbf{j} - 3\mathbf{j} \times \mathbf{k} - 4\mathbf{k} \times \mathbf{i} + 2\mathbf{k} \times \mathbf{j} - 3\mathbf{k} \times \mathbf{k} \\ &= -4.\mathbf{0} + 2\mathbf{k} - 3(-\mathbf{j}) - 4(-\mathbf{k}) + 2.\mathbf{0} - 3\mathbf{i} - 4\mathbf{j} + 2(-\mathbf{i}) - 3.\mathbf{0} \\ &= -5\mathbf{i} - \mathbf{j} + 6\mathbf{k}\end{aligned}$$

2 Find the area of the triangle with two adjacent sides given by

$$\mathbf{a} = 2\mathbf{i} + 3\mathbf{j},\ \mathbf{b} = -\mathbf{i} + 4\mathbf{j}$$

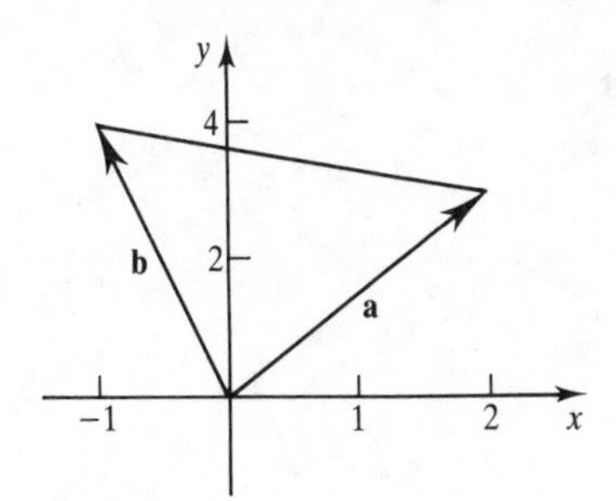

Figure 8.23

The vector product of **a** and **b** will give the vector area of the parallelogram with **a** and **b** as adjacent sides. The area of the triangle will be half that of the parallelogram.

$$\begin{aligned}\mathbf{a} \times \mathbf{b} &= (2\mathbf{i} + 3\mathbf{j}) \times (-\mathbf{i} + 4\mathbf{j})\\ &= -2\mathbf{i} \times \mathbf{i} + 8\mathbf{i} \times \mathbf{j}\\ &\quad - 3\mathbf{j} \times \mathbf{i} + 12\mathbf{j} \times \mathbf{j}\\ &= 8\mathbf{k} + 3\mathbf{k}\\ &= 11\mathbf{k}\end{aligned}$$

The area of the parallelogram is then 11 square units and the area of the triangle is 5.5 square units.

3 If $\mathbf{a} = -\mathbf{i} + 2\mathbf{j} - 3\mathbf{k}$, $\mathbf{b} = 4\mathbf{i} - 2\mathbf{j}$ and $\mathbf{c} = 2\mathbf{j} - \mathbf{k}$ show that

$$\mathbf{a}.(\mathbf{b} - \mathbf{c}) = (\mathbf{a}.\mathbf{b}) - (\mathbf{a}.\mathbf{c})$$

$$\begin{aligned}\mathbf{a}.(\mathbf{b} - \mathbf{c}) &= (-\mathbf{i} + 2\mathbf{j} - 3\mathbf{k}).[(4\mathbf{i} - 2\mathbf{j}) - (2\mathbf{j} - \mathbf{k})]\\ &= (-\mathbf{i} + 2\mathbf{j} - 3\mathbf{k}).(4\mathbf{i} - 4\mathbf{j} + \mathbf{k})\\ &= -\mathbf{i}.(4\mathbf{i} - 4\mathbf{j} + \mathbf{k}) + 2\mathbf{j}.(4\mathbf{i} - 4\mathbf{j} + \mathbf{k}) - 3\mathbf{k}.(4\mathbf{i} - 4\mathbf{j} + \mathbf{k})\\ &= -4 - 8 - 3\\ &= -15\end{aligned}$$

$$\begin{aligned}(\mathbf{a}.\mathbf{b}) - (\mathbf{a}.\mathbf{c}) &= [(-\mathbf{i} + 2\mathbf{j} - 3\mathbf{k}).(4\mathbf{i} - 2\mathbf{j})] - [(-\mathbf{i} + 2\mathbf{j} - 3\mathbf{k}).(2\mathbf{j} - \mathbf{k})]\\ &= (-4 - 4) - (4 + 3)\\ &= -15\end{aligned}$$

Therefore

$$\mathbf{a}.(\mathbf{b} - \mathbf{c}) = (\mathbf{a}.\mathbf{b}) - (\mathbf{a}.\mathbf{c})$$

4 Find the value of $\mathbf{i}.(\mathbf{j} \times \mathbf{k})$ and interpret your result geometrically.

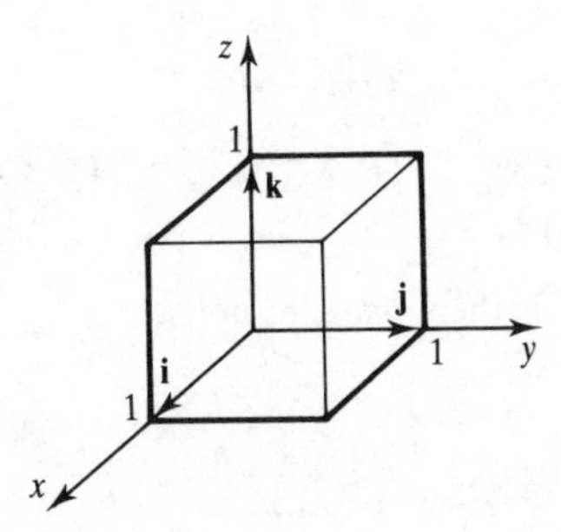

Figure 8.24

$$\mathbf{i}.(\mathbf{j} \times \mathbf{k}) = \mathbf{i}.(\mathbf{i}) = 1$$

$\mathbf{j} \times \mathbf{k}$ is the vector area with **j** and **k** as adjacent sides. This is a square of unit side length and area 1 with direction **i**. The scalar product of this vector area with **i** is then the volume of the cube of side length 1 given in Figure 8.24.

EXERCISES

1 If $\mathbf{a} = 5\mathbf{i} - \mathbf{j}$, $\mathbf{b} = -3\mathbf{i} - 2\mathbf{j}$, $\mathbf{c} = -2\mathbf{i} + \mathbf{j} + \mathbf{k}$ and $\mathbf{d} = \mathbf{i} + 2\mathbf{j} + 3\mathbf{k}$:

(a) Draw **a**, **b**, **a** + **b** and $-2\mathbf{a} + 4\mathbf{b}$

(b) Find: (i) $2\mathbf{c} - 4\mathbf{d}$ (ii) $\mathbf{a}.\mathbf{b}$ (iii) $\mathbf{c} \times \mathbf{d}$

(a)

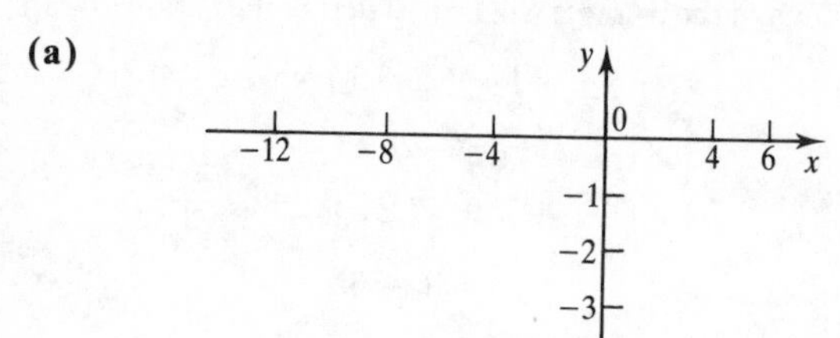

Figure 8.25

(b) (i) $2\mathbf{c} - 4\mathbf{d} = 2(-2\mathbf{i} + \mathbf{j} + \mathbf{k}) - *(*\mathbf{i} - *\mathbf{j} + *\mathbf{k})$

$= (-4 - *)\mathbf{i} + (2 + *)\mathbf{j} + (2 + *)\mathbf{k}$

$= *\mathbf{i} + *\mathbf{j} + *\mathbf{k}$

(ii) $\mathbf{a}.\mathbf{b} = (5\mathbf{i} - \mathbf{j}).(*\mathbf{i} + *\mathbf{j})$

$= *\mathbf{i}.\mathbf{i} + *\mathbf{i}.\mathbf{j} + *\mathbf{j}.\mathbf{i} + *\mathbf{j}.\mathbf{j}$

$= * + 0 + * + *$

$= *$

(iii) $\mathbf{c} \times \mathbf{d} = (-2\mathbf{i} + \mathbf{j} + \mathbf{k}) \times (*\mathbf{i} + *\mathbf{j} + *\mathbf{k})$

$= *\mathbf{i} \times \mathbf{i} + *\mathbf{i} \times \mathbf{j} + *\mathbf{i} \times \mathbf{k} + *\mathbf{j} \times \mathbf{i} + *\mathbf{j} \times \mathbf{j} + *\mathbf{j} \times \mathbf{k} + *\mathbf{k} \times \mathbf{i} + *\mathbf{k} \times \mathbf{j} + *\mathbf{k} \times \mathbf{k}$

$= *\mathbf{0} + *\mathbf{k} + *(-\mathbf{j}) + *(-*) + ** + ** + *\mathbf{j} + *(-\mathbf{i}) + *\mathbf{0}$

$= *\mathbf{i} + *\mathbf{j} + *\mathbf{k}$

2 Find the area of the rhombus with two adjacent sides given by

$$\mathbf{a} = 3\mathbf{i} - 4\mathbf{j}, \mathbf{b} = 4\mathbf{i} - 3\mathbf{j}$$

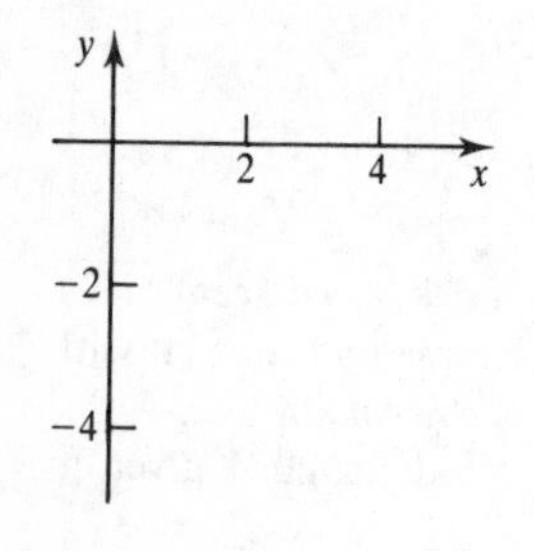

Figure 8.26

The vector product of **a** and **b** will give the vector area of the parallelogram with **a** and **b** as adjacent sides. Because $a = b = 5$ the parallelogram is a rhombus.

$\mathbf{a} \times \mathbf{b} = (3\mathbf{i} - 4\mathbf{j}) \times (*\mathbf{i} - *\mathbf{j})$

$= *\mathbf{i} \times \mathbf{i} - *\mathbf{i} \times \mathbf{j} - *\mathbf{j} \times \mathbf{i} + *\mathbf{j} \times \mathbf{j}$

$= *\mathbf{k}$

The area of the rhombus is then * square units.

3 If $\mathbf{a} = \mathbf{i} + \mathbf{j}$, $\mathbf{b} = \mathbf{i} - \mathbf{j}$ and $\mathbf{c} = \mathbf{k}$ show that

$$\mathbf{a} \times (\mathbf{b} - \mathbf{c}) = (\mathbf{a} \times \mathbf{b}) - (\mathbf{a} \times \mathbf{c})$$

$$
\begin{aligned}
\mathbf{a} \times (\mathbf{b} - \mathbf{c}) &= (\mathbf{i} + \mathbf{j}) \times (\mathbf{i} - \mathbf{j} - \mathbf{k}) \\
&= * \times (\mathbf{i} - \mathbf{j} - \mathbf{k}) + * \times (\mathbf{i} - \mathbf{j} - \mathbf{k}) \\
&= * \times \mathbf{i} - * \times \mathbf{j} - * \times \mathbf{k} + * \times \mathbf{i} - * \times \mathbf{j} - * \times \mathbf{k} \\
&= -* + * - * - * \\
&= -\mathbf{i} + \mathbf{j} - 2\mathbf{k}
\end{aligned}
$$

$$
\begin{aligned}
(\mathbf{a} \times \mathbf{b}) - (\mathbf{a} \times \mathbf{c}) &= (\mathbf{i} + \mathbf{j}) \times (* - *) \\
&= (\mathbf{i} + \mathbf{j}) \times * \\
&= -\mathbf{i} \times * + \mathbf{j} \times * - \mathbf{i} \times * - \mathbf{j} \times * \\
&= -* - * + * - * \\
&= *\mathbf{i} + *\mathbf{j} + *\mathbf{k}
\end{aligned}
$$

Therefore

$$\mathbf{a} \times (\mathbf{b} - \mathbf{c}) = (\mathbf{a} \times \mathbf{b}) - (\mathbf{a} \times \mathbf{c})$$

4 If $\mathbf{a} = \mathbf{i} + 2\mathbf{j}$, $\mathbf{b} = 2\mathbf{i} + \mathbf{j}$ and $\mathbf{c} = \mathbf{k}$ find

$$\mathbf{a} . (\mathbf{b} \times \mathbf{c})$$

and interpret your result geometrically.

$$
\begin{aligned}
\mathbf{a} . (\mathbf{b} \times \mathbf{c}) &= (\mathbf{i} + 2\mathbf{j}) . [(2\mathbf{i} + \mathbf{j}) \times \mathbf{k}] \\
&= (\mathbf{i} + 2\mathbf{j}) . [*\mathbf{j} + *\mathbf{i}] \\
&= *
\end{aligned}
$$

$\mathbf{b} \times \mathbf{c}$ is the vector area with * and * as adjacent sides. This is a * with a direction perpendicular to both * and *. The scalar product of this vector area with $\mathbf{a}$ is then the volume of the parallelepiped given in Figure 8.27.

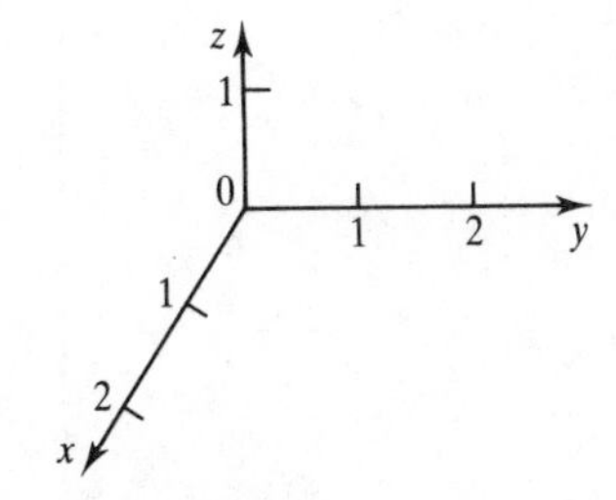

Figure 8.27

Module 8: Further exercises

1 Describe the type of each of the following matrices:

(a) $\begin{bmatrix} -4 \\ 3 \end{bmatrix}$ (b) $[1 \quad 2 \quad 3 \quad 4]$ (c) $\begin{bmatrix} 1 & 2 \\ -5 & 1 \end{bmatrix}$ (d) $\begin{bmatrix} 1 & 0 & 0 \\ 0 & 1 & 0 \end{bmatrix}$

2 Given matrices

$$A = \begin{bmatrix} -7 & -3 \\ 2 & 5 \end{bmatrix} \quad B = \begin{bmatrix} 8 & 1 \\ 0 & -3 \end{bmatrix} \quad C = [-2 \quad 4] \quad D = \begin{bmatrix} 9 \\ 4 \end{bmatrix}$$

Calculate each of the following:

(a) $A + B$ (c) $-3A$ (e) AD
(b) $B - A$ (d) AB (f) CB

3 Is $AB = BA$ for the matrices A and B in Question 2?

4 Find AA^T where:

(a) $A = \begin{bmatrix} 7 & -9 & 0 \\ 0 & 11 & -12 \end{bmatrix}$

(b) $A = \begin{bmatrix} 10 \\ 4 \\ 0 \\ -9 \end{bmatrix}$

(c) $A = \begin{bmatrix} 3 & 7 & -1 \\ 6 & 0 & 0 \\ -12 & -9 & -2 \end{bmatrix}$

5 If possible, find the inverse of each of the following matrices:

(a) $A = \begin{bmatrix} 11 & 2 \\ 55 & 10 \end{bmatrix}$ (b) $B = \begin{bmatrix} -1 & 2 \\ 4 & 3 \end{bmatrix}$

6 Write each pair of simultaneous equations in matrix form:

(a) $-x - 2y = 4$
$x + 6y = -8$

(b) $8x + y = 24$
$-y + 2x = 6$

(c) $-9x + 2y = -14$
$3x = 4y - 2$

(d) $-2x + 3y = 7$
$4y - 6 = 14$

7 Use matrix inversion to solve each pair of simultaneous equations in Question 6.

8 Describe the balance of forces in Figure 8.28 in a vector diagram.

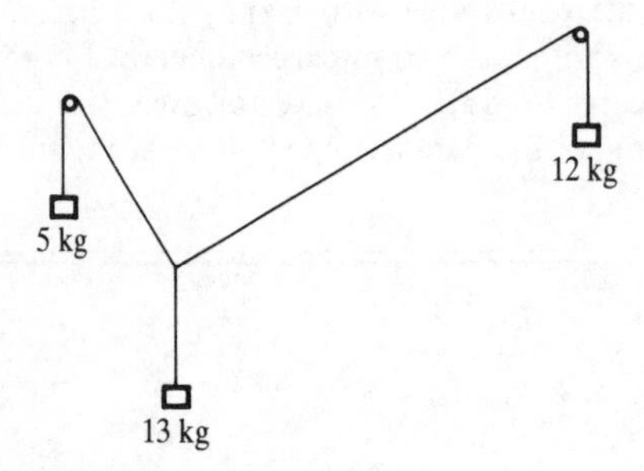

Figure 8.28

9 A swimmer can swim at 0.5 metres per second. The swimmer wishes to arrive at point P on the north bank of a river when the river is flowing at 1.5 metres per second due east. From where on the south bank should the swimmer begin if the river is 150 metres wide and how long will the swim last?

10 If $\mathbf{a} = 4\mathbf{i} - 5\mathbf{j}$, $\mathbf{b} = -\mathbf{i} - 8\mathbf{j}$, $\mathbf{c} = -2\mathbf{i} + 4\mathbf{j} - 8\mathbf{k}$ and $\mathbf{d} = \mathbf{i} - 3\mathbf{j} + 5\mathbf{k}$:

(a) Draw $\mathbf{a}$, $\mathbf{b}$, $\mathbf{a} + \mathbf{b}$ and $2\mathbf{a} - 5\mathbf{b}$

(b) Find: (i) $-3\mathbf{c} + 8\mathbf{d}$ (ii) $\mathbf{a}.\mathbf{b}$ (iii) $\mathbf{d} \times \mathbf{c}$

11 Find the area of the triangle with two adjacent sides given by

$$\mathbf{a} = -\mathbf{i} + 5\mathbf{j},\ \mathbf{b} = 2\mathbf{i} + 6\mathbf{j}$$

12 If $\mathbf{a} = \mathbf{i} + 4\mathbf{j} - \mathbf{k}$, $\mathbf{b} = 2\mathbf{i} + \mathbf{j}$ and $\mathbf{c} = -3\mathbf{j} + \mathbf{k}$ show that

$$\mathbf{a} \times (\mathbf{b} + \mathbf{c}) = (\mathbf{a} \times \mathbf{b}) + (\mathbf{a} \times \mathbf{c})$$

13 Find the value of $(\mathbf{i} + \mathbf{j} + \mathbf{k}).[(\mathbf{i} + \mathbf{j}) \times (\mathbf{j} + \mathbf{k})]$ and interpret your result geometrically.

Module 9: The Right-angled Triangle

Aim: To review the elements of trigonometry and thereby lay the groundwork for an extended description of the trigonometric functions in Module 13 of Chapter 3.

Objectives: When you have read this module you will be able to:

- ▶▶ Appreciate the geometric features of the right-angled triangle.
- ▶▶ Use Pythagoras' theorem.
- ▶▶ Use and manipulate expressions and equations involving the trigonometric ratios.

Unit 1 The Triangle

Test yourself with the following:

1 Without the aid of a calculator, convert the following angles, measured in degrees, into radians:

(a) $135°$ (b) $15°$ (c) $225°$ (d) $75°$ (e) $108°$

2 Convert the following angles, measured in radians, into degrees:

(a) $\pi/9$ radians (c) 0.537 radians
(b) $7\pi/8$ radians (d) 2.47 radians

3 Find the unknown angles in each of the following triangles:

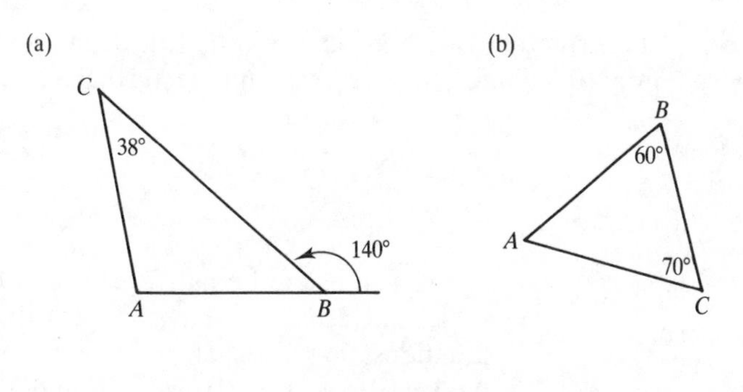

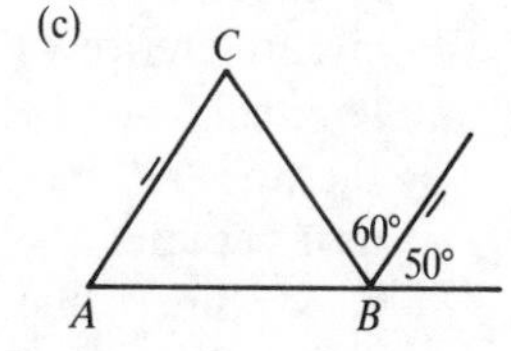

Figure 9.1

4 Identify the type of each of the following triangles:

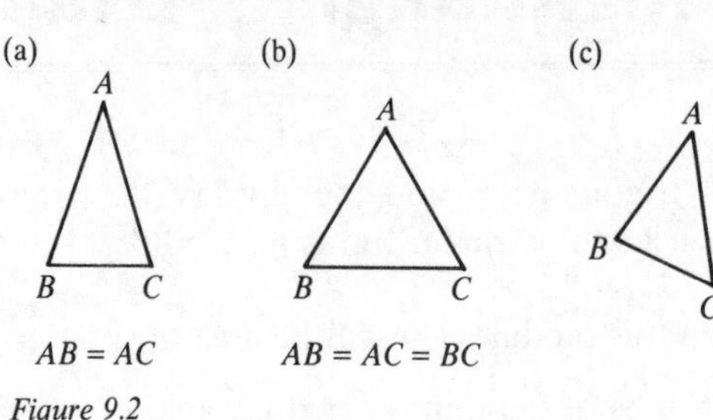

Figure 9.2

The Triangle

The triangle is a three-sided, rectilinear figure. The point where any two sides meet is called a **vertex** of the triangle and at each vertex is an **angle**.

Angles

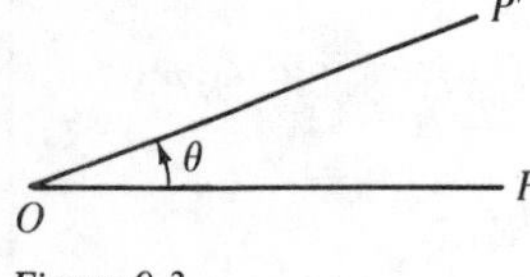

Figure 9.3

An angle is a measure of rotation. For example, the line OP, rotated about the point O to OP' in an anticlockwise direction, is said to have rotated through the angle θ. The angle can be measured in units of degrees or radians.

Degrees

A rotation through a complete revolution is defined as a rotation through 360 degrees (°). Each degree is subdivided into 60 minutes (′) and each minute further subdivided into 60 seconds (″).

Radians

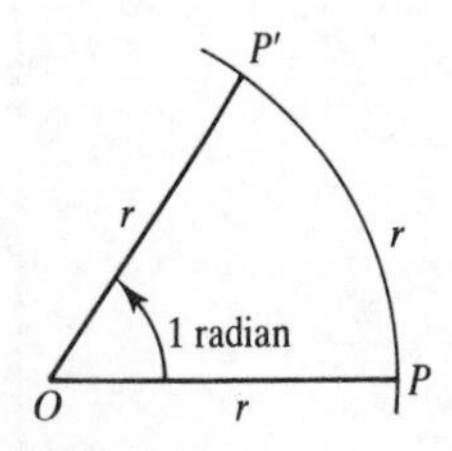

Figure 9.4

The radian is an alternative unit of measure of an angle, the use of which is imperative in certain situations that we shall meet later. When OP is rotated about O to OP' both P and P' lie on the circumference of a circle of radius $r = OP$. If the length of arc PP' is equal to the radius OP then the line OP' has rotated through an angle defined as 1 radian.

Since one complete revolution causes the point P' to move the full circumference $2\pi r$ then one complete revolution is equal to 2π radians.

Converting degrees to radians

Angles measured in degrees can be converted to angles measured in radians:

$$360 \text{ degrees} = 2\pi \text{ radians}$$

so that

$$\begin{aligned} 1 \text{ degree} &= \frac{2\pi}{360} \text{ radians} \\ &= \frac{\pi}{180} \text{ radians} \\ &= 0.017\,453\ldots \text{ radians} \end{aligned}$$

and

$$\begin{aligned} 1 \text{ radian} &= \frac{360}{2\pi} \text{ degrees} \\ &= \frac{180}{\pi} \text{ degrees} \\ &= 57.295\,7\ldots \text{ degrees} \end{aligned}$$

The Angles of a Triangle

The three angles of a triangle add up to a straight angle – 180 degrees or π radians. There are many different types of triangle and this fact is true for all of them.

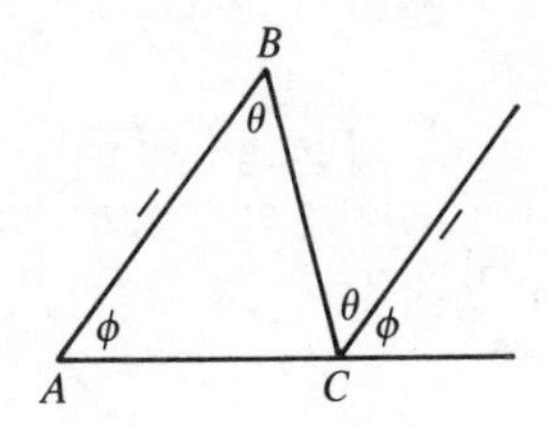

Angle $B = \theta$ *z*-angles
Angle $A = \phi$ corresponding angles
Angle $C + \theta + \phi = 180°$

Figure 9.5

Scalene Triangles

A triangle with unequal sides is called a **scalene triangle**.

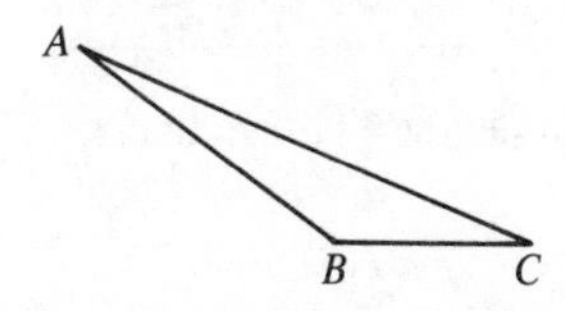

Figure 9.6

Acute-angled Triangle

A triangle, all of whose angles are acute, is called an **acute-angled triangle**.

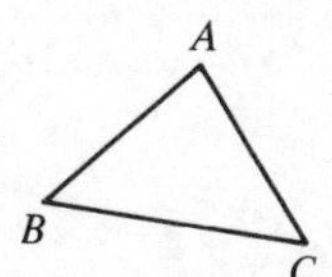

Figure 9.7

Right-angled Triangle

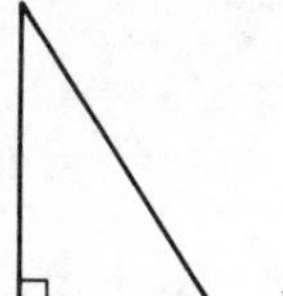

Figure 9.8

A triangle with one right angle is called a **right-angled triangle**. The side opposite to the right angle is referred to as the **hypotenuse**.

Obtuse-angled Triangle

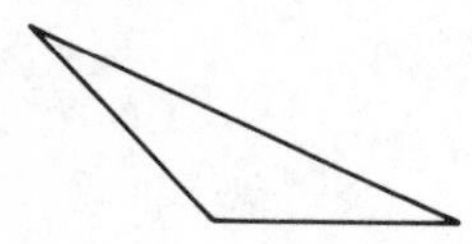

Figure 9.9

A triangle with one angle greater than 90° is called an **obtuse-angled triangle**.

Equilateral Triangles

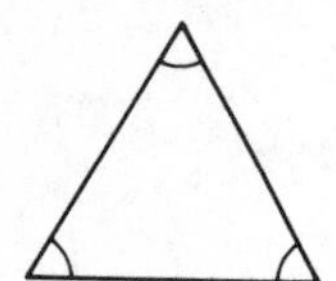

Figure 9.10

A triangle with all three sides equal is called an **equilateral triangle**. As well as all three sides being equal all three angles are equal – each angle being 60° or $\pi/3$ radians.

Isosceles Triangles

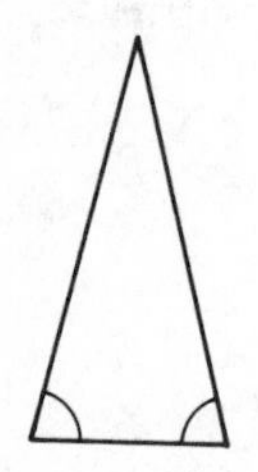

Figure 9.11

A triangle with two sides of equal length is called an **isosceles triangle**. The two angles at the feet of the two equal sides are equal.

EXAMPLES

1 Without the aid of a calculator, convert the following angles, measured in degrees, into radians:

(a) 90° **(b)** 30° **(c)** 270° **(d)** 25° **(e)** 132°

(a) $360° = 2\pi$ radians so $90° = 360°/4 = 2\pi/4 = \pi/2$ radians
(b) $30° = 360°/12 = 2\pi/12 = \pi/6$ radians
(c) $270° = 360° \times (3/4) = 2\pi \times 3/4 = 3\pi/2$ radians
(d) $25° = (\pi/180) \times 25 = 5\pi/36$ radians
(e) $132° = (\pi/180) \times 132 = 11\pi/15$ radians

2 Convert the following angles, measured in radians, into degrees:

(a) $\pi/5$ radians **(c)** 2.036 radians
(b) $5\pi/2$ radians **(d)** 34.56 radians

(a) $\pi/5$ radians $= (\pi/5)(180/\pi) = 36°$
(b) $5\pi/2$ radians $= (5\pi/2)(180/\pi) = 450°$
(c) 2.036 radians $= 2.036(180/\pi) = 116.654\,207\ldots° = 116°\,39'\,15''$
(d) 34.56 radians $= 34.56(180/\pi) = 1980.14°$ to two decimal places

3 Find the unknown angles in each of the following triangles:

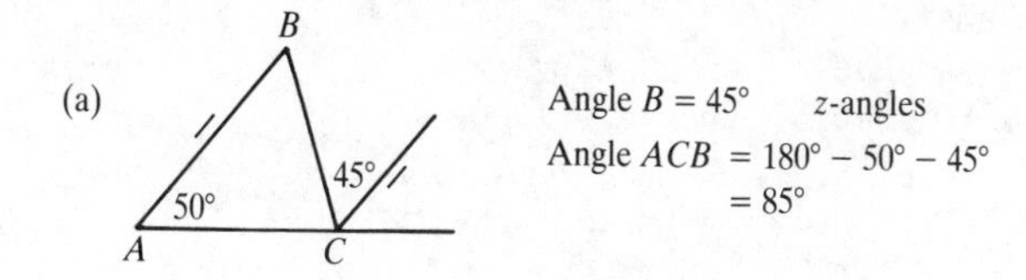

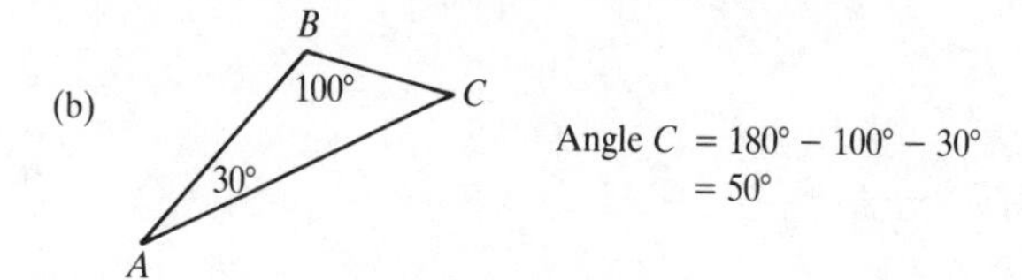

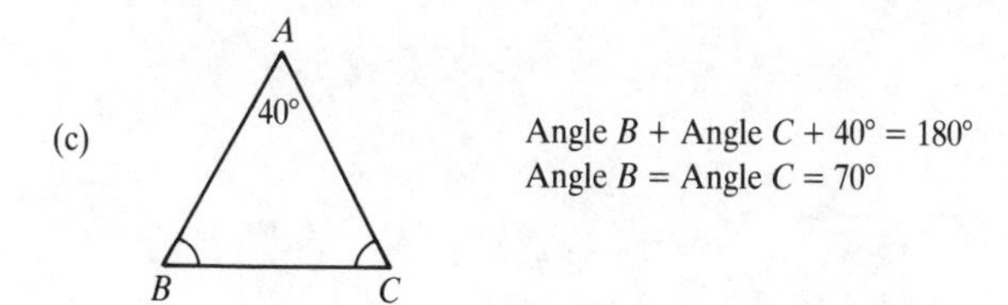

Figure 9.12

4 Identify the type of each of the following triangles:

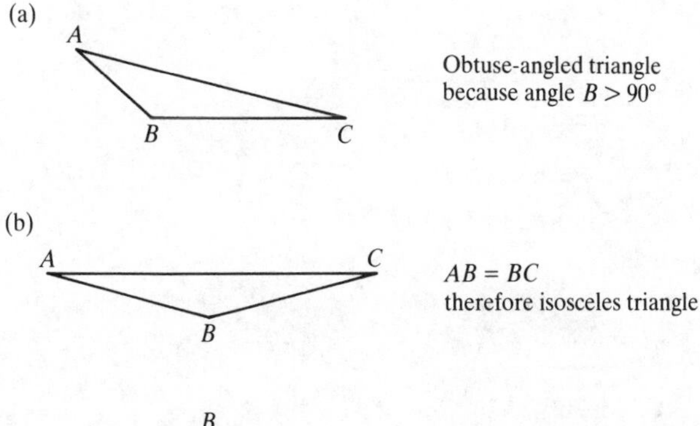

Figure 9.13

EXERCISES

1 Without the aid of a calculator, convert the following angles, measured in degrees, into radians:

(a) 45° **(b)** 60° **(c)** 270° **(d)** 36° **(e)** 204°

(a) 360° = * radians, so 45° = 360°/* = */* radians
(b) 60° = 360°/* = */* radians
(c) 270° = *°/* = */* radians
(d) 36° = (*/180) × 36 = */* radians
(e) 204° = (*/*) × 204 = */* radians

2 Convert the following angles, measured in radians, into degrees:

(a) $\pi/8$ radians **(c)** 1.125 radians
(b) $3\pi/5$ radians **(d)** 23.65 radians

(a) $\pi/8$ radians = $(\pi/8)(*/\pi)$ = *°
(b) $3\pi/5$ radians = $(*)(180/\pi)$ = *°
(c) 1.125 radians = (*)(180/*) = *° = *° *′ *″
(d) 34.56 radians = (*)(*/*) = *°

3 Find the unknown angles in each of the following triangles:

(a)

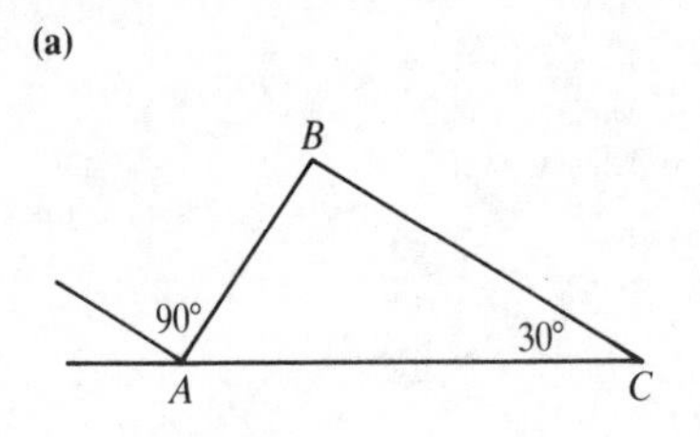

(b)

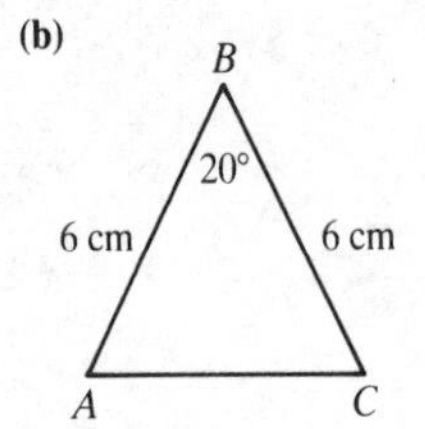

(c)

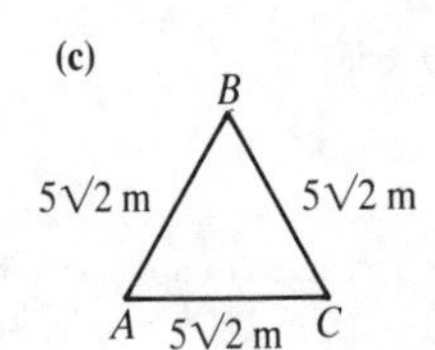

Figure 9.14

4 Identify the type of each of the following triangles:

(a)

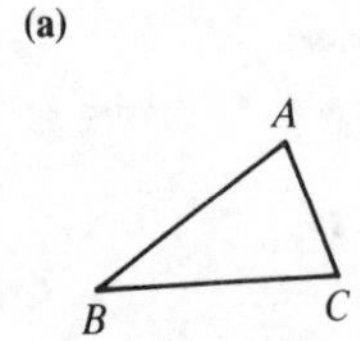

$BC > AB > AC$

(b)

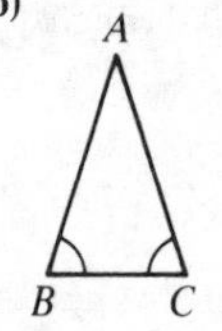

Angle B = Angle C = 75°

(c)

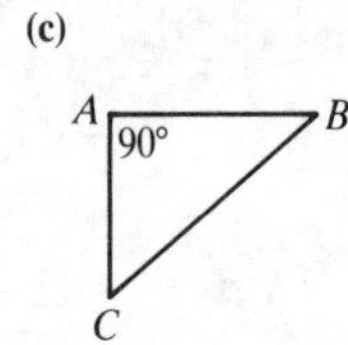

Figure 9.15

Unit 2 Pythagoras' theorem

Test yourself with the following:

1

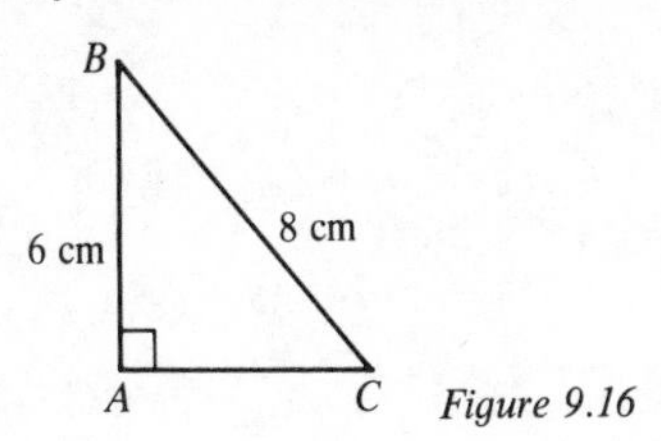

Figure 9.16

If ABC is a triangle right-angled at A and $AB = 6$ cm, $BC = 8$ cm calculate the length of AC.

2

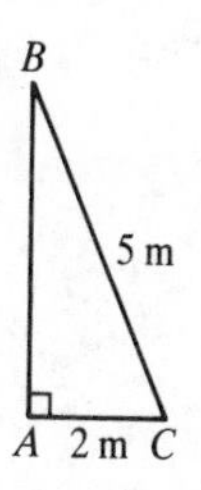

Figure 9.17

Calculate the area of a triangle ABC in which angle $A = 90°$, $BC = 5$ m and $AC = 2$ m.

3 Prove that a triangle with sides 2, 2 and $2\sqrt{2}$ cm long is a right-angled triangle.

4 Find the vertical height of an equilateral triangle with side length 4 cm.

5

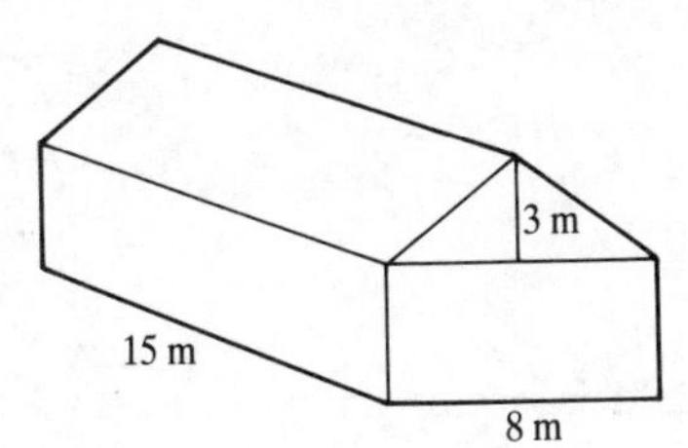

Figure 9.18

A symmetrical roof spans a 15 m by 8 m building and the apex of the roof is 3 m above the lower roof line. What is the roof area?

Right-angled Triangles

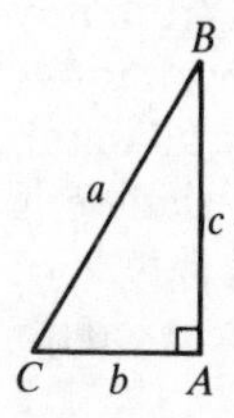

Figure 9.19

In a right-angled triangle the square on the hypotenuse is equal to the sum of the squares on the other two sides.

$$a^2 = b^2 + c^2$$

This statement, known as Pythagoras' theorem, is very easily demonstrated using Figure 9.20.

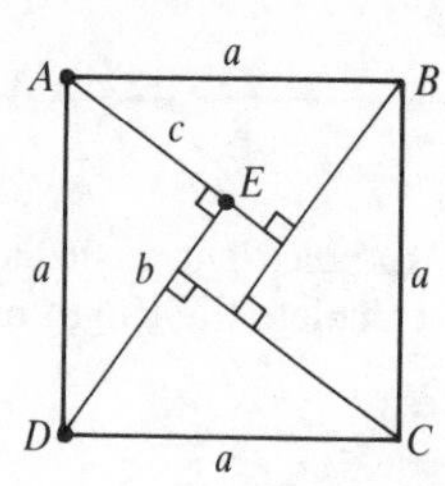

Square $ABCD$, side a
$DE = b$; $AE = c$

Area of four triangles + area of small square = area of large square

$$4(\tfrac{1}{2}b.c) + (b - c)(b - c) = a^2$$
$$2bc + b^2 - 2bc + c^2 = a^2$$
$$b^2 + c^2 = a^2$$

Figure 9.20

The 3:4:5 Triangle

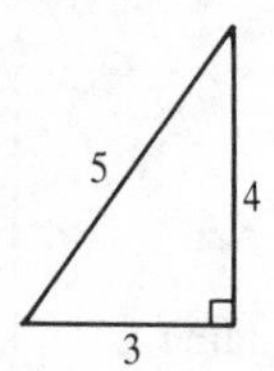

Figure 9.21

Since

$$3^2 + 4^2 = 9 + 16 = 25 = 5^2$$

Any triangle whose sides are in the ratio 3:4:5 is a right-angled triangle.

Right-angled Isosceles Triangle – the 1:1:$\sqrt{2}$ Triangle

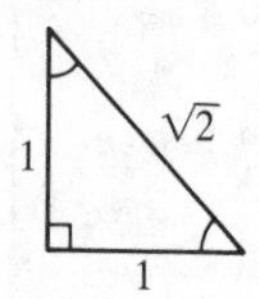

Figure 9.22

In a right-angled isosceles triangle the two other angles are equal to 45 degrees or $\pi/4$ radians. If the two equal sides are taken to have length 1 then the length of the hypotenuse is $\sqrt{2}$.

$$1^2 + 1^2 = 2 = (\sqrt{2})^2$$

Any triangle whose sides are in the ratio 1:1:$\sqrt{2}$ is a right-angled triangle.

Half Equilateral Triangle – the 1:$\sqrt{3}$:2 Triangle

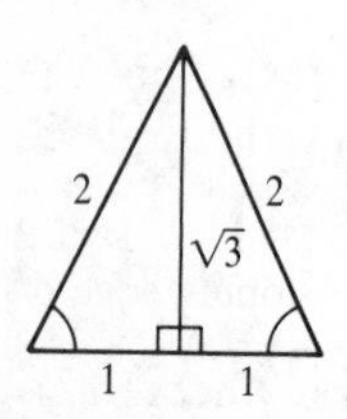

Figure 9.23

An equilateral triangle has three equal angles, each of $\pi/3$ radians. If a straight line is drawn from any vertex of an equilateral triangle to the opposite side, to bisect the side it forms two half equilateral, right-angled triangles. If the equilateral triangle is taken to have side length 2, the bisected side is of length 1 and the perpendicular bisector is of length $\sqrt{3}$. Any triangle whose sides are in the ratio 1:$\sqrt{3}$:2 is a right-angled triangle.

EXAMPLES

1

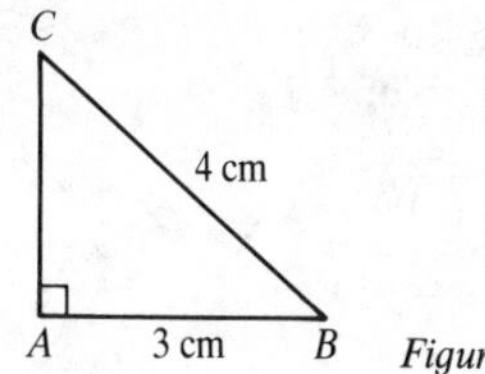

Figure 9.24

If ABC is a triangle right-angled at A and $AB = 3$ cm, $BC = 4$ cm calculate the length of AC.

By Pythagoras, $(BC)^2 = (AB)^2 + (AC)^2$ so that $4^2 = 3^2 + (AC)^2$. Hence, $(AC)^2 = 7$ therefore $AC = \sqrt{7}$ cm.

2

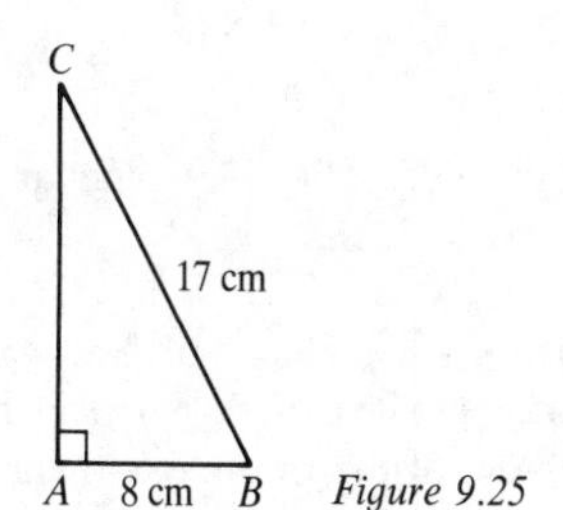

Figure 9.25

Calculate the area of a triangle ABC in which angle $A = 90°$, $BC = 17$ m and $AB = 8$ m.

By Pythagoras the length of AC is 15 m. The area of the triangle is half the area of the rectangle with sides AB and AC. That is, the area of the triangle is

$$\tfrac{1}{2} \times 15 \times 8 = 60 \text{ m}^2$$

3 Prove that a triangle with sides 9, 12 and 15 cm long is a right-angled triangle.

$15^2 = 225$, $9^2 = 81$ and $12^2 = 144$; $225 = 144 + 81$ and so

$$15^2 = 12^2 + 9^2$$

Because the side lengths of the triangle satisfy Pythagoras' theorem the triangle is right-angled. Alternatively, we could have simply stated that it is a right-angled triangle because the sides are in the ratio 3:4:5.

4 Find the vertical height of an equilateral triangle with side length 5 cm.

The vertical height of an equilateral triangle is equal to the length of the perpendicular bisector of any side drawn from the opposite vertex. The sides of a half equilateral triangle are in the ratio $1:\sqrt{3}:2$ where the 2 refers to the side of the equilateral triangle. Therefore, if the side of an equilateral triangle is of length 5 the vertical height is

$$\tfrac{5}{2}\sqrt{3} \text{ cm}$$

5

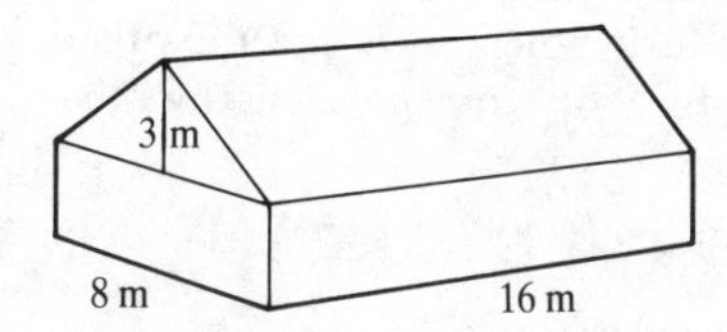

Figure 9.26

A symmetrical roof spans a 16 m by 8 m building and the apex of the roof is 3 m above the lower roof line. How long is the sloping side of the roof?

From Figure 9.26 it can be seen that the sloping edge of the roof forms the hypotenuse of a right-angled triangle with the other two sides of lengths 3 m and 4 m. The length of the hypotenuse is, therefore, 5 m.

EXERCISES

1 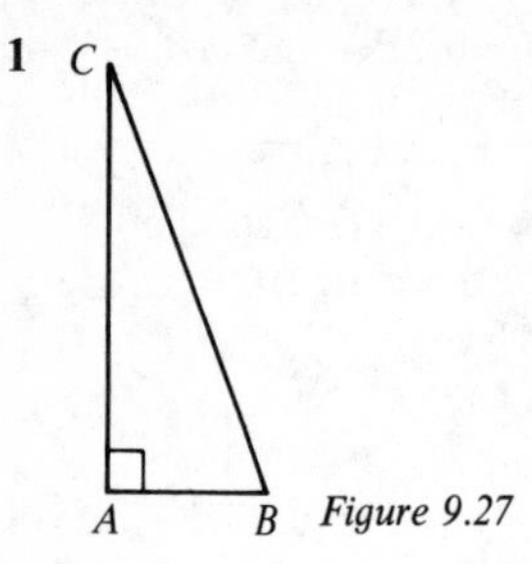

Figure 9.27

If ABC is a triangle right-angled at A and $AB = 5$ cm, $BC = 12$ cm calculate the length of AC.

By Pythagoras, $(*)^2 = (*)^2 + (*)^2$. That is

$$(*)^2 = *^2 - *^2 = 119$$

Therefore, $AC = *$ cm.

2 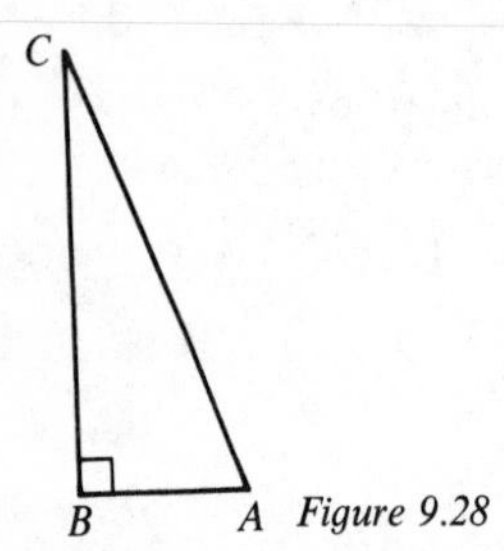

Figure 9.28

Calculate the area of a triangle ABC in which angle $B = 90°$, $BC = 24$ m and $AC = 26$ m.

By Pythagoras, the length of side $AB = *$ m. Sides AB and BC form two sides of a * whose area is twice that of the triangle ABC. The area of the triangle is then

$$\tfrac{1}{2} * . * = 120 \text{ m}^2$$

3 Prove that a triangle with sides 5, 10 and $5\sqrt{3}$ cm long is a right-angled triangle.

$(5\sqrt{3})^2 = *$, $5^2 = *$ and $10^2 = *$; $* = * + *$ and so

$$*^2 = *^2 + *^2$$

Because the side lengths of the triangle satisfy Pythagoras' theorem the triangle is right-angled.

4 Find the vertical height of a right-angled isosceles triangle with base length 12 cm.

The base of the right-angled isosceles triangle is opposite the right angle. The other two sides are of equal length L so that

$$12^2 = * + * = 2*^2$$

Therefore, $L^2 = *\text{ cm}^2$. The vertical height is V where $V^2 = L^2 - 6^2$ so that $V = *$.

5 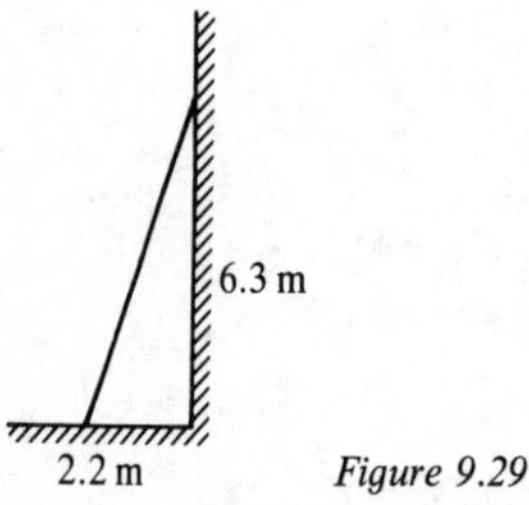

Figure 9.29

A ladder rests against a wall with its feet 2.2 m away from the wall. If the top of the ladder is 6.3 m up the wall, how long is the ladder?

The ladder, the ground and the side of the wall form a right-angled triangle with hypotenuse L and the other two sides * m and * m. By Pythagoras, the length of the ladder, L, is found from

$$*^2 = (*)^2 + (*)^2 = 44.53 \text{ m}^2$$

so that $L = *$ m to the nearest cm.

Unit 3 The Trigonometric Ratios

Test yourself with the following:

1 Find the sine, cosine and tangent of the following angles:
(a) 35° (b) 50° (c) $\pi/9$ radians (d) $\pi/36$ radians

2 Find the cosecant, secant and cotangent of the following angles:
(a) $\pi/7$ radians (b) $\pi/5$ radians (c) 40° (d) 28°

3 Find the trigonometric ratios of the two acute angles in a $2:1:\sqrt{3}$ triangle.

4 Find the reciprocal trigonometric ratios of the two acute angles in a $1:1:\sqrt{2}$ triangle.

5 Use a calculator to find the trigonometric ratios and their reciprocals for the following angles:
(a) 32.78° (b) $\pi/13$ radians

Trigonometry

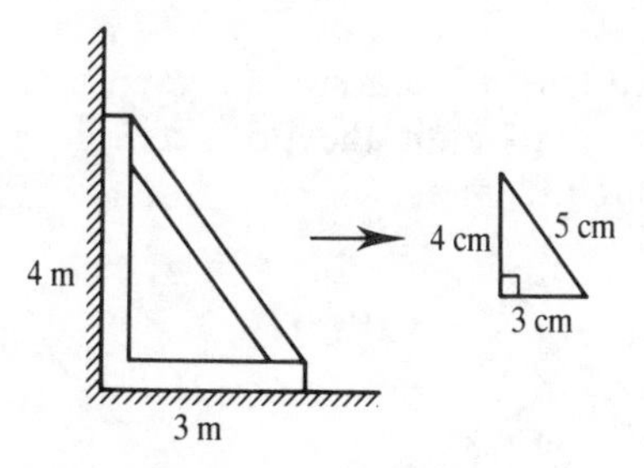

Figure 9.30

The aspect of mathematics called **trigonometry** deals with the properties of **similar triangles**. For example, a wall has to be supported by a triangular wooden prop as shown in Figure 9.30. It is known that the height of the vertical strut is 4 m and the length of the horizontal leg is 3 m but it is not known how long the hypotenuse strut is. If a diagram is drawn of this prop on a sheet of paper scaled down in the proportion 1 cm:1 m then it is found that measuring the hypotenuse gives 5 cm. From this it is deduced that the length of timber required for the hypotenuse strut will be 5 m. This deduction is made possible by the fact that the actual strut and the diagram are similar triangles whose sides are known to be in constant ratio to each other. Consequently, a knowledge of the ratios of the sides of a triangle will permit deductions to be made of all triangles that are similar to the original. These ratios are called the **trigonometric ratios** and are defined using the right-angled triangle.

The Right-angled Triangle

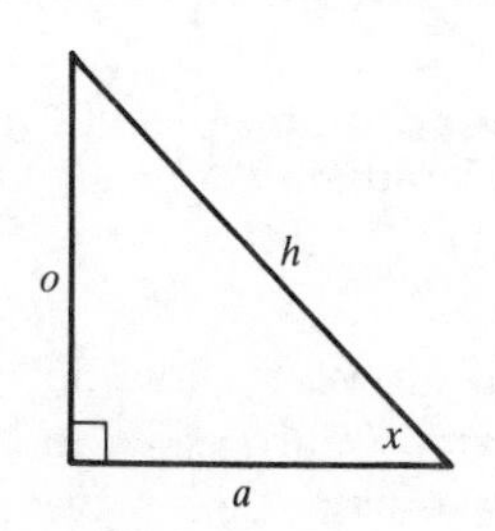

Figure 9.31

The right-angled triangle shown in Figure 9.31 possesses a right angle and an angle x. The side opposite to x is labelled o, the side adjacent to x is labelled a and the hypotenuse is labelled h. The three trigonometric ratios are then defined:

- **Sine** The sine of the angle x is defined as the ratio o/h:

$$\sin(x) = \frac{o}{h}$$

- **Cosine** The cosine of the angle x is defined as the ratio a/h:

$$\cos(x) = \frac{a}{h}$$

- **Tangent** The tangent of the angle x is defined as the ratio of the sine and the cosine:

$$\tan(x) = \frac{\sin(x)}{\cos(x)}$$
$$= \left(\frac{o}{h}\right) \Big/ \left(\frac{a}{h}\right)$$
$$= \frac{o}{a}$$

Notation

We shall see later that the use of brackets in the terms

$\sin(x)$, $\cos(x)$ and $\tan(x)$

is in full accordance with the later definitions of sin, cos and tan as functions. However, it is common practice to be inconsistent with the notation and to omit the brackets when considering these terms as simple trigonometric ratios, writing them as

$\sin x$, $\cos x$ and $\tan x$

Using a Calculator

You will find the three trigonometric ratios amongst the function keys on your calculator. These are used to find the trigonometric ratios of the angle that is displayed on the calculator screen. However, because angles can be measured in either degrees or radians you must first know whether the number displayed on the calculator screen is in degrees or radians. This is determined by the calculator mode which will be displayed on the screen as either DEG or RAD. The calculator manual will tell you how to change from one mode to another. As an example, put the calculator in degree mode (DEG) and enter the number 1. Press the **sin** key and the number

0.017 452 4...

will be displayed. This is the value of $\sin 1°$. Now clear the display, put the calculator in radian mode (RAD), enter the number 1 again and then press the **sin** key. This time the number

0.841 470 9...

will be displayed. This is the sine of 1 radian where, to 1 decimal place, 1 radian is 57.3°.

Reversing the Process

If we are given an angle then we can find its sine by using a calculator. However, how do we proceed if we are given, for example, that the sine of an angle is 0.5 and we are asked to find the angle. To find the angle we must reverse the process that produced the sine in the first place. On your calculator you will find an INV key which, when pressed, will enable you to reverse the process of any of the functions available. For example, enter the number 0.5, press INV followed by the sine function key and the displayed result is

30 in degree mode and 0.5235... in radian mode

What you have found is the angle whose sine is 0.5 – you have reversed the process of taking the sine of an angle. More about reversing such processes will be discussed in Chapter 3.

The Reciprocal Ratios

The reciprocal ratios are defined to make trigonometric expressions less cumbersome.

- **Cosecant** The cosecant of the angle x is the reciprocal of $\sin x$:

$$\operatorname{cosec}(x) = \frac{1}{\sin(x)}$$

- **Secant** The secant of the angle x is the reciprocal of $\cos x$:

$$\sec(x) = \frac{1}{\cos(x)}$$

- **Cotangent** The cotangent of the angle x is the reciprocal of $\tan x$:

$$\cot(x) = \frac{1}{\tan(x)}$$

The Graphs of the Trigonometric Ratios

Plotting the angle x against the corresponding value of one of the trigonometric ratios will produce a graph. The value of a trigonometric ratio for a given angle x can be found using the appropriate trigonometric function key on your calculator.

Remember that the ratios have only been defined for angles greater than zero and less than 90 degrees – those angles contained within a right-angled triangle. You will find, however, that your calculator will display values of the trigonometric ratios for angles greater than or equal to 90° and even for negative angles. To understand the meaning of such displayed values the definitions of the trigonometric ratios must be extended. This will be done in Module 13. What follows now are the graphs of the trigonometric ratios for angles greater than zero and less than 90°:

- **The sine graph**

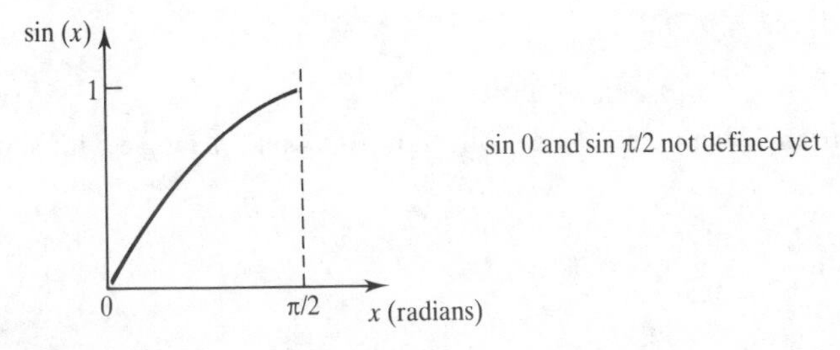

Figure 9.32

- **The cosine graph**

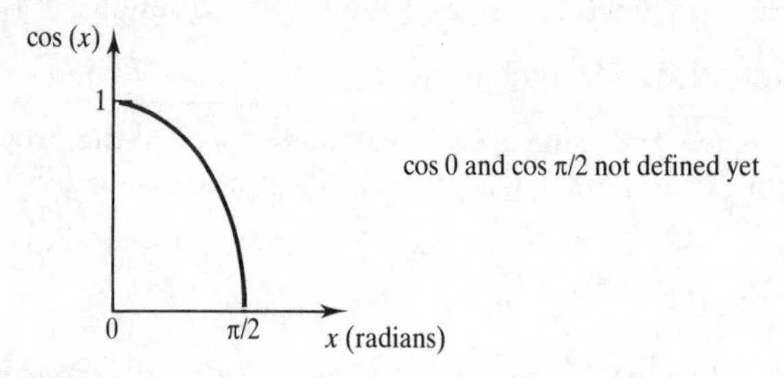

Figure 9.33

- **The tangent graph**

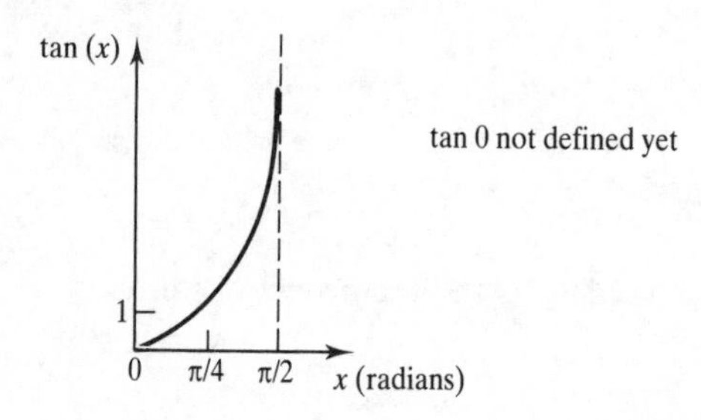

Figure 9.34

Notice that the horizontal axes in these three graphs are each graded in radians and not in degrees. It is essential that you be thoroughly familiar with both units of angle measure and that you can convert from one to the other with ease.

EXAMPLES

1 Find the sine, cosine and tangent of the following angles:

(a) 45° **(c)** $\pi/6$ radians
(b) 60° **(d)** $\pi/4$ radians

(a) The right-angled triangle with a 45° angle is an isosceles triangle with sides in the ratio $1{:}1{:}\sqrt{2}$. Consequently,

$$\sin 45° = 1/\sqrt{2} = 0.7071\ldots$$

$$\cos 45° = 1/\sqrt{2} = 0.7071\ldots$$

$$\tan 45° = 1$$

(b) The right-angled triangle with a 60° angle is the half equilateral triangle with sides in the ratio $1{:}\sqrt{3}{:}2$. Consequently,

$$\sin 60° = \sqrt{3}/2 = 0.8660\ldots$$

$$\cos 60° = 1/2 \quad = 0.5$$

$$\tan 60° = \sqrt{3} \quad = 1.7320\ldots$$

(c) $\pi/6$ radians = 30° which is the other angle in the half equilateral triangle. Thus,

$$\sin \pi/6 = 1/2 \quad = 0.5$$

$$\cos \pi/6 = \sqrt{3}/2 = 0.8660\ldots$$

$$\tan \pi/6 = 1/\sqrt{3} = 0.5773\ldots$$

(d) Using a calculator (in radian mode):

$$\sin \pi/24 = 0.1305\ldots$$

$$\cos \pi/24 = 0.9914\ldots$$

$$\tan \pi/24 = 0.1316\ldots$$

2 Find the cosecant, secant and cotangent of the following angles:

(a) $\pi/4$ radians **(c)** 30°
(b) $\pi/3$ radians **(d)** 15°

(a) $\operatorname{cosec} \pi/4 = \operatorname{cosec} 45° = \sqrt{2} = 1.4142\ldots$

$\sec \pi/4 = \sec 45° = \sqrt{2} = 1.4142\ldots$

$\cot \pi/4 = \cot 45° = 1$

(b) $\operatorname{cosec} \pi/3 = \operatorname{cosec} 60° = 2/\sqrt{3} = 1.1547\ldots$

$\sec \pi/3 = \sec 60° = 2$

$\cot \pi/3 = \cot 60° = 1/\sqrt{3} = 0.5773\ldots$

(c) $\operatorname{cosec} 30° = 2$

$\sec 30° = 2/\sqrt{3} = 1.1547\ldots$

$\cot 30° = \sqrt{3} = 1.7320\ldots$

(d) This will require the use of the $1/x$ function key:

$$\operatorname{cosec} 15° = \frac{1}{\sin 15°} = 3.8637\ldots$$

$$\sec 15° = \frac{1}{\cos 15°} = 1.0352\ldots$$

$$\cot 15° = \frac{1}{\tan 15°} = 3.7320\ldots$$

3

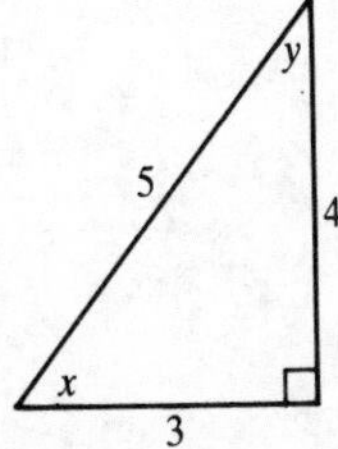

Figure 9.35

Find the trigonometric ratios of the two acute angles in a 3:4:5 triangle.

$$\sin x = 4/5 = \cos y$$

$$\cos x = 3/5 = \sin y$$

$$\tan x = 4/3$$

$$\tan y = 3/4$$

4

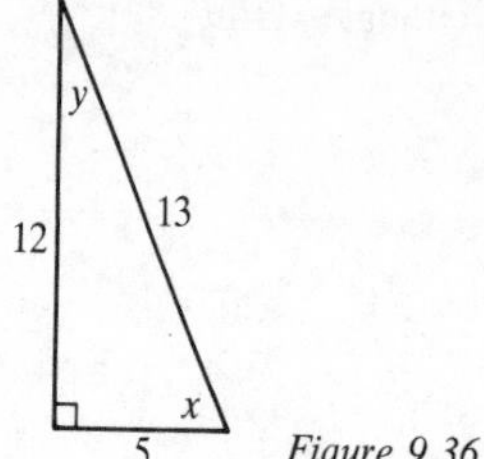

Figure 9.36

Find the reciprocal trigonometric ratios of the two acute angles in a 5:12:13 triangle.

$$\text{cosec } x = 13/12 = \sec y$$

$$\sec x = 13/5 = \text{cosec } y$$

$$\cot x = 5/12$$

$$\cot y = 12/5$$

5 Use a calculator to find the trigonometric ratios and their reciprocals of the following angles:

(a) 25.45° **(b)** 75°25′

(a) $\sin 25.45 = 0.4297\ldots$ $\text{cosec } 25.45 = 2.3270\ldots$
$\cos 25.45 = 0.9029\ldots$ $\sec 25.45 = 1.1074\ldots$
$\tan 25.45 = 0.4759\ldots$ $\cot 25.45 = 2.1012\ldots$

(b) $\sin 75°\,25' = 0.9677\ldots$ $\text{cosec } 75°\,25' = 1.0332\ldots$
$\cos 75°\,25' = 0.2517\ldots$ $\sec 75°\,25' = 3.9715\ldots$
$\tan 75°\,25' = 3.8436\ldots$ $\cot 75°\,25' = 0.2601\ldots$

EXERCISES

1 Find the sine, cosine and tangent of the following angles:

(a) 30° **(c)** $\pi/3$ radians
(b) 55° **(d)** $\pi/15$ radians

(a) The right-angled triangle with a 30° angle is the half equilateral triangle with sides in the ratio *:*:*. Consequently

$$\sin 30° = *$$

$$\cos 30° = *$$

$$\tan 30° = *$$

(b) Using a calculator (in degree mode):

$$\sin 55° = 0.8{*}9{*}\ldots$$

$$\cos 55° = 0.{*}7{*}5\ldots$$

$$\tan 55° = 1.{*}{*}81\ldots$$

(c) $\pi/3$ radians = 60°. Therefore

$$\sin 60° = *$$

$$\cos 60° = *$$

$$\tan 60° = *$$

(d) $\pi/15$ radians = 12°. Therefore

$$\sin 12° = *$$

$$\cos 12° = *$$

$$\tan 12° = *$$

2 Find the cosecant, secant and cotangent of the following angles:
(a) $\pi/6$ radians **(b)** $\pi/8$ radians **(c)** $60°$

(a) $\pi/6$ radians $= 30°$

cosec $30° = *$

sec $30° = *$

cot $30° = *$

(b) Using a calculator (in $*$ mode):

cosec $\pi/8 = *.6*3*\ldots$

sec $\pi/8 = 1.*8*3\ldots$

cot $\pi/8 = *.4*4*\ldots$

(c) cosec $60° = *$

sec $60° = *$

cot $60° = *$

3

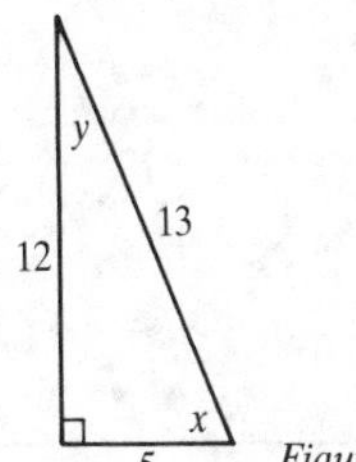

Figure 9.37

Find the trigonometric ratios of the two acute angles in a 5:12:13 triangle.

$\sin x = 12/* = \cos y$

$\cos x = */* = *$

$\tan x = */*$

$\tan y = */*$

4

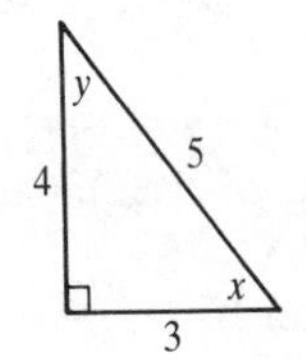

Figure 9.38

Find the reciprocal trigonometric ratios of the two acute angles in a 3:4:5 triangle.

$$\text{cosec } x = \frac{5}{*} = \sec y$$

$$\sec x = \frac{*}{*} = \text{cosec } y$$

$$\cot x = \frac{*}{*}$$

$$\cot y = \frac{*}{*}$$

5 Use a calculator to find the trigonometric ratios and their reciprocals of the following angles:
(a) $38.63°$ **(b)** $86°\,32'$

(a) $\sin 38.63° = 0.*2*2\ldots$ cosec $38.63° = *.6*1*\ldots$

$\cos 38.63° = 0.7*1*\ldots$ sec $38.63° = 1.*8*0\ldots$

$\tan 38.63° = 0.**91\ldots$ cot $38.63° = *.*5**\ldots$

(b) sin 86° 32′ = 0.9∗8∗... cosec 86° 32′ = ∗.0∗1∗...

cos 86° 32′ = 0.∗60∗... sec 86° 32′ = 1∗.5∗7∗...

tan 86° 32′ = 1∗.5∗7∗... cot 86° 32′ = ∗.0∗0∗...

Unit 4 Compound Angles and Trigonometric Identities

Test yourself with the following:

1 Expand: (a) $\cos(x + \pi/6)$ (c) $\sin(x - \pi/4)$
(b) $\tan(x + \pi/6)$

2 Without using a calculator find:
(a) $\sin 15°$ (b) $\tan 75°$

3 Expand $\cos 4x$ in terms of $\sin x$ and $\cos x$.

4 Show that:
(a) $\cos 45° + \cos 15° = \sqrt{3} \cos 15°$
(b) $\sin 45° - \sin 15° = \sqrt{3} \sin 15°$
(c) $\tan 45° + \tan 15° = \sqrt{3}(1 - \tan 15°)$

5 Establish the validity of each of the following identities:
(a) $\cos 2x \equiv 1 - 2 \sin^2 x$
(b) $(1 + \sin x)/\cos x \equiv \cos x/(1 - \sin x)$

Compound Angles

The trigonometric ratios of sums and differences of angles forms that aspect of trigonometry called the trigonometry of compound angles. Here we shall derive the sine of a sum and merely quote all the other combinations.

In Figure 9.39 PS and QT are perpendicular to OT and RQ is parallel to OT. Angle SOQ = angle OQR = angle $QPR = x$ and $RS = QT$. From Figure 9.39 we see that

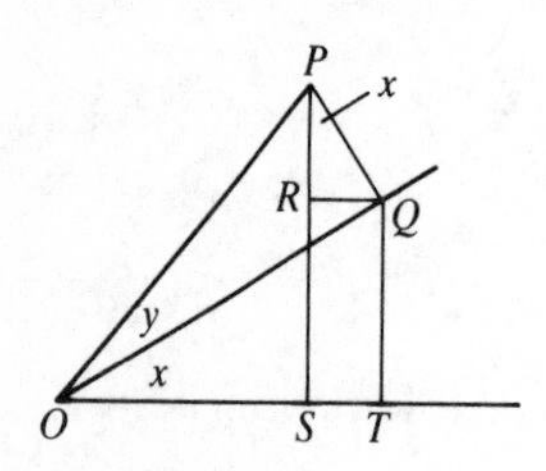

Figure 9.39

$$\begin{aligned}
\sin(x + y) &= \frac{SP}{OP} \\
&= \frac{SR + RP}{OP} \\
&= \frac{TQ + RP}{OP} \\
&= \frac{TQ}{OP} + \frac{RP}{OP} \\
&= \frac{TQ}{OQ}\cdot\frac{OQ}{OP} + \frac{RP}{QP}\cdot\frac{QP}{OP} \\
&= \sin x \cos y + \cos x \sin y
\end{aligned}$$

The Sine of a Sum or Difference

$$\sin(x + y) = \sin x \cos y + \sin y \cos x$$
$$\sin(x - y) = \sin x \cos y - \sin y \cos x$$

In particular, if $x = y$

$$\sin 2x = 2 \sin x \cos x$$

The Cosine of a Sum or Difference

$$\cos(x + y) = \cos x \cos y - \sin x \sin y$$
$$\cos(x - y) = \cos x \cos y + \sin x \sin y$$

In particular, if $x = y$

$$\cos 2x = \cos^2 x - \sin^2 x$$

The Tangent of a Sum or Difference

$$\tan(x + y) = \frac{\tan x + \tan y}{1 - \tan x \tan y}$$
$$\tan(x - y) = \frac{\tan x - \tan y}{1 + \tan x \tan y}$$

In particular, if $x = y$

$$\tan 2x = \frac{2 \tan x}{1 - \tan^2 x}$$

The Sum or Difference of Sines

$$\sin x + \sin y = 2 \sin\left(\frac{x + y}{2}\right) \cos\left(\frac{x - y}{2}\right)$$
$$\sin x - \sin y = 2 \sin\left(\frac{x - y}{2}\right) \cos\left(\frac{x + y}{2}\right)$$

The Sum or Difference of Cosines

$$\cos x + \cos y = 2 \cos\left(\frac{x + y}{2}\right) \cos\left(\frac{x - y}{2}\right)$$
$$\cos x - \cos y = -2 \sin\left(\frac{x - y}{2}\right) \sin\left(\frac{x + y}{2}\right)$$

Trigonometric Identities

Because there are six trigonometric ratios relating three triangular quantities there are many different trigonometric expressions that are equivalent to each other. To enable equivalent expressions to be deduced, extensive use of trigonometric identities is made. These identities stem from Pythagoras' theorem which is fundamental to the right-angled triangle.

The Fundamental Trigonometric Identity

From the following diagram, Pythagoras' theorem is expressed as

$$o^2 + a^2 \equiv h^2$$

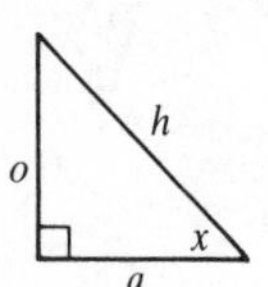

Figure 9.40

This is **Pythagoras' equation**. Dividing Pythagoras' equation throughout by h^2 yields

$$(o/h)^2 + (a/h)^2 \equiv 1$$

that is

$$\sin^2 x + \cos^2 x \equiv 1$$

Since this equation is true for any value of x it is an identity – it is called the **fundamental trigonometric identity**.

Two Derived Identities

From the fundamental trigonometric identity two further identities can be immediately derived.

(1) Dividing by $\sin^2 x$ throughout gives (note that $\sin^2 x$ means $[\sin x]^2$)

$$1 + \frac{\cos^2 x}{\sin^2 x} \equiv \frac{1}{\sin^2 x}$$

that is

$$1 + \cot^2 x \equiv \operatorname{cosec}^2 x$$

(2) Dividing the fundamental identity by $\cos^2 x$ throughout gives

$$\frac{\sin^2 x}{\cos^2 x} + 1 \equiv \frac{1}{\cos^2 x}$$

that is

$$\tan^2 x + 1 \equiv \sec^2 x$$

Proving the Validity of Trigonometric Identities

The validity of trigonometric identities can be proved using the fundamental identity and those identities derived from it. For example, to prove the following identity:

$$\tan x + \cot x \equiv \sec x \operatorname{cosec} x$$

the left-hand side must be shown to be equal to the right-hand side. Proceed as follows:

$$\begin{aligned} \text{LHS} &= \tan x + \cot x \\ &= \frac{\sin x}{\cos x} + \frac{\cos x}{\sin x} \\ &= \frac{\sin^2 x + \cos^2 x}{\sin x \cos x} \\ &= \frac{1}{\sin x \cos x} \\ &= \sec x \operatorname{cosec} x \\ &= \text{RHS} \end{aligned}$$

EXAMPLES

1 Expand: **(a)** $\cos(x + \pi/4)$ **(b)** $\tan(x - \pi/3)$ **(c)** $\sin(x - \pi/6)$

(a) $$\cos(x + \pi/4) = \cos x \cos \pi/4 - \sin x \sin \pi/4$$
$$= \frac{1}{\sqrt{2}}(\cos x - \sin x)$$

(b) $$\tan(x - \pi/3) = \frac{\tan x - \tan \pi/3}{1 + \tan x \tan \pi/3}$$
$$= \frac{\tan x - \sqrt{3}}{1 + \sqrt{3} \tan x}$$

(c) $$\sin(x - \pi/6) = \sin x \cos \pi/6 - \sin \pi/6 \cos x$$
$$= \frac{\sqrt{3}}{2} \sin x - \frac{1}{2} \cos x$$

2 Without using a calculator show that:

(a) $\cos 75^\circ = \dfrac{\sqrt{3} - 1}{2\sqrt{2}}$

(b) $\sin 75^\circ = \dfrac{\sqrt{3} + 1}{2\sqrt{2}}$

(a) $$\cos 75^\circ = \cos(30^\circ + 45^\circ)$$
$$= \cos 30^\circ \cos 45^\circ - \sin 30^\circ \sin 45^\circ$$
$$= \frac{\sqrt{3}}{2} \cdot \frac{1}{\sqrt{2}} - \frac{1}{2} \cdot \frac{1}{\sqrt{2}}$$
$$= \frac{\sqrt{3} - 1}{2\sqrt{2}}$$

(b) $$\sin 75^\circ = \sin(30^\circ + 45^\circ)$$
$$= \sin 30^\circ \cos 45^\circ + \sin 45^\circ \cos 30^\circ$$
$$= \frac{1}{2} \cdot \frac{1}{\sqrt{2}} + \frac{1}{\sqrt{2}} \cdot \frac{\sqrt{3}}{2}$$
$$= \frac{\sqrt{3} + 1}{2\sqrt{2}}$$

3 Expand $\cos 3x$ in terms of $\cos x$.

$$\cos 3x = \cos(2x + x)$$
$$= \cos 2x \cos x - \sin 2x \sin x$$
$$= (\cos^2 x - \sin^2 x) \cos x - 2 \sin x \cos x \sin x$$
$$= \cos^2 x \cos x - 3 \sin^2 x \cos x$$
$$= \cos^3 x - 3(1 - \cos^2 x) \cos x$$
$$= 4 \cos^3 x - 3 \cos x$$

4 Find the value of: **(a)** $\cos 75° - \cos 15°$
(b) $\sin 75° + \sin 15°$
(c) $\tan 60° - \tan 15°$, in terms of $\tan 15°$

(a)
$$\begin{aligned}\cos 75° - \cos 15° &= -2\sin\left(\frac{75° - 15°}{2}\right)\sin\left(\frac{75° + 15°}{2}\right)\\ &= -2\sin 30° \sin 45°\\ &= -2\left(\frac{1}{2}\right)\left(\frac{1}{\sqrt{2}}\right)\\ &= \frac{-1}{\sqrt{2}}\end{aligned}$$

(b)
$$\begin{aligned}\sin 75° + \sin 15° &= 2\sin\left(\frac{75° + 15°}{2}\right)\cos\left(\frac{75° - 15°}{2}\right)\\ &= 2\sin 45° \cos 30°\\ &= 2\left(\frac{1}{\sqrt{2}}\right)\left(\frac{\sqrt{3}}{2}\right)\\ &= \sqrt{\frac{3}{2}}\end{aligned}$$

(c)
$$\begin{aligned}\tan 60° - \tan 15° &= [\tan(60° - 15°)](1 + \tan 60° \tan 15°)\\ &= (\tan 45°)(1 + \tan 60° \tan 15°)\\ &= 1 + \sqrt{3}\tan 15°\end{aligned}$$

5 Establish the validity of each of the following identities:
(a) $\sin^3 x - \cos^3 x \equiv (\sin x - \cos x)(1 + \sin x \cos x)$
(b) $\tan^2 x - \sin^2 x \equiv \sin^4 x \sec^2 x$

(a)
$$\begin{aligned}\text{LHS} &= \sin^3 x - \cos^3 x\\ &\equiv (\sin x - \cos x)(\sin^2 x + \sin x \cos x + \cos^2 x)\\ &\equiv (\sin x - \cos x)(1 + \sin x \cos x)\\ &= \text{RHS}\end{aligned}$$

(b)
$$\begin{aligned}\text{RHS} &= \sin^4 x \sec^2 x\\ &\equiv (\sin^2 x \sin^2 x)/\cos^2 x\\ &\equiv \sin^2 x \tan^2 x\\ &\equiv \sin^2 x(\sec^2 x - 1)\\ &\equiv \tan^2 x - \sin^2 x\\ &= \text{LHS}\end{aligned}$$

EXERCISES

1 Expand **(a)** $\sin(x + \pi/6)$ **(b)** $\tan(x - \pi/4)$ **(c)** $\cos(x - \pi/3)$

(a) $\sin(x + \pi/6) = \sin x \cos * + \sin \pi/6 \cos *$

$= (*) \sin x + (1/2) \cos *$

$= (*)(* \sin x + \cos *)$

(b) $\tan(x - \pi/4) = \dfrac{\tan * - \tan *}{1 + \tan * \tan *}$

$= \dfrac{\tan * - *}{1 + \tan *}$

(c) $\cos(x - \pi/3) = \cos * \cos * + \sin * \sin *$

$= (*) \cos * + (*) \sin *$

$= (*)(\cos * + * \sin *)$

2 Without using a calculator find: **(a)** $\cos 15°$ **(b)** $\tan 15°$

(a) $\cos 15° = \cos(45° - *)$

$= \cos 45° \cos * + \sin 45° \sin *$

$= \dfrac{1}{\sqrt{2}}(*) + \dfrac{1}{\sqrt{2}}(*)$

$= *$

(b) $\tan 15° = \tan(* - *)$

$= \dfrac{\tan * - \tan *}{1 + \tan * \tan *}$

$= *$

3 Expand $\sin 4x$ in terms of $\sin x$ and $\cos x$.

$\sin 4x = * \sin * \cos *$

$= * \sin x \cos x(1 - 2*)$

4 Find the value of: **(a)** $\cos 75° + \cos 15°$
(b) $\sin 75° - \sin 15°$
(c) $\tan 30° + \tan 15°$, in terms of $\tan 15°$

(a) $\cos 75° + \cos 15° = 2 \cos\left(\dfrac{75° + *}{2}\right) \cos\left(\dfrac{75° - *}{2}\right)$

$= 2 \cos * \cos *$

$= *$

(b) $\sin 75^\circ - \sin 15^\circ = 2\sin\left(\frac{* - *}{2}\right)\cos\left(\frac{* + *}{2}\right)$

$= 2\sin * \cos *$

$= *$

(c) $\tan 30^\circ + \tan 15^\circ = [\tan(30^\circ * 15^\circ)](1 * \tan * \tan *)$

$= (\tan *)(1 * \tan * \tan *)$

$= *$

5 Establish the validity of each of the following identities:

(a) $\tan x + \cot x \equiv \sec x \operatorname{cosec} x$

(b) $\dfrac{\cos x - 1}{\sec x + \tan x} + \dfrac{\cos x + 1}{\sec x - \tan x} \equiv 2(1 + \tan x)$

(a) LHS $= \tan x + \cot x$

$\equiv \dfrac{\sin x}{\cos x} + \dfrac{*}{*}$

$\equiv \dfrac{\sin^2 x + \cos^2 x}{*.*}$

$\equiv \dfrac{*}{*.*}$

$\equiv \sec x \operatorname{cosec} x$

$=$ RHS

(b) LHS $= \dfrac{\cos x - 1}{\sec x + \tan x} + \dfrac{\cos x + 1}{\sec x - \tan x}$

$\equiv \dfrac{(\cos x - 1)(*) + (\cos x + 1)(*)}{\sec^2 x - \tan^2 x}$

$\equiv \dfrac{1 - * - * + \tan x + 1 + * + * + \tan x}{(1 - *^2)/*^2}$

$\equiv \dfrac{2 + 2\tan x}{*^2/*^2}$

$\equiv 2(1 + \tan x)$

$=$ RHS

Unit 5 The Sine and Cosine Rules

Test yourself with the following:

1

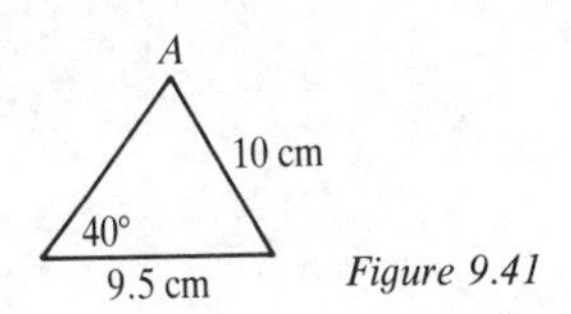

Figure 9.41

Given the triangle in Figure 9.41 find angle A using the sine rule.

2

12 cm
75°
14 cm

Figure 9.42

Given the triangle in Figure 9.42, find side a using the cosine rule.

3 Solve the following triangles:

(a) $a = 7$, $b = 6$, $c = 9$
(b) $A = 25°$, $b = 27$ cm, $c = 32.8$ cm
(c) $A = 74°$, $B = 10°$, $a = 14$
(d) $a = 33$, $B = 59°$, $C = \pi/6$ radians

4 Find the angles of the triangle whose sides are in the ratio 2:3:4.

Solving Triangles

All the properties of a right-angled triangle can be deduced when the length of any one of the sides and the size of one of the acute angles are given. However, this is not the case with a triangle that is not right-angled. To assist in the deduction of the properties of any triangle, the sine and cosine rules have been devised.

The Sine Rule

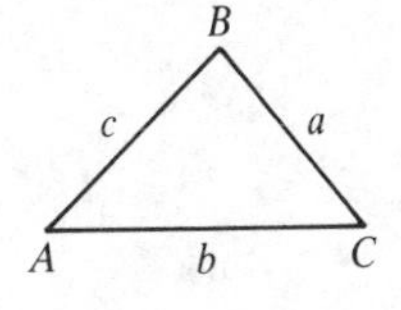

Figure 9.43

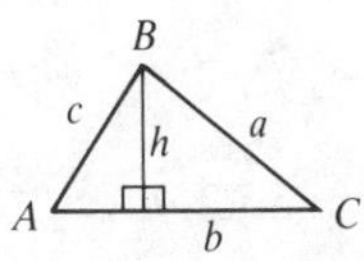

$\frac{h}{c} = \sin A \quad \frac{h}{a} = \sin C$ *Figure 9.44*

The area of a triangle is $\frac{1}{2}$(base)(height). Given the triangle shown in Figure 9.43 where sides a, b and c are respectively opposite angles A, B and C then the area of the triangle is given as

$$\tfrac{1}{2}bc \sin A = \tfrac{1}{2}ac \sin B = \tfrac{1}{2}ab \sin C$$

If each of these expressions for the area is divided by $\frac{1}{2}abc$ then what results is called the **sine rule**:

$$\frac{\sin A}{a} = \frac{\sin B}{b} = \frac{\sin C}{c}$$

For example, if, in triangle ABC, angle $A = 30°$, angle $B = 45°$ and side $a = 10$ cm then we can find side b by application of the sine rule:

$$\frac{\sin 30°}{10} = \frac{\sin 45°}{b}$$

and hence

$$b = \frac{10 \sin 45^\circ}{\sin 30^\circ}$$

$$= \frac{10/\sqrt{2}}{1/2}$$

$$= 10\sqrt{2}$$

$$= 14.14 \text{ cm to two decimal places}$$

By subtraction

$$C = 180^\circ - A - B$$

$$= 105^\circ$$

therefore

$$\frac{\sin 105^\circ}{c} = \frac{\sin 30^\circ}{10}$$

hence

$$c = \frac{10 \sin 105^\circ}{\sin 30^\circ}$$

$$= 19.32 \text{ cm to two dec. pl.}$$

This has solved the triangle because we now know all three angles and all three sides.

The Cosine Rule

Given the triangle shown in Figure 9.45 where h is an altitude:

$$h^2 = a^2 - x^2$$

$$= b^2 - (c - x)^2$$

$$= b^2 - c^2 + 2cx - x^2$$

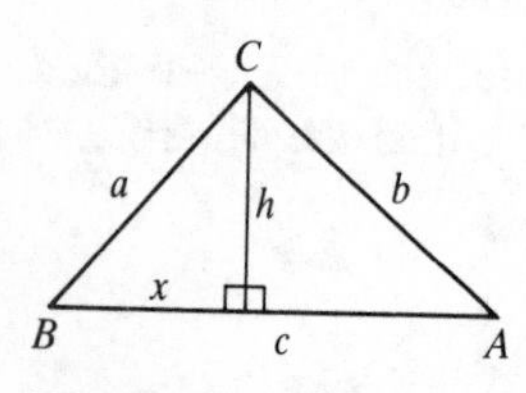

Figure 9.45

Therefore

$$a^2 - x^2 = b^2 - c^2 + 2cx - x^2$$

Adding x^2 to both sides of this equation gives

$$a^2 = b^2 - c^2 + 2cx$$

$$b^2 = a^2 + c^2 - 2ac \cos B$$

This is known as the **cosine rule**. Considering the other two altitudes of the triangle we see that

$$a^2 = b^2 + c^2 - 2bc \cos A$$

and

$$c^2 = a^2 + b^2 - 2ab \cos C$$

For example, if, in triangle ABC, side $b = 10$ cm, side $c = 12$ cm and angle $A = 60°$ then we can find side a by application of the cosine rule:

$$\begin{aligned} a^2 &= b^2 + c^2 - 2bc \cos A \\ &= 100 + 144 - 240 \cos 60° \\ &= 244 - 120 \\ &= 124 \end{aligned}$$

Therefore

$$a = 11.14 \text{ to two dec. pl.}$$

We now apply the sine rule to find angle B:

$$\frac{\sin 60°}{11.14} = \frac{\sin B}{10}$$

therefore

$$\begin{aligned} \sin B &= \frac{10 \sin 60°}{11.14} \\ &= 0.78 \text{ to two dec. pl.} \end{aligned}$$

therefore

$$B = 51.02° \text{ to two dec. pl.}$$

Subtracting:

$$\begin{aligned} A &= 180° - 51.02° - 60° \\ &= 68.98° \text{ to two dec. pl.} \end{aligned}$$

This completes the solution of the triangle.

Application of the Rules

Every triangle contains three sides and three angles. To find the values of all six items requires a prior knowledge of:

(1) two angles and one side length, or
(2) two side lengths and an angle opposite one of the given sides, or
(3) two side lengths and the included angle, or
(4) all three side lengths.

If the information given about a particular triangle is (1) or (2) then the sine rule can be applied to find the unknown side lengths and angles. If the information is (3) or (4) then the cosine rule can be applied.

EXAMPLES

1

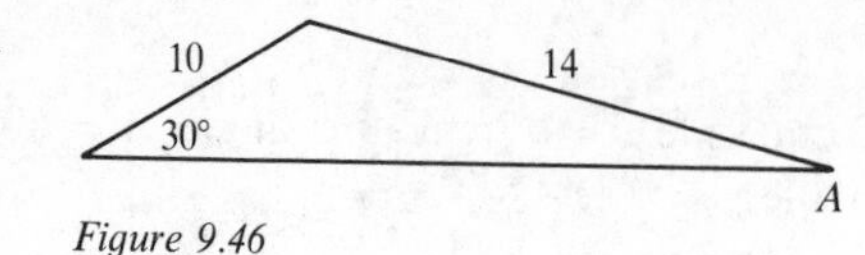

Figure 9.46

Given the triangle of Figure 9.46, find angle A using the sine rule.

$$\frac{\sin A}{10} = \frac{\sin 30°}{14}$$

so that

$$\sin A = \frac{10 \sin 30°}{14}$$

$$= 0.3571 \text{ to four dec. pl.}$$

hence $A = 20.9°$ to one dec. pl.

2

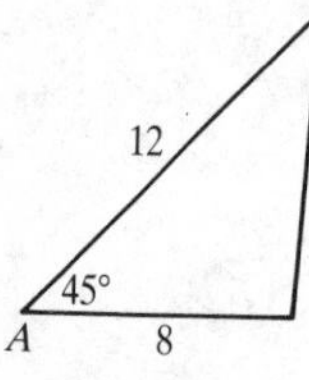

Figure 9.47

Given the triangle of Figure 9.47, find side a using the cosine rule.

$$a^2 = 8^2 + 12^2 - 2 \times 8 \times 12 \cos 45°$$

$$= 72.23 \text{ to two dec. pl.}$$

therefore $a = 8.50$ to two dec. pl.

3 Find all three sides and all three angles from the information given in each of the following triangles:

(a) $a = 13$, $b = 10$, $c = 16$
(b) $A = 85°$, $b = 13$ cm, $c = 18.5$ cm

(a) $13^2 = 10^2 + 16^2 - 2 \times 10 \times 16 \cos A$

that is

$$169 = 356 - 320 \cos A$$

$$\cos A = 0.5843\ldots$$

therefore $A = 54.24°$ to two dec. pl.

To find B we use the sine rule:

$$\frac{\sin 54.24°}{13} = \frac{\sin B}{10}$$

therefore

$$\sin B = \frac{10}{13} \sin 54.24$$

$$= 0.6242\ldots$$

hence $B = 38.62°$ to two dec. pl.

Finally, C can be obtained by subtracting the sum of A and B from 180° – the sum of all the angles in a triangle. Therefore

$$C = 180 - 54.24 - 38.62$$

$$= 87.14 \text{ to two dec. pl.}$$

(b) $a^2 = 13^2 + 18.5^2 - 2 \times 13 \times 18.5 \cos 85°$
$= 469.33 \text{ cm}^2$

so that $a = 21.66$ cm to two dec. pl. Also,

$$\frac{\sin 85°}{21.66} = \frac{\sin B}{13}$$

$$\sin B = \frac{13 \sin 85°}{21.66}$$

$$= 0.5979$$

so that $B = 36.72°$ to two dec. pl.

Finally,

$$C = 180° - 85° - 36.72°$$

$$= 58.28° \text{ to two dec. pl.}$$

4 Find the angles of the triangle whose sides are in the ratio 4:2:3.

Take the sides of the triangle to be

$$a = 2,\ b = 3 \text{ and } c = 4$$

Then

$$2^2 = 3^2 + 4^2 - 2 \times 3 \times 4 \cos A$$

so that

$$A = 28.955° \text{ to three dec. pl.}$$

Also,

$$\frac{\sin 28.955°}{2} = \frac{\sin B}{3}$$

so that $B = 46.567°$ to three dec. pl.

Finally,

$$C = 180° - A - B$$

$$= 104.478° \text{ to three dec. pl.}$$

EXERCISES

1 Given the triangle of Figure 9.48, find angle A using the sine rule.

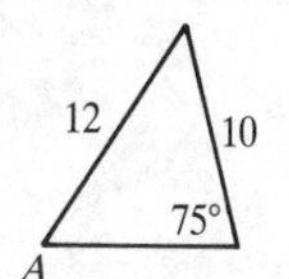

Figure 9.48

2 Given the triangle of Figure 9.49, find side a using the cosine rule.

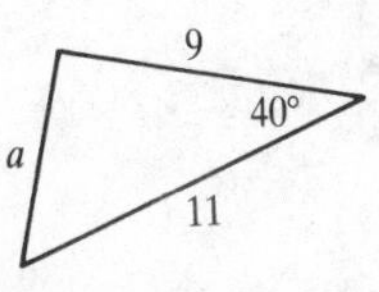

Figure 9.49

3 Solve the following triangles:

(a) $A = 35°$, $B = 64°$, $a = 19$ cm

(b) $A = 76°$, $b = 16.2$ cm, $c = 7.3$ cm

(a) $C = 180° - * - *$

$= *$

Also

$$\frac{\sin 35°}{*} = \frac{\sin *}{b}$$

therefore

$$b = \frac{* \sin *}{\sin 35°}$$

$= *$

Finally,

$$\frac{\sin 35°}{*} = \frac{\sin *}{c}$$

so that

$$c = \frac{* \sin *}{\sin 35°}$$

$= *$

(b) $a^2 = *^2 + *^2 - 2 \times * \times * \cos 76$

$= *$

Therefore

$a = *$

Also

$$\frac{\sin 76°}{*} = \frac{\sin B}{*}$$

so that

$$\sin B = \frac{* \sin 76°}{*}$$

$= *$

therefore $B = *$.

Finally,

$C = * - * - *$

$= *$

4 Find the angles of the triangle whose sides are in the ratio 5:6:2.

Take the sides of the triangle to be $a = 5$, $b = 6$ and $c = 2$. Then,

$5^2 = *^2 + *^2 - * \cos A$

so that $A = *°$. Also

$$\frac{\sin *}{5} = \frac{\sin B}{*}$$

so that $B = *°$.

Finally,

$$C = * - * - *$$
$$= *$$

Module 9: Further exercises

1 Without the aid of a calculator, convert the following angles, measured in degrees, into radians:
(a) 315° (b) 120° (c) 240° (d) 5° (e) 85°

2 Convert the following angles, measured in radians, into degrees:
(a) $\pi/12$ radians (b) $4\pi/5$ radians (c) 3.142 radians (d) 0.024 radians

3 Find the sine, cosine and tangent of the following angles:
(a) 70° (b) 85° (c) $\pi/8$ radians (d) $\pi/10$ radians

4 Find the cosecant, secant and cotangent of the following angles:
(a) $\pi/9$ radians (b) 20°

5 Find the trigonometric ratios of the two acute angles in a 9:12:15 triangle.

6 Find the reciprocal trigonometric ratios of the two acute angles in a 9:12:15 triangle.

7 Use a calculator to find the trigonometric ratios and their reciprocals of the following angles:
(a) 41.64° (b) $\pi/15$ radians

8 Expand: (a) $\cos(x - \pi/3)$ (b) $\tan(x + \pi/4)$ (c) $\sin(x - \pi/3)$

9 Expand $\sin 3x$ in terms of $\sin x$.

10 Expand $\cos 3x$ in terms of $\cos x$.

11 Solve the following trigonometric equations:

(a) $\cos x = \dfrac{1}{2}$ (c) $\tan x = 1$

(b) $\sin x = \dfrac{1}{\sqrt{3}}$ (d) $2 \sin x + 2 \cos x = \sqrt{6}$

12 Prove that

$$(1 - \cos x)^{1/2}(1 + \cos x)^{1/2} \equiv \sin x$$

13 Show that

$$-(\tan x + \sec x) \equiv \frac{1}{\tan x - \sec x}$$

14 Prove that

$$\frac{\operatorname{cosec} x \sec x}{\cot x} \equiv 1 + \tan^2 x$$

15 Eliminate x from the equations

$$x = 8 \cos x \text{ and } y = 6 \sin x$$

16 Given the triangle of Figure 9.50, find angle A using the sine rule.

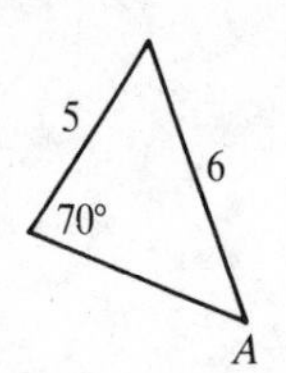

Figure 9.50

17 Given the triangle of Figure 9.51, find side a using the cosine rule.

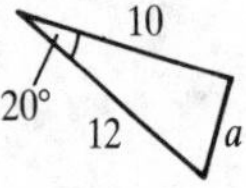

Figure 9.51

18 Solve the following triangles:

(a) $a = 9, b = 11, c = 8$

(b) $A = 40°, b = 15$ cm, $c = 25.6$ cm

(c) $A = 25°, B = 60°, a = 18$

(d) $a = 18, B = 32°, C = \pi/5$ radians

19 Find the angles of the triangle whose sides are in the ratio 10:12:8.

Module 10: Complex Numbers

Aims: To develop an awareness of the existence of complex numbers and to demonstrate their elementary properties in cartesian form.

Objectives: When you have read this module you will be able to:

▶▶ Plot complex numbers on an Argand diagram.

▶▶ Manipulate complex numbers in cartesian form by combining them with the arithmetic operations.

Unit 1 The Argand Diagram

Test yourself with the following:

1 Find the real and imaginary parts of:
(a) $1 - 2\mathrm{j}$ (b) -3 (c) $6\mathrm{j}$

2 Draw the following on an Argand diagram:
(a) $1 + \mathrm{j}$ (b) $-2 - \mathrm{j}$

3 Find the modulus and argument in each of the following:
(a) $-1/\sqrt{2} + \mathrm{j}/\sqrt{2}$ (b) $1/2 - \sqrt{3}\mathrm{j}/2$ (c) $-1.5 - 7.2\mathrm{j}$

Real, Imaginary and Complex Numbers

As we have seen, numbers of the form

$$a + \mathrm{j}b$$

where a and b are real numbers and $\mathrm{j}^2 = -1$, exist as solutions to those quadratic equations that have a negative discriminant. These numbers, being a mixture of real and imaginary numbers, are called **complex numbers**.

Real and Imaginary Parts

The traditional symbol for a complex number is z. We write

$$z = a + \mathrm{j}b$$

The real number a is called the **real part** of z and is denoted by

$$a = \operatorname{Re} z$$

The real number b is called the **imaginary part** of z and is denoted by

$$b = \operatorname{Im} z$$

Note that the imaginary part of z is the real number b and not the imaginary number $\mathrm{j}b$.

The Argand Diagram

Because every complex number contains two real numbers we can create a 'picture' of a complex number by using the cartesian coordinate system. Given the complex number $z = a + jb$ we plot a on the horizontal axis and b on the vertical axis to locate the point in the plane with coordinates (a, b). This point is now joined to the origin by a straight line and the straight line is taken as the graphical representation of the complex number. Notice that we have labelled the axes as the x-and y-axes respectively in keeping with the usual cartesian convention.

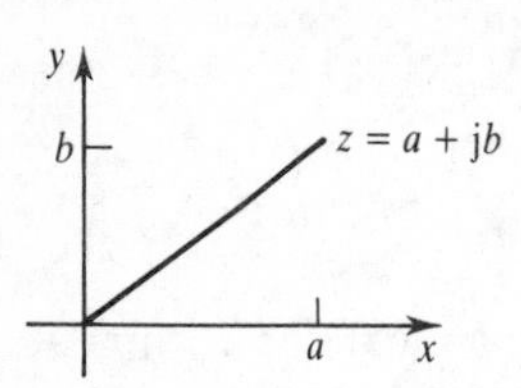

Figure 10.1

This representation of a complex number is called an **Argand diagram**.

Modulus and Argument

From the Argand diagram we can see that associated with every complex number is a length – the length of the line – and an inclination – the angle that the line makes with the x-axis. The length of the line is called the **modulus** of the complex number and is denoted by

$$|z| = \text{modulus } z$$

The length of the line can be found by employing Pythagoras' theorem:

$$|z| = \sqrt{(a^2 + b^2)}$$

The inclination of the line to the x-axis is called the **argument** of the complex number and is denoted by

$$\arg z$$

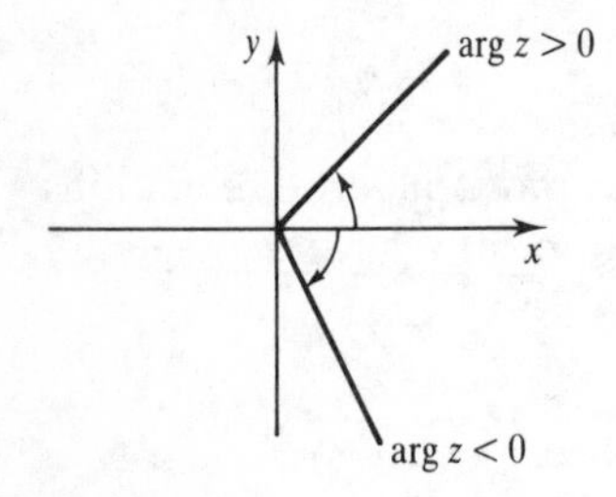

Figure 10.2

If, in the Argand diagram, the complex number is above the x-axis the argument is β where $0 \leqslant \beta \leqslant \pi$ measured anticlockwise. If the complex number is below the x-axis the argument is said to be negative with β measured clockwise where $-\pi < \beta < 0$.

EXAMPLES

1 Find the real and imaginary parts of the following:

(a) $3 + 4j$ **(b)** 2 **(c)** $-\sqrt{3}j$

(a) $\text{Re}(3 + 4j) = 3$ and $\text{Im}(3 + 4j) = 4$.
(b) We can write 2 as a complex number in the form $2 + 0j$. Then $\text{Re}(2 + 0j) = 2$ and $\text{Im}(2 + 0j) = 0$.
(c) We can write $-\sqrt{3}j$ as a complex number in the form $0 - \sqrt{3}j$. Then $\text{Re}(0 - \sqrt{3}j) = 0$ and $\text{Im}(0 - \sqrt{3}j) = -\sqrt{3}$.

2

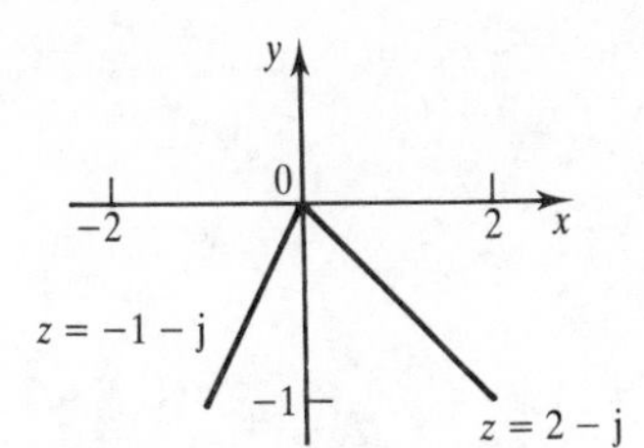

Figure 10.3

Draw each of the following on an Argand diagram:
(a) $2 - j$ **(b)** $-1 - j$

3

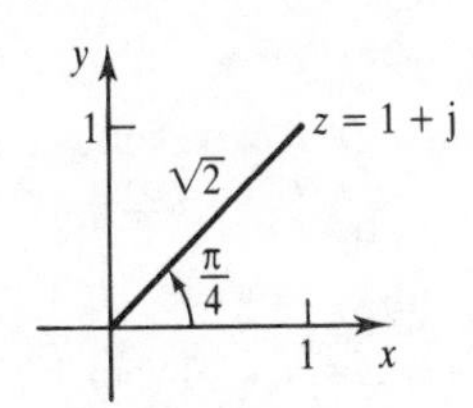

Figure 10.4

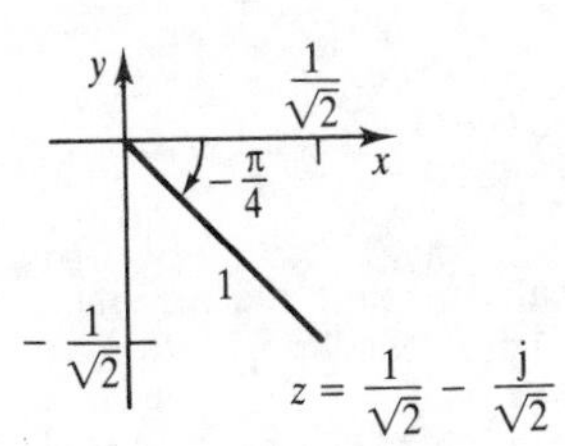

Figure 10.5

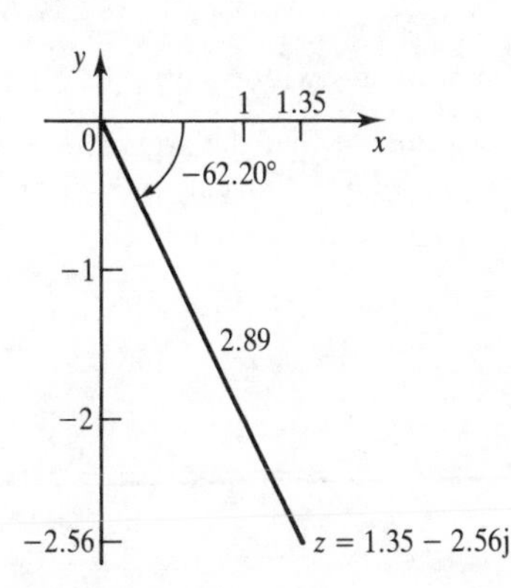

Figure 10.6

Find the modulus and the argument of each of the following:

(a) $1 + j$ **(c)** $1.35 - 2.56j$
(b) $1/\sqrt{2} - j\sqrt{2}$

(a) $|1 + j| = \sqrt{(1^2 + 1^2)} = \sqrt{2}$ From Figure 10.4 we can see that the argument is $\pi/4$ radians.

(b) $$|1/\sqrt{2} - j/\sqrt{2}| = \sqrt{[(1/\sqrt{2})^2 + (-1/\sqrt{2})^2]}$$
$$= \sqrt{(1/2 + 1/2)}$$
$$= 1$$

From Figure 10.5 we can see that the argument is $-\pi/4$ radians.

(c) $$|1.35 - 2.56j| = \sqrt{[(1.35)^2 + (-2.56)^2]}$$
$$= \sqrt{8.3761}$$
$$= 2.89 \text{ to two dec. pl.}$$

From Figure 10.6 we can see that the argument is the negative of the angle whose tangent is 2.56/1.35. That is, $-62.20°$ to two dec. pl.

EXERCISES

1 Find the real and imaginary parts of the following:
(a) $6 - 7j$ **(b)** -8 **(c)** j

(a) $\text{Re}(6 - 7j) = *$ and $\text{Im}(* - *) = *$.
(b) We can write -8 as a complex number in the form $* + *j$. Then $\text{Re}(* + *j) = *$ and $\text{Im}(* + *j) = *$.
(c) We can write j as a * number in the form *. Then $\text{Re}(*) = 0$ and $\text{Im}(*) = 1$.

2

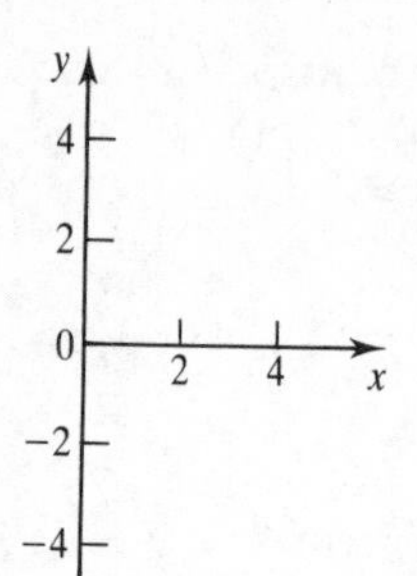

Figure 10.7

Draw each of the following on an Argand diagram:
(a) $3 + 4j$ **(b)** $3 - 4j$

3

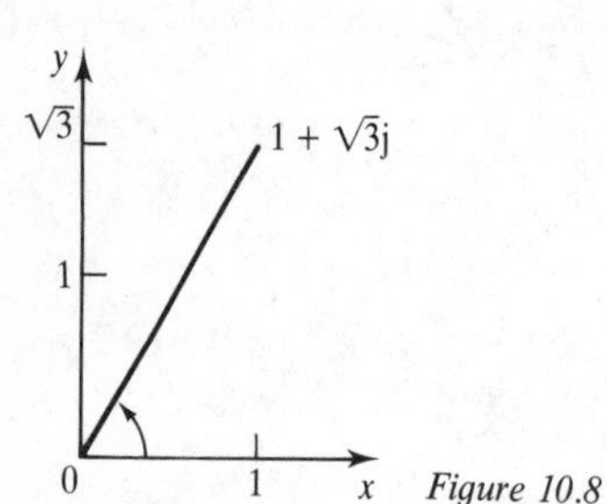

Figure 10.8

Find the modulus and argument of each of the following:
(a) $1 + \sqrt{3}j$ **(c)** $-3.24 + 5.16j$
(b) $-5\sqrt{2} + 5\sqrt{2}j$

(a) $|1 + \sqrt{3}j| = \sqrt{(1^2 + [*]^2)} = \sqrt{*} = *$ From Figure 10.8 we can see that the argument is $\pi/*$ radians.

(b) $|-5\sqrt{2} + 5\sqrt{2}j| = \sqrt{(*^2 + *^2)}$
$= \sqrt{(50 + *)}$
$= *$

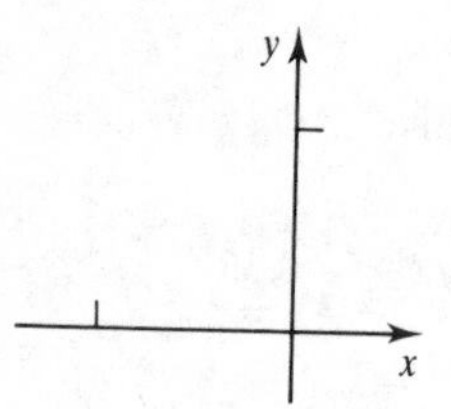

Figure 10.9

From Figure 10.9 we can see that the argument is * radians.

(c) $|-3.24 + 5.16j| = \sqrt{(*^2 + *^2)}$
$= \sqrt{*}$
$= 6.09$ to two dec. pl.

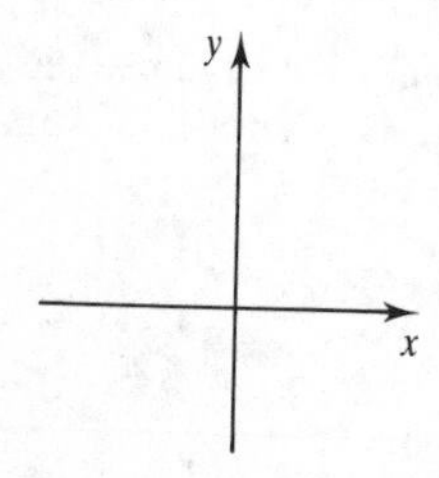

Figure 10.10

From Figure 10.10 we can see that the argument is $\pi/2$ radians plus the angle whose tangent is *. That is, 1*2.*2*°.

Unit 2 The Arithmetic of Complex Numbers

Test yourself with the following:

1 Given $z_1 = -1 + j$ and $z_2 = 3 - 4j$:
(a) draw z_1, z_2, and $z_1 - z_2$ on an Argand diagram
(b) find $3z_1 + 2z_2$, $z_1 . z_2$, z_1^* and z_1/z_2

2 By using the Argand diagram show that the argument of $(1 + j)^2$ is twice the argument of $(1 + j)$.

Arithmetic Operations

All the familiar operations of arithmetic apply to complex numbers.

Notation

Because we use the letter z to denote a complex number we resort to the use of subscripts to distinguish one complex number from another. For example,

$$z_1 = 3 - 2j \quad \text{and} \quad z_2 = 5 + j$$

Addition and Subtraction

To add or subtract two complex numbers we add or subtract their real parts and their imaginary parts. For example, if

$$z_1 = 3 - 2j \quad \text{and} \quad z_2 = 5 + j$$

then

$$\begin{aligned} z_1 - z_2 &= (3 - 2j) - (5 + j) \\ &= 3 - 2j - 5 - j \\ &= (3 - 5) + (-2j - j) \\ &= 8 - j \end{aligned}$$

and

$$\begin{aligned} z_1 - z_2 &= (3 - 2j) - (5 + j) \\ &= 3 - 2j - 5 - j \\ &= (3 - 5) + (-2j - j) \\ &= -2 - 3j \end{aligned}$$

If we plot two complex numbers of an Argand diagram then we see that they form two adjacent sides of a parallelogram and their sum forms the diagonal (see Figure 10.11). Their difference produces a similar geometric result (see Figure 10.12).

Multiplication

To multiply two complex numbers we proceed as in the following example:
If

$$z_1 = 2 - 3j \quad \text{and} \quad z_2 = 4 + j$$

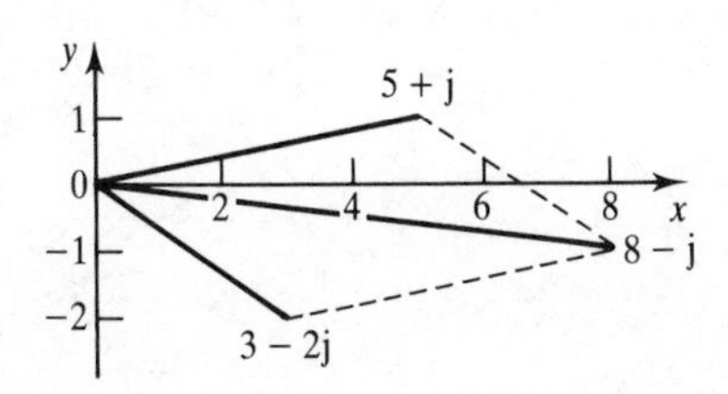

Figure 10.11

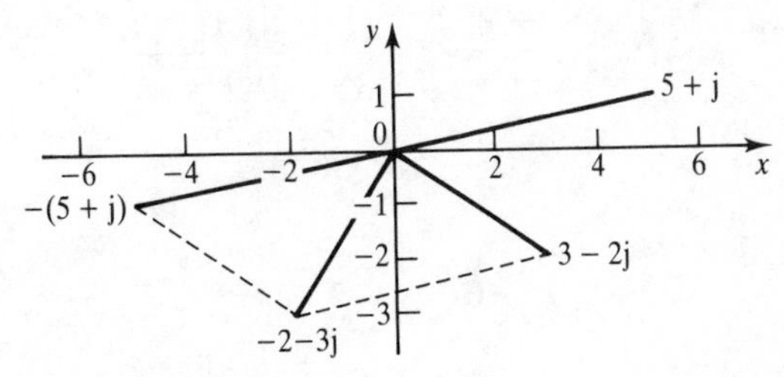

Figure 10.12

then

$$
\begin{aligned}
z_1 z_2 = (2 - 3\mathrm{j})(4 + \mathrm{j}) &= 2(4 + \mathrm{j}) - 3\mathrm{j}(4 + \mathrm{j}) \\
&= 8 + 2\mathrm{j} - 12\mathrm{j} - 3\mathrm{j}^2 \\
&= 8 - 10\mathrm{j} + 3 \\
&= 11 - 10\mathrm{j}
\end{aligned}
$$

Complex Conjugation

Complex conjugation is an operation that has no counterpart in real number arithmetic. Unlike addition, subtraction and multiplication which act on two numbers, complex conjugation acts on a single number.

If $z = a + \mathrm{j}b$ then the **complex conjugate**, denoted by z^* is defined to be

$$z^* = a - \mathrm{j}b$$

The imaginary part of z changes sign to form the complex conjugate. For example,

$(1 + \mathrm{j})^* = 1 - \mathrm{j}$ and $(5 - 4\mathrm{j})^* = 5 + 4\mathrm{j}$

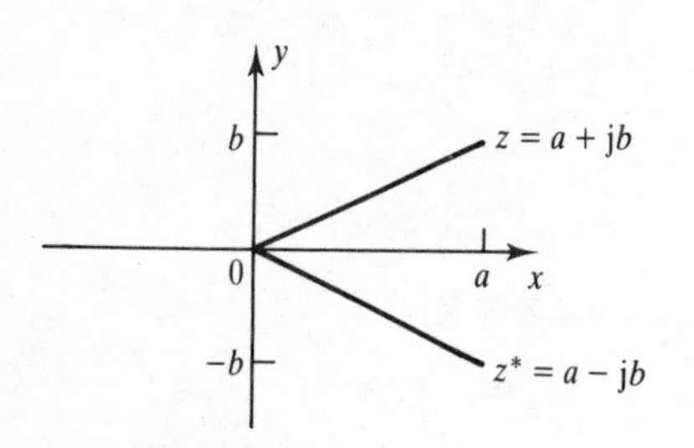

Figure 10.13

Note that in the Argand diagram the complex conjugate of a complex number is the mirror image of the original complex number with the mirror taken to lie along the x-axis. Note also that if $z = a + \mathrm{j}b$

$$
\begin{aligned}
zz^* &= (a + \mathrm{j}b)(a - \mathrm{j}b) \\
&= a^2 - \mathrm{j}^2 b^2 \\
&= a^2 + b^2
\end{aligned}
$$

zz^* is a real number equal to the square of the length of z in the Argand diagram. Also note that

$$(z^*)^* = z$$

Division

To effect the division of one complex number by another we multiply both the numerator and the denominator by the complex conjugate of the denominator. For example:
If

$$z_1 = 1 + 2\mathrm{j} \qquad \text{and} \qquad z_2 = 1 + \mathrm{j}$$

then

$$
\begin{aligned}
z_1/z_2 = \frac{\rightarrow + 2\mathrm{j}}{1 + \mathrm{j}} &= \frac{(1 + 2\mathrm{j})(1 - \mathrm{j})}{(1 + \mathrm{j})(1 - \mathrm{j})} \\
&= \frac{1 + \mathrm{j} - 2\mathrm{j}^2}{1 - \mathrm{j}^2} \\
&= \frac{3 + \mathrm{j}}{2} \\
&= \frac{3}{2} + \frac{\mathrm{j}}{2}
\end{aligned}
$$

EXAMPLES

1 Given $z_1 = 4 - 2j$ and $z_2 = 6 + 3j$:

(a) draw z_1, z_2 and $z_1 + z_2$ on an Argand diagram

(b) find $-z_1 + 4z_2$, $z_1 z_2$, z_2^* and z_2/z_1

(a)

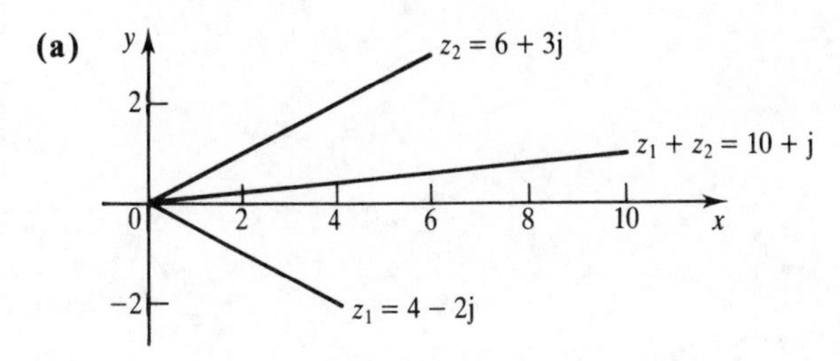

Figure 10.14

(b)
$$\begin{aligned}-z_1 + 4z_2 &= -(4 - 2j) + 4(6 + 3j)\\ &= -4 + 2j + 24 + 12j\\ &= 20 + 14j\end{aligned}$$

$$\begin{aligned}z_1 z_2 &= (4 - 2j)(6 + 3j)\\ &= 4(6 + 3j) - 2j(6 + 3j)\\ &= 24 + 12j - 12j - 6j^2\\ &= 24 + 6\\ &= 30\end{aligned}$$

$$\begin{aligned}z_2^* &= (6 + 3j)^*\\ &= 6 - 3j\end{aligned}$$

$$\begin{aligned}\frac{z_2}{z_1} &= \frac{6 + 3j}{4 - 2j}\\ &= \frac{(6 + 3j)(4 + 2j)}{(4 - 2j)(4 + 2j)}\\ &= \frac{24 + 12j + 12j + 6j^2}{16 - 4j^2}\\ &= \frac{18 + 24j}{20}\\ &= 0.9 + 1.2j\end{aligned}$$

2

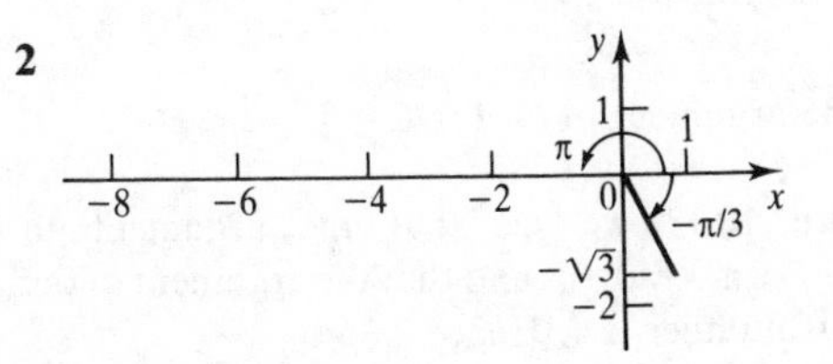

Figure 10.15

By using the Argand diagram show that the argument of $(1 - \sqrt{3}j)^3$ is three times the absolute value of the argument of $1 - \sqrt{3}j$. From Figure 10.15 we see that the argument of $1 - \sqrt{3}j$ is $-\pi/3$ radians and that the argument of the cube of this number, which is -8, is $\pi = 3|-\pi/3|$.

EXERCISES

1 Given $z_1 = 3 + 4j$ and $z_2 = -2 - j$:

(a) draw z_1, z_2 and $z_1 - z_2$ on an Argand diagram

(b) find $3z_1 + 2z_2$, $z_1 z_2$, z_1^* and z_2/z_1

(a)

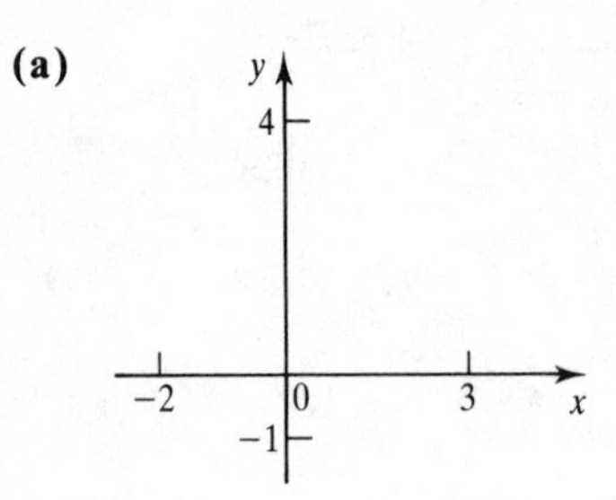

Figure 10.16

(b) $3z_1 + 2z_2 = 3(* + 4k) + 2(*)$

$= * + 12j + *$

$= 5 + 10j$

$z_1 z_2 = (3 + *)(* - j)$

$= 3(* - j) + *(* - j)$

$= * - 3j + * - *$

$= * - *j$

$z_2^* = (*)^*$

$= * + j$

$\frac{z_2}{z_1} = \frac{-2 - j}{3 + 4j}$

$= \frac{(-2 - j)(*)}{(3 + 4j)(*)}$

$= \frac{*}{*}$

$= \frac{-2}{5} + \frac{j}{5}$

2

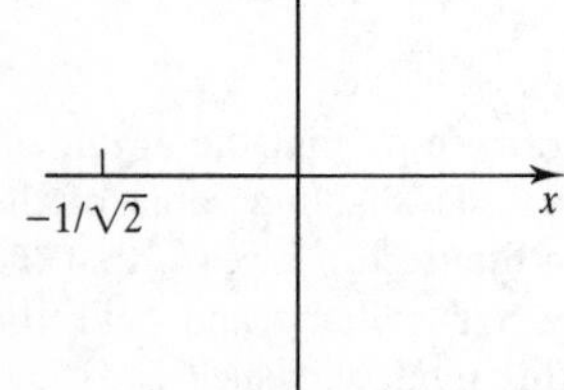

Figure 10.17

By using the Argand diagram show that twice the argument of

$$\frac{-1}{\sqrt{2}} + \frac{j}{\sqrt{2}}$$

is 2π plus the argument of $(-1/\sqrt{2} + j/\sqrt{2})^2$.

From Figure 10.17 we see that the argument of $-1/\sqrt{2} + j/\sqrt{2}$ is $*$ radians and that the argument of the square of this number is $*$. Hence $2\pi + * = *$.

Module 10: Further exercises

1 Find the real and imaginary parts of the following:
(a) $8-4j$ (b) -4 (c) $-9j$

2 Draw each of the following on an Argand diagram:
(a) $-6-j$ (b) $-6+j$

3 Find the modulus and argument of each of the following:
(a) $\frac{-1}{2}+\frac{\sqrt{3}j}{2}$ (b) $-1-j$ (c) $6.11+4.23j$

4 Given $z_1 = 5-4j$ and $z_2 = -6+7j$:
(a) draw z_1, z_2 and z_1+z_2 on an Argand diagram
(b) find $-4z_1+5z_2$, z_1z_2, z_2^* and z_2/z_1

5 By using the Argand diagram show that the argument of

$$\left(\frac{1}{2}+\frac{\sqrt{3}j}{2}\right)^3$$

is three times the argument of $(1/2+\sqrt{3}j/2)$.

Chapter 2: Miscellaneous exercises

1 If the radius of a circle is increased by 1 unit, find the ratio of the new circumference to the new diameter.

2 A square and an equilateral triangle have equal perimeters. The area of the triangle is $9\sqrt{3}$ cm^2. Find the length of the diagonal of the square.

3 A man on his way to dinner shortly after 6.00p.m. notices that the hands on his watch form an angle of 110°. Returning before 7.00p.m. he notices again the hands form an angle of 110°. How many minutes has he been away?

4 The dimensions of a rectangle R are a and b, $a<b$. It is required to obtain a rectangle with dimensions x and y, $x<a$ and $y<b$, so that its perimeter is $1/3$ that of R. How many different such rectangles are there?

5 Find S as a polynomial in x where

$$S=(x-1)^4+4(x-1)^3+6(x-1)^2+4(x-1)+1$$

6 Simplify the third term in the expansion of

$$\left(\frac{a}{\sqrt{x}}-\frac{\sqrt{x}}{a}\right)^6$$

7 Show that

$$\sqrt{(t^4+t^2)}=|t|\sqrt{(1+t^2)}$$

8 If x men working x hours a day for x days produce x articles (not necessarily a whole number of articles) find how many articles are produced by y men working y hours a day for y days.

9 If $3x^3-9x^2+kx-12$ is divisible by $x-3$, find k.

10 Let the roots of $ax^2+bx+c=0$ be r and s. Find the equation with roots $ar+b$ and $as+b$.

11 When x^9-x is factored as completely as possible into polynomials with real coefficients how many factors are there?

12 Find the difference between the larger and smaller roots of

$$x^2 - px + \frac{p^2 - 1}{4} = 0$$

13 R varies directly as S and inversely as T. When $R = 4/3$ and $T = 9/14$ then $S = 3/7$. Find S when $R = \sqrt{48}$ and $T = \sqrt{75}$.

14 If the parabola $y = ax^2 + bx + c$ passes through $(-1, 12)$, $(0, 5)$ and $(2, -3)$, find the value of $a + b + c$.

15 Let $A \uparrow B$ represent the operation on two numbers A and B which selects the larger of the two numbers, with $A \uparrow A = A$. Let $A \downarrow B$ represent the operation on two numbers A and B which selects the smaller of the two numbers, with $A \downarrow A = A$. Which of the following rules are correct?

(a) $A \uparrow B = B \uparrow A$
(b) $A \uparrow (B \uparrow C) = (A \uparrow B) \uparrow C$
(c) $A \downarrow (B \uparrow C) = (A \downarrow B) \uparrow (A \downarrow C)$

16 If x and y are both integers how many solutions are there to the equation

$$(x - 8)(x - 10) = 2^y$$

17 Find n when the values of $x = 2$ and $y = -2$ are substituted into the equation

$$n = x - y^{x-y}$$

18 For what value of k does this pair of equations have two identical solutions:

$$y = x^2 \text{ and } y = 3x + k$$

19 Find x if the reciprocal of $x + 1$ is $x - 1$.

20 Find the smallest positive integer x for which

$$1260x = N^2$$

where N is an integer.

21 Solve the following equation:

$$\frac{\dfrac{a}{a+x} + \dfrac{x}{a-x}}{\dfrac{x}{a+x} - \dfrac{a}{a-x}} = -1$$

22 Substitute $P = x + y$ and $Q = x - y$ into the following and simplify:

$$\frac{P+Q}{P-Q} - \frac{P-Q}{P+Q}$$

23 If

$$2^x = 8^{y-1} \text{ and } 9^y = 3^{x-9}$$

find x and y.

24 How many values of b and c are there such that the equations

$$3x + by + c = 0 \text{ and } cx - 2y + 12 = 0$$

have the same graph?

25 Find the sum of the reciprocals of the roots of the general quadratic equation

$$ax^2 + bx + c = 0$$

26 Show that the numbers 2, $-1 + \sqrt{3}j$ and $-1 - \sqrt{3}j$ satisfy the equation

$$x^3 - 8 = 0$$

27 If $z = \cos\theta + \mathrm{j}\sin\theta$ show that $z/z^* = \cos 2\theta + \mathrm{j}\sin 2\theta$.

28 If $\mathbf{a}.\mathbf{b} = \mathbf{a}.\mathbf{c}$ and $\mathbf{a} \neq 0$ is it necessarily true that $\mathbf{b} = \mathbf{c}$?

29 Find the interior angles of a triangle whose vertices are at

$$(-1, 0, 1), (1, 1, 0) \text{ and } (0, 0, -1)$$

30 Show that

$$(\mathbf{i} \times \mathbf{j}) \times \mathbf{k} = (\mathbf{i}.\mathbf{k})\mathbf{j} - (\mathbf{j}.\mathbf{k})\mathbf{i}$$

3

■ ■ ■ Functions

The aims of this chapter are to:

1. Develop the concept of a function from the idea of a system.
2. Demonstrate the graphical features of certain functions.
3. Extend the definitions of the trigonometric ratios to the trigonometric functions.
4. Investigate the properties of both exponential and logarithmic functions.
5. Consider both inverses of functions and inverse functions.

This chapter contains five modules:

Module 11: Systems and Functions
By considering inputs, processes and outputs, mathematical functions are defined. Algebraic functions and their composition are also discussed.

Module 12: Graphs of Functions
Graphs of functions and their related functions are discussed. The graphical description of even and odd functions is also given.

Module 13: Trigonometric Functions and Equations
The definitions of the trigonometric ratios are extended to define the circular trigonometric functions. Solutions of trigonometric equations are also considered.

Module 14: The Exponential and Logarithmic Functions
The exponential and logarithmic functions and their graphs are considered as are the hyperbolic trigonometric functions.

Module 15: Inverse Functions
The inverse of a function and inverse functions are distinguished. The use of inverse functions to solve trigonometric equations is discussed.

Module 11: Systems and Functions

Aim: To develop the concept of a function from the idea of a system.

Objectives: When you have read this module you will be able to:

- ▶▶ Distinguish between a system's input, process and output.
- ▶▶ Define a function.
- ▶▶ Separate a composition of functions into its composite parts.
- ▶▶ Recognize algebraic functions and find the domain and range of such functions.
- ▶▶ Combine functions using the arithmetic operations.

Unit 1 Input, Process and Output

Test yourself with the following:

1 Describe the processes that produce each of the following outputs:

(a) $f(x) = x + \dfrac{2}{3}$ (d) $f(x) = 12x + 4$

(b) $f(x) = 12.5x$ (e) $f(x) = 5x^2 - 2$

(c) $f(x) = x^{-3}$

2 Each of the following represents the output from a simple function. If this output were to form the input of a subsequent function that retrieves the original input x, describe the action of the subsequent process:

(a) $f(x) = 3 + x$ (c) $f(x) = x^{-3/4}$

(b) $f(x) = \dfrac{2x}{3}$

3 Which of the following defines a function?

(a) $f(x) = 7x^{-5}$ (c) $f(x) = 9x^{-5/3}$

(b) $f(x) = \pm 8x^{-5/2}$

4 Given that $f(x) = 1 + x$, $g(x) = x^{-2}$ and $h(x) = 3x/4$ find the output from each of the following:

(a) $f o g$ (b) $g o f$ (c) $f o g o h$ (d) $g o h o f$ (e) $f o f$

5 Each of the following is the output from a composition. Decompose each composition into its component functions where each component only involves one arithmetic operation:

(a) $f(x) = (1 - 2x)^5$ (c) $f(x) = (2x^2 + 5)$

(b) $f(x) = \left(9 - \dfrac{x}{4}\right)^{1/3}$

6 For which pair of functions is it true that $f \circ g = g \circ f$?
(a) $f{:}f(x) = 7 + x \quad g{:}g(x) = x - 7$
(b) $f{:}f(x) = 2x \quad g{:}g(x) = 4 - x$
(c) $f{:}f(x) = -x \quad g{:}g(x) = x + 2$

Systems

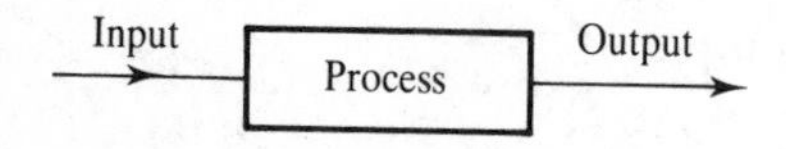

Figure 11.1 System diagram.

A **system** consists of an input, a process that acts on the input and an output that is the result of the processing. The processing unit of the system converts the input into an output. This notion of a system is used to define a mathematical function.

Functions

Figure 11.2

If the input and output of a system are numbers then provided that a single input number is processed into only one output number the processor is called a **function** (see Figure 11.2).

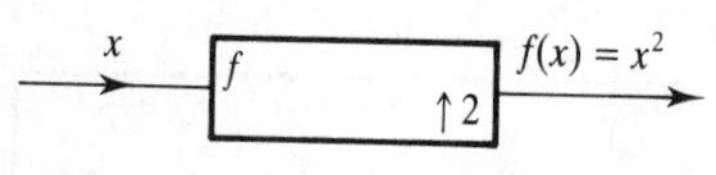

Figure 11.3

If the input number is labelled as x and the function labelled as f then the output is the effect of f acting on x and is accordingly labelled $f(x)$ – read as 'f of x'. For example, the process f that converts a number into its square is a function with output $f(x)$ (see Figure 11.3) where:

$$f{:}f(x) = x^2$$

This statement is to be read as f *is the function such that the output* $f(x) = x^2$.

Try this. Using a calculator, enter the number 4. The display is the value of x. Now press the $1/x$ function key and the display changes to 0.25. Pressing the $1/x$ function key is the signal to the calculator to process the number 4. The number 0.25 is the result of that process f – it is the output $f(4)$.

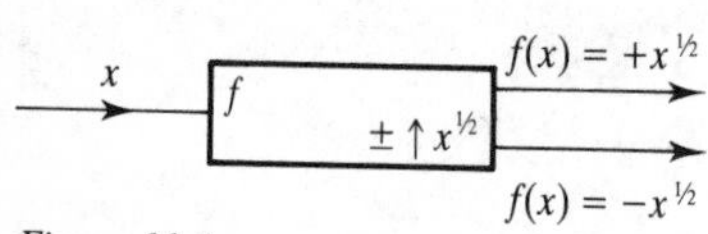

Figure 11.4

For a single value of the input x there is a single value for the output $f(x)$. The process that converts a number into both its positive and negative square roots is not a function (see Figure 11.4). Here:

$$f{:}f(x) = \pm x^{1/2}$$

For a single value of the input x there are two values for the output $f(x)$ – the positive and negative square roots. Accordingly, f in this example is not a function. If, however, the process is restricted to produce only a single value as in the case of a calculator then the process is a function. The calculator processes the input into the non-negative root.

The equation

$$f(x) = \text{some expression in } x$$

which forms the output from the function f is a **prescription** that *defines* the function in the sense that examination of the prescription dictates how the function processes the input. For example,

$$f(x) = x^2$$

is the prescription that defines the function that processes an input by squaring it.

The Composition of Functions

Many functions are composed of simpler functions connected together in such a way that the output from one forms the input for another. For example,

$$f{:}f(x) = (x + 1)^2$$

Here the process of obtaining the output $f(x)$ from the input x can be shown to be separated into two distinct processes g and h – we say f is *composed* of two simpler functions g and h.

To determine what processes g and h perform, think of how the value of $f(x)$ is obtained from the value of x using a calculator.

First put in a value for x – this is the input into g.

Then add 1 – this is the process g.

The output $x + 1$ from g then forms the input of h.

Now press the 'square' button to effect the process h to obtain the output of h as $(x + 1)^2$.

This is identical to the output from f.

Input	*Process*	*Output*
x	g	$g(x) = x + 1$
$x + 1$	h	$h(x + 1) = (x + 1)^2$

This is the same as:

x	f	$f(x) = (x + 1)^2$

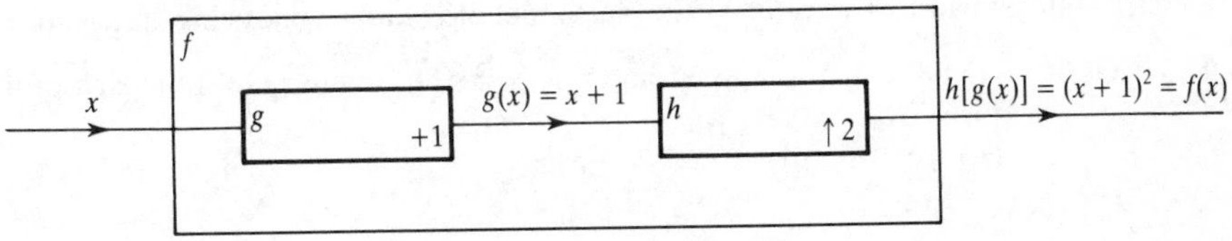

Figure 11.5

The joining of functions in this manner is referred to as **composition** and is represented algebraically as

$f = h \circ g$ read as 'h of g'

so

$$f(x) = h \circ g(x) = h[g(x)]$$

where

$$h: h(x) = x^2 \text{ and } g: g(x) = x + 1$$

Note that the order in which the functions are written in the composition $h \circ g$ is the reverse of the order in which they are activated in Figure 11.5.

EXAMPLES

1 Describe the process that produces each of the following outputs:

(a) $f(x) = x + 6$ **(d)** $f(x) = 4x - 2$
(b) $f(x) = 3x$ **(e)** $f(x) = 2x^2 + 6$
(c) $f(x) = x^2$

(a) The process adds 6 to the input.
(b) The process multiplies the input by 3.
(c) The process squares the input.
(d) The process multiplies the input by 4 and then subtracts 2.
(e) The process squares the input and multiplies it by 2. It then adds 6.

2 Each of the following represents the output from a simple function. If this output were to form the input of a subsequent function that retrieves the original input x, describe the action of the subsequent process:

(a) $f(x) = x + 10$ **(c)** $f(x) = x^{10}$
(b) $f(x) = 10x$

(a) Subtracting 10 retrieves the original input x.
(b) Dividing by 10 retrieves the original input x.
(c) Taking the 10th root retrieves the original input x.

3 Which of the following defines a function?

(a) $f(x) = \dfrac{2}{x^2}$ **(c)** $f(x) = \pm 2x^{1/4}$
(b) $f(x) = x^{1/3}$

(a) This is the output from a function as the process produces a single output for each input.
(b) This is the output from a function as the process produces a single output for each input.
(c) This is not the output from a function as the process produces more than a single output for each input – there are two values of the fourth root of a number, one being the negative of the other.

4 Given that $f(x) = x + 2$, $g(x) = x^2$ and $h(x) = 5x$ find the output from each of the following:

(a) $f \circ g$ **(c)** $f \circ g \circ h$ **(e)** $f \circ f$
(b) $g \circ f$ **(d)** $g \circ h \circ f$

(a) $f \circ g$

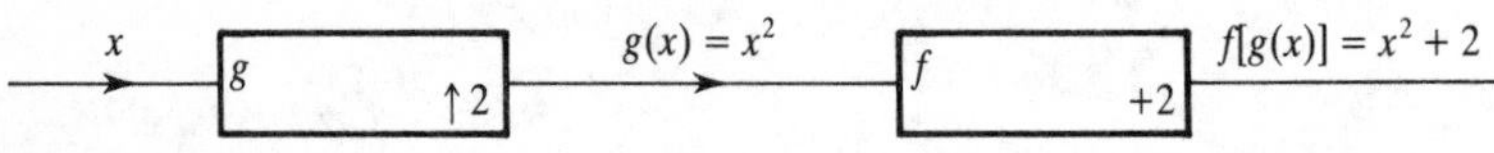

Figure 11.6

(b) $g \circ f$

x → [f +2] → $f(x) = x + 2$ → [g ↑2] → $g[f(x)] = (x + 2)^2$

Figure 11.7

(c) $f \circ g \circ h$

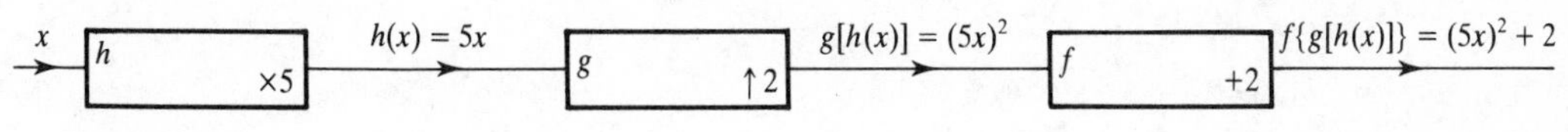

Figure 11.8

(d) $g \circ h \circ f$

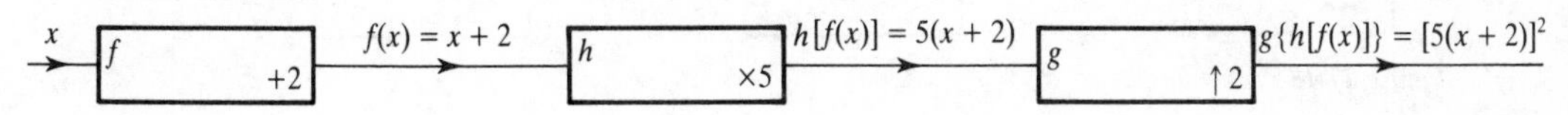

Figure i1.9

(e) $f \circ f$

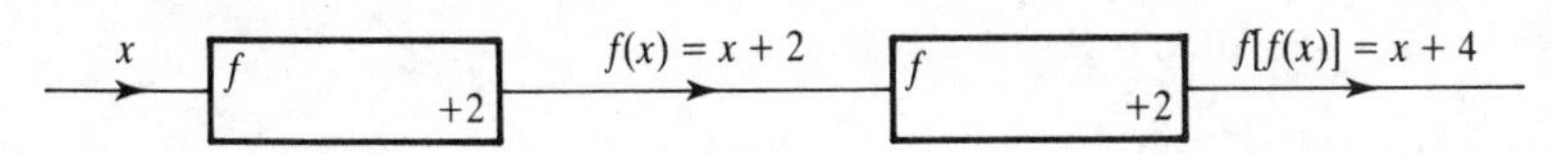

Figure 11.10

5 Each of the following is the output from a composition. Decompose each composition into its component functions where each component only involves one arithmetic operation:

(a) $f(x) = (2x + 1)^2$ **(c)** $f(x) = (x^2 - 2)^3$

(b) $f(x) = \left(\dfrac{x}{3} - 3\right)^{1/4}$

(a) $f(x) = c \circ b \circ a$ where

$$a(x) = 2x$$

$$b(x) = x + 1$$

$$c(x) = x^2$$

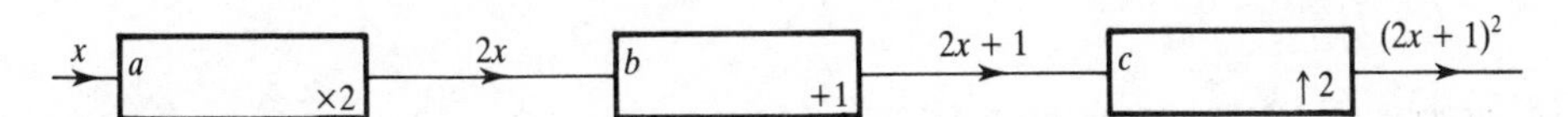

Figure 11.11

(b) $f = c \circ b \circ a$ where

$$a(x) = \frac{x}{3}$$

$$b(x) = x - 3$$

$c(x) = x^{1/4}$ (Notice that c is a function because the notation $x^{1/4}$ means the *positive* root of x only).

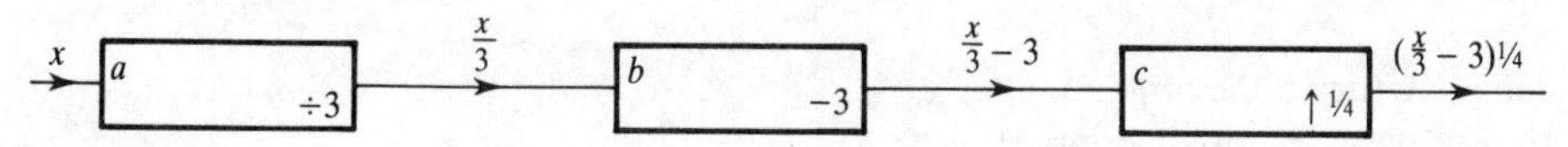

Figure 11.12

(c) $f = c \circ b \circ a$ where

$$a(x) = x^2$$
$$b(x) = x - 2$$
$$c(x) = x^3$$

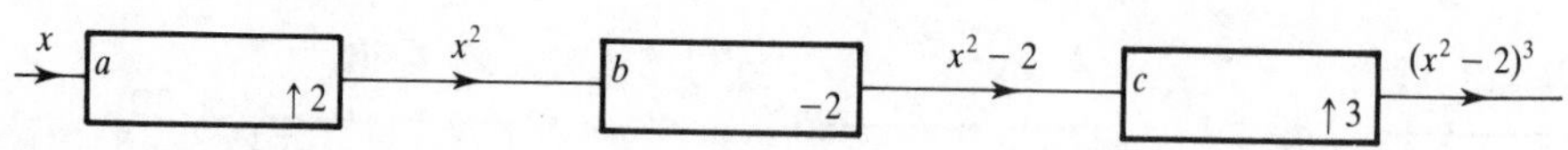

Figure 11.13

6 For which pair of functions is it true that $f \circ g = g \circ f$?

(a) $f(x) = x + 1, \quad g(x) = 2 - x$
(b) $f(x) = 6x, \quad g(x) = x + 3$
(c) $f(x) = \dfrac{x}{2}, \quad g(x) = 9x$

(a) $f[g(x)] = (2 - x) + 1 = 3 - x$
$g[f(x)] = 2 - (x + 1) = 1 - x$ so $f \circ g$ does not equal $g \circ f$
(b) $f[g(x)] = 6(x + 3) = 6x + 18$
$g[f(x)] = (6x) + 3 = 6x + 3$ so $f \circ g$ does not equal $g \circ f$
(c) $f[g(x)] = (9x)/2 = 4.5x$
$g[f(x)] = 9(x/2) = 4.5x$ so $f \circ g$ does equal $g \circ f$

EXERCISES

1 Describe the process that produces each of the following outputs:

(a) $f(x) = x - 2$ **(d)** $f(x) = \dfrac{x}{5} + 6$
(b) $f(x) = \dfrac{x}{4}$ **(e)** $f(x) = \dfrac{7}{x^2}$
(c) $f(x) = x^{1/3}$

(a) The process subtracts * from the *.
(b) The process * the input by *.
(c) The process takes the * of the input.
(d) The process * the input by * and then adds *.
(e) The process * the input and divides this into *.

2 Each of the following represents the output from a simple function. If this output were to form the input of a subsequent function that retrieves the original input x, describe the action of the subsequent process:

(a) $f(x) = x - 2$ **(c)** $f(x) = x^{1/4}$
(b) $f(x) = \dfrac{x}{3}$

(a) Adding * retrieves the original input x.
(b) * by * retrieves the original input x.
(c) * retrieves the original input x.

3 Which of the following are outputs from functions?

(a) $f(x) = 5x^3$ **(c)** $f(x) = \dfrac{1}{x^3}$

(b) $f(x) = \pm 2x^{-1/2}$

(a) This is/is not the output from a function as the process does produce a single output for each input.
(b) This is/is not the output from a function as the process does not produce a single output for each input.
(c) This is/is not the output from a function as the process does/does not produce a single output for each input.

4 Given that $f(x) = x - 6$, $g(x) = x^3$ and $h(x) = x/2$ find the output from each of the following:
(a) $f \circ g$ **(c)** $f \circ g \circ h$ **(e)** $f \circ f$
(b) $g \circ f$ **(d)** $g \circ h \circ f$

(a) $f \circ g$

Figure 11.14

(b) $g \circ f$

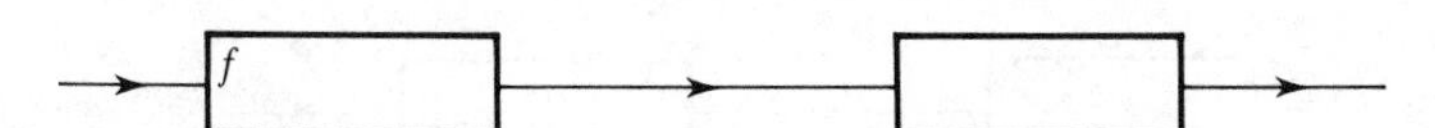

Figure 11.15

(c) $f \circ g \circ h$

Figure 11.16

(d) $g \circ h \circ f$

Figure 11.17

(e) $f \circ f$

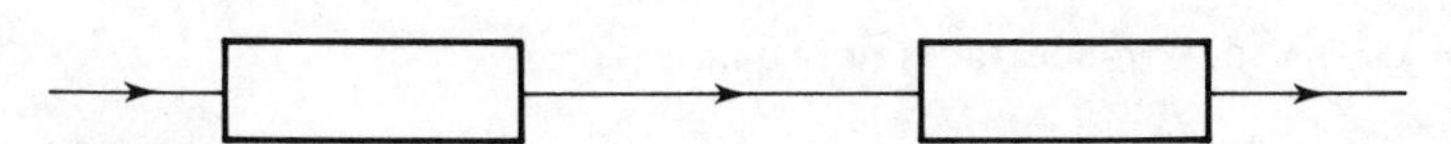

Figure 11.18

5 Each of the following is the output from a composition of functions. Decompose each composition into its component functions where each component only involves one arithmetic operation:

(a) $f(x) = (4x - 1)^3$ **(c)** $f(x) = (x - 2)(x + 2)$

(b) $f(x) = \sqrt{\left(\frac{x}{\pi} - 2\right)}$

(a) $f = c \circ b \circ a$ where

$$a(x) = *$$

$$b(c) = * - *$$

$$c(x) = x^*$$

Figure 11.19

(b) $f = c \circ b \circ a$ where

$$a(x) = \frac{x}{*}$$

$$b(x) = *$$

$$c(x) = x^*$$

Figure 11.20

(c) $f(x) = (x + 2)(x - 2) = * - *$. Consequently, $f = a \circ b$ where

$$a(x) = *$$

$$b(x) = *$$

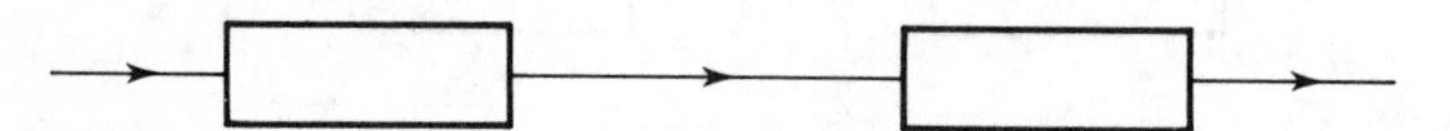

Figure 11.21

6 For which pair of functions is it true that $f \circ g = g \circ f$?

(a) $f(x) = x - 3, \quad g(x) = x + 6$

(b) $f(x) = \frac{3x}{4}, \quad g(x) = x - 1$

(c) $f(x) = 2x, \quad g(x) = x/5$

(a) $f[g(x)] = (*) - 3 = *$
$g[f(x)] = (*) + 6 = *$ so $f \circ g$ does/does not equal $g \circ f$

(b) $f[g(x)] = (*)(x - 1) = *x - *$
$g[f(x)] = (*) - 1 = *x - *$ so $f \circ g$ does/does not equal $g \circ f$

(c) $f[g(x)] = *(*) = *x$
$g[f(x)] = (*)(*) = *$ so $f \circ g$ does/does not equal $g \circ f$

Unit 2 Algebraic Functions

Test yourself with the following:

1 Write down the domain and range of each of the following functions:

(a) $f{:}f(x) = \left|\dfrac{1}{(x-5)^3}\right|$ (c) $f{:}f(x) = 5 \quad 0 < x \leqslant 5$

(b) $f{:}f(x) = 4x^2 \quad -1 \leqslant x < 1$

2 In each of the following form $f + g$, $f - g$, fg and f/g:

(a) $f{:}f(x) = 2x - 1$ domain $-2 \leqslant x \leqslant 10$
$g{:}g(x) = x - x^2$ domain $-1 \leqslant x \leqslant 8$

(b) $f{:}f(x) = 2 + x^2$ domain $0 \leqslant x \leqslant 9$
$g{:}g(x) = 1 + x + x^2$ domain $-4 \leqslant x \leqslant 4$

3 Which of the following are outputs from a one-to-one and which are outputs from a many-to-one function:

(a) $f(x) = 3x + 2$ (c) $f(x) = x^3 - 5|x|$

(b) $f(x) = (4x - 6)^2$ (d) $f(x) = |x^{-1}|$

Algebraic Expressions

We are already familiar with algebraic expressions; they consist of symbols in the form of numbers, variables and constants connected together by the arithmetic operations. We are also familiar with the process of evaluating expressions by substituting symbols with numbers. If the process of evaluating an algebraic expression produces a single number then the process of evaluation is a function. For example, the expression

$$\frac{x^2(x-1)}{x^2 - x + 1}$$

has a single value for every value of x substituted into the expression. Accordingly, we can define the function f where

$$f{:}f(x) = \frac{x^2(x-1)}{x^2 - x + 1}$$

Previously, when we evaluated an expression we assigned its value to another variable y, say. This was a natural thing to do, after all y is the next letter in the alphabet. Now, however, we have a superior notation. In the case of expressions that define functions we shall abandon the notation

$$y = \text{some expression in } x$$

in favour of

$$f(x) = \text{some expression in } x$$

where f denotes the process that evaluates the expression starting from some allowed value for the variable x.

Domain and Range of a Function

A function processes an input number to produce an output number. Sometimes a number can be input into a function that the function cannot process. For example, if a calculator displays the number 0 as an input then pressing the $1/x$ function key produces an error display. The function that produces the reciprocal of a number cannot process the number 0 because we have not defined what is meant by division by zero. Clearly, we need to know which numbers a function can process and which numbers it cannot.

Domain

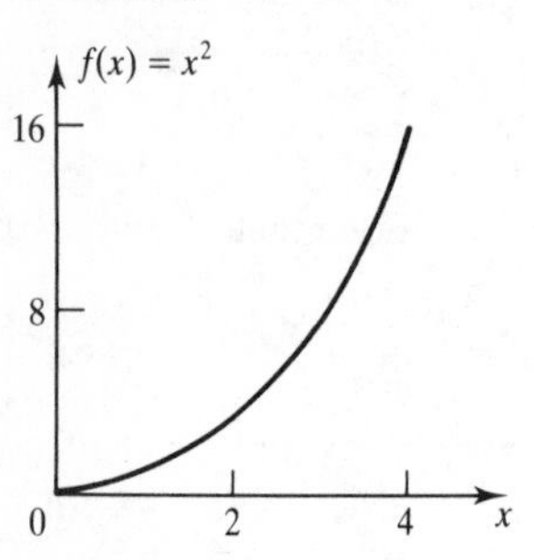

Figure 11.22

All the numbers that a function f can process are collectively called the **domain** of the function. Sometimes we may wish to restrict the domain to a smaller collection of numbers than the totality of the numbers that it can process. This smaller collection of numbers is called a **restricted domain**. For example, the function f with output

$$f(x) = x^2 \qquad 0 \leqslant x \leqslant 4$$

has a domain restricted to all the real numbers within the stated interval despite the fact that f without the restriction could process all the real numbers. The graph of this function is as shown in Figure 11.22.

Range

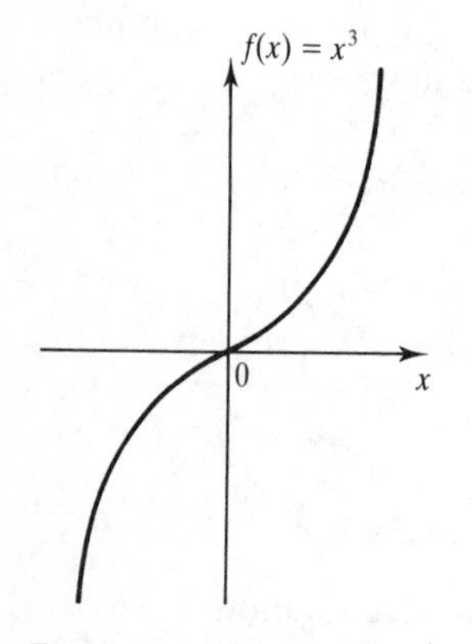

Figure 11.23

The range of a function consists of all those numbers $f(x)$ that correspond to each and every number in the domain. For example, the function f, where

$$f{:}f(x) = x^3$$

has a domain and a range that both consist of all the real numbers. We represent this domain by

$$-\infty < x < \infty$$

where these inequalities simply state that x is positive or negative and finite. The function f, where

$$f{:}f(x) = |x|$$

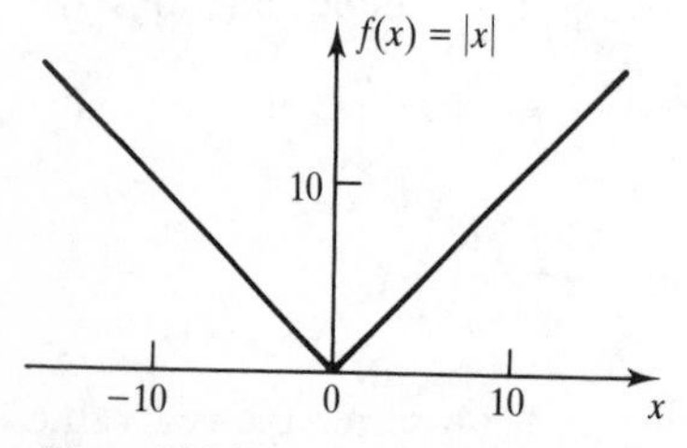

Figure 11.24

has a domain that consists of all the real numbers but a range that consists of only the non-negative real numbers, we write

$$0 \leqslant f(x) < \infty$$

$f(x)$ is non-negative and finite.

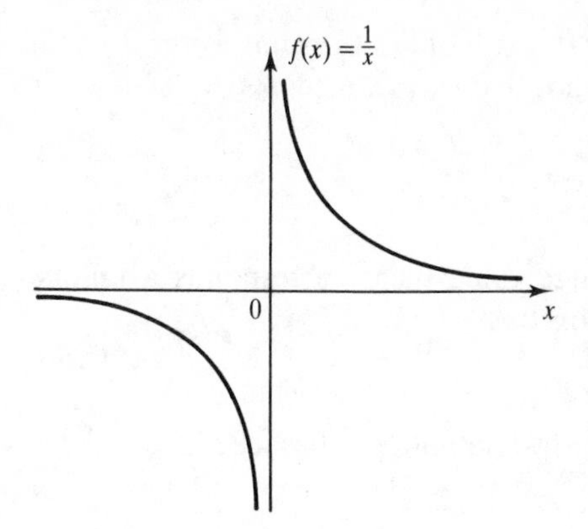

Figure 11.25

The function f, where

$$f{:}f(x) = \frac{1}{x}$$

has a domain and a range that both consist of all the real numbers excluding 0. Here, the domain is written as

$$-\infty < x < 0 \text{ and } 0 < x < \infty$$

Functions and the Arithmetic Operations

Functions can be subjected to the ordinary rules of arithmetic provided care is taken to define their common domain.

Sum and Difference

Two functions f and g can be added or subtracted to form their sum or difference $f \pm g$. For example,

$$f{:}f(x) = x^2 \qquad \text{domain } 0 \leqslant x \leqslant 4$$
$$g{:}g(x) = 3x \qquad \text{domain } -2 \leqslant x \leqslant 6$$

then if $h = f + g$ and $k = f - g$:

$$h{:}h(x) = f(x) + g(x) = x^2 + 3x \qquad \text{and} \qquad k{:}k(x) = f(x) - g(x) = x^2 - 3x$$

However, it should be noted that function f is not defined for certain points in the domain of g. That is, for $-2 \leqslant x < 0$ and for $4 < x \leqslant 6$. As a consequence h is not defined for this part of g's domain either. Indeed, h is only defined on the domain of f which is entirely within the domain of g.

Product and Quotient

A product or quotient can be formed from two functions. For example,

$$f{:}f(x) = 2x + 5 \qquad \text{domain } -5 \leqslant x \leqslant 5$$
$$g{:}g(x) = 3 - x \qquad \text{domain } 0 \leqslant x \leqslant 10$$

then if $h = fg$ and $k = f/g$:

$$h = fg{:}h(x) = f(x)g(x) = (2x + 5)(3 - x)$$

with domain $0 \leqslant x \leqslant 5$. It is only within this restricted domain that f and g are simultaneously defined. Similarly,

$$k = \frac{f}{g}{:}k(x) = \frac{f(x)}{g(x)} = \frac{2x + 5}{3 - x}$$

with domain $0 \leqslant x < 3$ and $3 < x \leqslant 5$. Notice that the domain of k is the same as the domain of h with the exception of the point $x = 3$ where k is not defined because $g(3) = 0$.

One-to-one Functions

A one-to-one function is such that for any given range value there corresponds a unique domain value. We say that if f is a one-to-one function then

$$f(x_1) = f(x_2) \text{ implies that } x_1 = x_2$$

For example, the prescription $f(x) = 2x$ defines a one-to-one function because if

$$f(x_1) = f(x_2) \text{ then } 2x_1 = 2x_2 \text{ so that } x_1 = x_2$$

Many-to-one Functions

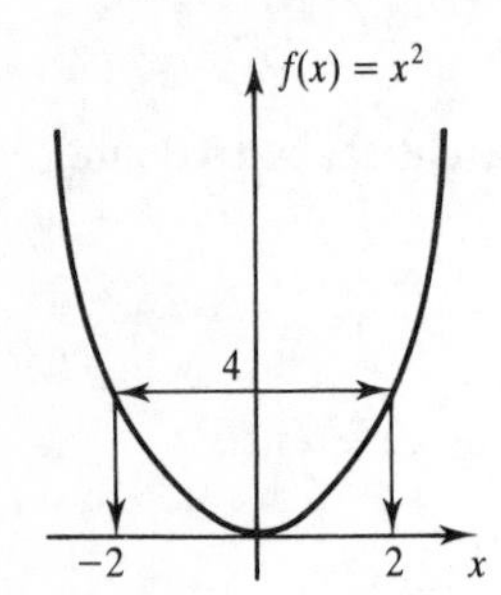

Figure 11.26

A many-to-one function is such that for any given range value there may be more than one corresponding domain value. For example, the prescription $f(x) = x^2$ defines a many-to-one function as can be seen from the graph of Figure 11.26. For example, $f(\pm 2) = 4$.

EXAMPLES

1 Write down the domain and the range of each of the following functions:

(a) $f{:}f(x) = 2x^5 \quad 0 \leqslant x < \infty$

(b) $f{:}f(x) = \dfrac{6x^2}{|x-2|}$

(c) $f{:}f(x) = |\sqrt{x}| \quad 1 < x \leqslant 2$

(a) $f{:}f(x) = 2x^5 \quad 0 \leqslant x < \infty$
Domain: $0 \leqslant x < \infty$. Range: $0 \leqslant f(x) < \infty$

(b) $f{:}f(x) = \dfrac{6x^2}{|x-2|}$
Domain: $-\infty < x < \infty$ but $x \neq 2$. Range: $0 \leqslant f(x) < \infty$

(c) $f{:}f(x) = |\sqrt{x}| \quad 1 < x \leqslant 2$
Domain: $1 < x \leqslant 2$. Range: $1 < f(x) \leqslant +\sqrt{2}$

2 In each of the following form $f + g$, fg and f/g:

(a) $f{:}f(x) = x + 1$ domain $0 \leqslant x \leqslant 25$
$g{:}g(x) = x^2 + x$ domain $0 \leqslant x \leqslant 15$

(b) $f{:}f(x) = 3x^2 - 4$ domain $-2 \leqslant x \leqslant 5$
$g{:}g(x) = \dfrac{1}{x^2} - 3x$ domain $0 < x \leqslant 8$

(a) $f + g : f(x) + g(x) = (x + 1) + (x^2 + x)$

$$= x^2 + 2x + 1$$

$$= (x + 1)^2$$

$$fg : f(x)g(x) = (x + 1)(x^2 + x)$$

$$= x(x + 1)^2$$

Domain $0 \leqslant x \leqslant 15$

$$\frac{f}{g} : \frac{f(x)}{g(x)} = \frac{x + 1}{x^2 + x}$$

$$= \frac{x + 1}{x(x + 1)}$$

$$= \frac{1}{x}$$

Domain $0 < x \leqslant 15$

(b) $f + g : f(x) + g(x) = (3x^2 - 4) + \left(\frac{1}{x^2} - 3x\right)$

$$= 3x^2 - 3x + \frac{1}{x^2} - 4$$

$$fg : f(x)g(x) = (3x^2 - 4)\left(\frac{1}{x^2} - 3x\right)$$

$$= 3x^2\left(\frac{1}{x^2} - 3x\right) - 4\left(\frac{1}{x^2} - 3x\right)$$

$$= 3 - 9x^3 - \frac{4}{x^2} + 12x$$

$$= 3 - \frac{4}{x^2} + 12x - 9x^2$$

Domain $0 < x \leqslant 5$

$$\frac{f}{g} : \frac{f(x)}{g(x)} = \frac{3x^2 - 4}{(1/x^2 - 3x)}$$

$$= \frac{x^2(3x^2 - 4)}{1 - 3x^3}$$

Domain $0 < x < 3^{-1/3}$ and $3^{-1/3} < x \leqslant 5$

3 Which of the following are outputs from a one-to-one and which are outputs from a many-to-one function:

(a) $f(x) = x - 5$ **(c)** $f(x) = x^2 - 5$
(b) $f(x) = (x - 5)^2$ **(d)** $f(x) = |x^3|$

(a) $f(x) = x - 5$ one to one
(b) $f(x) = (x - 5)^2$ many to one
(c) $f(x) = x^2 - 5$ many to one
(d) $f(x) = |x^3|$ many to one

EXERCISES

1 Write down the domain and the range of each of the following functions:

(a) $f{:}f(x) = \dfrac{12(x^4 - 3)}{(x+1)^3}$

(b) $f{:}f(x) = |x^2| \qquad -3 < x < 3$

(c) $f{:}f(x) = -2 \qquad -5 \leqslant x \leqslant -4$

(a) $f{:}f(x) = \dfrac{12(x^4 - 3)}{(x+1)^3}$

Domain: $-* < x < \infty$ provided $x \neq -1$. Range: $-* < f(x) < *$

(b) $f{:}f(x) = |x^2| \qquad -3 < x < 3$

Domain: $-* < x < *$. Range: $0 \leqslant f(x) < *$

(c) $f{:}f(x) = -2 \qquad -5 \leqslant x \leqslant -4$

Domain: $* \leqslant x \leqslant *$. Range: $f(x) = -*$

2 In each of the following form $f + g$, $f - g$, fg and f/g:

(a) $f{:}f(x) = 4x + 2 \qquad$ domain $-3 \leqslant x \leqslant 3$

$g{:}g(x) = x^2 \qquad$ domain $-2 \leqslant x \leqslant 2$

(b) $f{:}f(x) = 1 - 4x^2 \qquad$ domain $-2 \leqslant x \leqslant 12$

$g{:}g(x) = 1/x \qquad$ domain $-6 \leqslant x < 0$ and $0 < x \leqslant 6$

(a) $f + g{:}f(x) + g(x) = (4x + 2) + *$

$$= * + 4x + 2$$

$fg{:}f(x)g(x) = (*)x^2$

$$= *$$

Domain $-2 \leqslant x \leqslant *$

$\dfrac{f}{g}{:}\dfrac{f(x)}{g(x)} = \dfrac{*}{*}$

Domain $* \leqslant x < 0$ and $0 < x \leqslant *$

(b) $f + g{:}f(x) + g(x) = (*) + \dfrac{1}{x}$

$$= * + \frac{1}{x}$$

$fg{:}f(x)g(x) = (*)(*)$

$$= * - 4x$$

Domain $* \leqslant x < *$ and $0 < x \leqslant *$

$\dfrac{f}{g}{:}\dfrac{f(x)}{g(x)} = \dfrac{(*)}{(*)}$

$$= *$$

Domain $* \leqslant x \leqslant *$ provided $x \neq *$

3 Which of the following defines a one-to-one and which a many-to-one function?

(a) $f(x) = 2x + 3$ **(c)** $f(x) = (2x + 3)^2$

(b) $f(x) = 2x^3 + 3$ **(d)** $f(x) = |x^2|$

Module 11: Further exercises

1 Describe the process that produces each of the following outputs:

(a) $f(x) = x - \pi$ (c) $f(x) = \dfrac{1}{x^2}$ (e) $f(x) = 5x^2 - 1$

(b) $f(x) = -7x$ (d) $f(x) = 6 - 3x$

2 Each of the following represents the output from a simple function. If this output were to form the input of a subsequent function that retrieves the original input x, describe the action of the subsequent process:

(a) $f(x) = x - \pi$ (b) $f(x) = -6x$ (c) $f(x) = x^{-7}$

3 Which of the following defines a function?

(a) $f(x) = \dfrac{-3}{x^3}$ (b) $f(x) = x^{1/5}$ (c) $f(x) = \pm 8x^{1/6}$

4 Given that $f(x) = x - 5$, $g(x) = 1/x^2$ and $h(x) = -3x$ find the output from each of the following:

(a) $f \circ g$ (b) $g \circ f$ (c) $f \circ g \circ h$ (d) $g \circ h \circ f$ (e) $f \circ f$

5 Each of the following is the output from a composition. Decompose each composition into its component functions where each component only involves one arithmetic operation:

(a) $f(x) = (-3x - \pi)^2$ (c) $f(x) = 8(x^{-4} + 6)^{-2}$

(b) $f(x) = \left(\dfrac{x^2}{4} + 5\right)^3$

6 For which pair of functions is it true that $f \circ g = g \circ f$?

(a) $f(x) = 8 - 3x$ $g(x) = 2x + 4$

(b) $f(x) = -4x$ $g(x) = 6 - 2x$

(c) $f(x) = \dfrac{5}{x}$ $g(x) = 12x^2$

7 Write down the domain and the range of each of the following functions:

(a) $f{:}f(x) = 3x^{-3}$ $0 < x < 5$

(b) $f{:}f(x) = \dfrac{5}{|x - 2|}$ $x > -2$

(c) $f{:}f(x) = |x|^2$

8 In each of the following form $f + g$, $f - g$, fg and f/g:

(a) $f{:}f(x) = 6 - 2x^2$ domain $-1 \leqslant x \leqslant 1$
$g{:}g(x) = 9 + 3x$ domain $-2 \leqslant x \leqslant 4$

(b) $f{:}f(x) = 6 + 7x^2$ domain $-8 \leqslant x \leqslant 2$
$g{:}g(x) = -8/x$ domain $-10 \leqslant x < 0$ and $0 < x \leqslant 1$

9 Which of the following are outputs from a one-to-one and which are from a many-to-one function:

(a) $f(x) = 2 + 8x^2$ (c) $f(x) = |(x^{-3/4})|$

(b) $f(x) = (9x^{1/2} + 1)^2$ (d) $f(x) = |\sqrt{x}|$

Module 12: Graphs of Functions

Aim: To demonstrate the graphical features of certain functions.

Objectives: When you have read this module you will be able to:

▶▶ Generate a set of ordered pairs from a function.

▶▶ Plot the ordered pairs to obtain the graph of the function.

▶▶ Obtain the graphs of related functions.

▶▶ Derive the even and odd parts of a function where they exist.

Unit 1 Ordered Pairs

Test yourself with the following:

1 Generate a set of four ordered pairs from each of the following:

(a) $f(x) = x - 1$ (c) $f(x) = 2x(x - 1)^2$
(b) $f(x) = x^2 + x + 1$

2 Plot the graphs of each of the following:

(a) $f(x) = 3x + 4$ for $-3 \leqslant x \leqslant +3$ (d) $f(x) = x^{1/3}$ for $-27 \leqslant x \leqslant 27$
(b) $f(x) = 5 - 6x$ for $-5 \leqslant x \leqslant +9$ (e) $f(x) = x^2 - 3x + 2$ for $-2 \leqslant x \leqslant +6$
(c) $f(x) = x^3$ for $-3 \leqslant x \leqslant +3$

Ordered Pairs

Every function f relates an input value x to an output value $f(x)$. By virtue of this the function can be used to generate ordered pairs of numbers in the form

$$(x, f(x))$$

Each ordered pair generated from the function can then be plotted against a pair of cartesian axes to produce the graph of the function. For example, the function f with output

$$f(x) = x^2$$

can be used to generate the collection of ordered pairs of the form

$$(x, x^2)$$

When these are plotted they produce the graph of Figure 12.1. We have already seen that this graph is called a **parabola** and it is obtained by plotting a number of isolated points and joining the points together by the smooth curve shown.

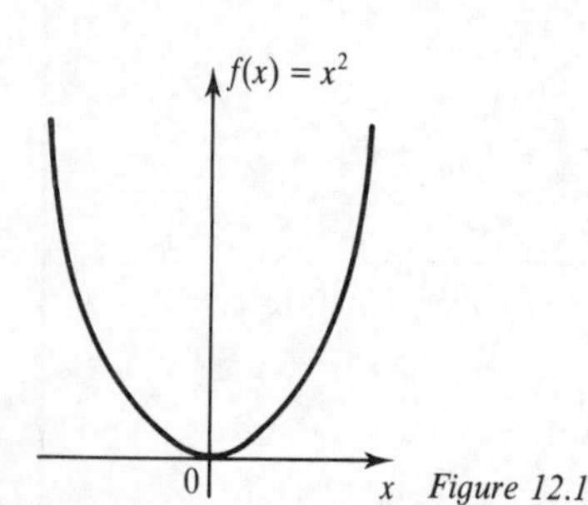

Figure 12.1

Graphs of Functions

One-to-one Functions

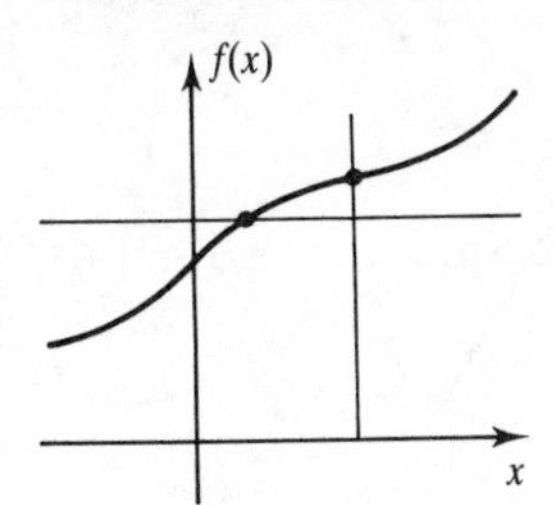

Figure 12.2

The graph of a one-to-one function is such that any vertical line through a domain value and any horizontal line through a range value will only intersect with the graph once at most. For example, see Figure 12.2.

Many-to-one Functions

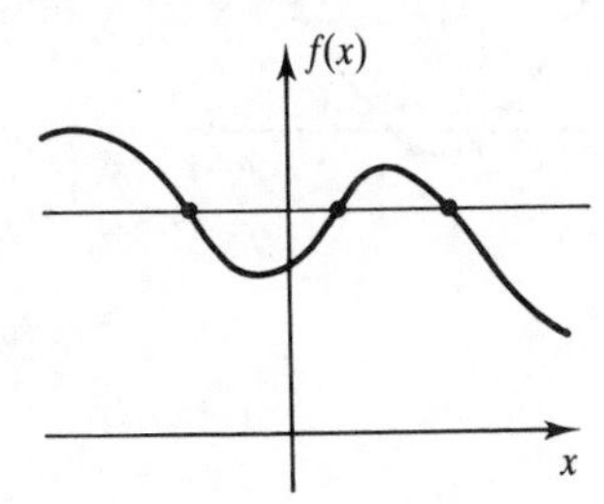

Figure 12.3

The graph of a many-to-one function is such that any vertical line through a domain value will only intersect with the graph once at most but a horizontal line through a range value may cut the graph more than once. For example, see Figure 12.3.

EXAMPLES

1 Generate a set of four ordered pairs from each of the following:

(a) $f(x) = 3x + 2$ **(c)** $f(x) = \dfrac{x+1}{x-1}$

(b) $f(x) = x^2 + 2x + 3$

(a) Four typical ordered pairs are: $(-2, -4)$, $(0, 2)$, $(3, 11)$ and $(-2/3, 0)$.
(b) Four typical ordered pairs are: $(-1, 2)$, $(0, 3)$, $(2, 11)$ and $(-1/2, 9/4)$.
(c) Four typical ordered pairs are: $(-1, 0)$, $(2, 3)$, $(0, -1)$ and $(-1/5, -2/3)$. Note that $x = 1$ is not allowed.

2 Plot the graphs of each of the following:
(a) $f(x) = x + 1$ for $-6 \leqslant x \leqslant 6$
(b) $f(x) = 2 - 3x$ for $-2 \leqslant x \leqslant 2$
(c) $f(x) = x^2$ for $-4 \leqslant x \leqslant 4$
(d) $f(x) = x^{1/2}$ for $0 \leqslant x \leqslant 16$
(e) $f(x) = x^2 - 5x + 6$ for $0 \leqslant x \leqslant 5$

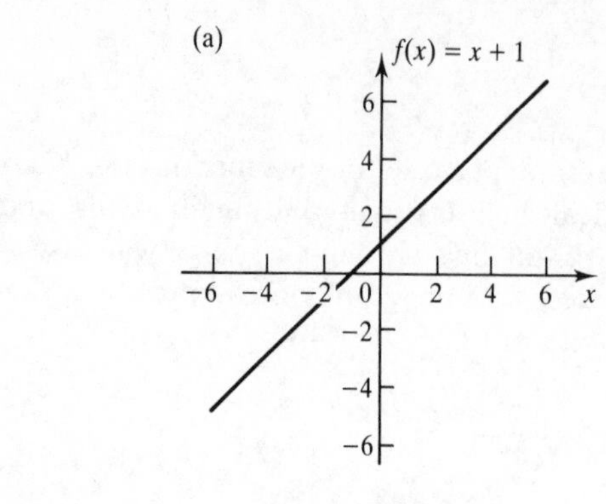

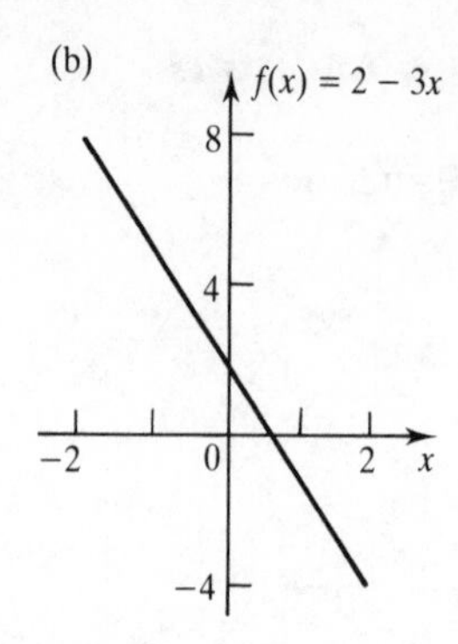

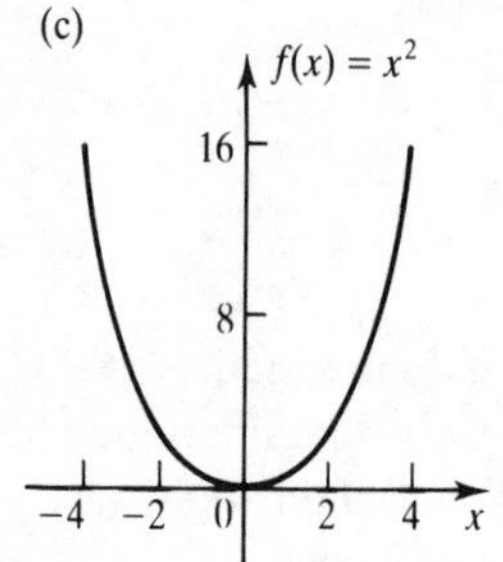

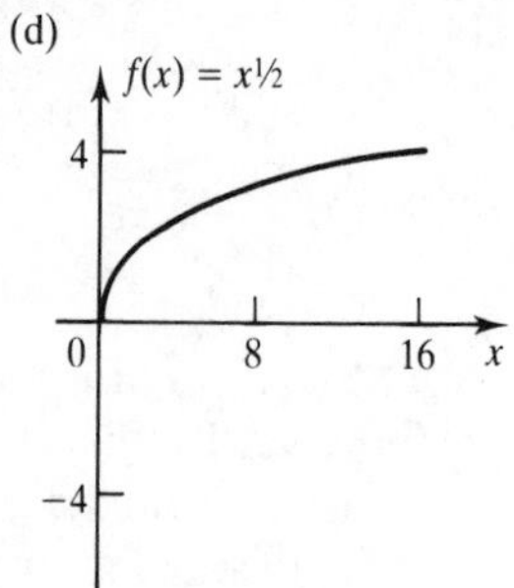

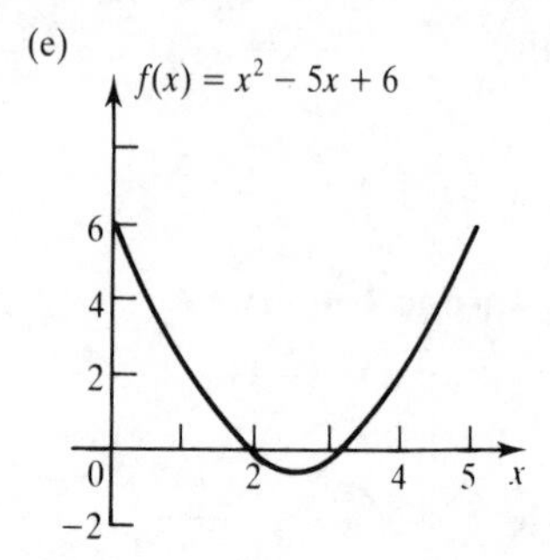

Figure 12.4

EXERCISES

1 Generate a set of four ordered pairs from each of the following:

(a) $f(x) = 5x - 6$ **(c)** $f(x) = \dfrac{x}{2x - 1}$

(b) $f(x) = x^3 - 8$

(a) Four typical ordered pairs are: $(-1, *)$, $(0, *)$, $(2, *)$ and $(*, *)$.
(b) Four typical ordered pairs are: $(-2, *)$, $(0, *)$, $(2, *)$ and $(*, *)$.
(c) Four typical ordered pairs are: $(-1, *)$, $(0, *)$, $(1, *)$ and $(*, *)$. Note that $x = *$ is not allowed.

2 Plot the graphs of each of the following:

(a) $f(x) = 2x - 1$ for $-8 \leqslant x \leqslant 8$
(b) $f(x) = 1 - 2x$ for $-5 \leqslant x \leqslant 6$
(c) $f(x) = -x^2$ for $-4 \leqslant x \leqslant 4$
(d) $f(x) = -x^{1/2}$ for $0 \leqslant x \leqslant 16$
(e) $f(x) = x^2 - 1$ for $-4 \leqslant x \leqslant 4$

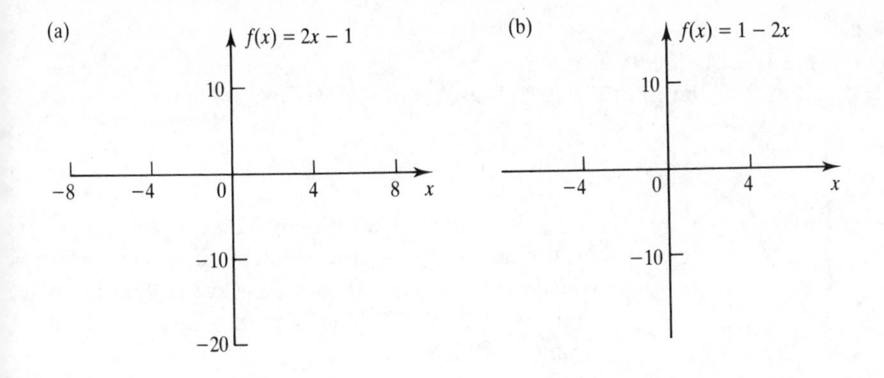

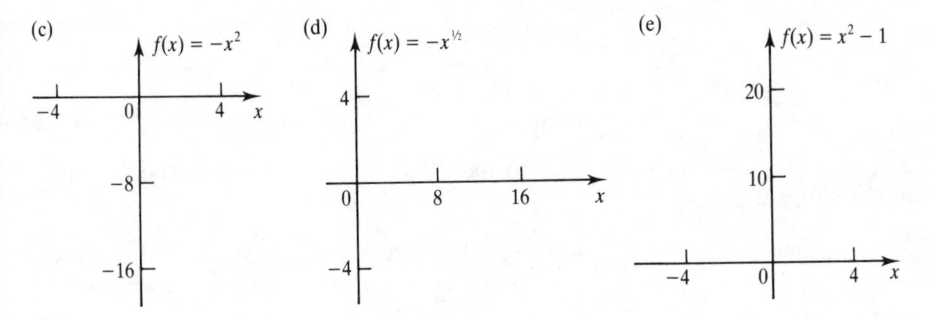

Figure 12.5

Unit 2 Related Graphs

Test yourself with the following:

1 Plot the graph of $f(x) = 2x - 3$ and deduce the graph of $f(x + 6)$.

2 Plot the graph of $f(x) = x^2 - x$ and deduce the graph of $f(-4x)$.

3 Plot the graph of $f(x) = x^2 - 3x + 2$ and deduce the graph of $f(x) - 3$.

4 Transform the graph of $f(x) = (x + 2)^3$ into a graph that passes through the origin.

Related Functions

For any function with output $f(x)$ there are a number of related functions whose graphs are similar to the original.

Graph of $f(x + a)$

The graph of $g(x) = f(x + a)$ is identical in shape to the graph of $f(x)$ but it is shifted a units to the left if $a > 0$ and to the right if $a < 0$. For example, if $f(x) = x^2$, then

$$g(x) = f(x + 3) = (x + 3)^2$$

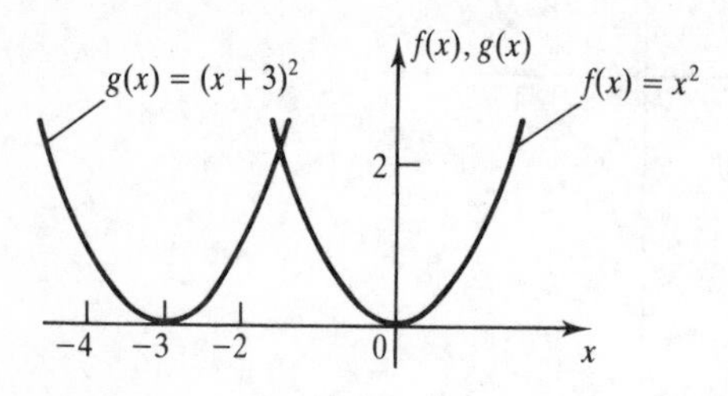

Figure 12.6

We can see from Figure 12.6 that the shapes of the two graphs are identical but $f(x) = x^2$ touches the x-axis at $x = 0$ whereas the other touches at $x = -3$. The second graph is shifted 3 units to the left.

Graph of $f(ax)$

The graph of $g(x) = f(ax)$ is similar in shape to the graph of $f(x)$. The difference in shape depends upon the value of a:

- $a > 1$ Here, the graph is stretched vertically. For example, if $f(x) = x^2$, the graph of

$$g(x) = f(2x) = (2x)^2 = 4x^2$$

is a vertically stretched version of $f(x) = x^2$ as can be seen in Figure 12.7 where it is a narrower parabola.

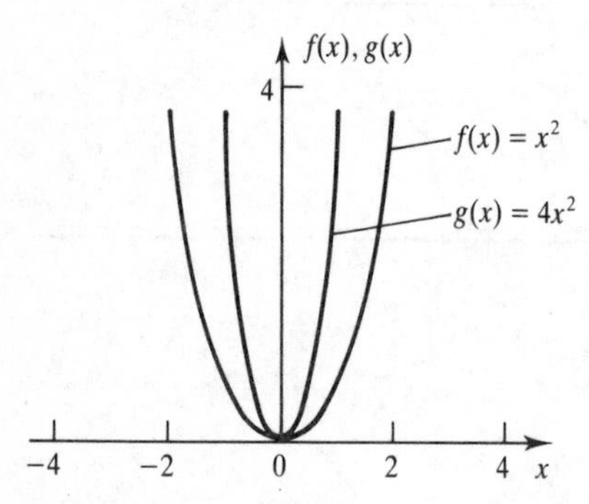

Figure 12.7

- $0 < a < 1$ Here, the graph is stretched horizontally. For example, if $f(x) = x^2$, the graph of

$$g(x) = f\left(\frac{x}{2}\right) = \left(\frac{x}{2}\right)^2 = \frac{x^2}{4}$$

is a horizontally stretched version of $f(x) = x^2$ as can be seen in Figure 12.8 where it is a wider parabola.

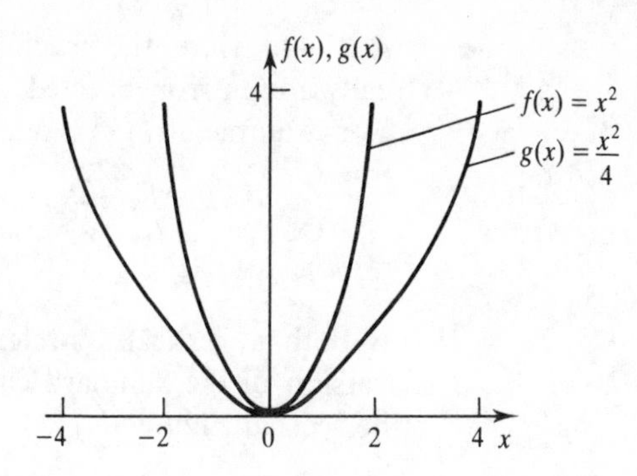

Figure 12.8

- $-1 < a < 0$ Here, the graph is stretched horizontally and reversed as though reflected in a mirror that lies along the vertical axis. For example, if $f(x) = x^3$, the graph of

$$\begin{aligned} g(x) &= f\left(\frac{-x}{2}\right) \\ &= \left(\frac{-x}{2}\right)^3 \\ &= \frac{-x^3}{8} \end{aligned}$$

which is both a horizontally stretched and reflected version of the graph of the standard cubic as can be seen in Figure 12.9.

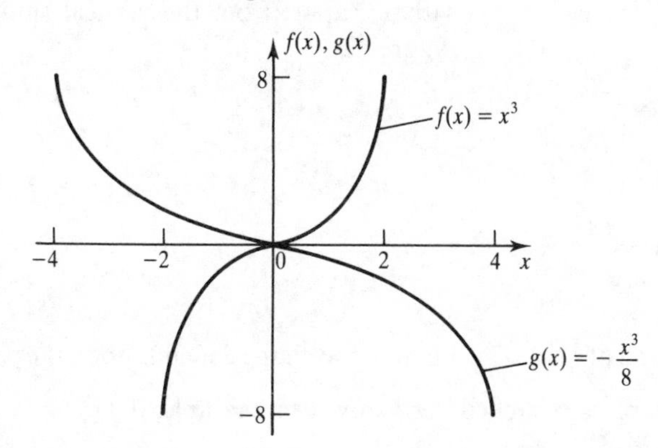

Figure 12.9

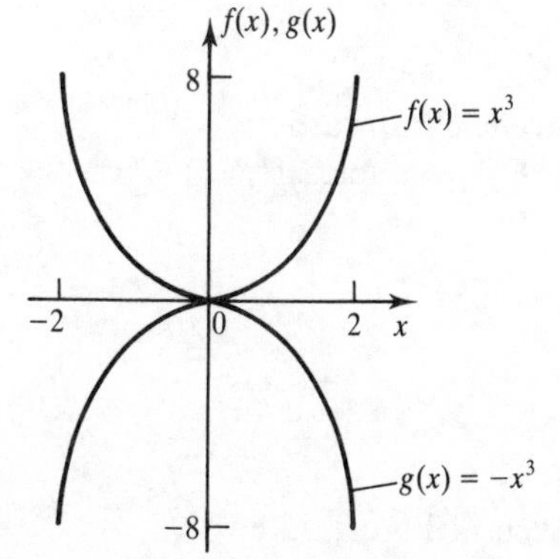

Figure 12.10

- $a = -1$ Here, the graph is simply mirror reflected in the vertical axis. For example, if $f(x) = x^3$, the graph of

$$\begin{aligned} g(x) &= f(-x) \\ &= (-x^3) \\ &= -x^3 \end{aligned}$$

is a mirror-reflected version of the standard cubic as can be seen in Figure 12.10.

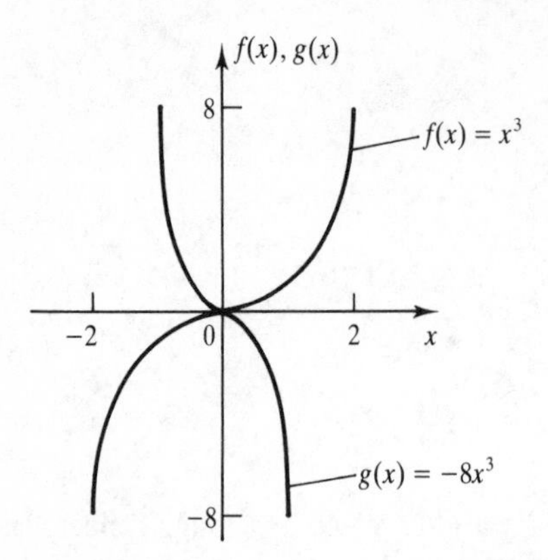

Figure 12.11

- $a < -1$ Here, the graph is stretched vertically and mirror reflected in the vertical axis. For example, if $f(x) = x^3$, the graph of

$$\begin{aligned} g(x) &= f(-2x) \\ &= (-2x)^3 \\ &= -8x^3 \end{aligned}$$

is both a vertically stretched and reflected version of the standard cubic again, as can be seen in Figure 12.11.

Graph of $f(x) + a$

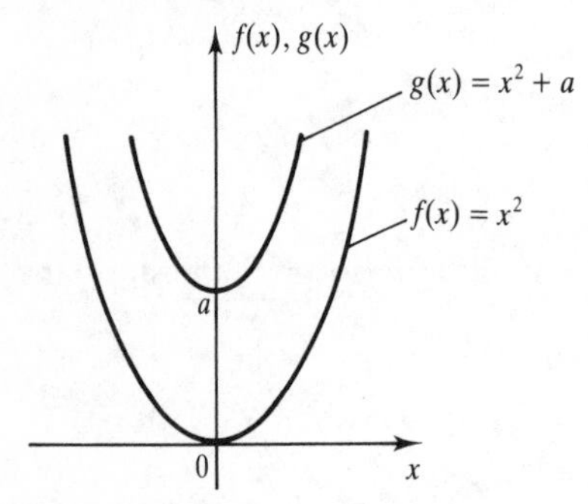

Figure 12.12

The graph of

$$g(x) = f(x) + a$$

is identical in shape to the graph of $f(x)$ except that it is shifted upwards if $a > 0$ or downwards if $a < 0$. For example, the graph associated with $x^2 + a$ is identical in shape to the graph associated with x^2, apart from the vertical shift as shown in Figure 12.12.

Graph of $af(x)$

The graph of

$$g(x) = af(x)$$

is similar in shape to the graph of $f(x)$. The difference in shape depends upon the value of a:

- $a > 1$ Here, the graph is stretched vertically. For example, if $f(x) = x^2$ then

$$\begin{aligned} g(x) &= 4f(x) \\ &= 4x^2 \end{aligned}$$

as can be seen in Figure 12.7 where the parabola is stretched upwards.

- $0 < a < 1$ Here, the graph is stretched horizontally. For example, if $f(x) = x^2$, the graph of

$$\begin{aligned} g(x) &= \frac{1}{4}f(x) \\ &= \frac{x^2}{4} \end{aligned}$$

is a flattened version of the graph of $f(x)$ as can be seen in Figure 12.8.

- $-1 < a < 0$ Here, the graph is stretched horizontally and reversed as though reflected in a mirror that lies along the horizontal axis. For example, if $f(x) = x^3$, the graph of

$$g(x) = \frac{-1}{8} f(x) = \frac{-x^3}{8}$$

is both a horizontally stretched and reflected version of the graph of the standard cubic as can be seen in Figure 12.9.

- $a = -1$ Here, the graph is simply mirror reflected in the horizontal axis. For example, if $f(x) = x^3$, the graph of

$$g(x) = -f(x) = -x^3$$

is a mirror-reflected version of the standard cubic as can be seen in Figure 12.10.

- $a < -1$ Here, the graph is stretched vertically and mirror reflected in the vertical axis. For example, if $f(x) = x^3$ the graph of

$$g(x) = -8f(x) = -8x^3$$

is both a vertically stretched and reflected version of the standard cubic again, as can be seen in Figure 12.11.

EXAMPLES

1 Plot the graph of $f(x) = x + 1$ and hence the graph of $g(x) = f(x + 2)$.

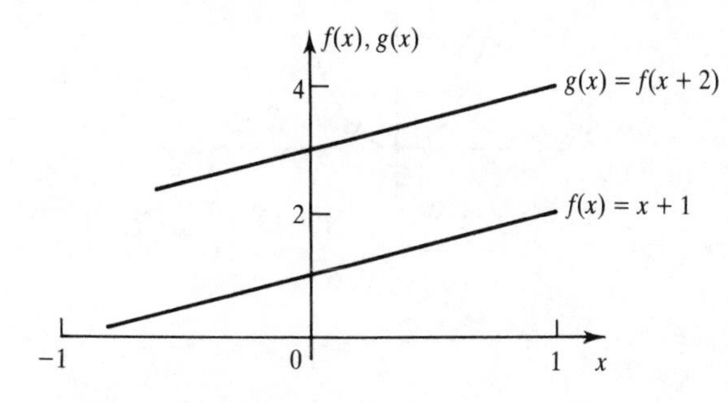

Figure 12.13

2 Plot the graph of $f(x) = x^2$ and hence the graph of $g(x) = f(3x)$.

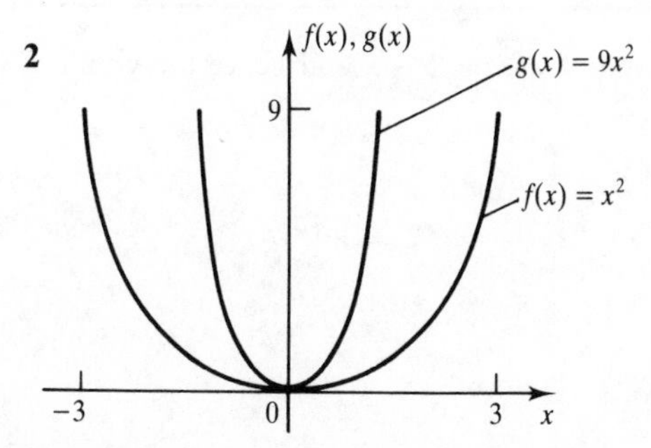

Figure 12.14

3 Plot the graph of $f(x) = x^2 - 5x + 6$ and hence the graph of $g(x) = f(x) - 6$.

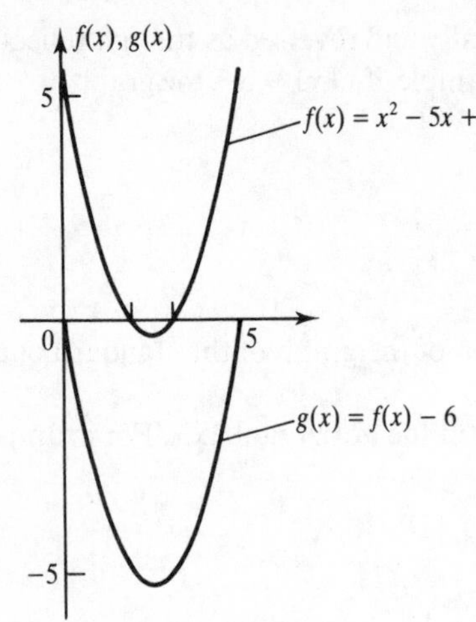

Figure 12.15

4 Transform the graph of $f(x) = x^2 - 2x + 1$ into a graph that is symmetric about the vertical axis.

$f(x) = (x - 1)^2$ so that $g(x) = f(x + 1)$ is the required function.

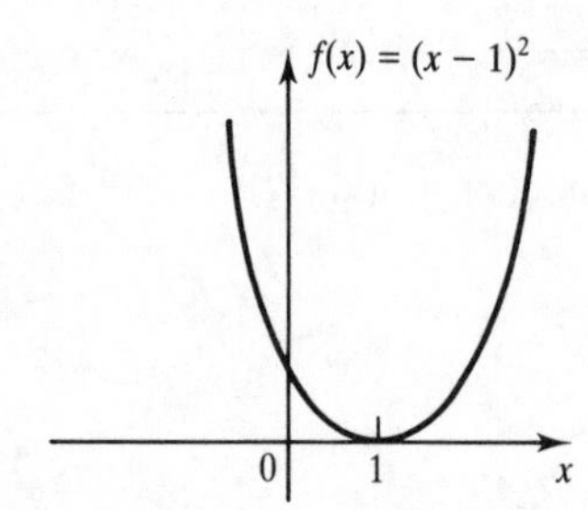

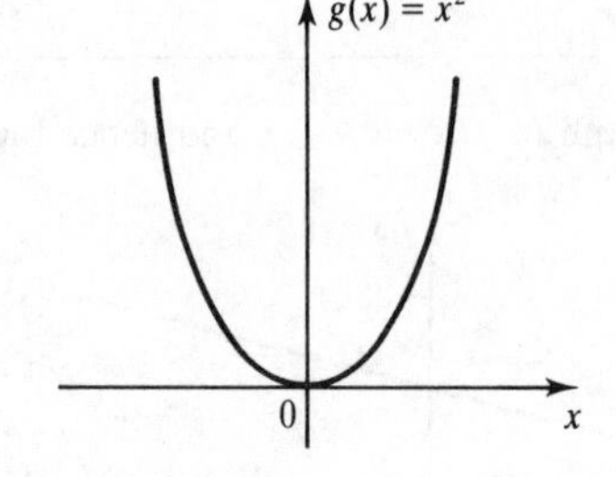

Figure 12.16

EXERCISES

1 Plot the graph of $f(x) = 1 - x$ and hence the graph of $g(x) = f(x - 4)$.

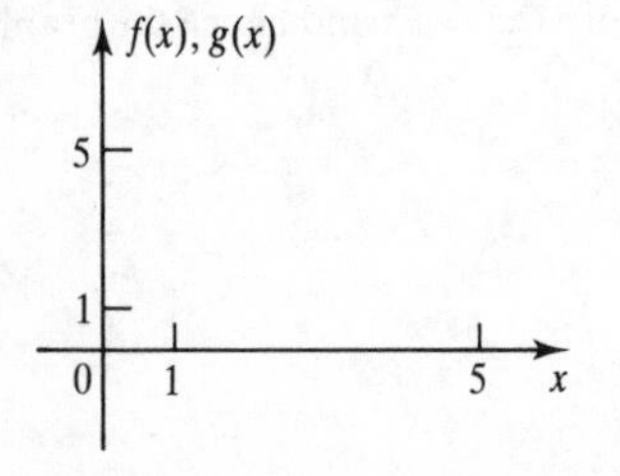

Figure 12.17

2

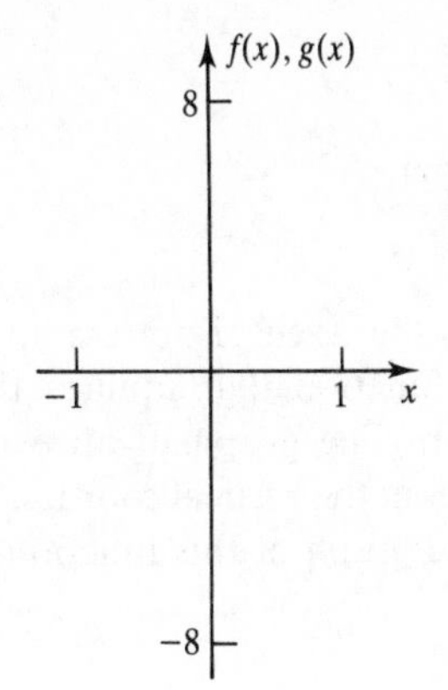

Figure 12.18

Plot the graph of $f(x) = -x^3$ and hence the graph of $g(x) = f(-2x)$.

3

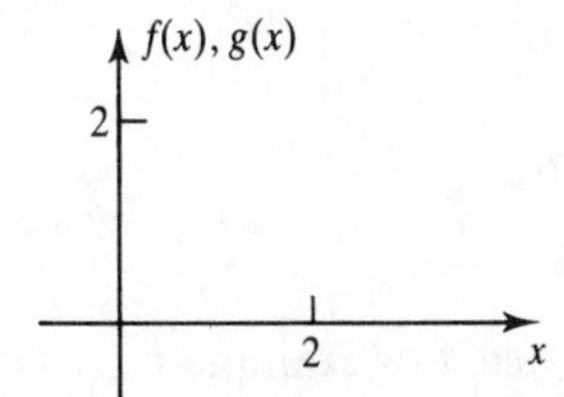

Figure 12.19

Plot the graph of $f(x) = (x - 2)^2$ and hence the graph of $g(x) = f(x) + 2$.

4

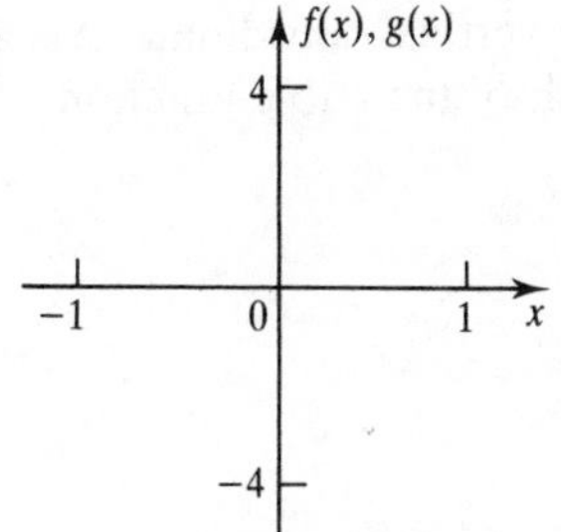

Figure 12.20

Transform the graph of $f(x) = 3x - 4$ into a graph that is parallel but intercepts the vertical axis at $x = 4$.

Unit 3 Even and Odd Functions

Test yourself with the following:

1 Which of the following functions are even, which are odd and which are neither?

(a) $f(x) = 3x$ (c) $f(x) = 2 - x$

(b) $f(x) = 4x^2$ (d) $f(x) = x^{-1/4}$

2 Generate an even function from each of the following:

(a) $f(x) = 3x + 4$ (b) $f(x) = x^3 + x^4$

3 Generate an odd function from each of the following:

(a) $f(x) = 5 - 3x$ (b) $f(x) = \dfrac{x + 1}{x}$

Even Function

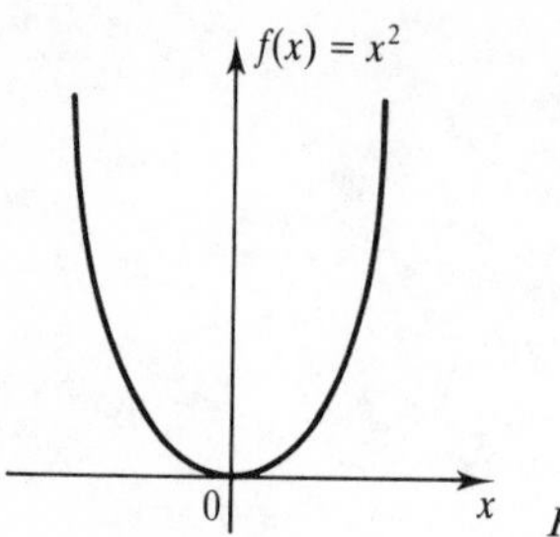

Figure 12.21

If a function f is such that

$$f(-x) = f(x)$$

the function is said to be **even**. For example, $(-x)^2 = x^2$ and so the function that squares the input is an even function. The graph of an even function is symmetrical about the vertical coordinate axis as is displayed in the graph of this function.

Odd Function

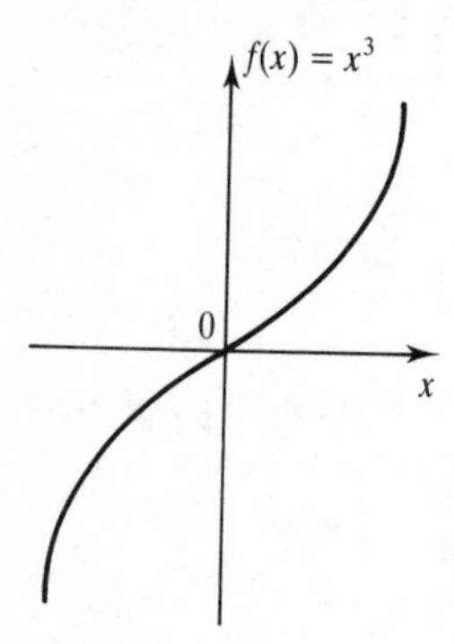

Figure 12.22

If a function f is such that

$$f(-x) = -f(x)$$

the function is said to be **odd**. For example, $(-x)^3 = -(x)^3$ and so the function that cubes the input is an odd function. The graph of an odd function is asymmetrical about the vertical coordinate axis as is displayed in the graph of the cubic function.

Even and Odd Parts of a Function

Not all functions are either odd or even. However, it may be possible to create an even and an odd function from one that is neither – we call them the even and odd parts of the function respectively. This is done as follows. Given the function f then if both $f(x)$ and $f(-x)$ are defined:

$$f_e(x) = \frac{f(x) + f(-x)}{2}$$ defines an even function called the even part of $f(x)$

$$f_o(x) = \frac{f(x) - f(-x)}{2}$$ defines an odd function called the odd part of $f(x)$

Note that

$$f(x) = f_e(x) + f_o(x)$$

For example, the function defined by

$$f(x) = (x - 2)^2$$

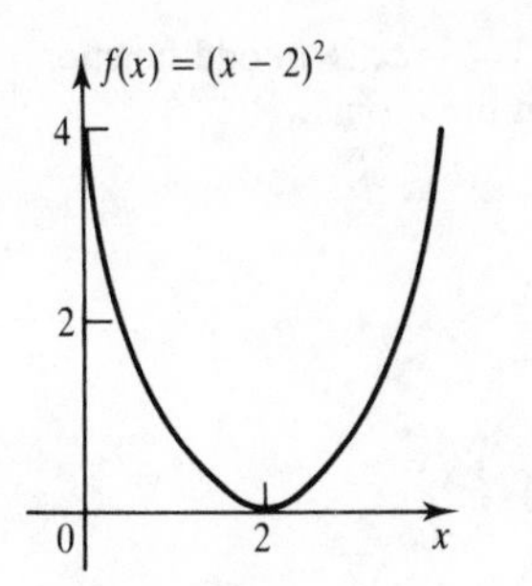

Figure 12.23

has the graph shown in Figure 12.23. As can be seen it is neither odd nor even. However, defining

$$
\begin{aligned}
f_e(x) &= \frac{f(x) + f(-x)}{2} \\
&= \frac{(x-2)^2 + (-x-2)^2}{2} \\
&= \frac{x^2 - 4x + 4 + x^2 + 4x + 4}{2} \\
&= x^2 + 4
\end{aligned}
$$

and

$$
\begin{aligned}
f_o(x) &= \frac{f(x) - f(-x)}{2} \\
&= \frac{(x-2)^2 - (-x-2)^2}{2} \\
&= \frac{x^2 - 4x + 4 - x^2 - 4x - 4}{2} \\
&= -4x
\end{aligned}
$$

So that $f_e(x) = x^2 + 4$ and $f_o(x) = -4x$ are the even and odd parts of $f(x)$ respectively.

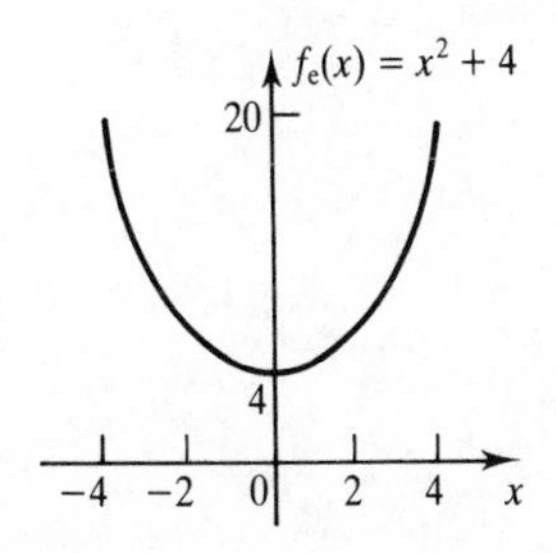

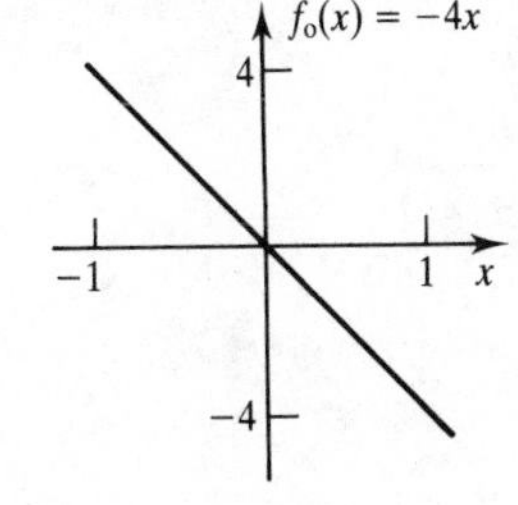

Figure 12.24

EXAMPLES

1 Which of the following functions are even, which are odd and which are neither?

(a) $f(x) = -2x$ **(c)** $f(x) = 5x^2 + 6$
(b) $f(x) = -8x^4$ **(d)** $f(x) = x^3$

(a) $f(-x) = -2(-x) = 2x = -f(x)$ so that the function defined by $f(x) = -2x$ is an odd function.
(b) $f(-x) = -8(-x)^4 = -8x^4 = f(x)$ so that this function is an even function.
(c) $f(-x) = 5(-x)^2 + 6 = 5x^2 + 6 = f(x)$ so this function is even.
(d) $f(-x) = (-x)^3 = -f(x)$ so this cubic is an odd function.

2 Generate an even function from each of the following:

(a) $f(x) = 2x + 1$ **(b)** $f(x) = x^3 - x^2$

(a) $f(-x) = -2x + 1$ Hence

$$\begin{aligned} f_e(x) &= \frac{f(x) + f(-x)}{2} \\ &= \frac{(2x+1) + (-2x+1)}{2} \\ &= 1 \end{aligned}$$

Consequently $f_e(x) = 1$ is the even function generated from $f(x)$.

(b) $f(-x) = -x^3 - x^2$ Hence

$$\begin{aligned} f_e(x) &= \frac{f(x) + f(-x)}{2} \\ &= \frac{(x^3 - x^2) + (-x^3 - x^2)}{2} \\ &= -x^2 \end{aligned}$$

3 Generate an odd function from each of the following:

(a) $f(x) = 1 - 2x$
(b) $f(x) = x - x^2$

(a) $f(-x) = 1 + 2x$ Hence

$$\begin{aligned} f_o(x) &= \frac{f(x) - f(-x)}{2} \\ &= \frac{(1-2x) - (1+2x)}{2} \\ &= -2x \end{aligned}$$

(b) $f(-x) = -x - x^2$ Hence

$$\begin{aligned} f_o(x) &= \frac{f(x) - f(-x)}{2} \\ &= \frac{(x - x^2) - (-x - x^2)}{2} \\ &= x \end{aligned}$$

EXERCISES

1 Which of the following functions are even, which are odd and which are neither?

(a) $f(x) = x/4$ **(c)** $f(x) = x^{1/3}$

(b) $f(x) = -x^3 + x$ **(d)** $f(x) = x^4$

(a) $f(-x) = (-x)/4 = *(x/4) = *f(x)$ so that the function defined by $f(x) = x/4$ is *.

(b) $f(-x) = -(-x)^3 + (-x) = *$ so that the function defined by $f(x)$ is *.

(c) $* = (-x)^{1/3} = *$ so that the function defined by $f(x)$ is *.

(d) $* = (-x)^4 = *(x^4) = *$ so that the function defined by $f(x) = x^4$ is an * function.

2 Generate an even function from each of the following:

(a) $f(x) = 1 - 2x$ **(b)** $f(x) = x - x^2$

(a) $f(-x) = *$, so that

$$\frac{f(x) + f(-x)}{2} = \frac{(*) + (*)}{2}$$

$$= *$$

$$= f_e(x)$$

(b) $f(-x) = -x - x^2$, so that

$$\frac{f(x) * f(-x)}{2} = \frac{(*) * (*)}{2}$$

$$= *$$

$$= f_e(x)$$

3 Generate an odd function from each of the following:

(a) $f(x) = 2x + 1$ **(b)** $f(x) = x^3 - x^2$

(a) $f(-x) = *$, so that

$$\frac{f(x) - f(-x)}{2} = \frac{(*) - (*)}{2}$$

$$= *$$

$$= f_o(x)$$

(b) $f(-x) = -x^3 - x^2$, so that

$$\frac{f(x) * f(-x)}{2} = \frac{(*) * (*)}{2}$$

$$= *$$

$$= f_o(x)$$

Module 12: Further exercises

1 Generate a set of four ordered pairs from each of the following:

(a) $f(x) = 7 - 5x$ (c) $f(x) = \sqrt{\dfrac{(x^2 - 1)}{(x^2 + 1)}}$

(b) $f(x) = 5 - 4x - 3x^2$

2 Plot the graphs of each of the following:

(a) $f(x) = 8 - x/4$ for $-8 \leqslant x \leqslant 4$ (d) $f(x) = |x^{3/2}|$ for $0 \leqslant x \leqslant 8$

(b) $f(x) = 9x + 4$ for $0 \leqslant x \leqslant 3$ (e) $f(x) = x^2 - x + 1$ for $-2 \leqslant x \leqslant 2$

(c) $f(x) = 5x^{-2}$ for $-1 \leqslant x \leqslant 1$

3 Plot the graph of $f(x) = 3x - 2$ and hence the graph of $g(x) = f(x - 4)$.

4 Plot the graph of $f(x) = 2x^4$ and hence the graph of $g(x) = f(-2x)$.

5 Plot the graph of $f(x) = x^2 - 6x + 8$ and hence the graph of $g(x) = f(x) + 2$.

6 Transform the graph of $f(x) = x^2 - 6x + 8$ into a graph that is symmetric about the vertical axis.

7 Transform the graph of $f(x) = 8 - 9x$ into a graph that is parallel but intercepts the vertical axis at $x = -2$.

8 Which of the following functions are even, which are odd and which are neither?

(a) $f(x) = \dfrac{-3x^2}{4}$ (c) $f(x) = x(x + 1)$

(b) $f(x) = x^3 + 1$ (d) $f(x) = x^{1/5}$

9 Generate an even function from each of the following:

(a) $f(x) = 7x - 2$ (b) $f(x) = x + \dfrac{1}{x}$

10 Generate an odd function from each of the following:

(a) $f(x) = 3 + 2x$ (b) $f(x) = \dfrac{x}{x^2 - 1}$

Module 13: Trigonometric Functions and Equations

Aims: To extend the definitions of the trigonometric ratios to the trigonometric functions.

Objectives: When you have read this module you will be able to:

▶▶ Define the trigonometric functions as related to angles of rotation.

▶▶ Derive the period, amplitude and phase of trigonometric functions.

▶▶ Solve simple trigonometric equations.

Unit 1 The Trigonometric Functions

Test yourself with the following:

1 Use a calculator to find the value of each of the following:
(a) $\sin 95°$ (b) $\cos 162°$ (c) $\tan 5\pi/3$ (d) $\sin(-\pi/4)$

2 In each of the following show that the quantity in square brackets is the period of the function:
(a) $f(x) = \sin 8x$ $[\pi/4]$
(b) $f(x) = \cos(9x + 5)$ $[2\pi/9]$
(c) $f(x) = 8\tan\left(\dfrac{5x+2}{6}\right)$ $[6\pi/5]$

3 Find the amplitude of each of the following:
(a) $f(x) = 6\sin 4x$
(b) $f(4 - 2x)$ where $f(x) = 3\cos 6x$
(c) $f(x) = 2\sin(2x - 4)\cos(2x - 4)$

4 Find the phase of each of the following:
(a) $f(x) = \sin(2 - x)$
(b) $f(x) = \cos(x + 7)$
(c) $f(x) = \cos^2 6x - \sin^2 6x$

The Trigonometric Ratios

The trigonometric ratios have been defined for a right-angled triangle and as such are only valid for angles $0 < \beta < \pi/2$. We can, however, extend the definitions of these quantities to

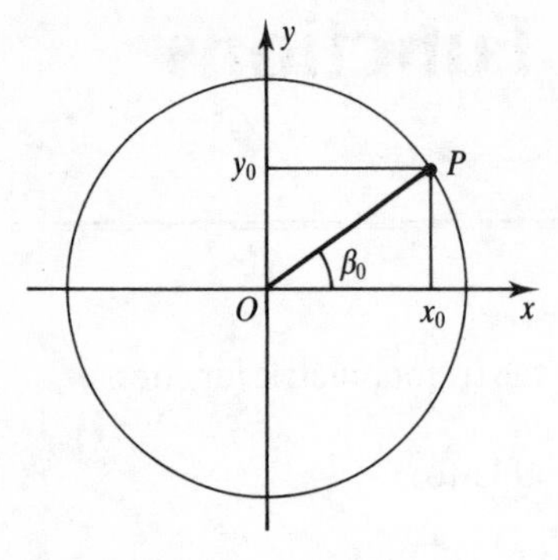

Figure 13.1

relate to any angle and thereby define the extended trigonometric functions.

Consider the circle centred at the origin of a cartesian coordinate system and having unit radius. If the radius is rotated through β_o radians to OP as shown in Figure 13.1, then the coordinates of P, (x_o, y_o), can be given as

$$x_o = \cos(\beta_o)$$
$$y_o = \sin(\beta_o)$$

These equations relate the angle rotated through by the radius to the coordinates of P.

The Sine Function

The equation

$$y = \sin(\beta)$$

relates the height of P above the x-axis to the angle β rotated through when β lies within the restricted range of values 0° to 90°. Unlike the restriction on the size of β when confined to being an angle in a right-angled triangle, there is no restriction on the value of β as an angle of rotation. Accordingly, we can create ordered pairs:

$$(\beta, y)$$

for any value of β, positive or negative – negative β means that the rotation is clockwise. When we plot these ordered pairs we obtain the graph of Figure 13.2.

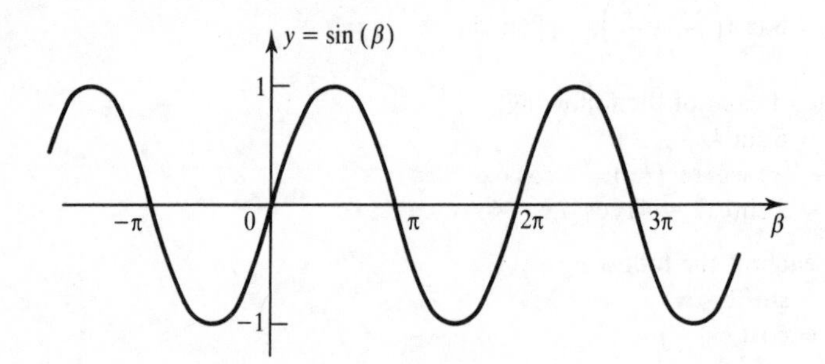

Figure 13.2

If the value of β lies within the range 0° to 90°, the value of y for a given value of β is the same as the value of $\sin(\beta)$. For example, if $\beta = 30°$ then $y = \frac{1}{2}$ and $\sin 30° = \frac{1}{2}$. If, however, β lies outside the range of values 0° to 90°, y has a value but $\sin \beta$ does not because $\sin \beta$ has not been defined outside the restricted range. To overcome this we define the values of $\sin \beta$ for β outside the restricted range to be equal to the corresponding values of y. In this way we extend the definition of the sine ratio. This extended sine ratio defines the **sine function** and, accordingly, the graph is called the graph of the sine function.

$f{:}f(\beta) = \sin(\beta)$ where $-\infty < \beta < \infty$

From the graph of Figure 13.2 it can be seen that

$$\sin(-\beta) = -\sin(\beta)$$

so that the sine function is odd. Also

$$\sin(\beta) = \sin(\pi - \beta)$$

and

$$\sin(\beta + 2\pi) = \sin(\beta) \text{ for } -\infty < \beta < \infty$$

Note the brackets in the term $\sin(\beta)$. The purpose of the brackets is to separate the name of the function, in this case, sin, from the input variable β. However, the inclusion of these brackets can often make simple trigonometric expressions look unwieldy. For this reason the brackets will be omitted from the trigonometric functions where it is appropriate to do so for the sake of simplicity.

The Cosine Function

The equation

$$x = \cos(\beta)$$

relates the horizontal distance of P from the y-axis to the angle β rotated through for angles within the range 0° to 90°. Again, unlike the restriction on the size of β within a triangle, there is no such restriction on the value of β as an angle of rotation. Accordingly, we can create ordered pairs:

$$(\beta, x)$$

for any value of β, positive or negative. When we plot these ordered pairs we obtain the graph of Figure 13.3.

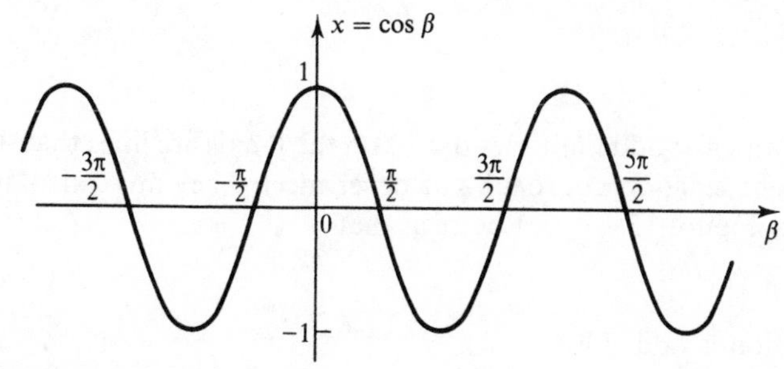

Figure 13.3

The values of x for any value of β outside the restricted range can then be used to give meaning to $\cos \beta$. Accordingly, the graph is called the graph of the extended cosine function f where

$f{:}f(\beta) = \cos(\beta)$ where $-\infty < \beta < \infty$

From the graph of Figure 13.3 it can be seen that

$$\cos(-\beta) = \cos(\beta)$$

so that the cosine function is even. Also

$$\cos(\beta) = -\cos(\pi - \beta)$$

and

$$\cos(\beta + 2\pi) = \cos(\beta) \text{ for } -\infty < \beta < \infty$$

We can also see a connection between the sine and cosine functions:

$$\sin(\pi/2 + \beta) = \cos(\beta) \text{ and } \cos(\pi/2 + \beta) = -\sin(\beta)$$

The Tangent Function

By defining

$$\tan\beta = \frac{\sin\beta}{\cos\beta} \text{ for } \cos\beta \neq 0$$

we are able to define and draw the graph of the extended tangent function.

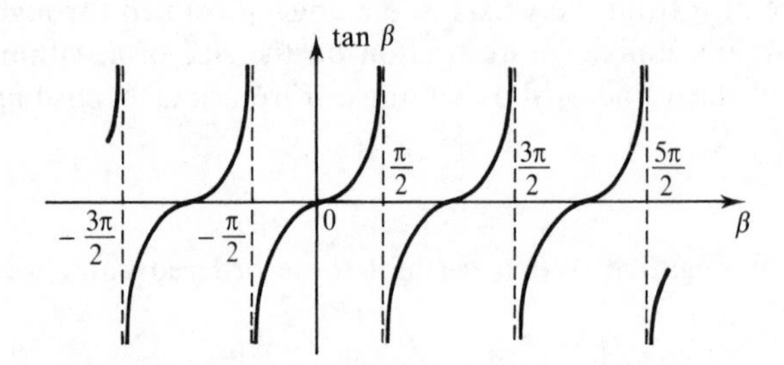

Figure 13.4

The vertical lines drawn at odd multiples of $\pi/2$ on the β-axis are lines that the branches of the graph of the tangent function approach but never meet. They are called **asymptotes**.

From the graph of Figure 13.4 it can be seen that

$$\tan(-\beta) = -\tan\beta$$

so that the tangent function is odd. Also

$$\tan\beta = -\tan(\pi - \beta)$$

and

$$\tan(\beta + \pi) = \tan\beta \text{ for } -\infty < \beta < \infty$$

The extended reciprocal trigonometric functions can be similarly defined.

It should also be noted that despite the fact that the original trigonometric ratios were defined from within the confines of a right-angled triangle, all the properties of the original ratios still apply to the extended definitions. To illustrate this, consider Pythagoras' theorem:

$$\sin^2 \beta + \cos^2 \beta \equiv 1$$

This was originally shown to be true for angles β where $0 < \beta < \pi/2$.

To show that it is true for angles β where $\pi/2 < \beta < \pi$ we shall define angle β to be

$$\beta = \pi/2 + \Phi \text{ where } 0 < \Phi < \pi/2$$

We then find that

$$\begin{aligned} \sin^2 \beta + \cos^2 \beta &= \sin^2(\pi/2 + \Phi) + \cos^2(\pi/2 + \Phi) \\ &= \cos^2 \Phi + (-\sin \Phi)^2 \\ &= \cos^2 \Phi + \sin^2 \Phi \\ &= 1 \end{aligned}$$

because $0 < \Phi < \pi/2$. We can repeat this exercise to demonstrate the validity of Pythagoras' theorem for any positive angle. The theorem is also true for negative angles, formed by a clockwise rotation. Let $\beta = -\Phi$ where $0 < \Phi < \pi/2$. Then

$$\begin{aligned} \sin^2 \beta + \cos^2 \beta &= \sin^2(-\Phi) + \cos^2(-\Phi) \\ &= (-\sin \Phi)^2 + \cos^2 \Phi \\ &= \sin^2 \Phi + \cos^2 \Phi \\ &= 1 \end{aligned}$$

because $0 < \Phi < \pi/2$. Again, we could repeat this exercise to demonstrate the validity of Pythagoras' theorem for any negative angle.

Period

If a function f is such that

$$f(x + A) = f(x)$$

then the values of $f(x)$ for $0 \leqslant x < A$ are repeated for every A units of x. For example, the sine function is such that

$$\begin{aligned} f(x + 2\pi) &= \sin(x + 2\pi) \\ &= \sin x \cos 2\pi + \sin 2\pi \cos x \\ &= \sin x \\ &= f(x) \end{aligned}$$

Here the range values of the sine function for $0 \leqslant x \leqslant 2\pi$ are repeated for every 2π radians increase in x. Graphically, this is seen as the repetitive pattern displayed in Figure 13.2.

Any function that repeats itself in this manner is said to be a **periodic function** with the domain interval of repetition being called the **period**. Clearly, both the sine and cosine functions are periodic with period 2π as can be seen from Figures 13.2 and 13.3.

The tangent function is again a periodic function but this time the period is π radians. Also, the tangent function is not wavelike in character; rather it is said to be branchlike with the principal branch lying between $\pm\pi/2$ as can be seen from Figure 13.4.

Amplitude

The **amplitude** of a periodic function is defined as the difference between the average value and the maximum value of the output taken over a period. In the case of the sine and cosine functions the outputs over a single period for increasing input range from a minimum of -1 to a maximum of $+1$ with an average value of 0. Consequently, the amplitude of both of these functions is 1. As an illustration, the function

$$f(x) = 5\sin(3x - 4)$$

has a maximum value of 5 because $\sin(3x - 4)$ has a maximum value of 1 and a minimum value of -1 giving an average value of 0 over a single period. Thus the amplitude of $f(x)$ is 5. It is meaningless to talk of amplitude in relation to the tangent function as each branch rises from minus infinity to plus infinity.

Phase

The related graphs

$$f(\beta) = \sin\beta$$

and

$$\begin{aligned} g(\beta) &= f(\beta - a) \\ &= \sin(\beta - a) \end{aligned}$$

are identical in shape but the graph of $g(\beta)$ is shifted a units to the right of the graph of $f(\beta)$. This shift is called the **phase** and we speak of the two functions being identical but out of phase with each other.

Similarly, for $p(\beta) = \cos\beta$

$$\begin{aligned} q(\beta) &= p(\beta - b) \\ &= \cos(\beta - b) \end{aligned}$$

is identical to

$$p(\beta) = \cos\beta$$

but b radians out of phase.

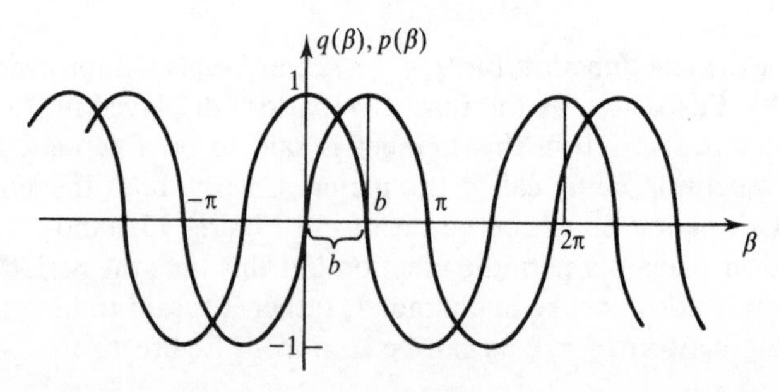

Figure 13.5

Also the phase of $f(x) = \sin(-\beta)$ is π as can be seen from Figure 13.6.

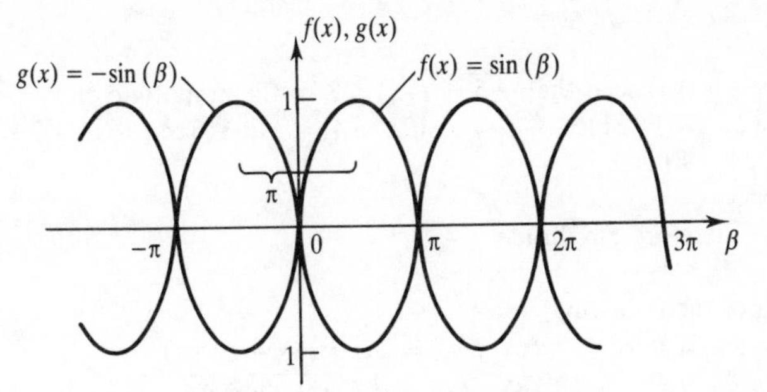

Figure 13.6

EXAMPLES

1 Use a calculator to find the value of each of the following:

(a) $\cos 380°$ **(b)** $\tan(-72°)$ **(c)** $\sin(8\pi/5)$ **(d)** $\cos(-3\pi/4)$

(a) $0.9396\ldots$ This is the same as $\cos 20°$.
(b) $-3.0776\ldots$ This is the same as $-\tan 72°$.
(c) $-0.9510\ldots$ This is the same as $-\sin(2\pi/5)$.
(d) $-0.7071\ldots$ This is the same as $\cos(3\pi/4) = -\cos(\pi/4)$.

2 In each of the following show that the quantity in square brackets is the period of the function:

(a) $f(x) = \sin 2x \quad [\pi]$
(b) $f(x) = \cos(x/2 + 1) \quad [4\pi]$
(c) $f(x) = 2\tan\left(\dfrac{3x - 4}{2}\right) \quad [2\pi/3]$

(a)
$$\begin{aligned} f(x+\pi) &= \sin 2(x+\pi) \\ &= \sin(2x + 2\pi) \\ &= \sin 2x \\ &= f(x) \quad \text{therefore period} = \pi \end{aligned}$$

(b)
$$\begin{aligned} f(x+4\pi) &= \cos\left(\frac{x+4\pi}{2} + 1\right) \\ &= \cos\left(\frac{x}{2} + 1 + 2\pi\right) \\ &= \cos\left(\frac{x}{2} + 1\right) \\ &= f(x) \quad \text{therefore period} = 4\pi \end{aligned}$$

(c)
$$\begin{aligned} f\left(x + \frac{2\pi}{3}\right) &= 2\tan\left[\frac{3(x + 2\pi/3) - 4}{2}\right] \\ &= 2\tan\left(\frac{3x + 2\pi - 4}{2}\right) \\ &= 2\tan\left(\frac{3x - 4}{2} + \pi\right) \\ &= 2\tan\left(\frac{3x - 4}{2}\right) \\ &= f(x) \quad \text{therefore period} = 2\pi/3 \end{aligned}$$

3 Find the amplitude of each of the following:

(a) $f(x) = 3 \sin 3x$ **(c)** $f(x) = 8 \sin 3x \cos 3x$
(b) $f(2x - 4)$ where $f(x) = 2 \cos x$

(a) Since $-1 \leqslant \sin 3x \leqslant 1$ it is seen that $-3 \leqslant f(x) \leqslant 3$ so the amplitude of $f(x)$ is 3.
(b) $f(2x - 4) = 2 \cos(2x - 4)$ Since $-1 \leqslant \cos(2x - 4) \leqslant 1$ it is seen that $-2 \leqslant f(2x - 4) \leqslant 2$ so the amplitude of $f(2x - 4)$ is 2.
(c) $f(x) = 8 \sin 3x \cos 3x$
$= 4 \sin 6x$ therefore amplitude $= 4$

4 Find the phase of each of the following:

(a) $f(x) = \sin(x + 5)$ **(c)** $f(x) = 2 \cos^2(3x - 5) - 1$
(b) $f(x) = \cos(2x - 6)$ **(d)** $f(x) = 10\sqrt{1 + \tan^2(9x - 1)}$

(a) $\sin(x + 5)$ is 5 radians behind $\sin x$, consequently the phase of $f(x)$ is -5 radians.
(b) Since $\cos(2x - 6) = \cos 2(x - 3)$ the phase of $f(x)$ is 3 radians.
(c) $2 \cos^2(3x - 5) - 1 = \cos 2(3x - 5)$
$= \cos 6(x - 5/3)$

Accordingly, the phase is 5/3 radians.
(d) $10\sqrt{1 + \tan^2(9x - 1)} = 10\sqrt{\sec^2(9x - 1)}$
$= \pm 10 \sec(9x - 1)$
$= \pm 10 \sec 9(x - \frac{1}{9})$

Consequently the phase of $f(x)$ is $\frac{1}{9}$ radians.

EXERCISES

1 Use a calculator to find the value of each of the following:

(a) $\tan 273°$ **(b)** $\sin 108°$ **(c)** $\cos(9\pi/2)$ **(d)** $\tan(-\pi/6)$

(a) * this is the same as tan *°.
(b) * this is the same as sin *.
(c) * this is the same as cos(*).
(d) * this is the same as −tan(*).

2 In each of the following show that the quantity in square brackets is the period of the function:

(a) $f(x) = \cos 3x$ $[2\pi/3]$
(b) $f(x) = \sin(4x - 3)$ $[\pi/2]$
(c) $f(x) = 5 \tan\left(\frac{2x - 6}{3}\right)$ $[3\pi/2]$
(d) $f(x) = 1 - 2 \sin^2(8x + 5)$ $[\pi/8]$

(a) $f(x + 2\pi/3) = \cos 3(x + *\pi)$
$= \cos(3x + *)$
$= \cos 3x$
$= f(x)$ therefore the period is $*\pi$
(b) $f(x + \pi/2) = \sin[4(*) - 3]$
$= \sin(4x - 3 + *)$
$= \sin(*)$
$= f(x)$ therefore the period is *

(c) $f(*) = 5 \tan\left[\dfrac{2(*) - 6}{3}\right]$

$= 5 \tan\left(\dfrac{2x - 6 + *}{3}\right)$

$= 5 \tan\left(\dfrac{2x - 6}{3} + *\right)$

$= 5 \tan(*)$

$= f(*)$ therefore the period is $*$

(d) $f(x + *) = 1 - 2\sin^2[8(x + *) + 5]$

$= \cos 2[8(x + *) + 5]$

$= \cos(* + 2\pi)$

$= \cos(*)$

$= \cos 2(*)$

$= 1 - 2\sin^2(*)$

$= f(x)$ therefore the period is $*$

3 Find the amplitude of each of the following:

(a) $f(x) = 4 \cos 5x$

(b) $f(5x + 1)$ where $f(x) = 5 \sin 2x$

(c) $f(x) = 4 \sin 4x \cos 4x$

(a) Since $-1 \leqslant * \leqslant 1$ it is seen that $-* \leqslant f(x) \leqslant *$ so the amplitude of $f(x)$ is $*$.

(b) $f(5x + 1) = 5 \sin 2(*)$ so the amplitude is $*$.

(c) $f(x) = 4 \sin 4x \cos 4x$

$= * \sin 8x$

The amplitude is equal to $*$.

4 Find the phase of each of the following:

(a) $f(x) = \cos(x - 3)$

(b) $f(x) = \sin(4x + 2)$

(c) $f(x) = 1 - 2\sin^2(4x + 3)$

(a) $\cos(x - 3)$ is $*$ radians ahead of $\cos x$ so the phase is $*$ radians.

(b) $\sin(4x + 2) = \sin *(x + *)$ so the phase is $*$ radians.

(c) $f(x) = 1 - 2\sin^2(4x + 3)$

$= \cos 2(*)$

$= \cos 8(*)$ so the phase is $*$ radians

Unit 2 Trigonometric Equations

Test yourself with the following:

1 Solve the following linear trigonometric equations:

(a) $\cos(x + \pi/3) + \cos(x - \pi/4) = 0$

(b) $\sin(x + \pi/4) - \sin(x - \pi/6) = 0$

(c) $\tan A - \tan B = 1.2$

$3 \tan A + 4 \tan B = 6.4$

2 If $\tan(A/2) = \sqrt{3}$ find the value of $\cos A$.

3 Solve the following equations:

(a) $\sin^2 x - 4 \sin x \cos x + 3 \cos^2 x = 0$

(b) $5 \sec x + \tan^2 x + 7 = 0$

Trigonometric Equations

Trigonometric equations are equations that involve the trigonometric functions. Having extended the definitions of the trigonometric ratios we find that we can now solve general trigonometric equations. We shall consider linear and quadratic equations only.

Linear Trigonometric Equations

There is no hard-and-fast rule for solving linear trigonometric equations, each equation is solved on its merits. The only common thread in their solution is the extensive use of the trigonometric identities. For example, to solve the equation

$$\sin(A + \pi/4) + \sin(A - \pi/4) = 1 \qquad \text{between 0 and } \pi/2$$

proceed as follows: expand the left-hand side to give

$$[\sin A \cos(\pi/4) + \cos A \sin(\pi/4)] + [\sin A \cos(\pi/4) - \cos A \sin(\pi/4)] = 1$$

that is

$$2 \sin A \cos(\pi/4) = 1$$

so that

$$2 \sin A\left(\frac{1}{\sqrt{2}}\right) = 1$$

that is

$$\sqrt{2} \sin A = 1 \text{ so that } \sin A = \frac{1}{\sqrt{2}}$$

If we use a calculator we find that $A = 45°$ or $0.7853\ldots = \pi/4$ radians and this is the principal solution. However, it is not the only solution. If we study the graph of $\sin x$ we see that there are an infinity of values for A that satisfy the equation

$$\sin A = \frac{1}{\sqrt{2}}$$

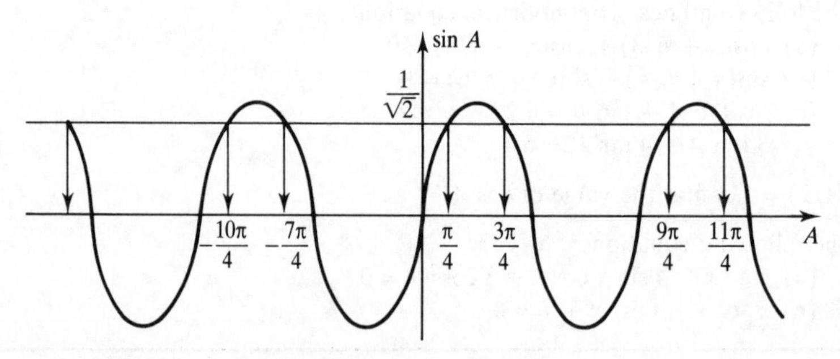

Figure 13.7

For example, $3\pi/4$ is another solution. Indeed, since

$$\sin(A + 2n\pi) = \sin A \text{ for any integer } n$$

the complete solution is

$$A = \pi/4 + 2n\pi \text{ or } 3\pi/4 \pm 2n\pi \qquad n = 0,1,2,\ldots$$

Note the remark about the values found by the calculator. When we are given the value of $\sin A$ then to find the value of A by calculator, we use the INV key to reverse the process of finding the sine. The sine function is a many-to-one function, and reversing the process could take a single value of the sine to many values of the angle as we have seen. The calculator can only take one value and that value is the **principal value**.

For the reverse sine and tangent processes the principal values lie in the range $-\pi/2$ to $\pi/2$, and for the reverse cosine process the principal values lie in the range 0 to π.

Quadratic Trigonometric Equations

Typical of quadratic trigonometric equations is the equation

$$2\sin^2\beta - 3\sin\beta + 1 = 0$$

This is solved by factorizing as follows:

$$(\sin\beta - 1)(2\sin\beta - 1) = 2(\sin\beta - 1)(\sin\beta - \tfrac{1}{2}) = 0$$

consequently

$\sin\beta = 1$ and $\sin\beta = \frac{1}{2}$ are solutions

If $\sin\beta = 1$ then the principal solution is $\beta = \pi/2$ radians, and if $\sin\beta = \frac{1}{2}$ the principal solution is $\beta = \pi/6$ radians. From the symmetry of the half-wave it can be seen that there is another solution at

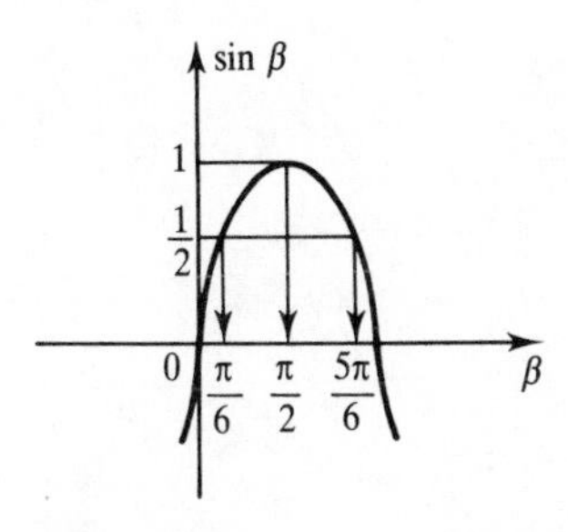

Figure 13.8

$$x = \pi - \pi/6$$
$$= 5\pi/6$$

These, however, are not the only solutions. From the graph of $\sin\beta$ it is seen that $\sin\beta = 1$ when β takes on the values

$$\ldots -7\pi/2,\ -3\pi/2,\ \pi/2,\ 5\pi/2,\ 9\pi/2,\ldots$$

This can be written as

$$\beta = (1 \pm 4n)\pi/2 \qquad n = 0,1,2,\ldots$$

Similarly, when $\sin\beta = \frac{1}{2}$ this is satisfied when β takes on the values

$$\pi/6 \pm 2n\pi \text{ and } 5\pi/6 \pm 2n\pi \text{ where } n = 0,1,2,\ldots$$

Be aware It is strongly recommended that you draw the graphs of the appropriate trigonometric functions whenever you attempt to solve trigonometric equations. In this way the symmetries of the graph will enable you to find all the solutions to the equation more confidently.

EXAMPLES

1 Solve the following linear trigonometric equations:

(a) $\cos(x + \pi/3) - \cos(x - \pi/3) = 1$
(b) $\tan(x + \pi/4) - \cot(x + \pi/4) = 0$
(c) Find the value of B given that:

$$\begin{aligned} \sin A - \sin B &= 0.3 \\ 2\sin A + 3\sin B &= 1.4 \end{aligned}$$

(a)

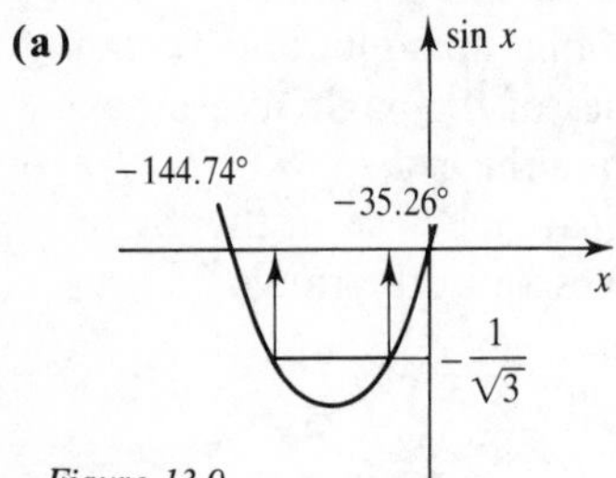

Figure 13.9

$$\begin{aligned} \cos(x + \pi/3) - \cos(x - \pi/3) &= -2\sin x \sin(\pi/3) \\ &= -2\left(\frac{\sqrt{3}}{2}\right)\sin x \\ &= 1 \end{aligned}$$

Therefore $\sin x = -1/\sqrt{3}$ hence $x = -35.26° \pm 360°n$ or $-144.74° \pm 360°n$.

(b)
$$\begin{aligned} \tan(x + \pi/4) - \cot(x + \pi/4) &= \tan(x + \pi/4) - 1/\tan(x + \pi/4) \\ &= \frac{\tan^2(x + \pi/4) - 1}{\tan(x + \pi/4)} \\ &= \frac{[\tan(x + \pi/4) - 1][\tan(x + \pi/4) + 1]}{\tan(x + \pi/4)} \\ &= 0 \end{aligned}$$

Figure 13.10

Therefore $\tan(x + \pi/4) = \pm 1$ so that $x + \pi/4 = \pm\pi/4 \pm n\pi$. That is, $x = \pm n\pi$ or $-\pi/2 \pm n\pi$.

(c) Find the value of B given that

$$\begin{aligned} \sin A - \sin B &= 0.3 \\ 2\sin A + 3\sin B &= 1.4 \end{aligned}$$

From the first equation

$$\sin A = 0.3 + \sin B$$

Substituting this into the second equation gives

$$2(0.3 + \sin B) + 3\sin B = 1.4$$

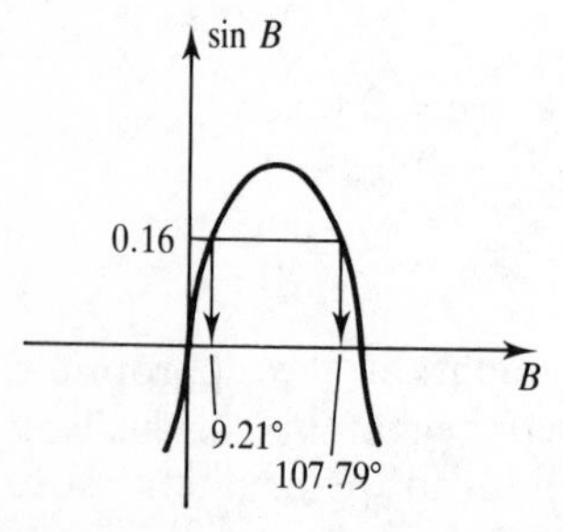

Figure 13.11

that is

$$\sin B = 0.16$$

The calculator gives the principal value of B as 9.21° to two decimal places. Inspection of the graph shows that another solution is given by

$$180° - 9.21° = 170.79°$$

to two decimal places

The final solution is then

$$9.21° \pm 360n° \text{ and } 170.79° \pm 360n°$$

2 If $\tan(A/2) = \sqrt{3}$ find the value of $\tan A$.

$$\tan A = \frac{2\tan(A/2)}{1 - \tan^2(A/2)}$$

therefore

$$\tan A = \frac{2\sqrt{3}}{1 - (\sqrt{3})^2}$$
$$= \frac{2\sqrt{3}}{-2}$$
$$= -\sqrt{3} \qquad \text{therefore } A = -60° \pm 180°n$$

3 Solve the following equations:

(a) $2\sin^2 x - 1 = 0$
(b) $2\sqrt{3}\cos^2 x - 5\cos x + \sqrt{3} = 0$
(c) $5\sec x + \tan^2 x + 7 = 0$

(a) $2\sin^2 x - 1 = (\sqrt{2}\sin x - 1)(\sqrt{2}\sin x + 1)$
$= 0$

Therefore $\sin x = \pm 1/\sqrt{2}$. Hence $x = \pm\pi/4 \pm 2n\pi$ or $\pm 3\pi/4 \pm 2n\pi$.

(b) $2\sqrt{3}\cos^2 x - 5\cos x + \sqrt{3} = 0$ therefore

$$\cos x = \frac{-(-5) \pm [(-5)^2 - 4(2\sqrt{3})(\sqrt{3})]^{1/2}}{2(2\sqrt{3})}$$
$$= \frac{5 \pm 1}{4\sqrt{3}}$$
$$= \frac{\sqrt{3}}{2} \text{ or } \frac{1}{\sqrt{3}}$$

Therefore $x = 30° \pm 360°n$ or $-30° \pm 360n°$ or $54.7° \pm 360n°$ or $-54.7° \pm 360n°$.

(c) $5\sec x + \tan^2 x + 7 = 5\sec x + (\sec^2 x - 1) + 7$
$= \sec^2 x + 5\sec x + 6$
$= (\sec x + 3)(\sec x + 2)$
$= 0$

Therefore $\sec x = -3$, that is, $\cos x = -1/3$; hence $x = 109.5° \pm 360n°$ or $250.5° \pm 360n°$.
For $\sec x = -2$, $\cos x = -1/2$; hence $x = 2\pi/3 \pm 2n\pi$ or $x = 4\pi/3 \pm 2n\pi$.

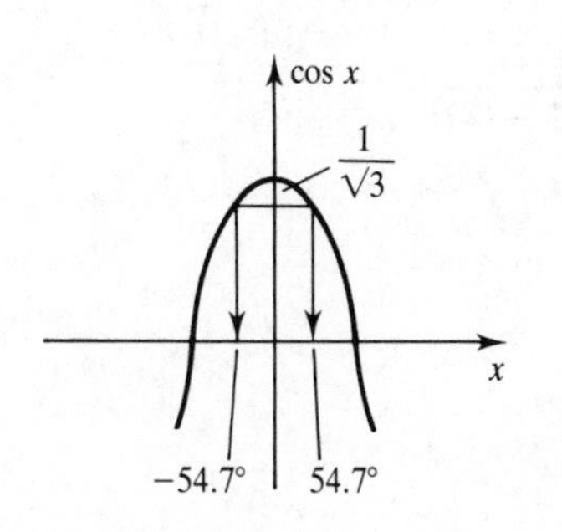

Figure 13.12

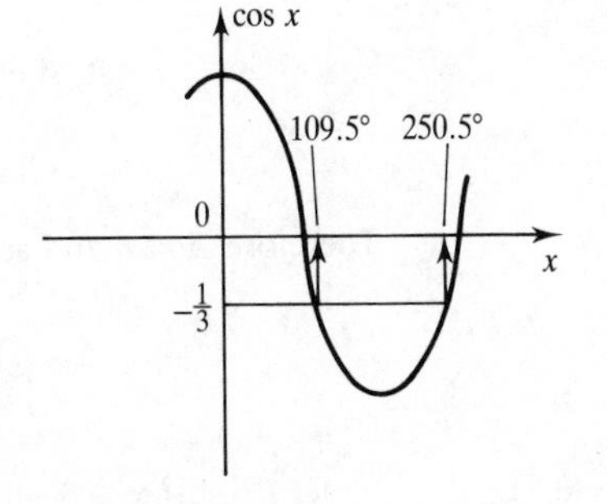

Figure 13.13

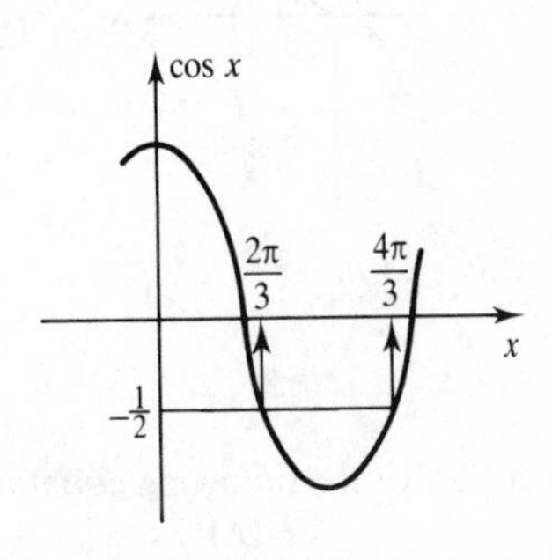

Figure 13.14

EXERCISES

1 Solve the following linear trigonometric equations:

(a) $\sin(x + \pi/6) + \sin(x - \pi/6) = \frac{1}{2}$

(b) $\tan(x + \pi) - \tan(x - \pi) = 0$

(c) Find the value of B given that

$\cos A + 2\cos B = 1.1$
$6\cos A - 5\cos B = 0.1$

(a) $\sin(x + \pi/6) + \sin(x - \pi/6) = 2\sin * \cos *$

$= 2(*)\sin *$

$= 0$

Therefore $\sin * = 0$ and so $x = *$ or $* \pm 2n\pi$.

(b) $\tan(x + \pi) - \tan(x - \pi) \equiv \dfrac{\tan x + \tan \pi}{1 - \tan x \tan \pi} - \dfrac{\tan x - \tan \pi}{1 + \tan x \tan \pi}$

$\equiv 0$ identically

Therefore this equation is an identity and as such is satisfied by * value of x for which the identity is valid. The identity is not valid at * where $\tan x$ is not defined.

(c) $\cos A + 2\cos B = 1.1$
$6\cos A - 5\cos B = 0.1$

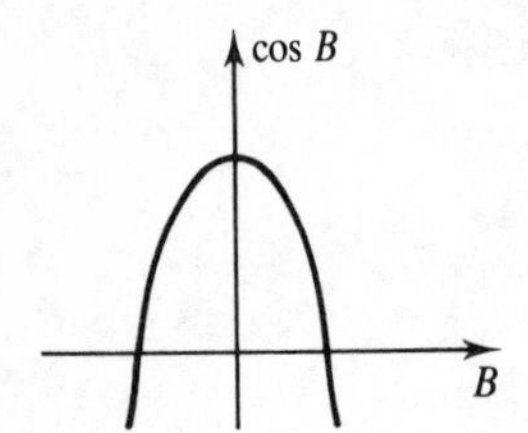

Figure 13.15

Multiply the first equation by * and * the second equation to give

$* \cos B = *$ so that $\cos B = *$

Therefore $x = *° + 360n°$ or $*° + *n°$

2 If $\tan(A/2) = 1/\sqrt{3}$ find the value of $\sin A$.

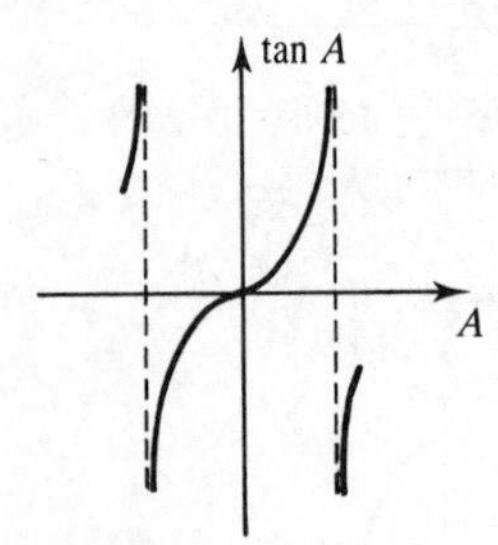

Figure 13.16

$$\tan A = \frac{2*}{1 - *^2}$$

therefore

$$\tan A = \frac{2(*)}{1 - (*)^2}$$

$$= *$$

Therefore $A = *$ and so $\sin A = *$.

3 Solve the following equations:

(a) $2\cos^2 x - 1 = 0.5$

(b) $\sec^2 x - 11\sec x + 30 = 0$

(c) $6\sin^2 x - 3\sin^2 2x + \cos^2 x = 0$

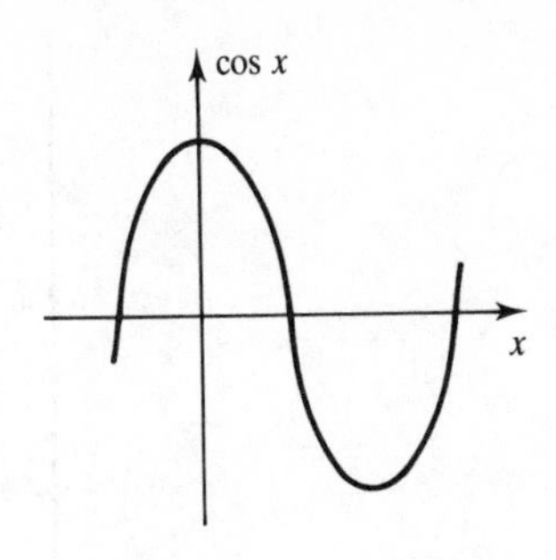

Figure 13.17

(a) $2\cos^2 x - 1 = 0.5$, therefore, $\cos 2x = *$. The solution is $*x = * \pm 2n\pi$ or $-* \pm 2n\pi$. Hence:

$$x = \pm * \pm *\pi$$

(b)
$$\begin{aligned}\sec^2 x - 11\sec x + 30 &= (*)(*)\\ &= 0\end{aligned}$$

Therefore

$$\sec x = * \text{ so } \cos x = *$$

hence

$$x = \pm *^\circ + *^\circ \text{ or } \sec x = * \text{ so } \cos x = *$$

giving

$$x = \pm * + *^\circ$$

(c)
$$\begin{aligned}6\sin^2 x - 3\sin^2 2x + \cos^2 x &= -*\sin^2 x\cos^2 x + \cos^2 x + *\\ &= -*\sin^2 x(1 - \sin^2 x) + *\\ &= *\sin^4 x - *\sin^2 x + *\\ &= (*\sin^2 x - *)(*\sin^2 x - *)\\ &= 0\end{aligned}$$

Therefore $\sin x = *$. In which case $x = *^\circ \pm *^\circ$ or $x = *^\circ \pm *^\circ$, or $\sin x = *$ in which case $x = \pm * + *$.

Figure 13.18

Module 13: Further exercises

1 Use a calculator to find the value of each of the following:
(a) $\sin(5\pi/2)$ (b) $\tan(-3\pi/4)$ (c) $\cos(-763^\circ)$ (d) $\sin 111^\circ$

2 Find the period of each of the following:
(a) $f(x) = \cos 11x$
(b) $f(x) = \sin(4 - 3x)$
(c) $f(x) = 10\cos\left(\dfrac{8x - 4}{16}\right)$

3 Find the amplitude of each of the following:
(a) $f(x) = 7\cos 2x$
(b) $f(3x + 3)$ where $f(x) = 2\sin 4x$
(c) $f(x) = 5\cos(3 - 2x)\sin(3 - 2x)$

4 Find the phase of each of the following:
(a) $f(x) = \cos(4 - 2x)$
(b) $f(x) = \tan(2x - 3)$
(c) $f(x) = 3\sin^2(4x - 1) - 3\cos^2(4x - 1)$

5 A function is defined by the following prescription:

$$\begin{aligned}f(t) &= -t + 3 \qquad 0 \leqslant t < 2\\ f(t + 2) &= f(t)\end{aligned}$$

Plot a graph of this function for $-9 \leqslant t \leqslant +9$ and find:

(a) the period of the function
(b) the amplitude of the function
(c) the phase of $f(t)+1$ with respect to $f(t)$

6 Solve the following linear trigonometric equations:

(a) $\sin(x+2\pi)+\sin(x-2\pi)=\frac{1}{2}$

(b) $\cos(x+\pi/6)+\cos(x-\pi/4)=0$

(c) $\cot A - \cot B = 0.2$
$2\cot A + 3\cot B = 1.4$

7 If $\tan(A/2)=\sqrt{3}$ find the value of $\sec A + \operatorname{cosec} A$.

8 Solve the following equations:

(a) $3\sin^2 x - \cos^2 x = \sin 2x$

(b) $18\cos^2 x + 3\cos x - 1 = 0$

(c) $2\sec^2 x + 3\tan x - 4 = 0$

Module 14: The Exponential and Logarithmic Functions

Aims: To investigate the properties of both exponential and logarithmic functions.

Objectives: When you have read this module you will be able to:

- ▶▶ Draw the graph of the exponential function and describe its general features.
- ▶▶ Understand and manipulate logarithms.
- ▶▶ Draw the graph of the logarithmic function and describe its general features.

Unit 1 The Exponential Function

Test yourself with the following:

1 Using a calculator complete the following table:

x	0.0	0.5	1.0	1.5	2.0	2.5
$f(x)$						

where $f(x) = e^{-x}$. Plot the graph of $f(x) = e^{-x}$ from this table and then use the graph to find the values of:

(a) $e^{-2.25}$ (b) $e^{-0.75}$

2 Find the value of x corresponding to each of the following:

(a) $3^{-x} = 1$ (b) $\exp(2x) = e$ (c) $e^x = 48.01$ (d) $\exp(-x) = 0.002$

3 Solve the following equations:

(a) $4^{x+2}5^{x+1} = 32\,000$ (c) $4^x + 1/4^x = 2$

(b) $e^{2x} - 5(e^x) + 6 = 0$

Raising to a Power

We have seen how the arithmetic operations can be used to create algebraic expressions and from these create functions. The only operation that we have not considered in detail yet is that of raising to a power. This operation gives rise to what are called **exponential functions**.

For example, the function f with output

$$f(x) = 2^x$$

is a typical exponential function. Notice that the form of this expression is somewhat different from other functional expressions in that the input variable is the power. Evaluating corresponding input and output values for f permits us to construct the following table:

x	-3	-2	-1	0	1	2	3
$f(x)$	$2^{-3} = 1/8$	$2^{-2} = 1/4$	$2^{-1} = 1/2$	$2^0 = 1$	$2^1 = 2$	$2^2 = 4$	$2^3 = 8$

Plotting x against $f(x)$ and joining the points with a smooth curve yields the following graph:

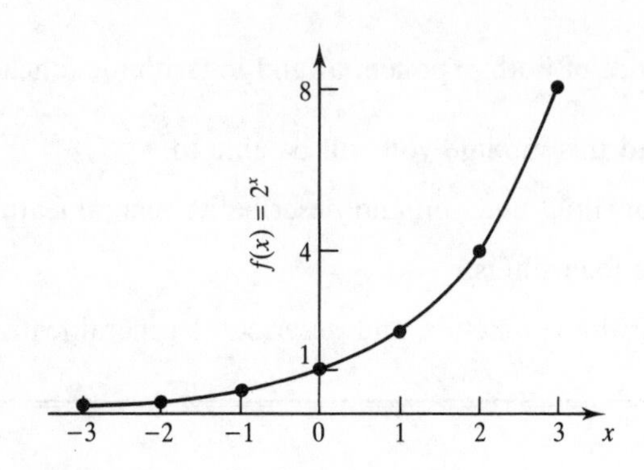

Figure 14.1

From the general features of this graph it can be seen that nowhere is the output zero or negative. The curve crosses the vertical axis at $f(0) = 1$, and as x increases the exponential curve also increases without bound. As x becomes increasingly negative the curve approaches the x-axis. However, the curve never actually meets the x-axis – we say that the x-axis is an **asymptote** to the curve and that the curve approaches the x-axis **asymptotically**.

The function with output

$$f(x) = (1/2)^x$$

has the following graph:

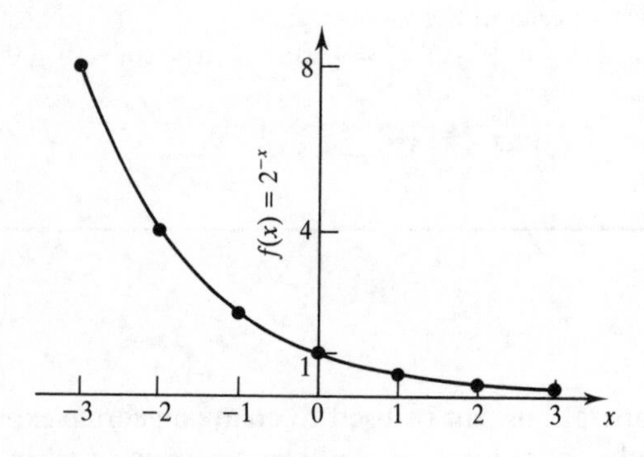

Figure 14.2

This graph is a mirror image of the graph in Figure 14.1, where the mirror is taken to lie along the vertical axis.

Any function f with output of the form

$$f(x) = a^x$$

is called an exponential function, where a is referred to as the **base** of the function. If base $a > 1$ the graph is similar in appearance to Figure 14.1, whereas if $0 < a < 1$ the graph is similar to Figure 14.2.

The Exponential Function

Of all the possible exponential functions there is one that is sufficiently special to be named *the* exponential function. This is the exponential function

$$f(x) = e^x$$

where the base is the irrational number e, where

$$e = 2.7182818\ldots$$

Some of the reasons why this exponential function holds such a special place in mathematics will become more clear as you progress through the book. Suffice it to say that because of its continual appearance throughout mathematics an alternative expression has been defined for it that is more in keeping with functional notation. This is

$$f(x) = \exp(x) \equiv e^x$$

The Reverse Process

When we considered the trigonometric functions, we found that by using the INV key on a calculator we could find the angle corresponding to a given trigonometric function value, thereby reversing the process of finding the trigonometric value of a given angle. This method can also be employed here. For example, if we are given that

$$e^x = 5$$

and we wish to find the value of x, proceed as follows:

Enter the number 5
Press the INV key
Press the e^x key

and the number 1.6094... is displayed. This is the required value of x.

Be aware On many calculators the exponential function key will be labelled e^x and $\ln x$. The $\ln x$ is called the logarithm function and we shall consider this in detail in the following unit. For now it is sufficient to know that the $\ln x$ process will reverse the exponential process.

EXAMPLES

1 Using a calculator complete the following table:

x	-2.0	-1.5	-1.0	-0.5	0.0	0.5	1.0
$f(x)$							

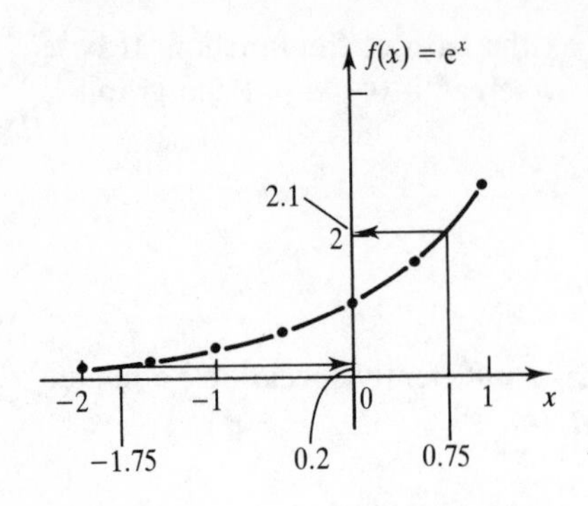

Figure 14.3

where $f(x) = e^x$. Plot the graph of $f(x) = e^x$ from this table and then use the graph to find the values of:

(a) $e^{0.75}$ **(b)** $e^{-1.75}$

Use your calculator to check the accuracy of your graph.

2 Find the value of x corresponding to each of the following:

(a) $e^x = 1$ **(b)** $\exp(x) = e$ **(c)** $e^x = 0.25$ (d) $\exp(x) = 6.3$

(a) Because any number raised to the power 0 is unity

$e^0 = 1$ therefore $x = 0$

(b) Any number raised to the power unity is itself. Therefore

$\exp(1) = e$, hence $x = 1$

(c) Using a calculator it is found that if $e^x = 0.25$ then $x = -1.3862\ldots$.
(d) Using a calculator it is found that if $\exp(x) = 6.3$ then $x = 1.8405\ldots$.

3 Solve the following equations:

(a) $3^{x-1}2^{2x} = 48$ **(c)** $2^x + 2^{1-x} = 9/2$
(b) $5^{2x} - 6(5^x) + 5 = 0$

(a) $3^{x-1}2^{2x} = 48$ This equation can be rewritten as

$$(3^x)(3^{-1})(4^x) = 48$$

that is

$$12^x\tfrac{1}{3} = 48$$

so that $12^x = 144$ and hence $x = 2$.

(b) $5^{2x} - 6(5^x) + 5 = 0$ This is a quadratic equation in 5^x with solution:

$$5^x = \frac{-(-6) \pm [(-6)^2 - 4(1)(5)]^{1/2}}{2(1)}$$

$$= \frac{6 \pm 4}{2}$$

$$= 1 \text{ or } 5$$

that is

$$5^{2x} - 6(5^x) + 5 = (5x - 1)(5^x - 5)$$

$$= 0$$

When $5^x = 1$ then $x = 0$ and when $5^x = 5$ then $x = 1$.

(c) $2^x + 2^{1-x} = 9/2$ By multiplying through by 2^x this equation can be rewritten as:

$$2^{2x} + 2 = \tfrac{9}{2}2^x$$

that is

$$2^{2x} - \tfrac{9}{2}2^x + 2 = 0$$

This is a quadratic equation in 2^x with solution:

$$2^x = \frac{-(-9/2) \pm [(-9/2)^2 - 4(1)(2)]^{1/2}}{2(1)}$$

$$= \frac{9/2 \pm 7/2}{2}$$

$$= 4 \text{ or } \frac{1}{2}$$

When $2^x = 4$ then $x = 2$ and when $2^x = 1/2$ then $x = -1$.

EXERCISES

1 Using a calculator complete the following table:

x	-2.0	-1.5	-1.0	-0.5	0.0	0.5	1.0
$f(x)$							

where $f(x) = 5^x$. Plot the graph of $f(x) = 5^x$ from this table and then use the graph to find the values of:

(a) $5^{0.25}$ **(b)** $5^{-1.25}$

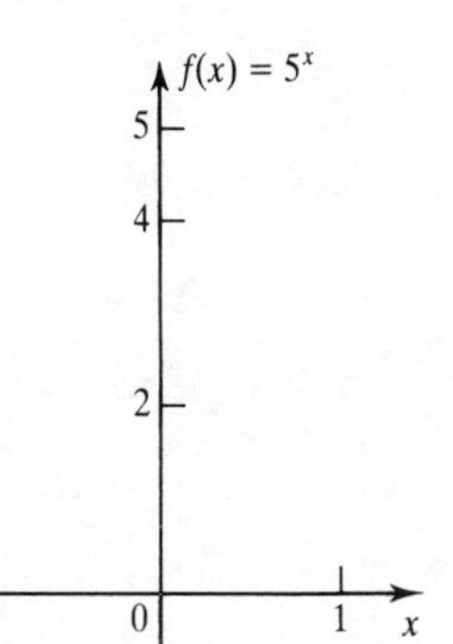

Figure 14.4

(a) $5^{0.25} = *$
(b) $5^{-1.25} = -*$

2 Find the value of x corresponding to each of the following:

(a) $e^x = 1/e$ **(b)** $\exp(x) = e^2$ **(c)** $e^x = 46.7$ **(d)** $\exp(x) = 0.002$

(a) Because any number raised to the power * is the reciprocal of the number,

$$e^* = \frac{1}{e} \text{ therefore } x = *$$

(b) The $\exp(x)$ and e^x notations are entirely equivalent so that $x = *$.
(c) Using a calculator it is found that if $e^x = 46.7$ then $x = *.8*3*\ldots$.
(d) Using a calculator it is found that if $\exp(x) = 0.002$ then $x = -*$.

3 Solve the following equations:

(a) $2^{x-3}5^{x+1} = 625$ (c) $4^x + 4^{-3-x} = 5/16$

(b) $10^{2x} - 11(10^x) + 10 = 0$

(a) $2^{x-3}5^{x+1} = 625$ This equation can be written as

$$(2^x)(2^*)(5^x)(5^*) = 625$$

that is

$$(*^x)\left(\frac{*}{*}\right) = 625$$

so that $*^x = 625(*/*) = *$, therefore $x = *$.

(b) $10^{2x} - 11(10^x) + 10 = 0$ This is a quadratic in $*^x$ with solution:

$$*^x = \frac{-(*) \pm [(*)^2 - 4(*)(*)]^{1/2}}{2(*)}$$

$$= \frac{* \pm *}{*}$$

$$= * \text{ or } 1$$

When $*^x = 1$ then $x = *$ and when $*^x = *$ then $x = *$.

(c) $4^x + 4^{-3-x} = 5/16$ This equation can be rewritten as

$$4^{*x} + *^{-3} = *$$

that is

$$4^{*x} - *4^x + 4^*3 = 0$$

which has solution

$$*^x = \frac{-(*) \pm [(*^2) - 4(*)(*)]^{1/2}}{2(*)}$$

$$= \frac{* \pm *}{*}$$

$$= * \text{ or } *$$

therefore $x = *$ or $*$.

Unit 2 The Logarithmic Function

Test yourself with the following:

1 Using a calculator complete the following table:

x	0.5	1.0	1.5	2.0	2.5	3.0	3.5
$f(x)$							

where $f(x) = \ln(x)$. Plot the graph of $f(x) = \ln(x)$ from this table and then use the graph to find the values of:

(a) $\ln(0.75)$ (b) $\ln(8)$

2 In each of the following find the value of x that satisfies the equation:

(a) $\frac{1}{2}\log(x^2) = 5\log(2) - 4\log(3)$ (d) $\log_x(36) = 2$

(b) $\log(x^{1/2}) = \log(x) + \log(3)$ (e) $\log(100) = x$

(c) $\log_4(x) = 5$

3 Given that

$$4\log(y) = 9\log(x) - 2\log(y^2)$$

find y in terms of x.

4 Solve for x:

$$\log_{10}(x - 5) = 0.55$$

5 Show that

$$5^{\log_5(x)} = x$$

Powers

Any real number can be written as a natural number raised to a power. For example,

$$9 = 3^2$$

and

$$27 = 3^3$$

By writing numbers in the form of another number raised to a power, some of the arithmetic operations can be performed in an alternative manner. For example,

$$\begin{aligned} 9 \times 27 &= 3^2 \times 3^3 \\ &= 3^{2+3} \\ &= 3^5 \\ &= 243 \end{aligned}$$

Provided we have a simple way of relating numbers such as 9 and 27 to powers of 3 and relating powers of 3 to numbers such as 243, we have converted the process of multiplication to the simpler process of adding powers.

In the past, one way of employing these relations was to construct tables of powers, for use in speeding-up calculations involving multiplication and division before the advent of electronic calculators. They were not called tables of powers but tables of **logarithms**. Nowadays, tables of logarithms have little value, but the logarithm remains as an essential concept.

Logarithms

If numbers a, b and c are such that

$$a = b^c$$

we call the power c the logarithm of the number a to the base b and we write

$$c = \log_b a$$

For example, because

$$25 = 5^2$$

we call the power 2 the logarithm of 25 to the base 5. It is written as

$$2 = \log_5 25$$

Notation

The only restriction that is placed on the base of the logarithm is that it be a positive real number. In practice, there are two bases that are commonly used, namely 10 and the irrational number e. When the base 10 is used the base is suppressed in the notation. For example,

$\log_{10} 4$ is written as $\log 4$

When the irrational number e is used as base then the notation employed is ln. For example,

$$\log_e 5 \equiv \ln 5$$

Also, in accordance with functional notation, we shall henceforth use brackets and write

$\log_a x$ as $\log_a(x)$

Rules of Logarithms

Just as powers are manipulated according to a set of rules so are logarithms.

Addition and Subtraction of Logarithms

If $x = a^b$ and $y = a^c$ so that $b = \log_a(x)$ and $c = \log_a(y)$, then

$$\begin{aligned} x . y &= a^b . a^c \\ &= a^{b+c} \end{aligned}$$

so that

$$\begin{aligned} b + c &= \log_a(xy) \\ &= \log_a(x) + \log_a(y) \end{aligned}$$

that is

$$\log_a(xy) \equiv \log_a(x) + \log_a(y)$$

The logarithm of a product is the sum of the logarithms. For example, using your calculator you will find that

$$\log_{10}(2) = 0.301\ldots \text{ and } \log_{10}(3) = 0.477\ldots$$

therefore

$$\begin{aligned}\log_{10}(2) + \log_{10}(3) &= 0.301\ldots + 0.477\ldots \\ &= 0.778\ldots \\ &= \log_{10}(6) \\ &= \log_{10}(2 \times 3)\end{aligned}$$

Similarly

$$\begin{aligned}\frac{x}{y} &= \frac{a^b}{a^a} \\ &= a^{b-c}\end{aligned}$$

so that

$$\begin{aligned}b - c &= \log_a\left(\frac{x}{y}\right) \\ &= \log_a(x) - \log_a(y)\end{aligned}$$

that is

$$\log_a\left(\frac{x}{y}\right) \equiv \log_a(x) - \log_a(y)$$

The logarithm of a quotient is the difference of the logarithms. For example, using your calculator you will find that

$$\begin{aligned}\log(3) - \log(2) &= 0.477\ldots - 0.301\ldots \\ &= 0.176\ldots \\ &= \log(1.5) \\ &= \log(\tfrac{3}{2})\end{aligned}$$

Multiplication of a Logarithm by a Number

If $x = a^b$ so that $b = \log_a(x)$, then

$$\begin{aligned}x^c &= (a^b)^c \\ &= a^{bc}\end{aligned}$$

so that

$$\begin{aligned}bc &= \log_a(x^c) \\ &= c\log_a(x)\end{aligned}$$

that is

$$\log_a(x^c) \equiv c\log_a(x)$$

For example, using your calculator you will find that

$$\begin{aligned}4\log(2) &= 1.204\ldots \\ &= \log(16) \\ &= \log(2^4)\end{aligned}$$

Change of Base of a Logarithm

If $x = p^a$ and $x = q^b$ where $p = q^c$ then

$$a = \log_p(x),\ b = \log_q(x) \text{ and } c = \log_q(p)$$

$$\begin{aligned} x &= p^a \\ &= (q^c)^a \\ &= q^{ca} \\ &= q^b \end{aligned}$$

so that

$$b = ca$$

that is

$$\log_q(x) \equiv \log_q(p)\log_p(x)$$

For example, to three decimal places, $\log_{10}(2) = 0.301$ and $\log_e(10) = 2.303$ so that

$$\begin{aligned} \log_{10}(2)\log_e(10) &= 0.301 \times 2.303 \\ &= 0.693 \\ &= \log_e(2) \text{ to three dec. pl.} \end{aligned}$$

The Logarithmic Function

The logarithmic function is defined by

$$f(x) = \log_a(x)$$

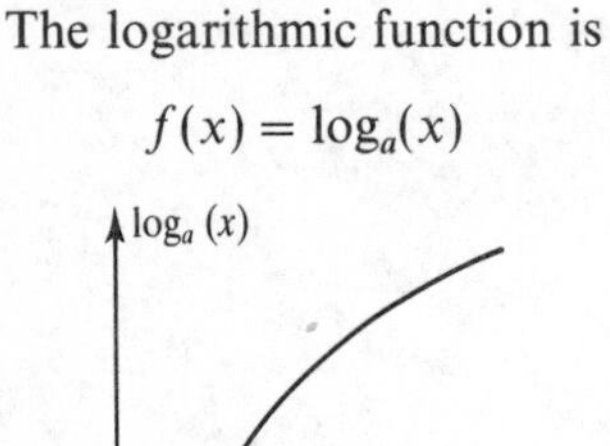

Figure 14.5

This function has the graph of Figure 14.5. The graph of

$$f(x) = \log_{10}(x) \equiv \log(x)$$

can be formed by computing $f(x)$ values for a selection of x values with a calculator and plotting the ordered pairs so formed. The isolated points plotted are then joined up with the smooth curve as shown.

From the general features of this graph it can be seen that the logarithmic function is not defined for negative values of x, and that as x increases so the logarithm increases without bound. The logarithmic curve passes through the x-axis at the point $x = 1$, which is a pictorial display of the statement that

$$\log_a 1 = 0 \text{ or alternatively } a^0 = 1$$

As x approaches zero so the curve approaches the vertical axis, the output becoming increasingly negative. The curve approaches the vertical axis **asymptotically** and the vertical axis is an **asymptote** to the curve.

Natural Logarithms

The natural logarithm has as base the irrational exponential number e and is denoted as $\ln(x)$, that is

$$\log_e(x) \equiv \ln(x)$$

Close inspection of the graph of the natural logarithmic function reveals that its shape is similar to that of the exponential function but with a different orientation. Indeed, its shape is identical, which indicates that it is somehow related to the exponential function. It is, in fact, the *inverse* function as we shall see in the next module.

EXAMPLES

1 Using a calculator complete the following table:

x	0.1	0.3	0.6	1.0	1.3	1.6	3.0
$f(x)$							

where $f(x) = \log(x)$. Plot the graph of $f(x) = \log(x)$ from this table and then use the graph to find the values of:

(a) $\log(2)$ **(b)** $\log(\frac{1}{2})$

Use your calculator to check the accuracy of your graph.

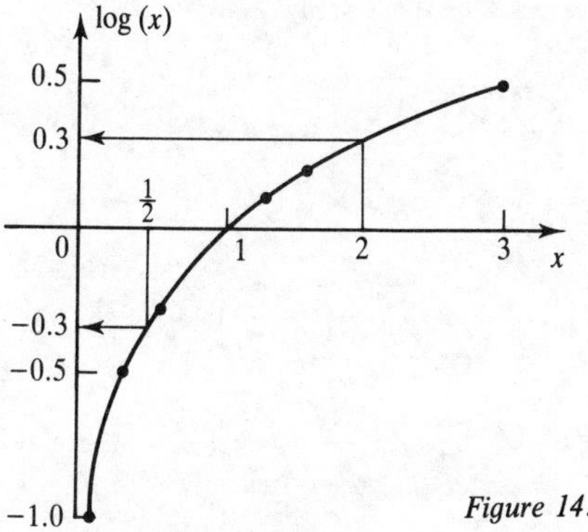

Figure 14.6

(a) $\log(2) = 0.3$ to one decimal place.
(b) $\log(\frac{1}{2}) = -0.3$ to one decimal place.
$= \log(1) - \log(2)$ by the rules of logarithms
$= -\log(2)$

2 In each of the following find the value of x that satisfies the equation:

(a) $\log(x) = 3\log(18) - 4\log(12)$
(b) $\log(x^4) = 2\log(x) + \log(4)$
(c) $\log_2(x) = 3$
(d) $\log_x(16) = 4$
(e) $\log_3(243) = x$

(a)
$$\begin{aligned}\log(x) &= 3\log(18) - 4\log(12)\\ &= \log(18^3) - \log(12^4)\\ &= \log\left(\frac{18^3}{12^4}\right)\\ &= \log\left(\frac{5832}{20\,736}\right)\\ &= \log\left(\frac{9}{32}\right)\end{aligned}$$

Therefore $x = 9/32$.

(b) $\log(x^4) = 2\log(x) + \log(4)$
$= \log(x^2) + \log(4)$

so that $\log(x^4) - \log(x^2) = \log(4)$ that is

$$\log\left(\frac{x^4}{x^2}\right) = \log(x^2)$$

$$= \log(4)$$

hence $x = \pm 2$.
(c) $\log_2(x) = 3$, that is, $2^3 = 8 = x$.
(d) $\log_x(16) = 4$, that is, $x^4 = 16$ so that $x = 2$.
(e) $\log_3(243) = x$, that is, $3^x = 243$ so that $x = 5$.

3 Given that

$$\log(x^3) = 4\log(x) - 3\log(y^2)$$

find y in terms of x.

$$\log(x^3) = 4\log(x) - 3\log(y^2)$$
$$= \log(x^4) + \log(y^{-6})$$

therefore

$$\log(x^3) - \log(x^4) = \log(x^{-1})$$
$$= \log(y^{-6})$$

Therefore $y^{-6} = x^{-1}$, that is, $y = x^{1/6}$.

4 Solve for x:

$$\ln(x^2 + \tfrac{1}{2}) = 2.27$$

Since $2.27 = \ln 9.68$ this equation can be rewritten as

$$x^2 + 0.5 = 9.68$$

that is, $x^2 = 9.18$, hence $x = \pm 3.03$.

5 Show that

$$\ln(e^x) \equiv x$$

By the rules of logarithms:

$$\text{LHS} = \ln(e^x)$$
$$= x\ln(e)$$
$$= x\ [\text{because } \ln(e) = 1]$$
$$= \text{RHS}$$

EXERCISES

1 Using a calculator complete the following table:

x	0.2	0.8	1.2	1.8	2.2	2.8	3.2
$f(x)$							

where $f(x) = -\log(x)$. Plot the graph of $f(x) = -\log(x)$ from this table and then use the graph to find the values of:

(a) $-\log(3)$ **(b)** $-\log(\frac{1}{3})$

Use your calculator to check the accuracy of your graph.

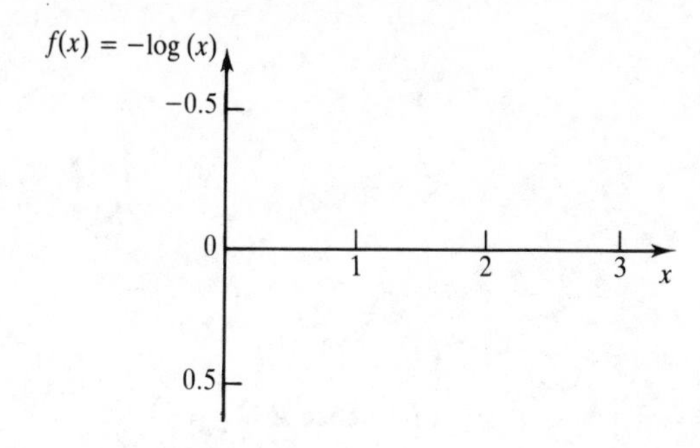

Figure 14.7

(a) $-\ln(3) = -*$
(b) $-\ln(\frac{1}{3}) = * = -\ln(*)$

2 In each of the following find the value of x that satisfies the equation:

(a) $2\log(x) = \frac{1}{5}\log(32) - 4\log(2)$ **(d)** $\log_x(125) = 3$
(b) $\log(x^2) = 3\log(x) + \log(4)$ **(e)** $\log_4(64) = x$
(c) $\log_3(x) = 4$

(a) $2\log(x) = \frac{1}{5}\log(32) - 4\log(2)$ This equation can be rewritten as

$$\log(x^2) = \log(32^*) - \log(2^*)$$

that is

$$\log(x^2) = \log(*) - \log(*)$$
$$= \log\left(\frac{*}{*}\right)$$

therefore $x = \pm *$.

(b) $\log(x^2) = 3\log(x) + \log(4)$ This equation can be rewritten as

$$\log(*^2) = \log(x^*) + \log(4)$$

that is

$$\log(*^2) - * = \log(4)$$

that is

$$\log(x^*) = \log(*)$$

therefore $x = *$.

(c) $\log_3(x) = 4$, that is, $*^* = x$ so that $x = *$.
(d) $\log_x(125) = 3$, that is, $x^* = *$ so that $x = *$.
(e) $\log_4(64) = x$, that is, $*^x = *$ so that $x = *$.

3 Given that

$$2\log(x) = 3\log(y) - 6\log(x^2)$$

find y in terms of x.

This equation can be rewritten as

$$\log(*^2) = \log(y^*) - \log(x^*)$$

that is

$$\log(*^2) + * = \log(*)$$

so that

$$\log(x^*) = \log(y^*)$$

therefore $y = x^*$.

4 Solve for x:

$$\ln(x^2 + \tfrac{3}{4}) = 2.35$$

Since $\ln(*) = 2.35$ then $x^2 + 0.75 = *$. That is, $x^2 = *$ and hence $x = \pm *$.

5 Show that

$$e^{\ln(x)} \equiv x$$

Taking the natural logarithms of both sides of this equation we find that

$$\begin{aligned}\ln(\text{LHS}) &= \ln e^{\ln(x)} \\ &= * \ln(*) \\ &= * \\ &= \ln(\text{RHS})\end{aligned}$$

Therefore $* = *$.

Module 14: Further exercises

1 Using a calculator complete the following table:

x	1	2	4	6	8	10
$f(x)$						

where $f(x) = \ln(x)$. Plot the graph of $f(x) = \ln(x)$ from this table and then use the graph to find the values of:

(a) $\ln(1.5)$ (b) $\ln(5.25)$

Use your calculator to check the accuracy of your graph.

2 Find the value of x corresponding to each of the following:

(a) $2^{-2x} = 1$ (b) $\exp(-3x) = e^2$ (c) $e^{1/x} = 5.43$ (d) $\exp\left(\frac{1}{x}\right) = 0.01$

3 In each of the following find the value of x that satisfies the equation:

(a) $\frac{1}{3}\log(x^6) = 5\log(3) - \log(27)$ (d) $\log_x(16) = 4$

(b) $\log(x^2) = \log\left(\frac{1}{x}\right) - \log(5)$ (e) $\log_9(81) = x$

(c) $\log_5(x) = 3$

4 Given that

$$12\log\left(\frac{1}{y}\right) = 4\log\left(\frac{1}{x}\right) - 3\log(y^2)$$

find y in terms of x.

5 Solve for x:

$$\log_4(3 - x^3) = -6.2$$

6 Using a calculator complete the following table:

x	0	1	2	3
$f(x)$				

where $f(x) = 7^x$. Plot the graph of $f(x) = 7^x$ from this table and then use the graph to find the values of:

(a) $7^{2.3}$ (b) $7^{1/0.56}$

7 Find the values of:

(a) $3^{\log_3(7)}$ (b) $10^{-\log(0.2)}$

8 Solve the following equations:

(a) $2^{x-3}5^{x+1} = 62.5$ (c) $4^x + 4^{2-x} = 17$

(b) $e^{2x} - e^{x+3} - e^{x+1} + e^4 = 0$

Module 15: Inverse Functions

Aims: To consider both inverses of functions and inverse functions.

Objectives: When you have read this module you will be able to:

- ▶▶ Derive the inverse of a function from the algebraic form of the function.
- ▶▶ Recognize that the inverse of a function is not necessarily the inverse function.
- ▶▶ Deduce the graph of the inverse of a function from the graph of a function.
- ▶▶ Use inverse functions to solve trigonometric equations.

Unit 1 The Inverse of a Function

Test yourself with the following:

1 Find the inverse of each of the functions with the following outputs:

(a) $f(x) = x - 3$ (c) $f(x) = x + \frac{1}{3}$ (e) $f(x) = x^2$

(b) $f(x) = 2x$ (d) $f(x) = \dfrac{x}{8}$

2 Give the inverse of each of the following:

(a) $f(x) = \tan\left(\dfrac{x}{4}\right)$ (b) $f(x) = \sin 3x$

3 Find the inverse of each of the following compositions:

(a) $f(x) = \dfrac{2x}{3} - 4$ (d) $f(x) = \log(2x + 1)$

(b) $f(x) = (x^3 - 8)^{1/2}$ (e) $f(x) = e^{2x}$

(c) $f(x) = \cos(x^2 - 1)$

4 Of the following, which has an inverse that is a function?

(a) $f(x) = x^3 - 27$ (b) $f(x) = \dfrac{2x^2}{4}$ (c) $f(x) = 3^{2x}$

5 Draw the graph of the inverse of each of the following:

(a) $f(x) = x - 1$ (b) $f(x) = x^3$ (c) $f(x) = \cos 3x$

6 Derive the inverse function of $f(x) = \tan 2x$.

Reversing a Process

A function is a process f that converts an input number x to an output number $f(x)$ (Figure 15.1).

If the input and output are reversed so that $f(x)$ now becomes the input and x becomes the output the action of the process f is reversed also. Because the process is reversed, it is a different process and it is labelled as **arc f** (Figure 15.2). Turning the diagram of the system around so that the input enters the process from the left gives Figure 15.3. As is seen from the diagram, the input is written as $f(x)$ and the output as x. So that $\text{arc} f(f(x)) = x$. This is now changed to read input x and output $\text{arc} f(x)$ so as to conform with the usual notation (Figure 15.4).

The process arc f is called the **inverse of the function f**. As an example, consider the function that adds 10 to a number (Figure 15.5). Now reverse the process, as in Figure 15.6. The process reversed now subtracts 10 from the input. Now turn the system around as in Figure 15.7. Now relabel the input and output as in Figure 15.8. The inverse of the function that adds 10 to an input is a process that subtracts 10 from an input. This is the inverse of the function.

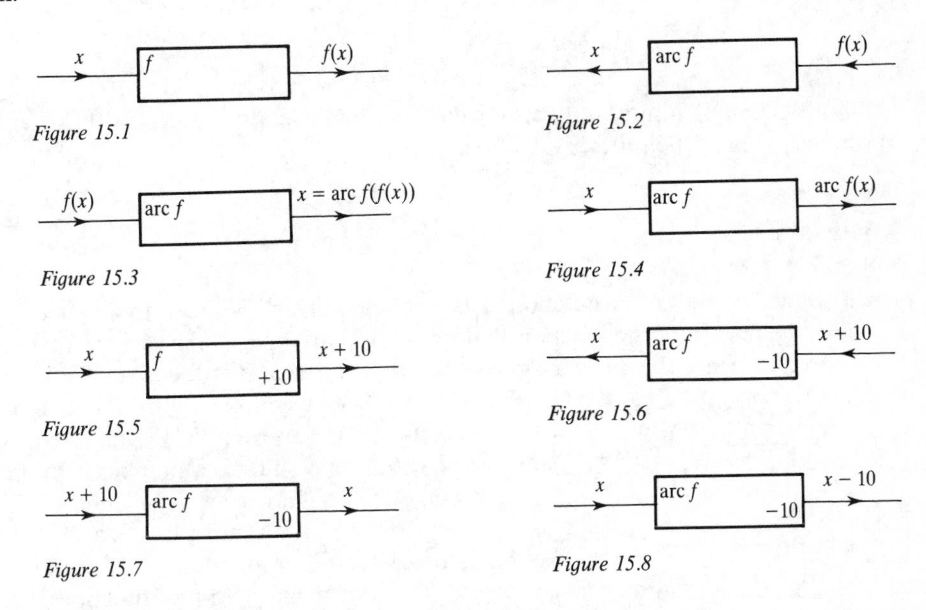

Figure 15.1

Figure 15.2

Figure 15.3

Figure 15.4

Figure 15.5

Figure 15.6

Figure 15.7

Figure 15.8

Inverse Processes

The inverse of the arithmetic operations are clear. Subtraction is the inverse process of addition and vice versa. Division is the inverse process of multiplication and vice versa. The inverse of raising to a power n is to raise to power $1/n$ and vice versa. For other, non-algebraic processes such as those available on a calculator we must appeal to the arc notation. For example, the inverse sine function is the arc sine function. On some calculators the inverse process is indicated by an INV key and on others it may be indicated by the symbol -1. For example, to find the value of

$$\arcsin(\tfrac{1}{2})$$

enter 0.5, press the INV key followed by the sin key to produce the display 30 in degree mode and 0.5236 to four dec. pl. in radian mode.

Notation

You will notice that we are using the notation

arc f

to denote the inverse of the function f. On your calculator you may notice the notation

f^{-1}

For example, the sin key may be labelled both $\sin x$ and $\sin^{-1} x$. Here the

$\sin^{-1} x$

is the calculator's equivalent of $\arcsin x$. Throughout this book the -1 notation will be avoided for two reasons. Firstly, its use creates a need for alternative notations. For example,

$$(\sin x)(\sin x) = \sin^2 x$$

but

$$\frac{1}{\sin x} = (\sin x)^{-1} \text{ and not } \sin^{-1} x$$

Secondly, and more importantly, its use creates a mind-set that somehow the inverse is related to the reciprocal, which it most definitely is not.

Inverse Functions

The inverse of a one-to-one function is itself a function, and is called the **inverse function**. For example, the inverse of the function that adds 10 to an input is the process that subtracts 10 from an input. Since this inverse process is also a function it is the inverse function.

The inverse of a many-to-one function is not a function. For example, the inverse process of the function that squares a number is the process that takes the square root of a number. Because the square root of a number has two values this inverse process is not a function.

For a many-to-one function the inverse *of the function* must be adjusted to form the *inverse function.* We consider this adjustment process graphically.

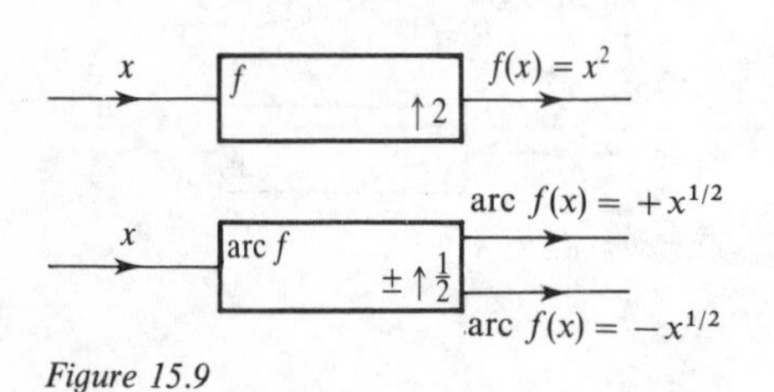

Figure 15.9

The Graphs of Inverse Processes

The cartesian graph of a function is traditionally drawn with the input values plotted on the horizontal axis and the output values plotted on the vertical axis. When the input and output are reversed the labelling on the graph must also be reversed. For example, the graph of the function defined by $f(x) = x^3$ is the familiar curve in Figure 15.10. If the input to and output from this function are reversed then the labelling of the graph must also be reversed as shown in Figure 15.11.

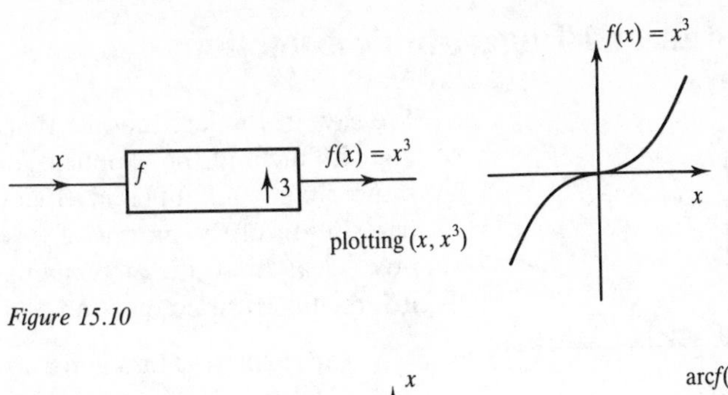

Figure 15.10

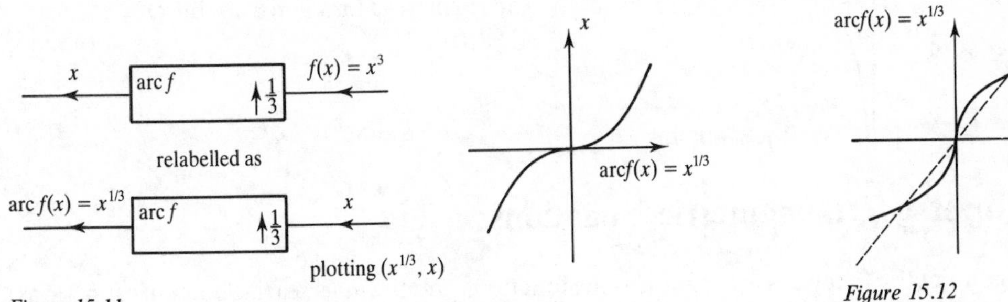

Figure 15.11

Figure 15.12

To maintain consistency with tradition we now rotate this entire graph – the axes and the parabola – out of the plane of the paper, about the line $f(x) = x$, to ensure that the input is plotted on the horizontal axis again. The effect, shown in Figures 15.12, 15.13, 15.14 and 15.15, illustrate the fact that the inverse of the function defined by

$$f(x) = x^2$$

is not itself a function. If, however, either the upper branch or the lower branch of the graph of the inverse is removed, the result is the graph of a function (Figure 15.16). Either choice can be taken to be the inverse function – we choose the removal of the lower branch (this is also the choice made by a calculator).

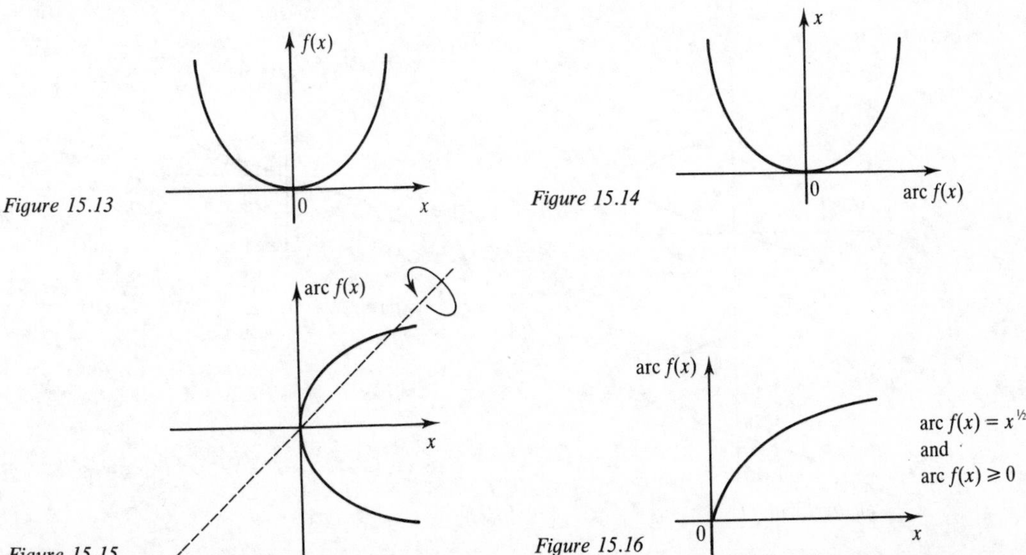

Figure 15.13

Figure 15.14

Figure 15.15

Figure 15.16

The Exponential and Logarithmic Functions

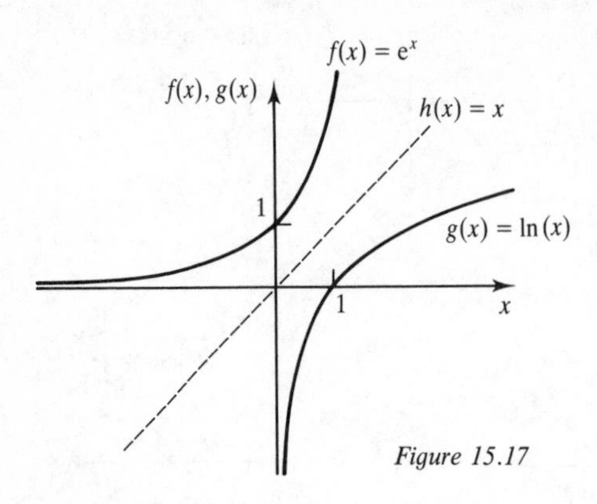

Figure 15.17

We saw in the last module that the graphs of the logarithmic and the exponential functions had the same shape but different orientations. From the description of the graphs of inverse functions it is now clear that the exponential function is the inverse logarithmic function and vice versa:

$$\text{arc}\exp(x) \equiv \ln(x) \text{ and } \text{arc}\ln(x) \equiv e^x$$

The Inverse Trigonometric Functions

The trigonometric functions are many-to-one functions and the inverse trigonometric functions are created by adjusting the graphs of the inverse processes.

The Inverse Sine Function

If the graph of the sine function is reflected in the line $f(x) = x$ and the axes appropriately relabelled, the graph of Figure 15.18 results. This is the graph of the inverse process of the sine function and is clearly not the graph of a function and hence, not the inverse function. To create the inverse sine function from this graph we restrict the range to the interval

$$-\pi/2 \leqslant \arcsin x \leqslant \pi/2$$

giving Figure 15.19. This is the inverse sine function and is defined by the prescription

$$f(x) = \arcsin x \qquad -\pi/2 \leqslant f(x) \leqslant \pi/2$$

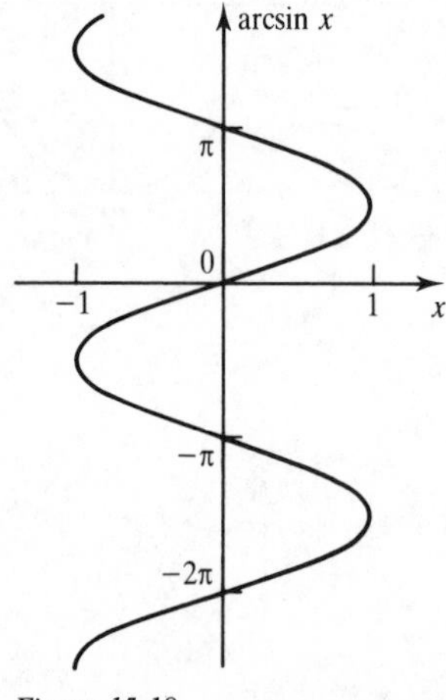

Figure 15.18

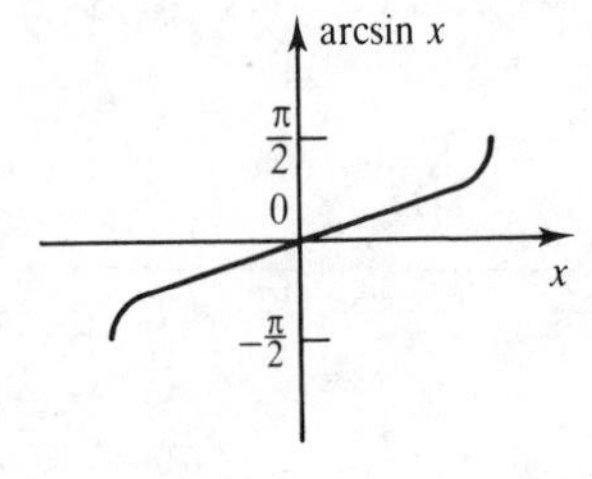

Figure 15.19

The Inverse Cosine Function

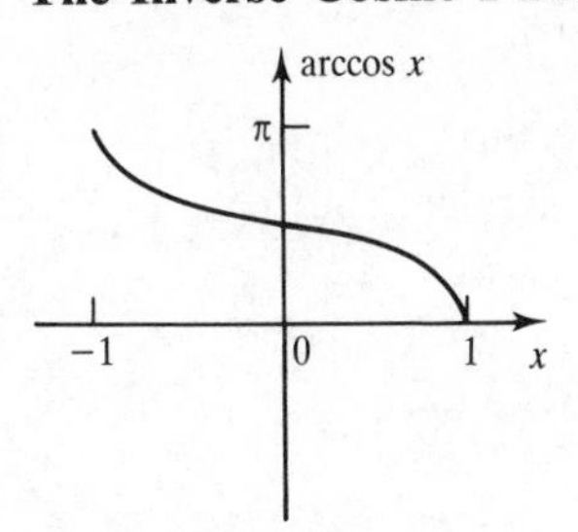

Figure 15.20

By a similar reasoning to that employed to create the inverse sine function, the inverse cosine function is defined by the prescription

$$f(x) = \arccos x \qquad 0 \leqslant f(x) \leqslant \pi$$

The Inverse Tangent Function

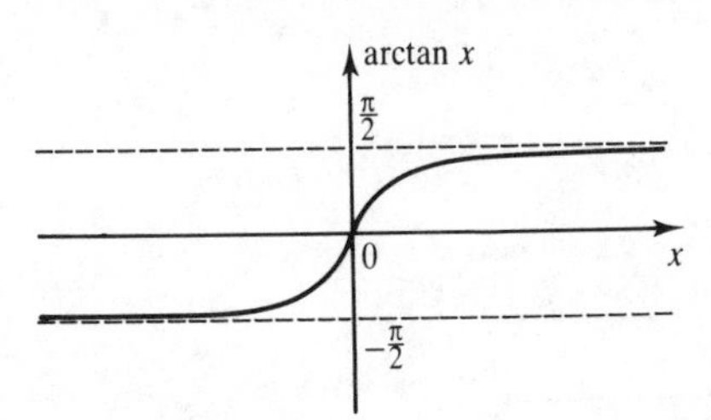

Figure 15.21

The inverse tangent function is defined by the prescription

$$f(x) = \arctan x \qquad -\pi/2 \leqslant f(x) \leqslant \pi/2$$

EXAMPLES

1 Find the inverse of each of the functions with the following outputs:

(a) $f(x) = x + 5$ **(d)** $f(x) = \dfrac{x}{4}$

(b) $f(x) = x - 6$ **(e)** $f(x) = x^3$

(c) $f(x) = 3x$

(a) x is retrieved from $x + 5$ by subtracting 5. Hence

$$\operatorname{arc} f(x) = x - 5$$

(b) x is retrieved from $x - 6$ by adding 6. Hence

$$\operatorname{arc} f(x) = x + 6$$

(c) x is retrieved from $3x$ by dividing by 3. Hence

$$\operatorname{arc} f(x) = x/3$$

(d) x is retrieved from $x/4$ by multiplying by 4. Hence

$$\operatorname{arc} f(x) = 4x$$

(e) x is retrieved from x^3 by taking the third root. Hence

$$\operatorname{arc} f(x) = x^{1/3}$$

2 Give the inverse of each of the following:

(a) $f(x) = \sin x$ **(b)** $f(x) = \mathrm{e}^x$

(a) Taking the sine of an angle is not a simple process and it is not possible to separate the process into a composition of simple processes. Instead, the arc notation is used:

$$\text{arc} f(x) = \arcsin x$$

(b) $\text{arc} f(x) = \text{arc}\, e^x$ Here the x is retrieved by taking the natural logarithm. Thus

$$\text{arc} f(x) = \ln(x)$$

3 Find the inverse of each of the following compositions:

(a) $f(x) = x^2 + 1$ **(c)** $f(x) = \sin(3x + 4)$

(b) $f(x) = (2x - 1)^2$ **(d)** $f(x) = e^x + 1$

(a) $\text{arc} f(x) = (x - 1)^{1/2}$ **(c)** $\text{arc} f(x) = \dfrac{(\arcsin x) - 4}{3}$

(b) $\text{arc} f(x) = \dfrac{\sqrt{x + 1}}{2}$ **(d)** $\text{arc} f(x) = \ln(x - 1)$

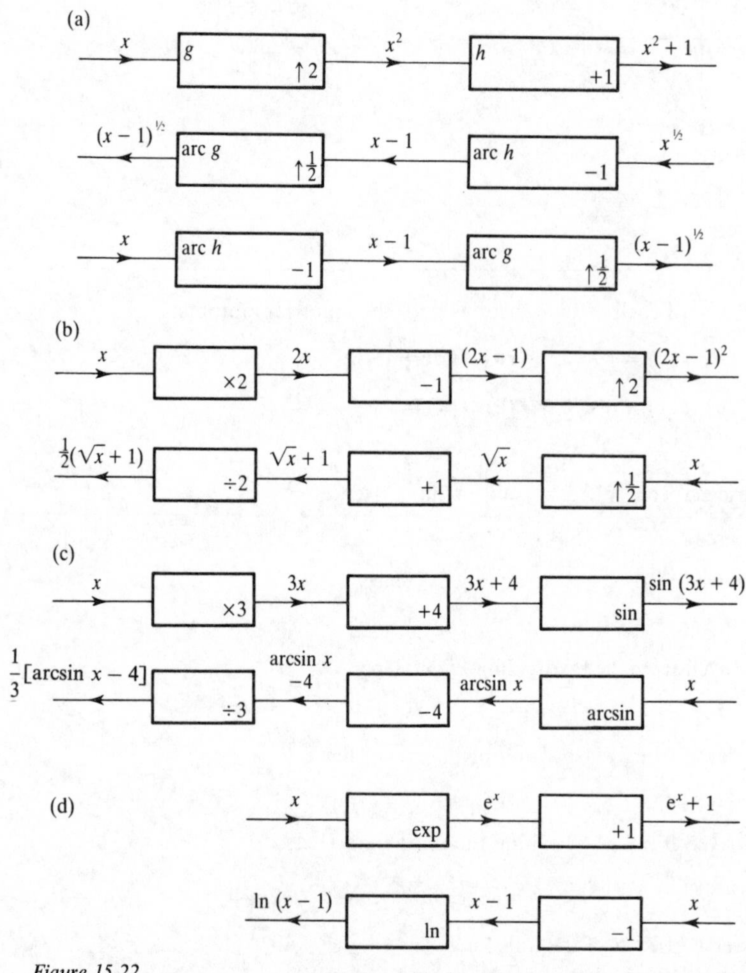

Figure 15.22

4 Of the following, which has an inverse that is a function?

(a) $f(x) = x^4 - 2$ **(c)** $f(x) = \ln(x)$

(b) $f(x) = \dfrac{5x}{3} + 5$

(a) $f(x) = x^4 - 2$

The graph is similar in shape to a parabola but it rises more steeply as can be seen from Figure 15.23. Its inverse is not a function.

(b) $f(x) = 5x/3 + 5$ $\quad \text{arc} f(x) = 3(x - 5)/5$ is a function.

(c) $f(x) = \ln(x)$ $\quad \text{arc} f(x) = e^x$ – the exponential function.

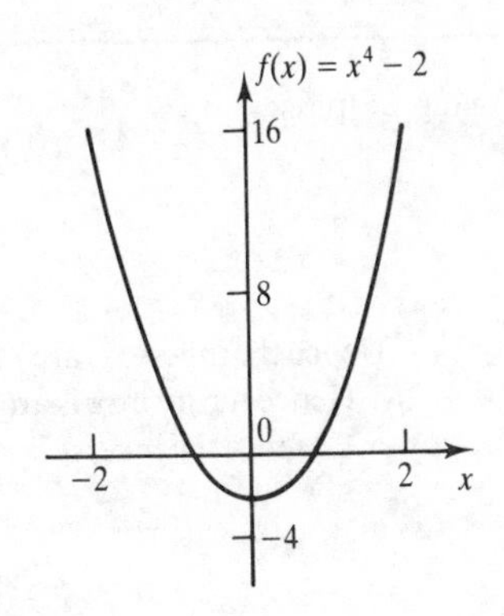

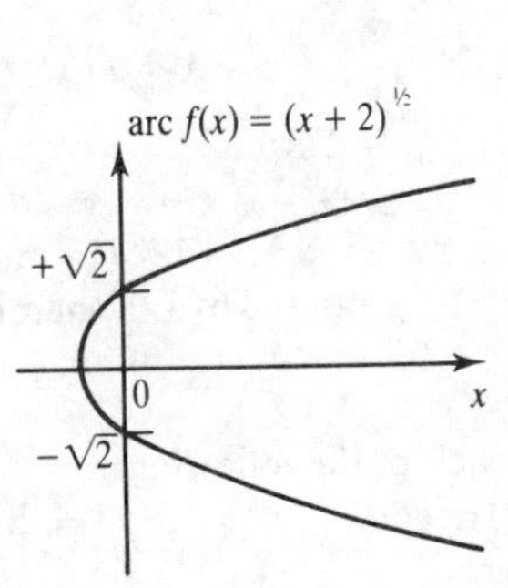

Figure 15.23

5 Draw the graph of the inverse of each of the following:

(a) $f(x) = 2x + 1$ **(b)** $f(x) = x^3$

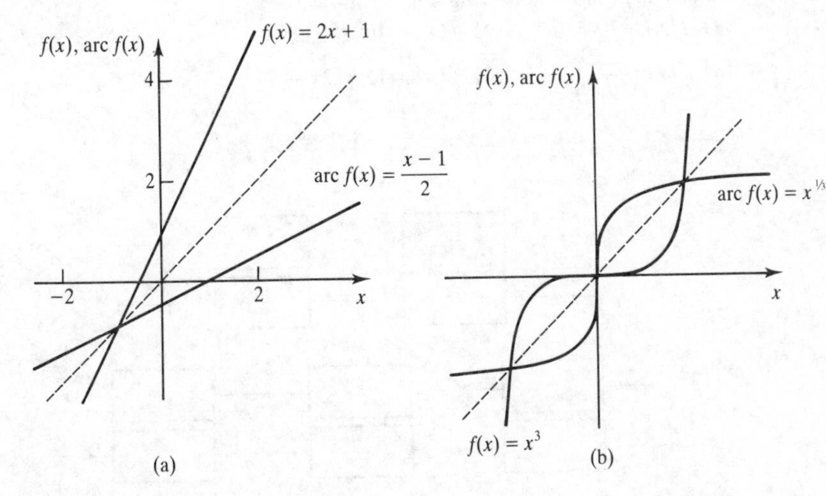

Figure 15.24

6 Derive the inverse function of $f(x) = \cos 2x$.

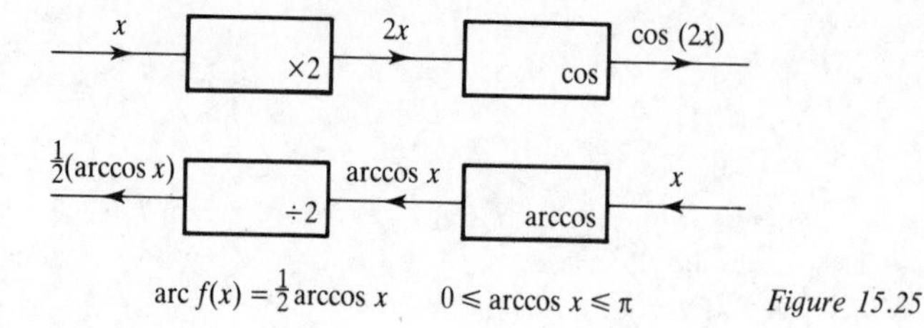

arc $f(x) = \frac{1}{2}$ arccos x $\quad 0 \leqslant \arccos x \leqslant \pi$ *Figure 15.25*

EXERCISES

1 Find the inverse of each of the functions with the following outputs:

(a) $f(x) = \dfrac{3x}{4}$ **(c)** $f(x) = x^{1/4}$

(b) $f(x) = x - 2/3$

(a) $f(x) = 3x/4$ x is retrieved from $3x/4$ by multiplying by $*$. Hence the inverse is arc $f(x) = *x$.
(b) $f(x) = x - 2/3$ x is retrieved by $*$ the number $*$ to $x - 2/3$. Hence the inverse is arc $f(x) = *$.
(c) $f(x) = x^{1/4}$ x is retrieved by $*$ the number $x^{1/4}$ to the power $*$. Hence the inverse is arc $f(x) = *$.

2 Give the inverse of each of the following:

(a) $f(x) = \cos x$ **(b)** $f(x) = \ln(x)$

(a) Taking the cosine of an angle is not a simple process and it is not possible to separate the process into a composition of simple processes. Instead, the arc notation is used:

arc $f(x) = *$

(b) arc $f(x) =$ arc ln x Here the x is retrieved by raising e to the power $*$:

arc $f(x) = *$

3 Find the inverse of each of the following compositions:

(a) $f(x) = 6x - 1$ **(c)** $f(x) = \tan(5x - 2)$

(b) $f(x) = \dfrac{1}{2x + 3}$ **(d)** $f(x) = \log_2(3x - 1)$

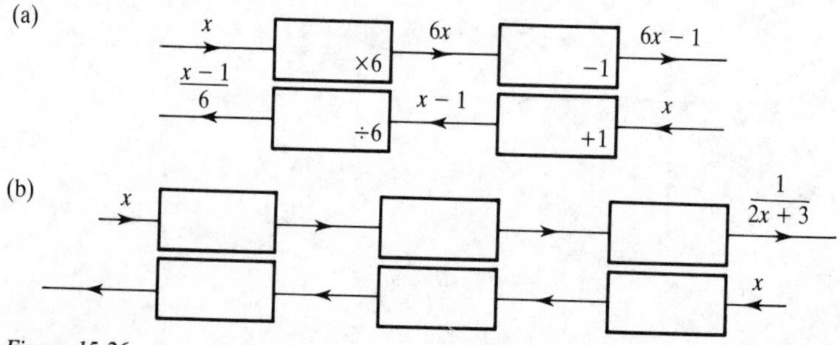

Figure 15.26

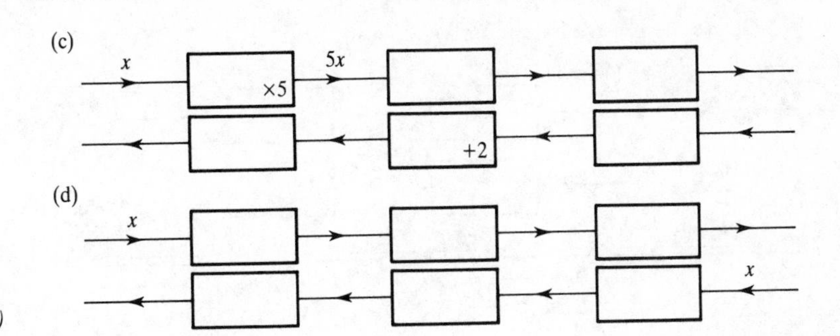

Figure 15.26 (continued)

4 Of the following, which has an inverse that is a function?

(a) $f(x) = 3x - 4$ **(c)** $f(x) = e^x$

(b) $f(x) = x^2 - 9$

(a) $f(x) = 3x - 4$

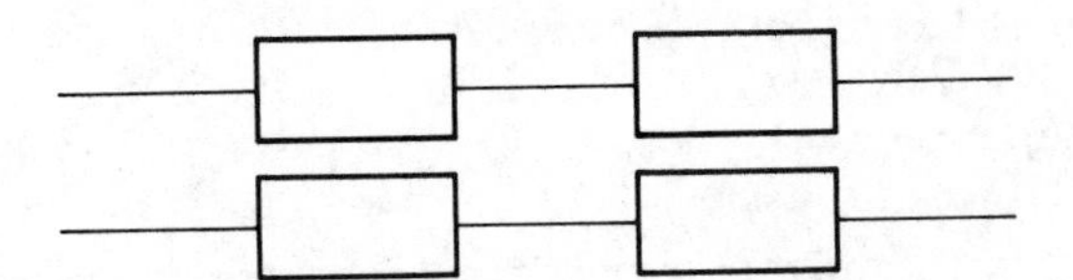

Figure 15.27

(b) $f(x) = x^2 - 9$

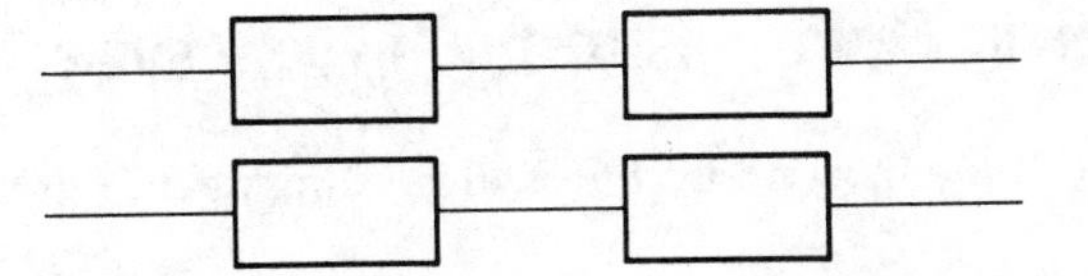

Figure 15.28

(c) $f(x) = e^x$ arc $f(x) = *$ which is/is not a function.

5 Draw the graph of the inverse of each of the following:

(a) $f(x) = 3x$ **(b)** $f(x) = \sin 3x$

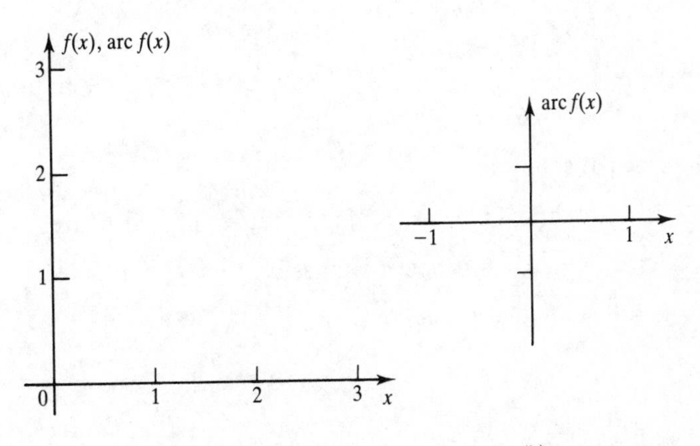

Figure 15.29

(b)

6 Derive the inverse function of $f(x) = \sin 2x$.

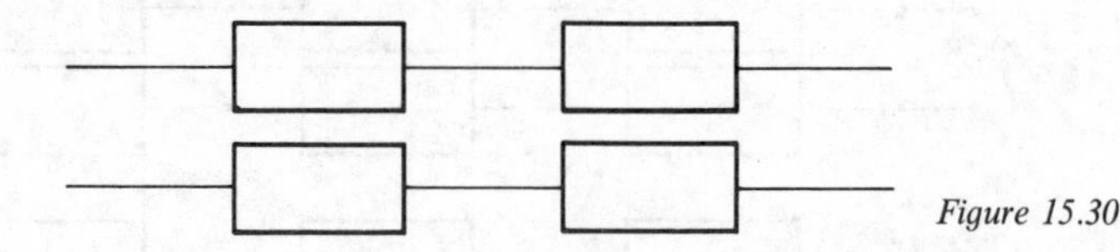

Figure 15.30

Unit 2 Solving Trigonometric Equations Using Inverse Functions

Test yourself with the following:

Solve the following equations:

(a) $4 \cos x = 1$

(b) $5 \sin 6x = 3$

(c) $9 \cos x - 4 \sin x = 2\sqrt{2}$ for $0 \leqslant x \leqslant \pi$

Trigonometric Equations of the Type $a \sin x + b \cos x = c$

Given any pair of numbers a and b it is always possible to find an angle β such that

$$\frac{b}{a} = \tan \beta$$

This is because $\tan \beta$ ranges from $-\infty$ to $+\infty$ as does any ratio b/a. For example, if $a = 5$ and $b = -23$ then

$$\frac{b}{a} = \frac{-23}{5} = -4.6$$

so that

$$\tan \beta = -4.6$$

with principal solution

$$\beta = \arctan(-4.6) = -77.74° \text{ to four significant figures}$$

If b/a is written as $\tan \beta$ then it is possible to say that

$$b = R \sin \beta \text{ and } a = R \cos \beta$$

because

$$\frac{b}{a} = \frac{R \sin \beta}{R \cos \beta} = \tan \beta$$

Furthermore,

$$a^2 + b^2 = R^2 \sin^2 \beta + R^2 \cos^2 \beta$$
$$= R^2$$

As a consequence, any equation of the form

$$a \sin x + b \cos x = c$$

can be rewritten as

$$R \cos \beta \sin x + R \sin \beta \cos x = c$$

that is

$$R \sin(x + \beta) = c$$

where

$$R^2 = a^2 + b^2 \text{ and } \beta = \arctan\left(\frac{b}{a}\right)$$

For example, to solve the equation

$$\sqrt{3} \sin x + \cos x = \sqrt{2}$$

we designate $\sqrt{3} = R \cos \beta$ and $1 = R \sin \beta$ thereby giving $\tan \beta = 1/\sqrt{3}$. This latter equation has the principal solution

$$\beta = \pi/6$$

Also, $R^2 = 3 + 1$ giving $R = 2$ so that the original equation can now be written as

$$\sqrt{3} \sin x + \cos x = 2 \sin(x + \pi/6)$$
$$= \sqrt{2}$$

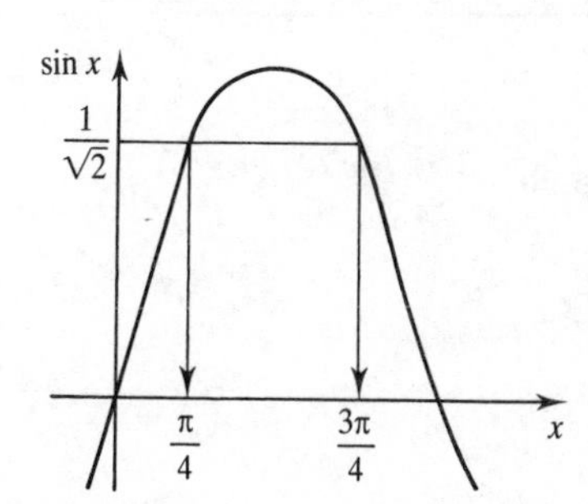

Figure 15.31

that is

$$\sin(x + \pi/6) = 1/\sqrt{2}$$

This has the principal solution

$$x + \pi/6 = \pi/4, \text{ therefore } x = \pi/12$$

From Figure 15.31 we can see that the complete solution to the equation is

$$x = \pi/12 \pm 2n\pi \text{ or } x = 7\pi/12 \pm 2n\pi$$

EXAMPLES

1 Solve the following equations: **(a)** $2 \sin x = 1$
(b) $2 \cos 2x = \sqrt{3}$
(c) $4 \sin x + 3 \cos x = 5$ for $0° \leqslant x \leqslant 180°$

(a) $2 \sin x = 1$
$\sin x = 1/2$

therefore

$$x = \arcsin(1/2) = \pi/6 \pm 2n\pi \text{ and } 5\pi/6 \pm 2n\pi$$

(b) $2\cos 2x = \sqrt{3}$
$\cos 2x = \sqrt{3}/2$
therefore

$$2x = \arccos(\sqrt{3}/2) = \pm\pi/6 \pm 2\pi$$

Therefore $x = \pm\pi/12 \pm n\pi$.
(c) $4\sin x + 3\cos x = 5$ This equation can be written as

$$R\sin(x + \beta) = 5$$

where $R = (4^2 + 3^2)^{1/2} = 5$ and $\beta = \arctan(3/4) = 0.64$ rad. Substituting into the above equation gives

$$5\sin(x + 0.64) = 5$$

therefore

$$\sin(x + 0.64) = 1$$

Hence

$$x + 0.64 = \arcsin 1$$
$$= \pi/2$$
$$= 1.57$$

giving

$$x = 0.93 \text{ rad}$$

EXERCISES

1 Solve the following equations: **(a)** $5\tan x = 10$
(b) $2\cot 3x = 1$
(c) $2\sin x - 5\cos x = 4$ for $0 \leqslant x \leqslant 2\pi$

(a) Since $5\tan x = 10$ then $\tan x = *$ giving

$$x = \arctan(*) = *^\circ$$

Therefore $x = *^\circ \pm 180n^\circ$.
(b) Since $2\cot 3x = 1$ then $\cot 3x = *$ so that $\tan 3x = *$ giving

$$3x = \arctan(*) = *^\circ \pm 180n^\circ$$

hence

$$x = *^\circ \pm 60n^\circ$$

(c) $2\sin x - 5\cos x = 4$ This equation can be written in the form

$$R\sin(x + \beta) = 4$$

where $R = (*^2 + *^2)^{1/2} = \sqrt{*}$ and $\beta = \arctan(-*/*) = -*$ rad. Substituting into the equation above gives

$$\sqrt{*}\sin(x - *) = 4$$

that is

$$x - * = \arcsin(4/\sqrt{*})$$
$$= * \text{ rad or } * \text{ rad}$$

therefore

$$x = * \pm 2n\pi \text{ or } * \pm 2n\pi$$

Module 15: Further exercises

1 Find the inverse of each of the functions with the following outputs:

(a) $f(x) = x - 1$ (c) $f(x) = \dfrac{5x}{6}$

(b) $f(x) = x + \dfrac{2}{5}$ (d) $f(x) = x^{1/6}$

2 Give the inverse of each of the following:

(a) $f(x) = \operatorname{cosec} x$ (b) $f(x) = 5^x$

3 Find the inverse of each of the following compositions:

(a) $f(x) = 9x - 7$ (d) $f(x) = \ln(3x + 4)$

(b) $f(x) = 2x^2 + 1$ (e) $f(x) = e^{x^2}$

(c) $f(x) = \sec(2x - 1)$

4 Of the following, which has an inverse that is a function?

(a) $f(x) = x^6 - 64$ (c) $f(x) = \log_2(3x)$

(b) $f(x) = \dfrac{3x^3}{5} - 1$

5 Draw the graph of the inverse of each of the following:

(a) $f(x) = 3x - 5$ (c) $f(x) = \tan x$

(b) $f(x) = \dfrac{2x}{3}$ (d) $f(x) = e^x$

6 Derive the inverse functions of:

(a) $f(x) = x^2$ (b) $f(x) = e^x$

7 Solve the following equations:

(a) $3 \operatorname{cosec} x = 4$ (c) $3 \sin x - 5 \cos x = 2$

(b) $2 \sec 5x = 8$

Chapter 3: Miscellaneous Exercises

1 Do the graphs of

$$f(x) = 2\log(x) \text{ and } g(x) = \log(2x)$$

intersect?

2 Which of the following points do not lie on the graph of

$$f(x) = \frac{x}{(x+1)}$$

$(0, 0)$, $(-\frac{1}{2}, -1)$, $(\frac{1}{2}, \frac{1}{3})$, $(-1, 1)$, $(-2, 2)$?

3 Let

$$f(x) = \log\left(\frac{1+x}{1-x}\right)$$

form a new function g by replacing x in $f(x)$ by

$$\frac{2x}{1+x^2}$$

Write g in terms of f.

4 Describe the graph of

$$x^2 - 4y^2 = 0$$

5 The two functions C and S are defined as the even and odd parts of the function f where

$$f(x) = a^x$$

Show that

$$C^2(x) - S^2(x) \equiv 1$$

6 Show, by using a diagram, that

$$\operatorname{arc}(f \circ g) \equiv (\operatorname{arc} g) \circ (\operatorname{arc} f)$$

7 Draw the graph of the inverse of the inverse of the sine function.

8 Which function is identical to its inverse?

9 Is it possible to find a value of x such that

$$\log_a(x) = a^x \text{ for } a > 1?$$

4

■ ■ ■ Sequences, Series and Limits

The aims of this chapter are to:

1. Introduce sequences as functions defined on the integers and develop the ideas of convergence and divergence.
2. Develop series as sequences of partial sums and extend the binomial theorem to any real index.
3. Consider limits of functions and thereby discuss the idea of continuity.

This chapter contains three modules:

Module 16: Sequences
Sequences are defined as functions and their limits are introduced, as are the concepts of convergence and divergence.

Module 17: Series
Series are defined, their limits considered and the issues of their convergence and divergence discussed. The binomial theorem is extended and the exponential number e is defined.

Module 18: Limits of Functions
Rational functions, compound prescriptions and limits of functions are considered, leading to a discussion of the continuity and discontinuity of functions.

Module 16: Sequences

Aims: To introduce sequences as functions defined on the integers and to develop the ideas of convergence and divergence.

Objectives: When you have read this module you will be able to:

▶▶ Recognize a sequence as a function.

▶▶ Plot the graphs of sequences and find the limits of sequences.

▶▶ Recognize specific types of sequence and derive their limits using the rules of limits.

▶▶ Use convergence tests to distinguish between convergent and divergent sequences.

Unit 1 Functions with Integer Input

Test yourself with the following:

Find the next two terms and the form of the general term for each of the following sequences:

(a) $0,5,10,15,\ldots$
(b) $5,25,125,625,\ldots$
(c) $-1/64,1/32,-1/16,1/8,\ldots$
(d) $4,-12,36,-108,\ldots$
(e) $1,2,3,2,0,-3,-5,\ldots$
(f) $-16,2,-14,-12,-26,\ldots$

Sequences

Any function f whose domain is restricted to positive or negative integer values n has a range in the form of a sequence of numbers. Accordingly, such a function is called a **sequence**. For example, the function defined by the prescription

$$f(n) = 5n - 2 \text{ where } n \text{ is a positive integer} \geqslant 1$$

is a sequence, the first three range values corresponding to successive domain values 1, 2 and 3 are

$$f(1) = 5 \times 1 - 2 = 3$$
$$f(2) = 5 \times 2 - 2 = 8$$
$$f(3) = 5 \times 3 - 2 = 13$$

Each range value $f(n)$ of the sequence is called a **term** of the sequence.

Graphs of Sequences

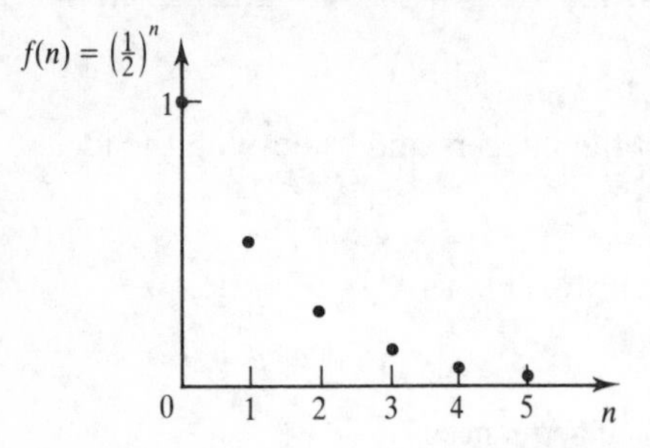

Figure 16.1

Since the range of a sequence consists of a sequence of discrete numbers, the graph of a sequence will take the form of a collection of isolated points in the cartesian plane. For example, the sequence defined by the prescription

$$f(n) = (\tfrac{1}{2})^n \qquad n = 0, 1, 2, \ldots$$

has the graph of Figure 16.1.

The Arithmetic Sequence

Any sequence defined by the prescription

$$f(n) = a + nd \qquad n = 0, 1, 2, \ldots$$

is called an **arithmetic sequence**. The number a is the first term as it is the output from the function when $n = 0$. The number d is called the **common difference** as it is the difference between any output other than the first and its predecessor.

For example, the sequence

$$f(n) = 3 - 2n \qquad n = 0, 1, 2, \ldots$$

has a sequence of range values:

$$3, 1, -1, -3, -5, \ldots$$

Here, the first term $a = 3$ and the common difference $d = -2$.

The Geometric Sequence

Any sequence f of the form

$$f{:}f(n) = ar^n$$

where n is an integer $\geqslant 0$, is called a **geometric sequence**. The number a is the first term as it is the output from the function when $n = 0$. The number r is called the **common ratio** as it is the ratio between any output other than the first and its predecessor. For example, the sequence

$$f(n) = 3(2^n) \qquad n = 0, 1, 2, \ldots$$

has a sequence of range values:

$$\begin{aligned} f(0) &= 3(2^0) = 3 \\ f(1) &= 3(2^1) = 6 \\ f(2) &= 3(2^2) = 12 \\ f(3) &= 3(2^3) = 24 \\ &\vdots \end{aligned}$$

Here, the first term $a = 3$ and the common ratio $r = 2$.

The Harmonic Sequence

The **harmonic sequence** is defined by the prescription

$$f(n) = \frac{1}{n} \qquad n = 1, 2, 3, \ldots$$

The sequence of range values is then:

$$1, \tfrac{1}{2}, \tfrac{1}{3}, \tfrac{1}{4}, \tfrac{1}{5}, \tfrac{1}{6}, \ldots$$

Note that in the geometric and arithmetic sequences the first term is given for $n = 0$ whereas in the harmonic sequence the first term is given for $n = 1$. There is no hard-and-fast rule to decide on the value of n for starting the sequence off – it is all a matter of circumstance.

Sequences Generated from Initial Terms

Sequences need not be formed by continually adding a certain number as in the arithmetic sequences, nor by multiplying by a certain number as in geometric sequences. A sequence can be generated by calculating a range value from earlier, specified range values in the sequence called **initial terms**. For example, each term in the **Fibonacci sequence** is obtained by adding the two preceding terms. Naturally, the first two terms, the initial terms, must be given; they are 0 and 1. This then permits the sequence to be generated:

$$0, 1, 1, 2, 3, 5, 8, 13, \ldots$$

The prescription for the Fibonacci sequence is

$$f(n+2) = f(n+1) + f(n) \qquad n = 0, 1, 2, \ldots$$

where the initial terms are:

$$\begin{aligned} f(0) &= 0 \\ f(1) &= 1 \end{aligned}$$

Sequences Generated from Other Prescriptions

A sequence can also be generated from other prescriptions. For example, the sequence

$$1, 2, 5, 10, 17, 26, \ldots$$

is generated by the prescription

$$f(n) = 1 + n^2 \qquad n = 0, 1, 2, 3, \ldots$$

Again, this sequence is neither an arithmetic sequence nor a geometric sequence. Determining the prescription for a sequence from the first few terms is a problem that is usually solved by experience and ingenuity.

EXAMPLES

Find the next two terms and the form of the general term for each of the following sequences:

(a) $1, 3, 5, 7, \ldots$ **(d)** $-3, 6, -12, 24\ldots$
(b) $-4, -1, 2, 5, \ldots$ **(e)** $1, 1, 1, 3, 5, 9, \ldots$
(c) $16, 8, 4, 2, \ldots$ **(f)** $1, 1, 4, 10, 28, 76, \ldots$

(a) $1, 3, 5, 7, 9, 11, \ldots$ general term $f(n) = 1 + 2n$.
(b) $-4, -1, 2, 5, 8, 11, \ldots$ general term $f(n) = -4 + 3n$.
(c) $16, 8, 4, 2, 1, 1/2, \ldots$ general term $f(n) = 16/2^n$.
(d) $-3, 6, -12, 24, -48, 96, \ldots$ general term $f(n) = -3(-2)^n$.
(e) $1, 1, 1, 3, 5, 9, 17, 31, \ldots$ general term:

$$f(n+3) = f(n+2) + f(n+1) + f(n) \qquad n \geqslant 0$$

given $f(0) = f(1) = f(2) = 1$.
(f) $1, 1, 4, 10, 28, 76, 208, 568, \ldots$ general term:

$$f(n+2) = 2[f(n+1) + f(n)] \qquad n \geqslant 0$$

given $f(0) = f(1) = 1$.

EXERCISES

Find the next two terms and the form of the general term for each of the following sequences:

(a) $1, 2, 4, 8, \ldots$ **(d)** $-2, 2, -2, 2, \ldots$
(b) $-18, -22, -26, -30, \ldots$ **(e)** $2, 1, 1, 2, 0, 3, \ldots$
(c) $5, 5/3, 5/9, 5/27, \ldots$ **(f)** $6, 8, 7, 7.5, 7.25, 7.375, \ldots$

(a) $1, 2, 4, 8, 16, *, \ldots$ general term $f(n) = *$.
(b) $-18, -22, -*, -30, \ldots$ general term $f(n) = -18 + *$.
(c) $5, \frac{5}{3}, \frac{5}{9}, \frac{5}{27}, \frac{5}{*}, \frac{5}{*}, \ldots$ general term $f(n) = \frac{5}{*}$.
(d) $-2, 2, -2, 2, *, *, \ldots$ general term $f(n) = *(*)^n$.
(e) $2, 1, 1, 2, 0, 3, *, 4, \ldots$ general term:

$$f(n) = f(n-2) * f(n-3) * f(n-1)$$

(f) $6, 8, 7, 7.5, 7.25, 7.375, *, *, \ldots$ general term:

$$f(n) = (*)[f(n-1) * f(n-2)]$$

Unit 2 Limits of Sequences

Test yourself with the following:

Evaluate the following limits:

(a) $\lim_{n\to\infty} (4 - 3n)$

(b) $\lim_{n\to\infty} (-2)^{2n}$

(c) $\lim_{n\to\infty} 0.999(1^n)$

(d) $\lim_{n\to\infty} \left(\frac{1}{100}\right)^{-n}$

(e) $\lim_{n\to\infty} \frac{n-4}{3-n}$

(f) $\lim_{n\to\infty} \frac{4-n^2}{3n^2+n+1}$

(g) $\lim_{n\to\infty} \cos\left(\frac{\pi n^2}{n^2-1}\right)$

(h) $\lim_{n\to\infty} \arccos\left(\frac{\sqrt{3n^3}}{3n^3-1}\right)$

(i) $\lim_{n\to\infty} \ln\left(\frac{3+4n}{5n}\right)$

(j) $\lim_{n\to\infty} \exp\left(\frac{-1}{n}\right)$

(k) $\lim_{n\to\infty} \sin\left(\frac{n\pi}{2}\right)$

Unbounded Input Values

Each successive term of a sequence is obtained by increasing the input integer n by unity. As a consequence it is natural to ask how the output behaves as the input integer n takes on arbitrarily large values. This introduces the idea of **infinity** and the concept of a **limit**.

Infinity

There is no largest integer; this fact is embodied in the statement that the integers increase without bound – no matter how large an integer you can think of you can always add 1 to it to obtain an even larger integer. An alternative description of this idea is to say that the integers increase to infinity where infinity is represented by the symbol ∞ (negative infinity is represented by $-\infty$). Unfortunately, this latter description tends to give the notion of infinity some quantifiable aspect that it does not possess. It must be clearly understood that although infinity is a well-established concept it is numerically indefinable and so it cannot be used as a number in any arithmetic calculations.

The Concept of a Limit

The idea that the integers increase without bound necessitates the introduction of the idea of a limit. For example, consider the sequence with output

$$f(n) = (\tfrac{1}{2})^n \qquad n = 0, 1, 2, \ldots$$

The successive outputs from this function that correspond to the inputs $0, 1, 2, 3, 4, \ldots$ are:

$$1, \tfrac{1}{2}, \tfrac{1}{4}, \tfrac{1}{8}, \tfrac{1}{16}, \ldots \text{ respectively}$$

As the input value n increases so the output value $(\frac{1}{2})^n$ decreases, in fact, as n increases by 1 the output is equal to half the previous output.

Put another way, as n increases without bound – becomes arbitrarily large – so $(\frac{1}{2})^n$ decreases and becomes closer to zero. The output can never be equal to zero because n can never be chosen to be large enough – there is no bound to n. However, it is possible to obtain a value of $(\frac{1}{2})^n$ as close to zero as we wish, by selecting a sufficiently large value of n. This statement is expressed as follows:

as n becomes arbitrarily large so $(\frac{1}{2})^n$ becomes arbitrarily close to zero

or, more succinctly

as n approaches infinity so $(\frac{1}{2})^n$ approaches zero

Here, zero is called the limit of $(\frac{1}{2})^n$. This is written in symbolic form as

$$\lim_{n \to \infty} \left(\frac{1}{2}\right)^n = 0$$

or, in words, the limit of $(\frac{1}{2})^n$ as n approaches ∞ is 0.

Infinite Limits

Sometimes, as n becomes arbitrarily large so does $f(n)$. For example, the sequence $f(n) = n^2$ becomes larger faster than n does. In this case, we write the limit as

$$\lim_{n \to \infty} n^2 = \infty$$

Be aware This notation can be misleading if it is not correctly understood. It does not mean what it appears to mean, namely, that the limit is equal to infinity. It cannot, because infinity is not defined so nothing can be said to be equal to it. What it does mean is that as n becomes arbitrarily large so does n^2. The notation is unfortunate but it is in common usage so we have to accept it in the absence of anything better.

Arithmetic Sequences

The limit of an arithmetic sequence $f(n) = a + nd$ is infinite because $a + nd$ increases without bound as n increases without bound. We write

$$\lim_{n \to \infty} (a + nd) = \infty \text{ or } \lim_{n \to \infty} (a + nd) = -\infty$$

depending on whether d is positive or negative. For example, the terms of the arithmetic sequence $2, 4, 6, 8, \ldots$ increase without bound. Again, beware of the ∞ on the right-hand side of the equation; its use suggests some quantifiable aspect that it does not possess.

Geometric Sequences

The limit of a geometric sequence $f(n) = ar^n$ depends upon the value of r:

(1) $\lim_{n \to \infty} ar^n$ for $r = 1$

Here the sequence has range values $a, a, a, a, \ldots$ and clearly their limit is a:

$$\lim_{n \to \infty} a = a$$

(2) $\lim_{n \to \infty} ar^n$ for $r > 1$

Take, for example, $r = 2$. Here the sequence has range values $a, 2a, 4a, 8a, \ldots$ and their limit is ∞.

(3) $\lim_{n \to \infty} ar^n$ for $-1 < r < 1$

Take, for example, $r = -\frac{1}{2}$. Here the sequence has range values $a, -a/2, a/4, -a/8, \ldots$ and their limit is 0 – they alternate in sign but their magnitudes decrease.

(4) $\lim_{n \to \infty} ar^n$ for $r = -1$

Here the sequence has range values $a, -a, a, -a, \ldots$. In this case the magnitude of successive terms stays the same but the signs alternate. Consequently, it is not possible to define a unique limit and so no limit exists at all. Limits must be uniquely defined:

$$\lim_{n \to \infty} a(-1)^n = \text{undefined}$$

(5) $\lim_{n \to \infty} ar^n$ for $r < -1$

Take, for example, $r = -2$. Here the sequence has range values $a, -2a, 4a, -8a, \ldots$. In this case the magnitude of successive terms increases but the signs alternate. Consequently, it is not possible to define a unique limit and so, again, no limit exists at all.

These five cases can be summarized as follows:

$$\lim_{n \to \infty} ar^n = \begin{cases} a \text{ if } r = 1 \\ 0 \text{ if } -1 < r < 1 \\ \infty \text{ if } r > 1 \\ \text{undefined if } r \leqslant -1 \end{cases}$$

Harmonic Sequence

The limit of the harmonic sequence $f(n) = 1/n$ is zero. As n increases without bound so $1/n$ approaches zero.

Sequences Generated from Initial Terms or Other Prescriptions

The limits of sequences generated from initial terms or other prescriptions depend upon each sequence individually. There is no overall rule, and to find their limits requires informed use of the rules of limits.

The Rules of Limits

Limits can be manipulated algebraically according to the following rules: assume

$$\lim_{n \to \infty} a_n = A \text{ and } \lim_{n \to \infty} b_n = B$$

Sums and Differences

The limit of a sum or difference is the sum or difference of the limits:

$$\lim_{n \to \infty} (a_n + b_n) = \lim_{n \to \infty} a_n + \lim_{n \to \infty} b_n = A + B$$

and

$$\lim_{n \to \infty} (a_n - b_n) = \lim_{n \to \infty} a_n - \lim_{n \to \infty} b_n = A - B$$

For example,

$$\lim_{n \to \infty} \frac{1}{n^2} = 0 \text{ and } \lim_{n \to \infty} 5 = 5$$

so that

$$\begin{aligned} \lim_{n \to \infty} \left(5 + \frac{1}{n^2}\right) &= \lim_{n \to \infty} 5 + \lim_{n \to \infty} \frac{1}{n^2} \\ &= 5 + 0 \\ &= 5 \end{aligned}$$

Multiplication by a Constant

$$\lim_{n \to \infty} (\beta a_n) = \beta \lim_{n \to \infty} (a_n)$$

where β is constant. In particular,

$$\lim_{n \to \infty} \beta = \beta \ (\beta \text{ a constant})$$

Products

The limit of a product is the product of the limits:

$$\lim_{n \to \infty} a_n . b_n = \lim_{n \to \infty} a_n . \lim_{n \to \infty} b_n = AB$$

For example,

$$\begin{aligned} \lim_{n \to \infty} \frac{5}{n^2} &= \lim_{n \to \infty} 5 . \lim_{n \to \infty} \frac{1}{n^2} \\ &= 5 \times 0 \\ &= 0 \end{aligned}$$

Quotients
The limit of a quotient is the quotient of the limits:

$$\lim_{n\to\infty} \frac{a_n}{b_n} = \frac{\lim\limits_{n\to\infty} a_n}{\lim\limits_{n\to\infty} b_n} = \frac{A}{B} \text{ provided } B \neq 0$$

For example,

$$\lim_{n\to\infty} \frac{5}{2 - 1/n^n} = \frac{\lim\limits_{n\to\infty} 5}{\lim\limits_{n\to\infty} (2 - 1/n^n)} = \frac{5}{2-0} = \frac{5}{2}$$

Powers

$$\lim_{n\to\infty} p^n = \begin{cases} 0 \text{ if } -1 < p < 1 \\ 1 \text{ if } p = 1 \\ \infty \text{ if } p > 1 \\ \text{undefined if } p \leqslant -1 \end{cases}$$

For example,

$$\lim_{n\to\infty} (\tfrac{1}{3})^n = 0 \text{ whereas } \lim_{n\to\infty} 3^n = \infty$$

Function of a Sequence
The limit of a function of a sequence is the function of the limit of the sequence:

$$\lim_{n\to\infty} f(a_n) = f(\lim_{n\to\infty} a_n) = f(A), \text{ provided } \lim_{x\to A} f(x) = f(A)$$

For example,

$$\lim_{n\to\infty} \cos\left(\frac{1}{n}\right) = \cos\left(\lim_{n\to\infty} \frac{1}{n}\right) = \cos 0 = 1$$

EXAMPLES

Evaluate the following limits:

(a) $\lim_{n\to\infty}\left(1-\frac{n}{100}\right)$ **(g)** $\lim_{n\to\infty}\sin\left(\frac{\pi n}{n-1}\right)$

(b) $\lim_{n\to\infty} 5^n$ **(h)** $\lim_{n\to\infty}\arcsin\left(\frac{\sqrt{2n^2}}{2n^2+3n+4}\right)$

(c) $\lim_{n\to\infty} 1000(0.999)^n$ **(i)** $\lim_{n\to\infty}\ln\left(\frac{1}{n}\right)$

(d) $\lim_{n\to\infty} 25(-1)^n$ **(j)** $\lim_{n\to\infty}\exp\left(\frac{n+1}{n-1}\right)$

(e) $\lim_{n\to\infty}\frac{n}{n+1}$ **(k)** $\lim_{n\to\infty}\cos 2n\pi$

(f) $\lim_{n\to\infty}\frac{n^2+n+1}{n^2-n-1}$

(a) $\lim_{n\to\infty}\left(1-\frac{n}{100}\right)=\lim_{n\to\infty}(1)-\left(\frac{1}{100}\right)\lim_{n\to\infty} n=-\infty$

Here, writing the limit as $-\infty$ means that $-n$ becomes arbitrarily large and negative.

(b) $\lim_{n\to\infty} 5^n=\infty$ because $5>1$

(c) $\lim_{n\to\infty} 1000(0.999)^n=0$ because $0.999<1$

(d) $\lim_{n\to\infty} 25(-1)^n=$ undefined

(e) $\lim_{n\to\infty}\frac{n}{n+1}=\lim_{n\to\infty}\frac{1}{1+1/n}=1$ (dividing numerator and denominator by n)

(f) $\lim_{n\to\infty}\frac{n^2+n+1}{n^2-n-1}=\lim_{n\to\infty}\frac{1+1/n+1/n^2}{1-1/n-1/n^2}=1$ (dividing numerator and denominator by n^2)

(g) $\lim_{n\to\infty}\sin\left(\frac{\pi n}{n-1}\right)=\lim_{n\to\infty}\sin\pi\left(\frac{1}{1-1/n}\right)=\sin\pi=0$

(h) $\lim_{n\to\infty}\arcsin\left(\frac{\sqrt{2n^2}}{2n^2+3n+4}\right)=\lim_{n\to\infty}\arcsin\left(\frac{\sqrt{2}}{2+3/n+4/n^2}\right)=\arcsin\frac{1}{\sqrt{2}}=\frac{\pi}{4}$

(i) $\lim_{n\to\infty}\ln\frac{1}{n}=\ln\left(\lim_{n\to\infty}\frac{1}{n}\right)=\ln 0=-\infty$

(j) $\lim_{n\to\infty}\exp\left(\frac{n+1}{n-1}\right)=\exp\left(\lim_{n\to\infty}\frac{n+1}{n-1}\right)=\exp\left(\lim_{n\to\infty}\frac{1+1/n}{1-1/n}\right)=\exp(1)=e$

(k) $\lim_{n\to\infty}\cos 2n\pi$ is undefined because $\cos\infty$ is not defined.

EXERCISES

Evaluate the following limits:

(a) $\lim_{n\to\infty} (8 + 2n)$

(b) $\lim_{n\to\infty} \left(\frac{1}{3}\right)^n$

(c) $\lim_{n\to\infty} 1000(0.999)^{-n}$

(d) $\lim_{n\to\infty} \left(\frac{-1}{10}\right)^n$

(e) $\lim_{n\to\infty} \frac{3n + 2}{4n - 5}$

(f) $\lim_{n\to\infty} \frac{2n^2 - 3n + 4}{5n^2 - 6n + 7}$

(g) $\lim_{n\to\infty} \tan\left(\frac{\pi n + 1}{4n + 3}\right)$

(h) $\lim_{n\to\infty} \operatorname{arccot}\left(\frac{n - 5}{n + 5}\right)$

(i) $\lim_{n\to\infty} \ln\left(\frac{3}{n} + e\right)$

(j) $\lim_{n\to\infty} \exp\left(\frac{n^2 + n + 1}{2n^2 + 3n + 4}\right)$

(k) $\lim_{n\to\infty} \tan\left(\frac{n\pi}{4}\right)$

(a) $\lim_{n\to\infty} (8 + 2n) = \lim_{n\to\infty} * + \lim_{n\to\infty} * = \infty$

(b) $\lim_{n\to\infty} \left(\frac{1}{3}\right)^n = *$

(c) $\lim_{n\to\infty} 1000(0.999)^{-n} = 1000 \lim_{n\to\infty} (0.999)^{-n} = *$

(d) $\lim_{n\to\infty} \left(\frac{-1}{10}\right)^n = *$

(e) $\lim_{n\to\infty} \frac{3n + 2}{4n - 5} = \frac{\lim_{n\to\infty} (*)}{\lim_{n\to\infty} (*)} = \frac{3}{4}$

(f) $\lim_{n\to\infty} \frac{2n^2 - 3n + 4}{5n^2 - 6n + 7} = \lim_{n\to\infty} \frac{2 - * + 4/*}{* - 6/n + */n^2} = \frac{2}{5}$

(g) $\lim_{n\to\infty} \tan \frac{\pi n + 1}{4n + 3} = \tan\left(\lim_{n\to\infty} \frac{\pi + *}{4 + *}\right)$

$$= \tan *$$
$$= 1$$

(h) $\lim_{n\to\infty} \operatorname{arccot} \frac{n - 5}{n + 5} = \operatorname{arccot}\left(\lim_{n\to\infty} \frac{1 - *}{1 + *}\right)$

$$= \operatorname{arccot} 1$$
$$= *$$

(i) $\lim_{n\to\infty} \ln\left(\frac{3}{n} + e\right) = \ln\left(\lim_{n\to\infty} \frac{3}{n} + e\right)$

$$= \ln *$$
$$= 1$$

(j) $\lim_{n\to\infty} \exp\left(\dfrac{n^2+n+1}{2n^2+3n+4}\right) = \exp\left(\lim_{n\to\infty} \dfrac{1+*+*}{*+*/n+*/n^2}\right)$
$= \exp(*)$
$= 1.6487\ldots$

(k) $\lim_{n\to\infty} \tan\left(\dfrac{n\pi}{4}\right) = *$

Unit 3 Convergence and Divergence

Test yourself with the following:

Which of the following sequences converge and which diverge?

(a) $f(n) = \dfrac{n-1}{n^2}$

(b) $f(n) = \dfrac{0.0001}{0.1^n}$

(c) $f(n) = \dfrac{10f(n-1)-2}{5}$, $f(0) = 10$

(d) $f(n) = \tan\left(\dfrac{\pi n^2}{4n^2-1}\right)$

(e) $f(n) = (0.1)f(n-1)$, $f(0) = 1000$

Finite Limits

If a sequence possesses a finite limit then it is said to **converge** to that limit. If a sequence does not converge then it is said to **diverge**. A sequence diverges if:

(1) It has an infinite limit. For example, the sequence defined by $f(n) = 3 - 4n$ has a limit of $-\infty$ and so diverges.

(2) It has an undefined limit. For example, the sequence defined by $f(n) = \cos(n\pi)$ has terms that alternate as ± 1. As a consequence it is not possible to define a unique limit so the limit is undefined. This sequence diverges because it does not converge.

EXAMPLES

Which of the following sequences converge and which diverge?

(a) $f(n) = 3 - 3n$

(b) $f(n) = (0.99)^{99n}$

(c) $f(n+1) = \dfrac{f(n)+2}{5}$, $f(0) = 1$

(d) $f(n) = \tan\left(\dfrac{\pi n}{2n+1}\right)$

(e) $f(n) = \dfrac{1}{f(n-1)}$, $f(0) = 1/2$

(a) $f(n) = 3 - 3n$ diverges as it is an arithmetic sequence.
(b) $f(n) = (0.99)^{99n}$ converges as it is a geometric sequence with common ratio 0.99.

(c) $f(n+1) = \dfrac{f(n)+2}{5}$, $f(0) = 1$

$$f(0) \qquad = 1$$

$$f(1) = \frac{3}{5} \quad = 0.6$$

$$f(2) = \frac{2.6}{5} \quad = 0.52$$

$$f(3) = \frac{2.52}{5} \quad = 0.504$$

$$f(4) = \frac{2.504}{5} \quad = 0.5008$$

$$f(5) = \frac{2.5008}{5} = 0.50016$$

The sequence appears to converge to 0.5. This result could also be obtained as follows:

$$\lim_{n\to\infty} f(n+1) = \lim_{n\to\infty} \frac{f(n)+2}{5}$$

$$= \frac{1}{5} \lim_{n\to\infty} f(n) + \frac{2}{5}$$

Now, if

$$\lim_{n\to\infty} f(n) = F \text{ then } \lim_{n\to\infty} f(n+1) = F$$

so that

$$F = \tfrac{1}{5}F + \tfrac{2}{5}$$

therefore

$$F = 0.5$$

(d) $f(n) = \tan(\pi n/(2n+1))$ has the limit $\tan \pi/2$ which is not defined. The sequence diverges.

(e) $f(n) = 1/f(n-1) \qquad f(0) = 1/2$

$$f(0) = 0.5$$

$$f(1) = 2$$

$$f(2) = 0.5$$

$$f(3) = 2$$

The sequence oscillates between 2 and 0.5 and so diverges because it does not converge to a unique limit.

EXERCISES

Which of the following sequences converge and which diverge?

(a) $f(n) = 1000 + \dfrac{1000}{n}$

(b) $f(n) = (0.0001)(0.1)^{-n}$

(c) $f(n+1) = \dfrac{3f(n) - 5}{6}$, $f(0) = 2$

(d) $f(n) = \cot\left(\dfrac{\pi n^2}{10 - 4n^2}\right)$

(e) $f(n) = \dfrac{f(n-1)}{0.1}$, $f(0) = 1$

(a) $f(n) = 1000 + 1000/n$ converges to *.

(b) $f(n) = (0.0001)(0.1)^{-n}$ is a geometric sequence with common ratio * and so the sequence *verges.

(c) $f(n+1) = \dfrac{3f(n) - 5}{6}$, $f(0) = 2$

$$f(0) = 2$$
$$f(1) = 0.*$$
$$f(2) = *$$
$$f(3) = *$$
$$f(4) = *$$
$$f(5) = *$$

The sequence *verges.

(d) $f(n) = \cot\left(\dfrac{\pi n^2}{10 - 4n^2}\right)$

$$\lim_{n \to \infty} \frac{\pi n^2}{10 - 4n^2} = \lim_{n \to \infty} \frac{\pi}{*/n^2 - *} = \pi/4$$

therefore

$$\lim_{n \to \infty} \cot\left(\frac{\pi n^2}{10 - 4n^2}\right) = \cot\left(\lim_{n \to \infty} \frac{\pi n^2}{10 - 4n^2}\right)$$
$$= \cot *$$
$$= * \qquad (*\text{verges})$$

(e) $f(n) = \dfrac{f(n-1)}{0.1} \qquad f(0) = 1$

$$f(0) = 1$$
$$f(1) = \frac{1}{0.1} = 10$$
$$f(2) = \frac{10}{0.1} = 100$$
$$\vdots$$
$$f(n) = 10^n$$

The sequence *verges.

Module 16: Further exercises

1 Find the next two terms and the form of the general term for each of the following sequences:

(a) $1, 6, 11, 16, \ldots$

(b) $-101, -99, -97, -95, \ldots$

(c) $10, 1, 0.1, 0.01, \ldots$

(d) $1234.5, 123.45, 12.345, \ldots$

(e) $2, 4, -2, 6, -8, \ldots$

(f) $5, 8, 13, 21, \ldots$

2 Evaluate the following limits:

(a) $\lim_{n \to \infty} (5 + 9n)$

(b) $\lim_{n \to \infty} \left(\frac{1}{4}\right)^{-3n}$

(c) $\lim_{n \to \infty} 0.999(-1)^n$

(d) $\lim_{n \to \infty} 152^{-2n}$

(e) $\lim_{n \to \infty} \frac{n + 7}{n - 7}$

(f) $\lim_{n \to \infty} \frac{1 - n - n^2}{5n^2 - 2n - 3}$

(g) $\lim_{n \to \infty} \sec\left(\frac{\pi n^2}{2n^2 - 1}\right)$

(h) $\lim_{n \to \infty} \text{arccosec}\left(\frac{6n}{3n + 4}\right)$

(i) $\lim_{n \to \infty} \ln\left(\frac{n^2 - 1}{n - 1}\right)$

(j) $\lim_{n \to \infty} \exp(-n)$

(k) $\lim_{n \to \infty} \sin 2n\pi$

3 Which of the following sequences converge and which diverge?

(a) $f(n) = 6 + n$

(b) $f(n) = (n + 1)^2$

(c) $f(n) = \frac{16f(n - 1) - 2}{32}, \ f(0) = 1$

(d) $f(n) = \sec\left(\frac{\pi n + 2}{3 - 6n}\right)$

(e) $f(n) = f^2(n - 1), \ -1 < f(n) < 1$

Module 17: Series

Aims: To develop series as sequences of partial sums and to extend the binomial theorem to any real index.

Objectives: When you have read this module you will be able to:

- ▶▶ Calculate the partial sums of a sequence and generate a series.
- ▶▶ Recognize and use the sigma notation.
- ▶▶ Manipulate arithmetic and geometric series.
- ▶▶ Use the binomial theorem for any real power.

Unit 1 Partial Sums and Series

Test yourself with the following:

1 Form the first four partial sums of each of the following sequences:

(a) $f(n) = \dfrac{1}{n!}$ (b) $f(n) = \dfrac{n}{n+1}$

2 Find the sum of the first 10 terms of each of the following:

(a) $f(n) = 5 + 2n$ (c) $f(n) = 8(3^n)$

(b) $f(n) = 2 - 4n$ (d) $f(n) = \dfrac{5}{4^n}$

3 If £P is invested at r% per annum compound interest, the amount £A accrued after n years is given by

$$A = P\left(1 + \frac{r}{100}\right)^n$$

Find the amount accrued after 20 years if £500 is invested every year for at 15% per annum.

Partial Sums

The partial sums of a sequence f with output $f(n)$ are as follows:

zeroth partial sum $s(0) = f(0)$
first partial sum $s(1) = f(0) + f(1)$
second partial sum $s(2) = f(0) + f(1) + f(2)$
⋮
Nth partial sum $s(N) = f(0) + f(1) + f(2) + \ldots + f(N)$

Series

The partial sums $s(0)$, $s(1)$, $s(2)$,... define a sequence s. However, because this sequence is created from the partial sums of an originating sequence it is called a **series**. For example, the sequence defined by

$$f(n) = (\tfrac{1}{2})^n$$

can be used to define the series:

$$\begin{aligned} s(0) &= 1 \\ s(1) = 1 + \tfrac{1}{2} &= \tfrac{3}{2} \\ s(2) = 1 + \tfrac{1}{2} + \tfrac{1}{4} &= \tfrac{7}{4} \\ s(4) = 1 + \tfrac{1}{2} + \tfrac{1}{4} + \tfrac{1}{8} &= \tfrac{15}{8} \end{aligned}$$

The Sigma Notation

To avoid writing out a long list of expressions added together, the sigma notation is used. This takes the form of an uppercase Greek sigma (Σ – a Greek equivalent of the letter S) followed by the general term of the originating sequence. For example, if $f(n)$ is the originating sequence then

$$\begin{aligned} s(N) &= f(0) + f(1) + f(2) + \ldots + f(N) \\ &= \sum_{n=0}^{N} f(n) \end{aligned}$$

Note the indication above and below the sigma that denotes the range of values of the variable n. Because the variable n does not appear in the final form of $s(N)$ it is called a **dummy variable** – any other symbol would do just as well. Also notice that the originating sequence $f(n)$ here starts with $n = 0$, but it may be that in some cases it starts with some other value of n such as 1, for example.

Arithmetic Series

The general term of the arithmetic series is given by the sum of the first N terms of the arithmetic sequence:

$$\begin{aligned} s(N-1) &= f(0) + f(1) + \ldots + f(N-1) \\ &= \sum_{n=0}^{N-1} (a + nd) \\ &= a + (a + d) + (a + 2d) + \ldots + [a + (N-1)d] \end{aligned}$$

Note that because the first term of this series is given when $n = 0$, the Nth term is given when $n = N - 1$.

This sum of terms is reversed to read:

$$s(N-1) = [a + (N-1)d] + [a + (N-2)d] + \ldots + a$$

and then both equations are added to give

$$\begin{aligned} 2s(N-1) &= [2a + (N-1)d] + [2a + (N-1)d] + \ldots + [2a + (N-1)d] \\ &= N[2a + (N-1)d] \end{aligned}$$

Because there are N terms of the form $[2a + (N-1)d]$. This means that we have obtained an alternative, algebraic expression to the Σ-notation for the $(N-1)$th partial sum:

$$s(N-1) = \frac{N}{2}[2a + (N-1)d]$$

This is the Nth term of the arithmetic series. Note that

$$\begin{aligned} \lim_{N\to\infty} s(N-1) &= \lim_{N\to\infty} \frac{N}{2}[2a + (N-1)d] \\ &= \lim_{N\to\infty} Na + \lim_{N\to\infty}\left[\frac{N(N-1)d}{2}\right] \\ &= \pm\infty \end{aligned}$$

so all arithmetic series diverge to $+\infty$ if $d > 0$ and $-\infty$ if $d < 0$.

Geometric Series

The general term of the geometric series is given by the sum of the first N terms of the geometric sequence:

$$\begin{aligned} s(N-1) &= f(0) + f(1) + \ldots + f(N-1) \\ &= \sum_{n=0}^{N-1} ar^n \\ &= a + ar + ar^2 + \ldots + ar^{N-1} \end{aligned}$$

If this sum is multiplied by the common ratio r the result reads as

$$rs(N-1) = ar + ar^2 + ar^3 + \ldots + ar^N$$

Now perform the subtraction:

$$s(N-1) - rs(N-1) = a - ar^N$$

Factorizing both sides of this equation gives

$$(1-r)s(N-1) = a(1-r^N)$$

so that

$$s(N-1) = \frac{a(1-r^N)}{1-r}$$

which is an alternative, algebraic expression to the Σ-notation for the $(N-1)$th partial sum of the geometric sequence – the Nth term of the geometric series. Note that

$$\begin{aligned}\lim_{N\to\infty} s(N-1) &= \lim_{N\to\infty} \frac{a(1-r^N)}{1-r}\\ &= \lim_{N\to\infty}\left[\left(\frac{a}{1-r}\right)-\left(\frac{r^N}{1-r}\right)\right]\\ &= \frac{a}{1-r}-\left(\frac{a}{1-r}\right)\left(\lim_{N\to\infty} r^N\right)\\ &= \begin{cases}\dfrac{a}{1-r} \text{ if } -1<r<1\\ \pm\infty \text{ if } r>1 \text{ (depending on whether } a \text{ is } -\text{ve or } +\text{ve)}\\ \text{undefined if } r<-1\end{cases}\end{aligned}$$

A geometric series converges to $a/(1-r)$ only if $-1<r<1$.

Note that this form for the general term of a geometric series is not valid if the common ratio $r=1$ – the denominator is zero when $r=1$. Indeed, a geometric sequence with common ratio equal to unity is an arithmetic sequence with common difference equal to zero. For example,

$$5+5+5+5+\ldots$$

Is both an arithmetic series with common difference $d=0$ and a geometric series with common ratio $r=1$.

EXAMPLES

1 Form the first four partial sums of each of the following sequences:

(a) $f(n)=n$ **(b)** $f(n)=n^2+1$

(a) $f(n)=n$

$$\begin{aligned}s(0)&=f(0) && &&=0\\ s(1)&=f(0)+f(1) &&=0+1 &&=1\\ s(2)&=f(0)+f(1)+f(2) &&=0+1+2 &&=3\\ s(3)&=f(0)+f(1)+f(2)+f(3) &&=0+1+2+3 &&=6\end{aligned}$$

(b) $f(n)=n^2+1$

$$\begin{aligned}s(0)&=f(0) && &&=0\\ s(1)&=f(0)+f(1) &&=0^2+1^2 &&=1\\ s(2)&=f(0)+f(1)+f(2) &&=0^2+1^2+2^2 &&=5\\ s(3)&=f(0)+f(1)+f(2)+f(3) &&=0^2+1^2+2^2+3^2 &&=14\end{aligned}$$

2 Find the sum of the first 10 terms of each of the following:

(a) $f(n)=3+n$ **(c)** $f(n)=2^n$
(b) $f(n)=2-4n$ **(d)** $f(n)=(1/4)^n$

(a) Arithmetic sequence: $s(N-1) = (N/2)[2a + (N-1)d]$ where $a = 3$, $d = 1$, $N = 10$ so that

$$s(9) = 5(6 + 9) = 75$$

(b) Arithmetic sequence: $s(N-1) = (N/2)[2a + (N-1)d]$ where $a = 2, d = -4, N = 10$ so that

$$s(9) = 5(4 - 36) = -160$$

(c) Geometric sequence: $s(N-1) = a(1 - r^N)/(1 - r)$ where $a = 1$, $r = 2$, $N = 10$ so that

$$s(9) = \frac{1 - 2^{10}}{1 - 2} = 1023$$

(d) Geometric sequence: $s(N-1) = a(1 - r^N)/(1 - r)$ where $a = 1$, $r = 1/4$, $N = 10$ so that

$$s(9) = \frac{1 - (1/4)^{10}}{1 - 1/4} = 1.333 \text{ to three dec. pl.}$$

3 If £P is invested at r% per annum compound interest, the amount £A accrued after n years is given by

$$A = P(1 + r/100)^n$$

Find the amount accrued after 10 years if £100 is invested every year for 10 years at 10% per annum.

After 10 years the first £100 accrues to $100(1.1)^{10}$
After 9 years the second £100 accrues to $100(1.1)^9$
After 8 years the third £100 accrues to $100(1.1)^8$
⋮
After 1 year the last £100 accrues to $100(1.1)$

These amounts form a geometric sequence of 10 terms with common ratio 1.1 and first term 100(1.1). Their sum is, therefore

$$\begin{aligned} s &= \frac{[100(1.1)][1 - (1.1)^{10}]}{1 - 1.1} \\ &= \frac{(110)(-1.593\,742\,5\ldots)}{-0.1} \\ &= 1753.12 \end{aligned}$$

EXERCISES

1 Form the first four partial sums of each of the following sequences:

(a) $f(n) = \dfrac{1}{n}$ **(b)** $f(n) = n!$

(a) $f(n) = 1/n$

$$\begin{aligned} s(1) &= f(1) &&= * &&= * \\ s(2) &= f(1) + f(2) &&= * + * &&= * \\ s(3) &= f(1) + f(2) + f(3) &&= * + * + * &&= 11/6 \\ s(4) &= f(1) + f(2) + f(3) + f(4) &&= * + * + * + * &&= */12 \end{aligned}$$

(b) $f(n) = n!$

$$s(0) = f(0) \qquad = *! \qquad = *$$

$$s(1) = f(0) + f(1) \qquad = * + 1! \qquad = * + * \qquad = *$$

$$s(2) = f(0) + f(1) + f(2) \qquad = * + 1! + * = * + 1 + * = 4$$

$$s(3) = f(0) + f(1) + f(2) + f(3) = 4 + * \qquad = *$$

2 Find the sum of the first 10 terms of each of the following:

(a) $f(n) = 1 - 2n$ **(c)** $f(n) = 4^{n+1}$
(b) $f(n) = n - 5$ **(d)** $f(n) = (0.1)^{2n}$

(a) Arithmetic sequence: $s(N-1) = (N/2)[2a + (N-1)d]$ where $a = *$, $b = *$, $N = 10$ so that

$$s(9) = 5(* - *) = -*$$

(b) * sequence: $s(N-1) = (*)$ where $a = *$, $d = *$, $N = 10$ so that

$$s(9) = *(*) = -5$$

(c) Geometric sequence: $s(N-1) = a(1 - r^N)/(1 - r)$ where $a = *$, $r = *$, $N = 10$ so that

$$s(*) = \frac{*(1 - *^{10})}{1 - *} = 1\,398\,100$$

(d) * sequence: $s(N-1) = (*)$ where $a = *$, $r = *$, $N = 10$ so that

$$s(*) = \frac{*(1 - *^{10})}{1 - *} = *$$

3 If £X is due to be paid in n years time, that amount of money has a present value of £Y where

$$Y = \frac{X}{(1 + r/100)^n}$$

and where money currently earns r% compound interest per annum. Find the present value of 20 annual payments of £100, the first instalment being paid one year from now. The current rate of interest is 10%.

The first instalment is currently worth *
The second instalment is currently worth *
The third instalment is currently worth *
⋮
The last instalment is currently worth *

These amounts form a * sequence of 20 terms with common * = * and first term *. Their sum is, therefore

$$s = \frac{(*)(* - *)}{* - *}$$

$$= 851.36$$

Unit 2 Limits of Series

Test yourself with the following:

1 Find the limits of the series formed from the partial sums of the following sequences:

$$\text{(a) } f(n) = \frac{5}{4^n} \qquad \text{(b) } f(n) = 6(0.1)^{3n}$$

2 Test the following series for convergence:

(a) $\sum_{n=0}^{\infty} \frac{1}{7^n}$ (d) $\sum_{n=1}^{\infty} \frac{n}{2^n}$

(b) $\sum_{n=0}^{\infty} \frac{1}{0.1^{-n}}$ (e) $\sum_{n=0}^{\infty} \frac{2^n}{n!}$

(c) $\sum_{n=0}^{\infty} \left(\frac{-99}{100}\right)^n$

The Harmonic Series

The arithmetic and geometric series have general Σ-notation terms that can be rewritten as an algebraic expression, so finding their limits is a relatively simple affair. Most series, however, are not capable of having their general, Σ-notation terms rewritten as an algebraic expression, which makes the problem of determining their limits more difficult, if not impossible. For example, the harmonic sequence

$$f(n) = \frac{1}{n} \qquad n = 1, 2, 3, \ldots$$

is convergent. The general term of the harmonic series written in Σ-notation is

$$s(N) = \sum_{n=1}^{N} \frac{1}{n}$$

Note that $f(0)$ does not exist for the harmonic sequence so the sum starts with $n = 1$. This sum cannot be rewritten as an algebraic expression and so we must use an alternative method to find its limit.

To find the limit of the harmonic series we proceed as follows. By grouping terms together, the harmonic series can be written as

$$s(N) = 1 + \frac{1}{2} + \left(\frac{1}{3} + \frac{1}{4}\right) + \left(\frac{1}{5} + \frac{1}{6} + \frac{1}{7} + \frac{1}{8}\right) + \left(\frac{1}{9} + \frac{1}{10} + \frac{1}{11} + \frac{1}{12} + \frac{1}{13} + \frac{1}{14} + \frac{1}{15} + \frac{1}{16}\right) + \ldots + \left(\ldots + \frac{1}{N}\right)$$

The total value of each bracket is greater than $1/2$. For example, the first bracket has a value

$$\frac{1}{3} + \frac{1}{4} = \frac{7}{12}$$
$$> \frac{1}{2}$$

If we define the series t where

$$t(N) = 1 + \frac{1}{2} + \left(\frac{1}{2}\right) + \left(\frac{1}{2}\right) + \ldots + \left(\frac{1}{2}\right)$$

where there are as many $(1/2)$s as there are brackets in the bracketed harmonic series, and where

$$s(N) > t(N)$$

$t(N)$ defines a divergent series t, because as N increases without bound so the number of $(1/2)$s increase without bound. The inequality states that the harmonic series term $s(N)$ is greater than the equivalent term of a diverging series. As a result, it can be shown that the limit of the harmonic series is greater than the limit of a known divergent sequence:

$$\lim_{N \to \infty} s(N) \geqslant \lim_{N \to \infty} t(N)$$

Therefore the harmonic series diverges.

Convergence Tests

The procedure used to show that the harmonic series diverges is known as a **convergence test**. The name of the test is the **comparison test**.

The Comparison Test

If each term $s(N)$ of a series s is *greater than* the corresponding term $t(N)$ of a series t that is known to diverge then the series s also diverges.

If each term $s(N)$ of a series s is *less than* the corresponding term $t(N)$ of a series t that is known to converge then the series s also converges.

The Alternating Sign Test

If a series is generated from an originating sequence, each of whose terms becomes progressively smaller and alternate in sign then the series converges. For example, the series

$$1 - \frac{1}{2} + \frac{1}{3} - \frac{1}{4} + \frac{1}{5} - \frac{1}{6} + \ldots$$

converges by the alternating sign test. This particular series is called the alternating harmonic series.

The Ratio Test

If

$$s(N) = f(1) + f(2) + \ldots + f(N)$$

and

$$\lim_{N \to \infty} \left| \frac{f(N+1)}{f(N)} \right| = L$$

then if

- $L < 1$ the series converges
- $L > 1$ the series diverges

and if

- $L = 1$ the test fails to produce a result

For example, the series

$$s(N) = 1 + \frac{1}{2!} + \frac{1}{3!} + \frac{1}{4!} + \ldots + \frac{1}{N!}$$

is such that

$$f(N) = \frac{1}{N!} \text{ and } f(N+1) = \frac{1}{(N+1)!}$$

so that

$$\begin{aligned}\frac{f(N+1)}{f(N)} &= \frac{1/(N+1)!}{1/N!} \\ &= \frac{N!}{(N+1)!} \\ &= \frac{1}{N+1} \text{ because } (N+1)! = (N+1)N!\end{aligned}$$

So that

$$\begin{aligned}\lim_{N\to\infty}\left|\frac{f(N+1)}{f(N)}\right| &= \lim_{N\to\infty}\frac{1}{N+1} \\ &= 0\end{aligned}$$

Here $L = 0$ so that $L < 1$ which means that the series converges by the ratio test.

Note that these tests will only test whether a series converges or not. If a series does converge the test will not reveal the limit to which the series converges. Indeed, it can often be very difficult, if not impossible, to find the limit of a known convergent series.

EXAMPLES

1 Find expressions for the limits of the series formed from the partial sums of the following sequences:

(a) $f(n) = \frac{2}{3^n}$ **(b)** $f(n) = 5(0.2^{-2n})$

(a) The series formed from the partial sums of the sequence

$$f(n) = \frac{2}{3^n}$$

is

$$s(N) = \sum_{n=0}^{N} \frac{2}{3^n}$$

This is a geometric series with common ratio 1/3 and first term 2. The alternative expression for $s(N)$ is then

$$s(N) = \frac{2[1 - (1/3)^N]}{1 - 1/3}$$

$$= \frac{2[1 - (1/3)^N]}{2/3}$$

$$= 3[1 - (1/3)^N]$$

The limit of this series is then

$$s(\infty) = \lim_{N \to \infty} 3[1 - (1/3)^N]$$

$$= 3\left[\lim_{N \to \infty} 1 - \lim_{N \to \infty} (1/3)^N\right]$$

$$= 3(1 - 0)$$

$$= 3 \qquad \text{(Note the use of } s(\infty) \text{ to denote } \lim_{N \to \infty} s(N).\text{)}$$

(b) The series formed from the partial sums of the sequence

$$f(n) = 5(0.2)^{-2n}$$

is

$$s(N) = \sum_{n=0}^{N} \frac{5}{(0.2^2)^n}$$

$$= 5 \sum_{n=0}^{N} (0.04^{-1})^n$$

$$= 5 \sum_{n=0}^{N} 25^n$$

This is a geometric series with common ratio 25 and first term 5. The alternative expression for $s(N)$ is then

$$s(N) = \frac{5(1 - 25^N)}{1 - 25}$$

$$= \frac{5(1 - 25^N)}{-24}$$

$$= \frac{-5}{24}(1 - 25^N)$$

The limit of this series is then

$$s(\infty) = \lim_{N \to \infty} \frac{-5}{24}(1 - 25^N)$$

$$= \frac{-5}{24}\left(\lim_{N \to \infty} 1 - \lim_{N \to \infty} 25^N\right)$$

$$= \frac{-5}{24}(1 - \infty)$$

$$= \infty$$

2 Test the following series for convergence:

(a) $\sum_{n=0}^{\infty} \frac{1}{3^n}$ (Use the comparison test)

(b) $\sum_{n=0}^{\infty} (1.1)^n$ (Use the comparison test)

(c) $\sum_{n=0}^{\infty} \left(\frac{-3}{5}\right)^n$ (Use the alternating sign test)

(d) $\sum_{n=0}^{\infty} \frac{10^n}{n!}$ (Use the ratio test)

(e) $\sum_{n=0}^{\infty} \frac{1}{2^n}$ (Use the ratio test)

(a) Since $1/3 < 1/2$ and $\sum_{n=0}^{\infty} 1/2^n$ converges then so $\sum_{n=0}^{\infty} 1/3^n$ converges by the comparison test.
(b) Since $1.1 > 1$ and $\sum_{n=0}^{\infty} 1^n$ diverges then so $\sum_{n=0}^{\infty} 1.1^n$ also diverges by the comparison test.
(c) Since $3/5 < 1$ and the sign alternates, the series converges by the alternating sign test.
(d) Writing $\sum_{n=0}^{\infty} f(n) = \sum_{n=0}^{\infty} 10^n/n!$ it is seen that

$$\frac{f(n+1)}{f(n)} = \frac{10^{n+1}/(n+1)!}{10^n/n!}$$

$$= \frac{10n!}{(n+1)!}$$

$$= \frac{10}{n+1}$$

then

$$\operatorname*{Lim}_{n \to \infty} \left|\frac{f(n+1)}{f(n)}\right| = \operatorname*{Lim}_{n \to \infty} \left(\frac{10}{n+1}\right)$$

$$= 0$$

Since $0 < 1$ the series converges by the ratio test.
(e) Writing $\sum_{n=0}^{\infty} f(n) = \sum_{n=0}^{\infty} (1/2)^n$ it is seen that

$$\frac{f(n+1)}{f(n)} = \frac{(1/2)^{n+1}}{(1/2)^n}$$

$$= 1/2$$

then

$$\operatorname*{Lim}_{n \to \infty} \left|\frac{f(n+1)}{f(n)}\right| = \operatorname*{Lim}_{n \to \infty} \frac{1}{2}$$

$$= \frac{1}{2}$$

Since $1/2 < 1$ the series converges by the ratio test.

EXERCISES

1 Find the limits of the series formed from the partial sums of the following sequences:

(a) $f(n) = (\frac{3}{4})^n$ **(b)** $f(n) = 1 - n$

(a) The series formed from the partial sums of the sequence

$$f(n) = (\tfrac{3}{4})^n$$

is

$$s(N-1) = \sum_{n=0}^{N-1} (*)^n$$

This is a geometric series with common ratio * and first term *. The alternative expression for $s(N)$ is then

$$s(N-1) = \frac{*(1 - *^N)}{1 - *}$$

$$= *(1 - *^N)$$

The limit of this series is then

$$s(\infty) = \lim_{N \to \infty} *(1 - *^N)$$

$$= *\left(\lim_{N \to \infty} 1 - \lim_{N \to \infty} *^N\right)$$

$$= *$$

(b) The series formed from the partial sums of the sequence

$$f(n) = 1 - n$$

is

$$s(N-1) = \sum_{n=0}^{N-1} *$$

This is an arithmetic series with common difference * and first term *. The alternative expression for $s(N-1)$ is then

$$s(N-1) = \frac{N}{2}[* + (N-1)*]$$

$$= \frac{N}{2}*$$

$$= \frac{N}{2}(* - N)$$

$$= \frac{*N}{2} - \frac{N^2}{2}$$

The limit of this series is then

$$s(\infty) = \lim_{N\to\infty}\left(\frac{*N}{2} - \frac{N^2}{2}\right)$$

$$= *$$

2 Test the following series for convergence:

(a) $\sum_{n=0}^{\infty}\left(\frac{1}{5}\right)^n$ (Use the comparison test)

(b) $\sum_{n=0}^{\infty}\left(\frac{4}{3}\right)^n$ (Use the comparison test)

(c) $\sum_{n=0}^{\infty}\left(\frac{-1}{6}\right)^{3n}$ (Use the alternating sign test)

(d) $\sum_{n=1}^{\infty}\frac{2^n}{n}$ (Use the ratio test)

(e) $\sum_{n=1}^{\infty}\frac{1}{n(n!)}$ (Use the ratio test)

(a) Since $1/5 < 1/2$ and $\sum_{n=0}^{\infty}(1/2)^n$ *verges then so $\sum_{n=0}^{\infty}(1/5)^n$ *verges by the comparison test. Each term of the series in question is * than the corresponding term of a series that is known to *verge.

(b) Since $4/3 > 1$ and $\sum_{n=0}^{\infty}1^n$ *verges then so $\sum_{n=0}^{\infty}(4/3)^n$ *verges by the comparison test. Each term of the series in question is * than the corresponding term of a series that is known to *verge.

(c) Since each term in the series * and the sign of each term *, the series converges by the alternating sign test.

(d) Writing $\sum_{n=1}^{\infty}f(n) = \sum_{n=1}^{\infty} *$ it is seen that

$$\frac{f(n+1)}{f(n)} = \frac{*}{*}$$

$$= *$$

then

$$\lim_{n\to\infty}\left|\frac{f(n+1)}{f(n)}\right| = \lim_{n\to\infty} *$$

$$= *$$

Since $* > 1$ the series *verges by the ratio test.

(e) Writing $\sum_{n=1}^{\infty}f(n) = \sum_{n=1}^{\infty} *$ it is seen that

$$\frac{f(n+1)}{f(n)} = \frac{*}{*}$$

$$= *$$

then

$$\lim_{n\to\infty}\left|\frac{f(n+1)}{f(n)}\right| = \lim_{n\to\infty} *$$

$$= *$$

Since $* < 1$ the series *verges by the ratio test.

Unit 3 The Binomial Theorem

Test yourself with the following:

1 Expand each of the following as far as the fourth term:

(a) $(1 + 4x)^{-3}$ (b) $(3 - x)^{2/3}$ (c) $\left(9 - \dfrac{1}{x}\right)^{-1/3}$

2 Find the nth term in the following expansion, and then use the ratio test to find those values of x for which the expansion converges:

$(1 - 2x)^{-1}$

The Binomial Expansion Revisited

In Unit 2 of Module 6 we saw that the binomial expansion of $(x + a)^n$, where n is a natural number is given as

$$(x + a)^n = x^n + nax^{n-1} + \frac{n(n-1)a^2x^{n-2}}{2!} + \frac{n(n-1)(n-2)a^3x^{n-3}}{3!} + \ldots + a^n$$

This is a sum of a finite number of terms – $n + 1$ of them in fact – and as such it converges because the sum of a finite number of finite terms is a finite number.

An alternative way of expressing this expansion is:

$$(x + a)^n = x^n + nax^{n-1} + \frac{n(n-1)a^2x^{n-2}}{2!} + \frac{n(n-1)(n-2)a^3x^{n-3}}{3!}$$

$$+ \ldots + n(n-1)(n-2)(\ldots)\frac{[n-(m-1)]a^m x^{n-m}}{m!} + \ldots \qquad m = 0, 1, 2, \ldots$$

where the last term a^n has been replaced by the general form of the $(m + 1)$th term. This is the form of the binomial expansion for any real number power n, not just a natural number n.

The Validity of the Binomial Expansion

Expressed in the alternative form, the binomial expansion takes on the appearance of a sum of an infinite number of terms – there is no stated restriction to a maximum value of m. The fact that the expansion terminates after a finite number of terms when n is a natural number is because when $m = n + 1$ the factor $n - (m - 1)$ in the general term equals zero. As this factor appears in all subsequent terms of the expansion, all subsequent terms are zero. The binomial expansion for natural number power n is a sum of a finite number of non-zero terms and an infinite number of zero terms.

If n is not a natural number it is not possible to find a natural number m that makes the factor $n - (m - 1)$ be equal to 0. The consequence is that for powers n that are not natural numbers, the binomial expansion is a sum of an infinite number of non-zero terms – it is the limit of a series and that brings into question its convergence.

Convergence of the Binomial Expansion

It can be shown by using the ratio test for convergence that the binomial series expansion of $(x + a)^n$, for values of n that are not positive integers, converges when the value of x is restricted to

$$-a < x < +a$$

that is

$$-1 < \frac{x}{a} < 1$$

For this reason we shall expand $(x + a)^n$ using the binomial theorem when n is not a positive integer as follows:

$$\begin{aligned}(x + a)^n &= (a + x)^n \\ &= \left[a\left(1 + \frac{x}{a}\right)\right]^n \\ &= a^n\left(1 + \frac{x}{a}\right)^n \\ &= a^n\left[1 + n\left(\frac{x}{a}\right) + \frac{n(n-1)}{z!}\left(\frac{x}{a}\right)^2 + \ldots\right]\end{aligned}$$

where, for the expansion to be valid (that is, convergent):

$$-1 < \frac{x}{a} < 1$$

For example, the expression

$$(1 - x)^{-1}$$

can be shown both by the binomial theorem and by direct division to produce the expansion

$$(1 - x)^{-1} = 1 + x + x^2 + x^3 + \ldots$$

However, by the ratio test, this expansion is only valid if

$$-1 < x < 1$$

The ratio test does not say anything for $x = \pm 1$. If $x = +1$ then the left-hand side is not defined and the right-hand side diverges. If, however, $x = -1$ then the left-hand side is equal to $1/2$ but the right-hand side becomes

$$1 - 1 + 1 - 1 + 1 - \ldots$$

which has no limit. In conclusion, the binomial series expansion of

$$(1 - x)^{-1}$$

is only valid for values of x where

$$-1 < x < +1$$

Note that the binomial expansion of $(1 - x)^{-1}$ was written in ascending powers of x. When we used the binomial expansion for expressions raised to positive integer powers such as $(1-x)^2$ then we wrote the expansion in descending powers of x:

$$\begin{aligned}(1-x)^2 &= [(-x)+1]^2\\ &= x^2-2x+1\end{aligned}$$

The reason we expand in ascending powers of x when the expression is raised to a non-positive integer power is because the magnitudes of successive terms in the expansion decrease. As a consequence, an approximation to the value of the complete expansion can be made by considering terms in ascending power order and cutting the expansion off after a specified number of terms.

EXAMPLES

1 Expand each of the following as far as the fourth term:

(a) $(1+2x)^{-1}$ **(b)** $(4-5x)^{1/2}$ **(c)** $\left(6+\dfrac{1}{x}\right)^{-1/4}$

(a) $$\begin{aligned}(1+2x)^{-1} &= 1+(-1)(2x)+\frac{(-1)(-2)(2x)^2}{2!}+\frac{(-1)(-2)(-3)(2x)^3}{3!}+\dots\\ &= 1-2x+4x^2-8x^3+\dots\end{aligned}$$

(b) $$\begin{aligned}(4-5x)^{1/2} &= [4(1-5x/4)]^{1/2}\\ &= 2[1+(-5x/4)]^{1/2}\\ &= 2\left[1+(\tfrac{1}{2})(-5x/4)+\frac{(\frac{1}{2})(-\frac{1}{2})(-5x/4)^2}{2!}+\frac{(\frac{1}{2})(-\frac{1}{2})(-\frac{3}{2})(-5x/4)^3}{3!}+\dots\right]\\ &= 2\left(1-\frac{5x}{8}-\frac{25x^2}{128}-\frac{125x^3}{1024}-\dots\right)\\ &= 2-\frac{5x}{4}-\frac{25x^2}{64}-\frac{125x^3}{512}-\dots\end{aligned}$$

(c) $$\begin{aligned}(6+1/x)^{-1/4} &= 6^{-1/4}\left(1+\frac{1}{6x}\right)^{-1/4}\\ &= 6^{-1/4}\left[1+\left(\frac{-1}{4}\right)\left(\frac{1}{6x}\right)\right.\\ &\quad+\left(\frac{-1}{4}\right)\left(\frac{-5}{4}\right)\left(\frac{1}{6x}\right)^2\left(\frac{1}{2!}\right)\\ &\quad\left.+\left(\frac{-1}{4}\right)\left(\frac{-5}{4}\right)\left(\frac{-9}{4}\right)\left(\frac{1}{6x}\right)^3\left(\frac{1}{3!}\right)+\dots\right]\\ &= 6^{-1/4}\left(1-\frac{1}{24x}+\frac{5}{1152x^2}-\frac{5}{9216x^3}\dots\right)\end{aligned}$$

2 Find the nth term in the following expansion and then use the ratio test to find those values for which the expansion converges:

$$(1-x)^{-1}$$

$$\begin{aligned}(1-x)^{-1} &= [1+(-x)]^{-1}\\ &= 1+(-1)(-x)+\frac{(-1)(-2)(-x)^2}{2!}\\ &\quad+\frac{(-1)(-2)(-3)(-x)^3}{3!}+\dots\\ &= 1+x+x^2+x^3+\dots+x^n+\dots\end{aligned}$$

Here the general term $f(n) = x^n$ so that $f(n+1) = x^{n+1}$. As a consequence

$$\frac{f(n+1)}{f(n)} = \frac{x^{n+1}}{x^n}$$
$$= x$$

So that

$$\lim_{n\to\infty}\left|\frac{f(n+1)}{f(n)}\right| = \lim_{n\to\infty}|x|$$
$$= |x|$$

Hence, by the ratio test for convergence, provided $|x| < 1$, that is, $-1 < x < 1$ the binomial expansion converges.

EXERCISES

1 Expand each of the following as far as the fourth term:

(a) $(1-3x)^{-2}$ **(b)** $(7+2x)^{1/3}$ **(c)** $\left(\frac{1}{x}-4\right)^{-1/5}$

(a) $(1-3x)^2 = [1+(*)]^{-2}$

$$= 1 + (-2)(*) + \frac{(-2)(-*)(*)^2}{2!} + \frac{(-2)(-*)(-*)(*)^3}{3!} + \ldots$$
$$= 1 + (-2)(*) - (-*)(*)^2 + (-*)(*)^3 + \ldots$$
$$= 1 + *x + *x^2 + *x^3 + \ldots$$

(b) $(7+2x)^{1/3} = 7^{1/3}[1+(*)]^{1/7}$

$$= 7^{1/3}\left[1 + (1/7)(*) + \frac{(1/7)(*)(*)^2}{2!} + \frac{(1/7)(*)(*)(*)^3}{3!} + \ldots\right]$$
$$= 7^{1/3}(1 + *x - *x^2 + *x^3 + \ldots)$$

(c) $(1/x-4)^{-1/5} = \left(-4+\frac{1}{x}\right)^{-1/5}$

$$= (-4)^{-1/5}[1+(*)]^{-1/5}$$
$$= (-4)^{-1/5}\left[1 + (-1/5)(*) + \frac{(-1/5)(*)(*)^2}{2!} + \frac{(-1/5)(*)(*)(*)^3}{3!} + \ldots\right]$$
$$= (-4)^{-1/5}(1 + *x + *x^2 + *x^3 + \ldots)$$

2 Find the nth term in the following expansion and then use the ratio test to find those values for which the expansion converges:

$$(1+5x)^{-1}$$

$$(1+5x)^{-1} = 1 + (-*)(*) + \frac{(-*)(-*)(*)^2}{*!} + \frac{(-*)(-*)(-*)(*)^3}{*!} + \ldots$$
$$= 1 + *x + *x^2 + *x^3 + \ldots + (*x)^n + \ldots$$

Here the general term $f(n) = (*)^n$ so that $f(n+1) = (*)^{n+1}$. As a consequence

$$\frac{f(n+1)}{f(n)} = \frac{*^{n+1}}{*^n}$$

$$= *$$

so that

$$\lim_{n \to \infty} \left| \frac{f(n+1)}{f(n)} \right| = \lim_{n \to \infty} *$$

$$= *$$

Hence, by the ratio test for convergence, provided $-1 < *x < 1$, that is, $-1/* < x < 1/*$, the binomial expansion *verges.

Unit 4 The Exponential Number e

Test yourself with the following:

1 Show that $\lim_{n \to \infty} \left(1 - \frac{1}{2n}\right)^n = e^{-1/2}$

2 If money earns compound interest at r% per annum and the interest is paid m times a year then the amount £A accrued from an investment of £P after n years is given by

$$A = P\left[1 + \frac{(r/m)}{100}\right]^{mn}$$

Find the amount accrued after 20 years from an investment of £200 at 15% per annum where the investment is paid:

(a) once a year
(b) four times a year
(c) twelve times a year
(d) continuously

The Binomial Expansion of $(1 + 1/n)^n$

As an example of the application of the binomial theorem, consider the binomial expansion of

$$(1 + 1/n)^n$$

where n is a natural number. The binomial expansion produces

$$(1 + 1/n)^n = 1 + n(1/n) + \frac{n(n-1)(1/n)^2}{2!} + \frac{n(n-1)(n-2)(1/n)^3}{3!} + \ldots + \frac{n(n-1)(n-2)(\ldots)[n-(n-1)](1/n)^n}{n!}$$

Since

$$n(1/n) = 1,\ n(n-1)/n^2 = 1 - 1/n \ldots$$

we can rewrite this equation as

$$(1 + 1/n)^n = 1 + 1 + \frac{(1 - 1/n)}{2!} + \frac{(1 - 1/n)(1 - 2/n)}{3!} + \ldots + \frac{(1 - 1/n)(1 - 2/n)(\ldots)(1/n)}{n!}$$

Taking limits term by term we then find that

$$\lim_{n\to\infty} [(1 + 1/n)^n] = \lim_{n\to\infty} \left[1 + 1 + \frac{(1 - 1/n)}{2!} + \frac{(1 - 1/n)(1 - 2/n)}{3!} + \ldots + \frac{(1 - 1/n)(1 - 2/n)(\ldots)(1/n)}{n!} \right]$$

$$= 1 + 1 + \frac{1}{2!} + \frac{1}{3!} + \ldots$$

It has already been shown by the ratio test for convergence that this series converges. It converges to an irrational number that is given the numeral e. It is called the exponential number. In sigma notation this reads as

$$\mathrm{e} = \sum_{r=0}^{\infty} \frac{1}{r!}$$

EXAMPLES

1 Show that $\lim_{n\to\infty} \left(1 + \frac{2}{n}\right)^n = \mathrm{e}^2$

By the binomial theorem the left-hand side is:

$$(1 + 2/n)^n = 1 + n(2/n) + \frac{n(n-1)(2/n)^2}{2!} + \frac{n(n-1)(n-2)(2/n)^3}{3!} + \ldots$$

$$= 1 + 2 + \frac{(1 - 1/n)2^2}{2!} + \frac{(1 - 1/n)(1 - 2/n)2^3}{3!} + \ldots$$

so that

$$\lim (1 + 2/n)^n = 1 + 2 + \frac{2^2}{2!} + \frac{2^3}{3!} + \ldots$$

The right-hand side is

$$\mathrm{e}^2 = \left[\lim_{n\to\infty} (1 + 1/n)^n \right]^2$$

$$= \lim_{n\to\infty} (1 + 1/n)^{2n}$$

By the binomial theorem:

$$(1 + 1/n)^{2n} = 1 + (2n)(1/n) + \frac{(2n)(2n-1)(1/n)^2}{2!} + \frac{(2n)(2n-1)(2n-2)(1/n)^3}{3!} + \ldots$$

$$= 1 + 2 + \frac{2(2 - 1/n)}{2!} + \frac{2(2 - 1/n)(2 - 2/n)}{3!} + \ldots$$

so that

$$\lim_{n\to\infty} (1 + 1/n)^{2n} = 1 + 2 + \frac{2^2}{2!} + \frac{2^3}{3!} + \ldots$$

$$= \lim_{n\to\infty} (1 + 2/n)^n$$

$$= e^2$$

2 If money earns compound interest at $r\%$ per annum and the interest is paid m times a year then the amount £A accrued from an investment of £P after n years is given by

$$A = P\left[1 + \frac{(r/m)}{100}\right]^{mn}$$

Find the amount accrued after 10 years from an investment of £100 at 12% per annum where the investment is paid:

(a) once a year
(b) four times a year
(c) twelve times a year
(d) continuously

(a) Once per year:

$$A = 100\left(1 + \frac{12}{100}\right)^{10}$$

$$= 100(1.12)^{10}$$

$$= 310.58$$

(b) Four times per year:

$$A = 100\left(1 + \frac{0.12}{4}\right)^{40}$$

$$= 100(1.03)^{40}$$

$$= 326.20$$

(c) Twelve times per year:

$$A = 100\left(1 + \frac{0.12}{12}\right)^{120}$$

$$= 100(1.01)^{120}$$

$$= 330.04$$

(d) Continuously:

$$A = \lim_{m \to \infty} 100\left(1 + \frac{0.12}{m}\right)^{10m}$$

$$= 100\left[\lim_{m \to \infty}\left(1 + \frac{0.12}{m}\right)^{m}\right]^{10}$$

$$= 100(e^{0.12})^{10}$$

$$= 100e^{1.2}$$

$$= 332.01$$

EXERCISES

1 Show that $\lim_{n \to \infty} (1 - 2/n)^n = e^{-2}$

By the binomial theorem the left-hand side is

$$(1 - 2/n)^n = 1 + n(-*) + \frac{n(n-1)(-*)^2}{2!} + \frac{n(n-1)(n-2)(-*)^3}{3!} + \ldots$$

$$= 1 - * + \frac{(1 - 1/n)*^2}{2!} - \frac{(1 - 1/n)(1 - 2/n)*^3}{3!} + \ldots$$

so that

$$\lim (1 - 2/n)^n = 1 - * + \frac{*^2}{2!} - \frac{*^3}{3!} + \ldots$$

The right-hand side is:

$$e^{-2} = \left[\lim_{n \to \infty} (1 + 1/n)^n\right]^{-2}$$

$$= \lim_{n \to \infty} (1 + 1/n)^{-2n}$$

By the binomial theorem:

$$(1 + 1/n)^{-2n} = 1 + (-*)(1/n) + \frac{(-*)(-*)(1/n)^2}{2!} + \frac{(-*)(-*)(-*)(1/n)^3}{3!} + \ldots$$

$$= 1 - * + \frac{*(* + 1/n)}{2!} - \frac{*(* + 1/n)(* + 2/n)}{3!} + \ldots$$

so that

$$\lim_{n \to \infty} (1 + 1/n)^{-2n} = 1 - * + \frac{*^2}{2!} - \frac{*^3}{3!} + \ldots$$

$$= \lim_{n \to \infty} (1 - 2/n)^n$$

$$= e^{-2}$$

2 If money earns compound interest at $r\%$ per annum and the interest is paid m times a year then the amount £A accrued from an investment of £P after n years is given by

$$A = P\left[1 + \frac{(r/m)}{100}\right]^{mn}$$

Find the amount accrued after 15 years from an investment of £500 at 10% per annum where the investment is paid:

(a) once a year
(b) four times a year
(c) twelve times a year
(d) continuously

(a) Once per year:

$$A = 500\left(1 + \frac{*}{100}\right)^{*}$$

$$= 500(*)^{*}$$

$$= 2088.62$$

(b) Four times per year:

$$A = 500\left(1 + \frac{*}{4}\right)^{*}$$

$$= 500(*)^{*}$$

$$= 2199.89$$

(c) Twelve times per year:

$$A = 500\left(1 + \frac{*}{12}\right)^{*}$$

$$= 500(*)^{*}$$

$$= 2226.96$$

(d) Continuously:

$$A = \lim_{m \to \infty} 500\left(1 + \frac{*}{m}\right)^{*}$$

$$= 500\left[\lim_{m \to \infty}\left(1 + \frac{*}{m}\right)^{*}\right]^{*}$$

$$= 500(e^{*})^{*}$$

$$= 500e^{*}$$

$$= 2240.84$$

Module 17: Further exercises

1 Form the first four partial sums of each of the following sequences:

(a) $f(n) = n! - (n-1)!$ (b) $f(n) = n^2 - 1$

2 Find the sum of the first 10 terms of each of the following:

(a) $f(n) = 7n - 2$ (c) $f(n) = \frac{1}{2}.5^n$

(b) $f(n) = 3 - 5n$ (d) $f(n) = \frac{7}{(-2)^n}$

3 If £x is due to be paid in n years time that amount of money has a present value of £Y where

$$Y = \frac{X}{(1 + r/100)^n}$$

and where money currently earns r% compound interest per annum. Find the present value of four annual payments of £50, the first instalment being paid one year from now. The current rate of interest is 12%.

4 Find the limits of the series formed from the partial sums of the following sequences:

(a) $f(n) = (\frac{2}{3})^n$ (b) $f(n) = n$

5 Test the following series for convergence:

(a) $\sum_{n=0}^{\infty} \left(\frac{6}{7}\right)^{2n}$ (d) $\sum_{n=0}^{\infty} \frac{1}{5^n n!}$

(b) $\sum_{n=0}^{\infty} \frac{1}{5^{-n}}$ (e) $\sum_{n=1}^{\infty} \frac{10^n}{n}$

(c) $\sum_{n=0}^{\infty} (-0.1)^n$

6 Expand each of the following as far as the fourth term:

(a) $(1 - 5x)^{-4}$ (c) $\left(\frac{2}{x} + 1\right)^{-1/2}$

(b) $(8 + 3x)^{3/4}$

7 Find the nth term in the following expansion and then use the ratio test to find those values for which the expansion converges:

$$\left(1 + \frac{x}{5}\right)^{-1}$$

8 Show that $\lim_{n \to \infty} \left(1 - \frac{1}{3n}\right)^n = e^{-1/3}$

9 If money earns compound interest at r% per annum and the interest is paid m times a year then the amount £A accrued from an investment of £P after n years is given by

$$A = P\left(1 + \frac{r/m}{100}\right)^{mn}$$

Find the amount accrued after 30 years from an investment of £400 at 16% per annum where the investment is paid:

(a) once a year (c) twelve times a year

(b) four times a year (d) continuously

Module 18: Limits of Functions

Aims: To consider limits of functions and thereby discuss the idea of continuity.

Objectives: When you have read this module you will be able to:

- ▶▶ Recognize situations where division by zero is implied.
- ▶▶ Understand that limits of functions exist as finite input points.
- ▶▶ Handle limits defined at finite input points.

Unit 1 Division by Zero

Test yourself with the following:

Find each of the following limits:

(a) $\lim_{x \to -5} \left(\dfrac{x^2 - 25}{x + 5} \right)$ (c) $\lim_{x \to 3\pi/4} \left(\dfrac{\cos 2x}{\sqrt{2} \cos x + 1} \right)$

(b) $\lim_{x \to -1} \left(\dfrac{x^4 - x^2}{x + 1} \right)$

Rational Expressions

Rational expressions cannot be numerically evaluated for values of a variable that make the denominator zero. For example, the expression

$$\frac{1}{x - 1}$$

can be evaluated for any real value of the variable x provided $x \neq 1$. When $x = 1$ the denominator is zero and division by zero is not allowed.

Compound Prescriptions

The problem of zero-valued denominators is one to be aware of when dealing with functions whose output is in the form of an expression that can be simplified by cancelling common factors. For example, if f is the function defined by

$$f(x) = \frac{x^2 + 2x + 1}{x^2 + 3x + 2}$$

it will be recognized that both the numerator and the denominator on the right-hand side can be factorized to give

$$f(x) = \frac{(x+1)^2}{(x+1)(x+2)}$$

After cancellation of the common factor $(x + 1)$ this equation becomes

$$f(x) = \frac{x+1}{x+2}$$

From this equation it would appear that $f(x)$ has an output for all values of x provided $x \neq -2$. However, because the equation was derived by cancelling $(x + 1)$ in the denominator – remember cancelling is division in disguise – $f(x)$ has no defined value when $x = -1$ either, even though this is not immediately apparent from the final form for $f(x)$. To clarify matters, the output from the function should be written as

$$f(x) = \frac{x+1}{x+2} \text{ provided } x \neq -1 \text{ or } -2$$

or by using the compound prescription

$$f(x) = \frac{x+1}{x+2}$$
$$f(-1), f(-2) \text{ undefined}$$

Limits of Functions

Even though a function may not be defined for a certain finite input value, it may have a finite limit at that value. For example, the function defined by the prescription

$$f(x) = \frac{x^2-4}{x-2} = \frac{(x+2)(x-2)}{x-2}$$

After cancellation of the common factor $(x - 2)$ the prescription becomes

$$f(x) = x + 2$$
$$f(2) \text{ undefined}$$

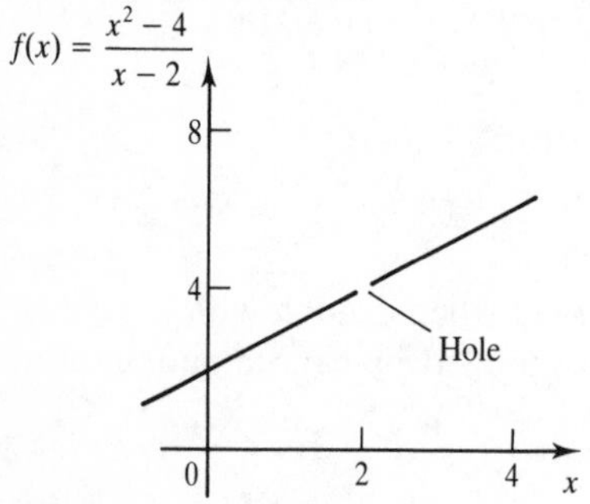

$f(2)$ is not defined because when $x = 2$ the denominator $x - 2 = 0$ and cancellation would then imply division by zero. The graph of this function displays a gap at the point $x = 2$ where there is no output – it is called a **discontinuous function**. However, it is clear from the graph that $f(x)$ does not become arbitrarily large as x approaches 2. On the contrary $f(x)$ approaches the value 4 as x approaches 2. As a consequence it is

Figure 18.1

deduced that

$$\lim_{x \to 2} f(x) = 4$$

even though $f(2)$ has not been defined.

Two-sided Limits

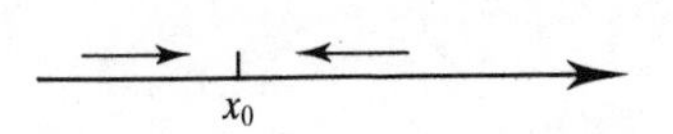

Figure 18.2

When we consider the limit of $f(x)$ as x approaches some fixed, finite point x_0 then it must be recognized that x_0 can be approached in either one of two directions – from the left or from the right. As a result of this we have two limits – the left-hand limit and the right-hand limit. For example,

$$f(x) = \frac{1}{2 - x}$$

is not defined at $x = 2$. However, a value for $f(x)$ can be found for any x arbitrarily close to $x = 2$. Indeed, let $x = 2 - \varepsilon$ where $\varepsilon > 0$. Then

$$f(2 - \varepsilon) = \frac{1}{\varepsilon}$$

Hence as x approaches 2 from values less than 2 – we say x approaches the number 2 from the left – then ε approaches 0 and

$$\lim_{\varepsilon \to 0} \frac{1}{\varepsilon} = \infty$$

If now, we choose $x = 2 + \varepsilon$ where $\varepsilon > 0$ then

$$f(2 + \varepsilon) = \frac{-1}{\varepsilon}$$

In this case, as x approaches 2 from values greater than 2 – we say x approaches the value 2 from the right – then again ε approaches 0, but this time:

$$\lim_{\varepsilon \to 0} \frac{-1}{\varepsilon} = -\infty$$

We can express these two facts using left- and right-hand limits.

$$\lim_{x \to 2\uparrow} \frac{1}{2 - x} = +\infty \qquad \text{the left-hand limit}$$

where the arrow up indicates x is approaching 2 by increasing and

$$\lim_{x \to 2\downarrow} \frac{1}{2 - x} = -\infty \qquad \text{the right-hand limit}$$

where the arrow down indicates x is approaching 2 by decreasing. For the limit to exist, it must do so regardless of direction – there must be no ambiguity in the limit. To ensure this both the left-hand and the right-hand limits must exist and be equal.

For example, consider the function f defined by

$$f(x) = \frac{x^2 - 1}{x - 1}$$

we see that we can write $f(x)$ as

$$f(x) = x + 1 \text{ provided } x \neq 1$$

Here,

$$\operatorname*{Lim}_{x \to 1\uparrow} f(x) = \operatorname*{Lim}_{x \to 1\uparrow} (x + 1) = 2$$

and

$$\operatorname*{Lim}_{x \to 1\downarrow} f(x) = \operatorname*{Lim}_{x \to 1\downarrow} (x + 1) = 2$$

Because these two limits exist and are equal, we can deduce that

$$\operatorname*{Lim}_{x \to 1} f(x) = 2$$

EXAMPLES

1 Find each of the following limits:

(a) $\displaystyle\operatorname*{Lim}_{x \to -1} \frac{x^2 - 1}{x + 1}$ **(b)** $\displaystyle\operatorname*{Lim}_{x \to 0} \frac{\sin 2x}{\sin x}$

(b) $\displaystyle\operatorname*{Lim}_{x \to -1} \frac{x^2 - x - 2}{x^2 + 4x + 3}$

(a) $$\frac{x^2 - 1}{x + 1} = \frac{(x + 1)(x - 1)}{x + 1}$$
$$= x - 1 \text{ provided } x \neq -1$$

therefore

$$\operatorname*{Lim}_{x \to -1} \frac{x^2 - 1}{x + 1} = \operatorname*{Lim}_{x \to -1} (x - 1)$$
$$= -2$$

(b) $$\frac{x^2 - x - 2}{x^2 + 4x + 3} = \frac{(x - 2)(x + 1)}{(x + 3)(x + 1)}$$
$$= \frac{x - 2}{x + 3} \text{ provided } x \neq -1$$

so that

$$\lim_{x\to -1}\frac{x^2-x-2}{x^2+4x+3} = \lim_{x\to -1}\frac{x-2}{x+3}$$

$$= \frac{-3}{2}$$

(c) $\dfrac{\sin 2x}{\sin x} = \dfrac{2\sin x\cos x}{\sin x}$

$= 2\cos x$ provided $\sin x \neq 0$, that is, provided $x \neq 0$ (or any multiple of π)

so that

$$\lim_{x\to 0}\frac{\sin 2x}{\sin x} = \lim_{x\to 0} 2\cos x$$

$$= 2\cos 0$$

$$= 2$$

EXERCISES

1 Find each of the following limits:

(a) $\displaystyle\lim_{x\to 3/2}\frac{4x^2-9}{2x-3}$ **(c)** $\displaystyle\lim_{\beta\to\pi/4}\frac{1-\tan\beta}{\cot 2\beta}$

(b) $\displaystyle\lim_{x\to 5}\frac{x^2-9x+20}{x^2-3x-10}$

(a) $\dfrac{4x^2-9}{2x-3} = \dfrac{(*x-*)(*x+*)}{2x-3}$

$= *x + *$ provided $2x \neq 3$, that is $x \neq 3/2$

so that

$$\lim_{x\to 3/2}\frac{4x^2-9}{2x-3} = \lim_{x\to 3/2}(*x+*)$$

$$= 6$$

(b) $\dfrac{x^2-9x+20}{x^2-3x-10} = \dfrac{(x-*)(x-*)}{(x-*)(x+*)}$

$= \dfrac{x-*}{x+*}$ provided $x \neq 5$

so that

$$\lim_{x\to 5}\frac{x^2-9x+20}{x^2-3x-10} = \lim_{x\to 5}\frac{x-*}{x+*}$$

$$= \frac{5-*}{5+*}$$

$$= \frac{1}{7}$$

(c) $\dfrac{1-\tan\beta}{\cot 2\beta} = (1-\tan\beta)\tan 2\beta$

$$= (1-\tan\beta)\left(\frac{2*}{1-*^2}\right)$$

$$= \frac{2*(1-\tan\beta)}{(1-*)(1+*)}$$

$$= \frac{2*}{*+*} \text{ provided } * \neq *,$$

that is $\beta \neq *$

so that

$$\lim_{\beta\to\pi/4} \frac{1-\tan\beta}{\cot 2\beta} = \lim_{\beta\to\pi/4} \frac{2*}{*+*}$$

$$= 1$$

Unit 2 Continuity and Discontinuity

Test yourself with the following:

Locate and describe any discontinuities in each of the following:

(a) $f(x) = \cot x$

(b) $f(x) = \begin{cases} -1 & x \leqslant 0 \\ 3 & x > 0 \end{cases}$

(c) $f(x) = \dfrac{x^2-25}{x-5}$

(d) $f(x) = \begin{cases} \sin x & x \leqslant \pi/2 \\ 2x/\pi & x > \pi/2 \end{cases}$

Continuity

A function is said to be continuous at $x = x_0$ if

$$\lim_{x\to x_0} f(x) = f(x_0)$$

and discontinuous otherwise. There are two types of discontinuity, the **hole** and the **break**, and at each type of discontinuity this equation is not valid. A discontinuity in a function is most easily appreciated by its effect on the graph of the function.

Break Discontinuities

A break discontinuity occurs in the graph of a function when an otherwise continuous graph undergoes a literal break. For example, the function defined by

$$f(x) = \begin{cases} -x^2 & 0 \leqslant x \leqslant 2 \\ x^2 & 2 < x \leqslant 4 \end{cases}$$

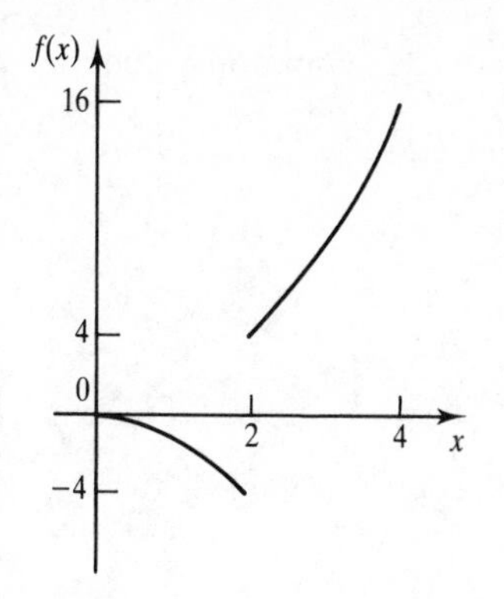

Figure 18.3

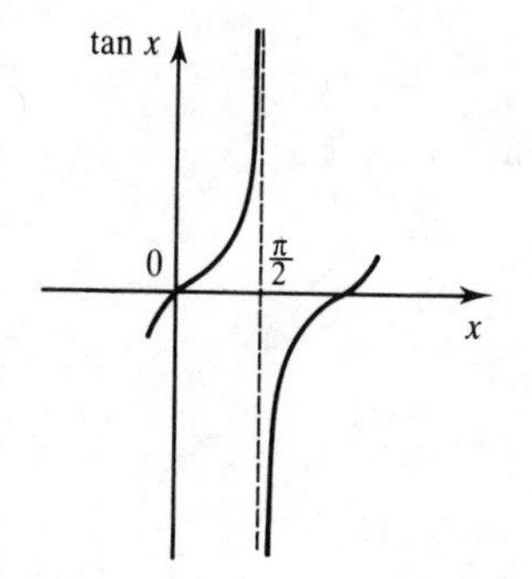

Figure 18.4

has the graph of Figure 18.3. Here the break occurs at $x = 2$. The left-hand limit has a value -4:

$$\lim_{x\to 2\uparrow} f(x) = \lim_{x\to 2\uparrow} (-x^2) = -4$$

and the right-hand limit has a value $+4$:

$$\lim_{x\to 2\downarrow} f(x) = \lim_{x\to 2\downarrow} (x^2) = 4$$

Because these two limits are not equal, the limit regardless of direction does not exist:

$$\lim_{x\to 2} f(x) \text{ does not exist}$$

consequently,

$$\lim_{x\to 2} f(x) \neq f(2)$$

A further example of a break discontinuity is given by the tangent function. The function defined by $f(x) = \tan x$ is not defined at $x = \pi/2$ and so is not continuous there. This is an **infinite break** discontinuity.

Hole Discontinuities

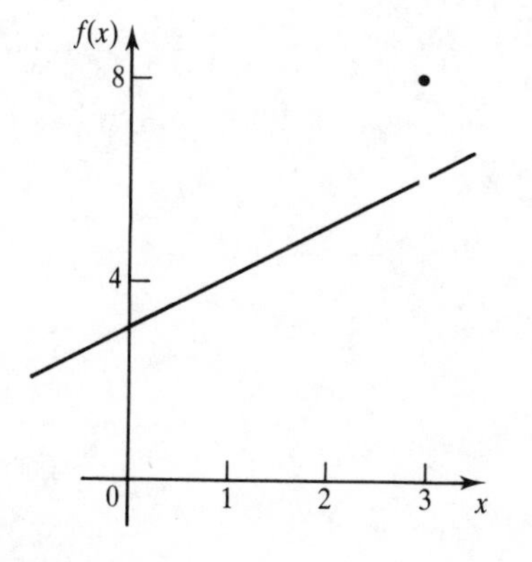

Figure 18.5

A hole discontinuity occurs in the graph of a function when an otherwise continuous graph contains a literal hole or gap. For example, the function defined by

$$f(x) = \frac{x^2 - 9}{x - 3} \qquad x \neq 3$$
$$f(3) = 8$$

Here, the hole occurs at $x = 3$. In this case both the left- and right-hand limits exist and are equal:

$$\lim_{x\to 3\uparrow} f(x) = \lim (x + 3) = 6$$

and

$$\lim_{x\to 3\downarrow} f(x) = \lim (x + 3) = 6$$

As a result, the limit regardless of direction also exists and takes on this common value:

$$\lim_{x \to 3} f(x) = 6$$

However, because $f(3)$ is defined to be equal to 8:

$$\lim_{x \to 3} f(x) \neq f(3)$$

EXAMPLES

1 Locate and describe any discontinuities in each of the following:

(a) $f(x) = \tan 2x$ **(c)** $f(x) = \dfrac{x^2 - 1}{x - 1}$

(b) $f(x) = \begin{cases} 1 & x \leqslant 5 \\ 0 & x > 5 \end{cases}$ **(d)** $f(x) = \begin{cases} e^x & x \leqslant 1 \\ (e + 1) - x & x > 1 \end{cases}$

(a)

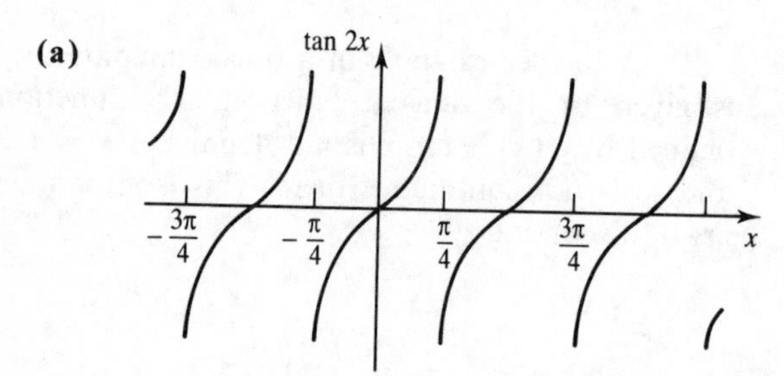

$f(x) = \tan 2x$

Figure 18.6

(b)

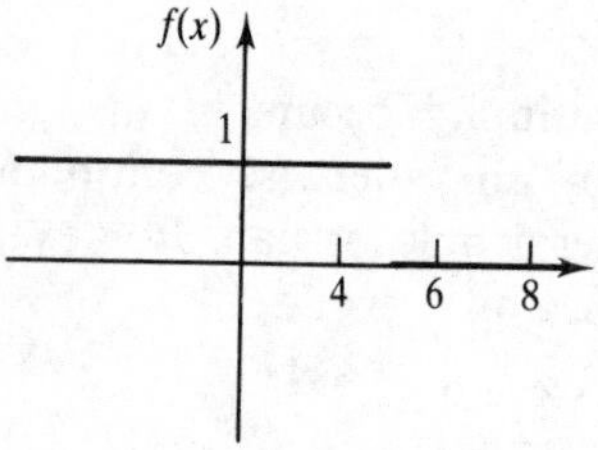

$f(x) = \begin{cases} 1 & x \leqslant 5 \\ 0 & x > 5 \end{cases}$

Figure 18.7

(c)

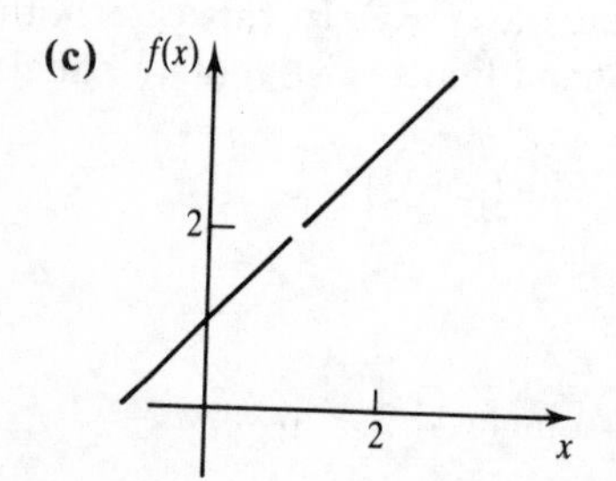

$f(x) = \dfrac{x^2 - 1}{x - 1}$

Figure 18.8

(d)

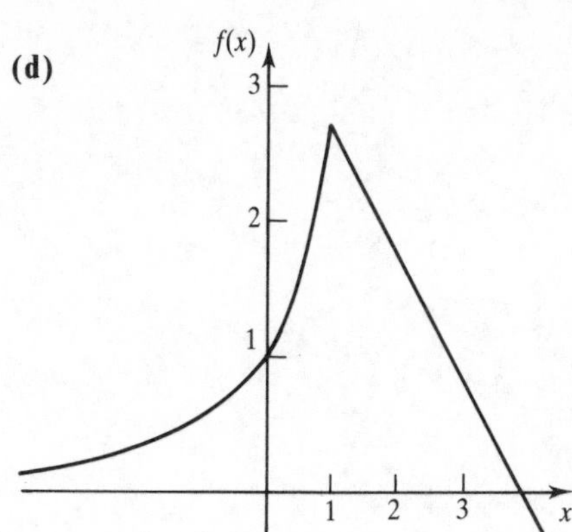

$$f(x) = \begin{cases} e^x & x \leqslant 1 \\ (e+1) - x & x > 1 \end{cases}$$

This function is continuous.

Figure 18.9

EXERCISES

1 Locate and describe any discontinuities in each of the following:

(a) $f(x) = \operatorname{cosec} x$

(b) $f(x) = \begin{cases} \sin x & x \leqslant \pi/2 \\ x - 1 & x > \pi/2 \end{cases}$

(c) $f(x) = \dfrac{4x^2 - 9}{2x + 3}$

(d) $f(x) = \begin{cases} 1 - x & x \leqslant 1 \\ \ln x & x > 1 \end{cases}$

(a)

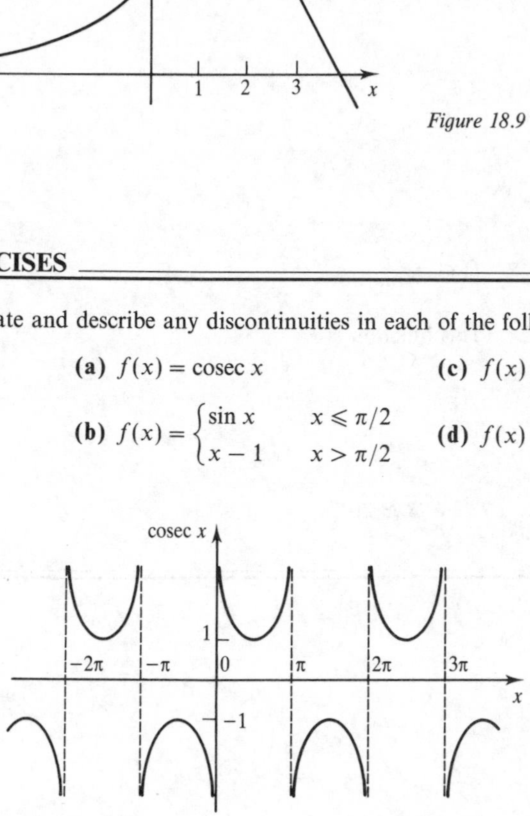

$$f(x) = \operatorname{cosec} x$$

Figure 18.10

(b)

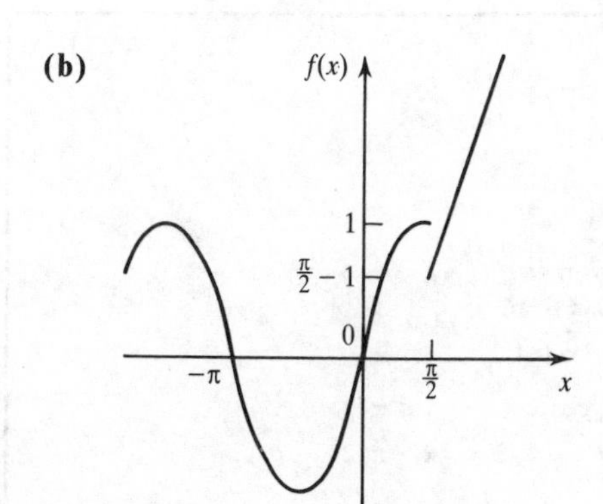

$$f(x) = \begin{cases} \sin x & x \leqslant \pi/2 \\ x - 1 & x > \pi/2 \end{cases}$$

Figure 18.11

(c)

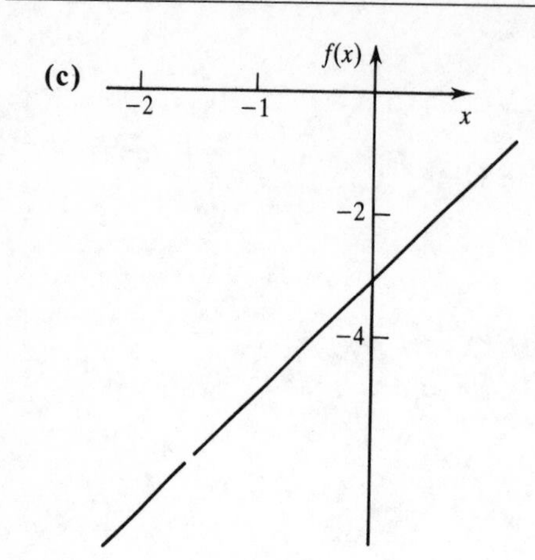

$$f(x) = \frac{4x^2 - 9}{2x + 3}$$

Figure 18.12

(d)

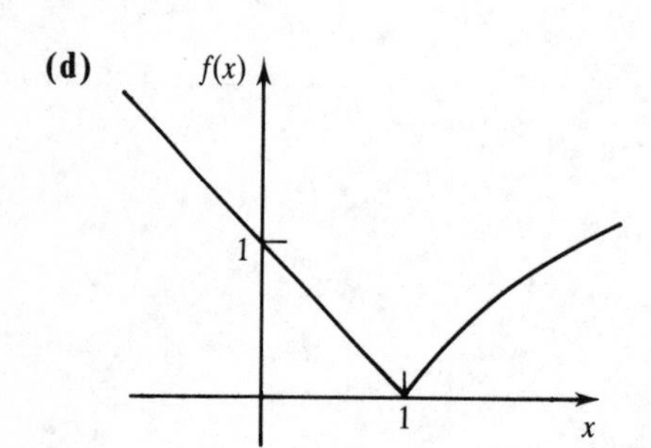

$$f(x) = \begin{cases} 1 - x & x \leqslant 1 \\ \ln x & x > 1 \end{cases}$$

This function is *.

Figure 18.13

Module 18: Further exercises

1 Find each of the following limits:

(a) $\displaystyle\lim_{x \to 4/3} \frac{9x^2 - 16}{3x - 4}$ (c) $\displaystyle\lim_{x \to \pi/4} \frac{\cos 2x}{\sqrt{2}\cos x - 1}$

(b) $\displaystyle\lim_{x \to -7} \frac{x^2 + 2x - 35}{x^2 + 9x + 14}$

2 Locate and describe any discontinuities in each of the following:

(a) $f(x) = \sec 5x$ (c) $f(x) = \dfrac{9x^2 - 16}{3x + 4}$

(b) $f(x) = \begin{cases} \cos x & x \leqslant \pi/2 \\ x & x > \pi/2 \end{cases}$ (d) $f(x) = \begin{cases} \cos x & x \leqslant \pi/2 \\ x - \pi/2 & x > \pi/2 \end{cases}$

Chapter 4: Miscellaneous exercises

1 Find the first six terms of the sequence defined by

$$f(n+1) = \frac{1}{1+f(n)} \quad \text{where } f(0) = 0$$

What is special about this sequence?

2 Newton's method for approximating to the square root of a number x is given by

$$f(n+1) = \frac{1}{2}\left[f(n) + \frac{x}{f(n)}\right] \qquad f(0) = 1$$

Use Newton's method to find the square root of 5 to three decimal places.

3 Does $f(n)$ converge when:

$$f(n) = \begin{cases} 1/n^n & \text{for } n \text{ odd} \\ 0 & \text{for } n \text{ even} \end{cases}$$

4 Find the limits of each of the following:

(a) $f(n) = \dfrac{(6 + 1/n)^2 - 36}{1/n}$

(b) $f(n) = \dfrac{(-2 + 1/n)^3 + 8}{1/n}$

(c) $f(n) = \dfrac{(x + 1/n)^4 - x^4}{1/n}$ $\qquad$ x is a real number

5 For what values of x does the following sequence converge?

$$1 + 4x^2 + 16x^4 + 64x^6 + \ldots$$

6 Find the value of

$$\frac{1}{\sqrt{2}} + \left(\frac{1}{\sqrt{2}}\right)^2 + \left(\frac{1}{\sqrt{2}}\right)^3 + \left(\frac{1}{\sqrt{2}}\right)^4 + \ldots$$

7 Find

$$\lim_{n \to \infty} \frac{\sqrt{(n+1)} - \sqrt{n}}{\sqrt{(n+1)} + \sqrt{n}}$$

8 If a whole is divided into four and one quarter is retained then that quarter represents one third of that not retained. If this quarter is further subdivided into four and one quarter of it is retained then, again, the quarter retained is one third of that not retained. This process is repeated indefinitely, with the result that the total amount retained tends to one third of that not retained. We can express this as

$$\sum_{n=1}^{\infty} \left(\frac{1}{4^n}\right) = \frac{1}{3}$$

By a similar reasoning find the values of:

(a) $\dfrac{1}{5} + \dfrac{1}{25} + \dfrac{1}{125} + \ldots$

(b) $\dfrac{1}{8} + \dfrac{1}{64} + \dfrac{1}{512} + \ldots$

(c) $\dfrac{2}{7} + \dfrac{4}{49} + \dfrac{8}{343} + \ldots$

9 Find the value of

$$\sum_{n=1}^{\infty}\left[\frac{1}{n(n+1)}\right]$$

Hint Break $1/n(n+1)$ into partial fractions to create a 'telescoping' series.

10 Discuss the continuity of each of the following at the point given:

$$\text{(a) } f(x)=\begin{cases} 1 & x \leqslant 2 \\ 4-x & 2 < x \leqslant 5 \\ -1 & x > 5 \end{cases}$$

at $x = 2$ and $x = 5$.

$$\text{(b) } f(x)=\frac{|x|}{x}$$

$$f(0) = 1$$

at $x = 0$.

5
■ ■ ■ The Differential Calculus

The aims of this chapter are to:

1. Develop the derivative of a function by considering the gradient of a tangent to the graph of the function.
2. Develop the ability to differentiate expressions using the rules of differentiation, demonstrate characteristic features of graphs and demonstrate how to locate and identify stationary points.
3. Examine the geometric significance of the differential and apply it to the approximation of solutions to equations.

This chapter contains three modules:

Module 19: Rates of Change
Because we use expressions to represent physical quantities and because we are interested in how physical quantities change, we consider the changes in outputs from functions as the input changes. This gives rise to the idea of an instantaneous rate of change of a function – the derivative.

Module 20: The Rules of Differentiation and Their Application
Expressions consist of symbols combined with various operations. To differentiate compound expressions requires a knowledge of the rules of differentiation.

Module 21: The Differential and Its Uses
The differential is described and it is seen that the ratio of differentials forms an alternative notation for the derivative. The differential is further used to find approximate solutions to equations.

Module 19: Rates of Change

Aim: To develop the derivative of a function by considering the gradient of a tangent to the graph of the function.

Objectives: When you have read this module you will be able to:

- ▶▶ Calculate average rates of change of a dependent variable with respect to an independent variable.
- ▶▶ Relate the average rate of change of a function to the gradient of a chord to the graph of the function.
- ▶▶ Use limits to obtain gradients of tangents to a curve.
- ▶▶ Define the derivative as the gradient of a tangent.
- ▶▶ Differentiate simple expressions from first principles.

Unit 1 Average Rates of Change

Test yourself with the following:

1 If a car travels 210 miles in 5 hours find its average speed over the journey.

2 A particle changes its velocity from $15\,\text{cm}\,\text{s}^{-1}$ to $25\,\text{cm}\,\text{s}^{-1}$ in 6 seconds. Find the average acceleration during that time.

3 A chord cuts the curve

$$f(x) = x^4$$

at $(2, 16)$ and $(-1, 1)$. Find the gradient of the chord.

4 What is the average rate of change with respect to x of

$$f(x) = x^4$$

between $x = 1$ and $x = 2$?

Speed and Average Speed

Imagine that you are driving a car. You start your journey from rest at midday and drive through the town keeping your speed below 30 mph so as to negotiate the traffic with safety. Some time later you drive on to the motorway where you increase your speed to a maximum of 70 mph. As you drive, you keep an eye on the speedometer to make sure that you do not exceed the speed limit. Eventually you arrive at your destination at 2.00 p.m. to find that you have travelled 80 miles – you have averaged 40 miles per hour during the journey.

If you recount your journey to a friend you would not say that you travelled at 40 mph for the entire journey but that you averaged 40 mph – your **average speed** was 40 mph.

Whenever you looked at your speedometer to check the car's speed you would see the speed of the car at that particular instant – the car's **instantaneous speed**.

The average speed of the car during the 80 mile journey is measured by dividing the distance it has travelled by the time it has taken to travel that distance. As the car travelled 80 miles in two hours we say that its average speed during the journey was

$$\frac{\text{Distance travelled}}{\text{Time taken}} = \frac{80}{2}$$

$$= 40 \text{ miles per hour}$$

Average speed is the rate of change of distance with respect to time and, as we see, it is calculated from the ratio of distance travelled to the time taken. Instantaneous speed, as opposed to average speed, is the rate of change of distance with respect to time *at a specific time* and because of this we cannot calculate instantaneous speed from a ratio because the denominator – at a specific time – is zero.

Functions

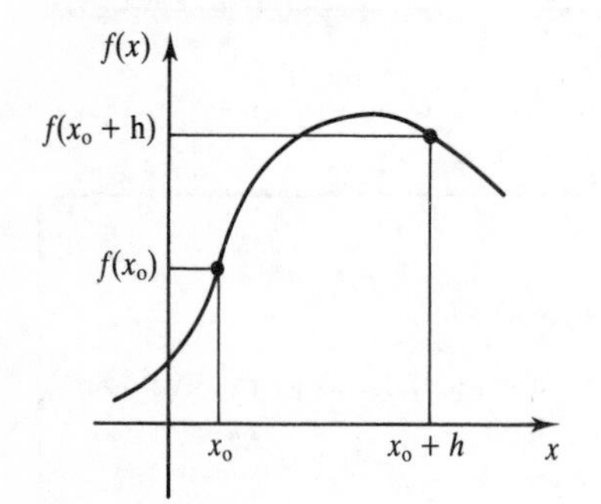

Figure 19.1

If a function f has output $f(x)$ then the average rate of change of output with respect to input between inputs x_0 and $x_0 + h$, where $h > 0$, is given by the ratio

$$\frac{\text{Change in output}}{\text{Change in input}} = \frac{f(x_0 + h) - f(x_0)}{(x_0 + h) - x_0}$$

$$= \frac{f(x_0 + h) - f(x_0)}{h}$$

This ratio represents the average rate of change of $f(x)$ with respect to x over the interval from x_0 to $x_0 + h$.

Gradients of Chords

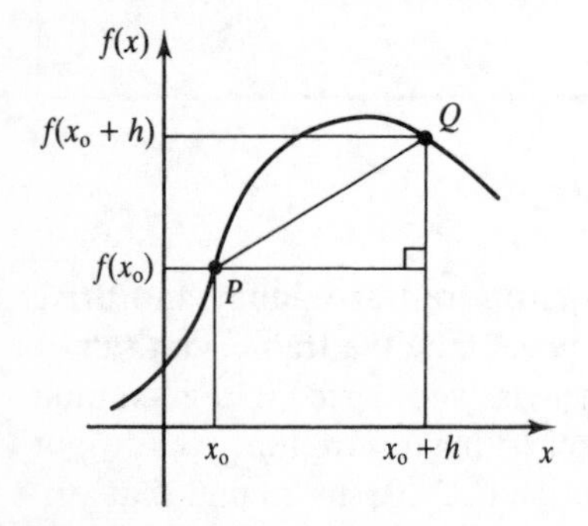

Figure 19.2

If, in the graph of Figure 19.1 we label the points with coordinates $(x_0, f(x_0))$ as P and $(x_0 + h, f(x_0 + h))$ as Q, then the straight line that joins P to Q has a gradient given by

$$\frac{f(x_0 + h) - f(x_0)}{h}$$

The straight line joining P and Q is called a **chord** and, as we see by comparing Figures 19.1 and 19.2, its gradient is equal to the average rate of change of $f(x)$ with respect to x between x_0 and $x_0 + h$.

EXAMPLES

1 If a car travels 150 miles in 4 hours, find its average speed over the journey.

$$\text{Average speed} = \frac{\text{distance travelled}}{\text{time taken}}$$

$$= \frac{150}{4} \text{ miles per hour}$$

$$= 37.5 \text{ mph}$$

2 A particle changes its velocity from 10 cm s^{-1} to 15 cm s^{-1} in 5 seconds. Find the average acceleration during that time.

$$\text{Average acceleration} = \frac{\text{change in velocity}}{\text{time taken}}$$

$$= \frac{\text{final velocity} - \text{initial velocity}}{\text{time}}$$

$$= \frac{15 - 10}{5}$$

$$= \frac{5}{5} \text{ cm per second per second}$$

$$= 1 \text{ cm s}^{-2}$$

3 A chord cuts the curve

$$f(x) = x^2$$

at (1, 1) and (2, 4). Find the gradient of the chord.

$$\text{Gradient of the chord} = \frac{\text{vertical distance between intersection points}}{\text{horizontal distance between intersection points}}$$

$$= \frac{f(2) - f(1)}{2 - 1}$$

$$= \frac{4 - 1}{2 - 1}$$

$$= 3$$

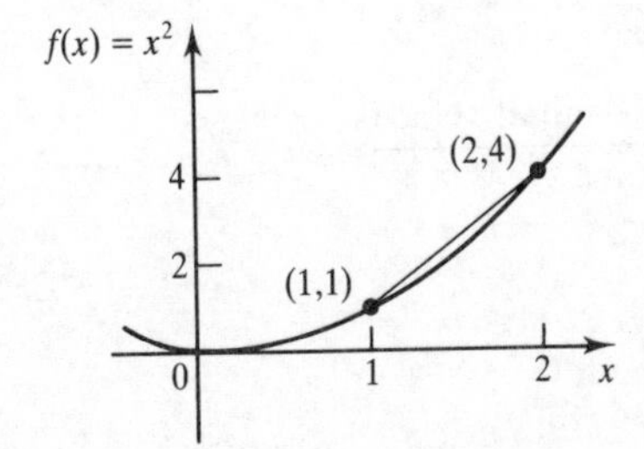

Figure 19.3

4 What is the average rate of change of

$$f(x) = 2x^2$$

between $x = 0$ and $x = 3$?

$$\text{Average rate of change} = \frac{\text{total change of } f(x)}{\text{range of } x \text{ values over which change has taken place}}$$

$$= \frac{f(3) - f(0)}{3 - 0}$$

$$= \frac{(2)(9) - 0}{3}$$

$$= 6$$

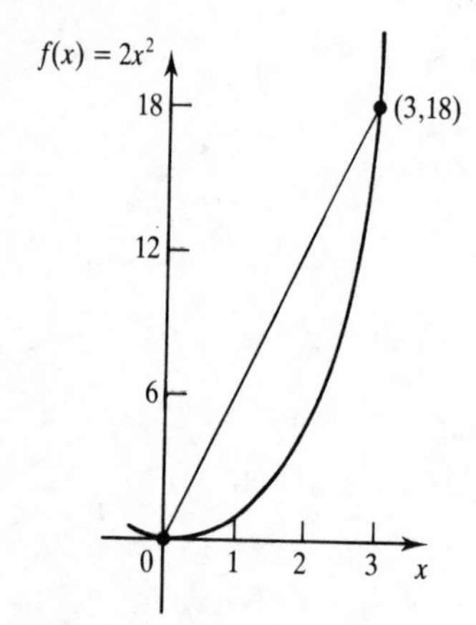

Figure 19.4

EXERCISES

1 If a car travels 340 miles in 3 hours find its average speed over the journey.

$$\text{Average speed} = \frac{\text{distance travelled}}{\text{time taken}}$$

$$= \frac{*}{*} \text{ miles per hour}$$

$$= 113.33\ldots \text{ mph}$$

2 A particle changes its velocity from 20 cm s^{-1} to 10 cm s^{-1} in 4 seconds. Find the average acceleration during that time.

$$\text{Average acceleration} = \frac{\text{change in velocity}}{\text{time taken}}$$

$$= \frac{\text{final velocity} - \text{initial velocity}}{\text{time}}$$

$$= \frac{* - *}{*}$$

$$= -* \text{ cm s}^{-1}$$

$$= -2.5 \text{ cm s}^{-2}$$

3 A chord cuts the curve

$$f(x) = -x^3$$

at (2, −8) and (4, −64). Find the gradient of the chord.

$$\text{Gradient of the chord} = \frac{\text{vertical distance between intersection points}}{\text{horizontal distance between intersection points}}$$

$$= \frac{f(*) - f(*)}{* - *}$$

$$= \frac{* - *}{* - *}$$

$$= *$$

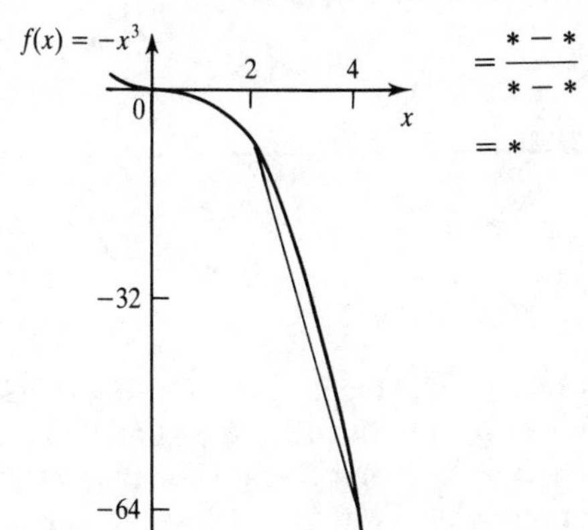

Figure 19.5

4 What is the average rate of change of

$$f(x) = -3x^5$$

between $x = -1$ and $x = 1$?

$$\text{Average rate of change} = \frac{\text{total change of } f(x)}{\text{range of } x \text{ values over which change has taken place}}$$

$$= \frac{f(*) - f(*)}{1 - (-1)}$$

$$= \frac{* - *}{*}$$

$$= *$$

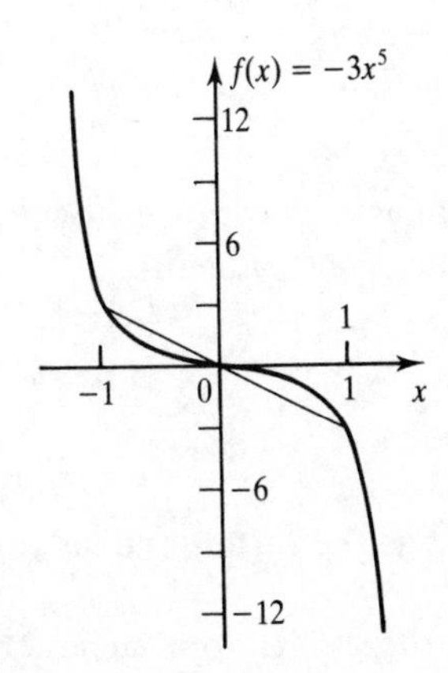

Figure 19.6

Unit 2 Gradients of Tangents

Test yourself with the following:

1 Differentiate from first principles:

(a) $f(x) = 4x^2$ (b) $f(x) = (1 - 3x)^3$

2 Find the gradient of the tangent to:

$$f(x) = -x^4 \quad \text{at} \quad x = 2$$

3 Using the table of standard forms write down the derivative of each of the following:

(a) $f(x) = x^9$ (b) $f(x) = 2$ (c) $f(x) = \dfrac{1}{x^3}$ (d) $f(x) = -x^{-2}$

Tangents to a Curve

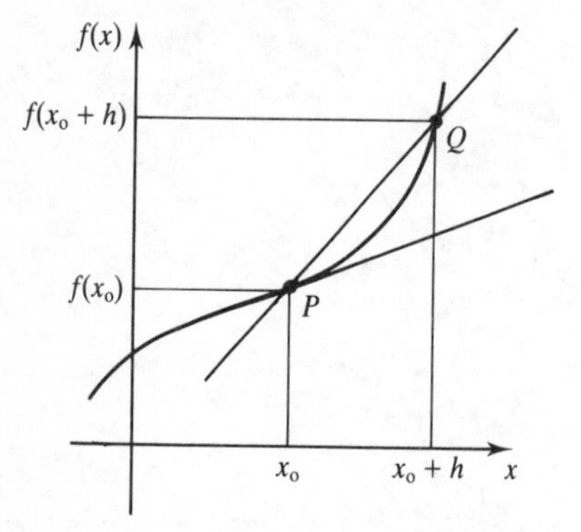

Figure 19.7

Consider the graph of the function f in Figure 19.7. Points P and Q are joined by the chord and the chord is extended beyond P and Q. A **tangent** to a curve is a straight line that touches the curve at one point. In Figure 19.7, the other line at P is a tangent to the curve at P.

If the point Q were to move down the curve towards P, the chord would rotate about P and the gradient of the chord would change – its value would approach a value equal to the gradient of the tangent at P. As Q approaches P both $f(x_0 + h) - f(x_0)$ and the interval h become closer to zero. However, no matter how close to zero these two quantities are, provided they are not zero, the gradient of the chord can still be calculated from their ratio:

$$\frac{f(x_0 + h) - f(x_0)}{h}$$

Because the gradient of the chord approaches the gradient of the tangent as Q moves towards P, this ratio has a limit, and this limit we define to be the gradient of the tangent:

$$\text{Gradient of tangent} = \lim_{h \to 0} \frac{f(x_0 + h) - f(x_0)}{h}$$

The expression on the right-hand side of this equation is identified as the instantaneous rate of change of $f(x)$ with respect to x at the point x_0.

It is important to understand clearly what this limit is. As the chord approaches the tangent, the gradient of the chord approaches the gradient of the tangent. As the chord

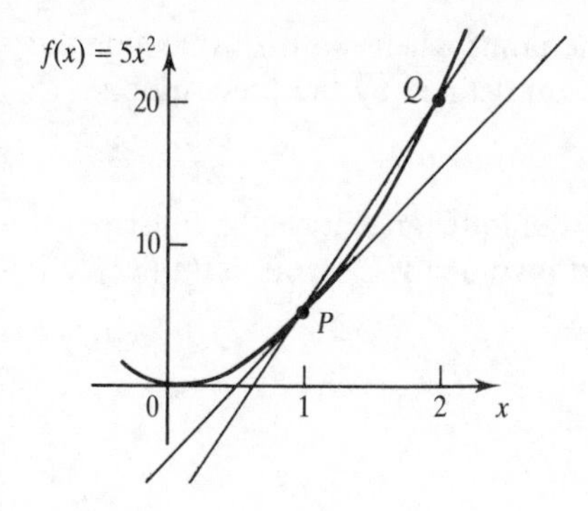

Figure 19.8

approaches the tangent, the ratio that is used to calculate the gradient of the chord approaches a limiting value. This limiting value is the gradient of the tangent. For example, consider the graph of the function f with output $f(x)$ where

$$f(x) = 5x^2$$

Let $(1, 5)$ and $(2, 20)$ be two points P and Q respectively on this curve. The following table records the gradients of successive chords as Q approaches P (the numbers in the table are not rounded and are just the numbers that appear on a calculator display):

x_0	$x_0 + h$	$f(x_0)$	$f(x_0 + h)$	$f(x_0 + h) - f(x_0)$	h	$\frac{f(x_0 + h) - f(x_0)}{h}$
1	2	5	20	15	1	15
1	1.5	5	11.25	6.25	0.5	12.5
1	1.05	5	5.512 5	0.512 5	0.05	10.25
1	1.005	5	5.050 125	0.050 125	0.005	10.025
1	1.0005	5	5.005 001 3	0.005 001 3	0.0005	10.0025
1	1.0001	5	5.001 000 1	0.001 000 1	0.0001	10.0005

From this table it can be seen that as h approaches 0 so the gradient of the chord approaches the limiting value of 10. This is the value of the tangent's gradient.

The Derivative and Differentiation

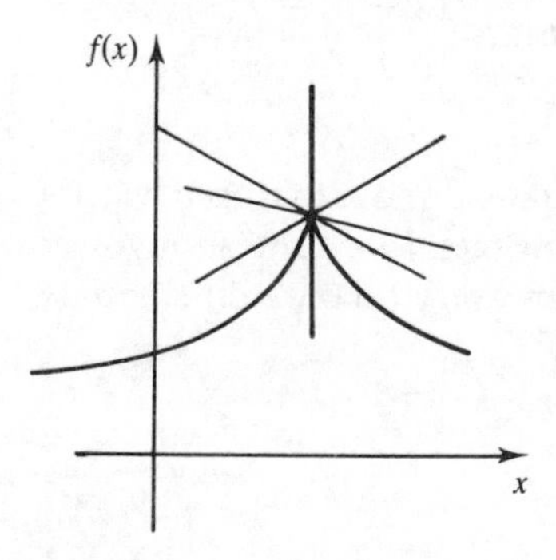

Figure 19.9

At every point $(x, f(x))$ on the graph of a function f where a unique tangent can be constructed, the tangent will have a gradient given by

$$\lim_{h \to 0} \frac{f(x + h) - f(x)}{h}$$

Limits are unique. If there is any ambiguity in the evaluation of the limit then the limit does not exist. We have seen this non-existence of limits in cases where left- and right-hand limits exist but differ in value. Because it is impossible to construct a unique tangent at the sharp peak of the graph of Figure 19.9, it is impossible to define a unique gradient and hence impossible to define a unique limit. The consequence is that at this point the limit does not exist.

The problem is that the graph is not smooth at the point where we wish to define the limit. For example, the graph in Figure 19.8 of the function defined by the prescription

$$f(x) = 5x^2$$

is smooth and continuous everywhere. As a consequence, a unique tangent can be constructed at any point on the curve. The gradient of the tangent drawn at the general point $(x, f(x))$ will, therefore, be

$$\begin{aligned}
\text{Gradient of tangent} &= \lim_{h\to 0} \frac{f(x+h) - f(x)}{h} \\
&= \lim_{h\to 0} \frac{5(x+h)^2 - 5x^2}{h} \\
&= \lim_{h\to 0} \frac{(5x^2 + 10xh + 5h^2) - 5x^2}{h} \\
&= \lim_{h\to 0} \frac{10xh + 5h^2}{h} \\
&= \lim_{h\to 0} (10x + 5h) \\
&= 10x
\end{aligned}$$

We see that the gradient of the tangent at the general point is an expression in x. This is to be expected, as the gradient of the tangent is different at different points on the curve. Indeed, the expression for the gradient of the tangent is the output from a function. Because this function is derived from the function f, it is called the **derivative** of f and is represented by the symbol

f' (read as 'f primed')

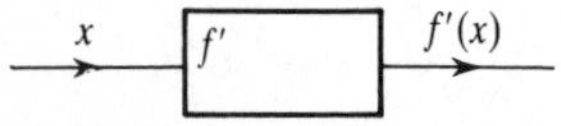

Figure 19.10

that is

$$f' : f'(x) = \lim_{h\to 0} \frac{f(x+h) - f(x)}{h}$$

The process of deriving the derivative is called **differentiation** and deriving it via the limit, as we have just done, is called **differentiation from first principles**. It is quite an involved process, and to avoid having to differentiate from first principles every time we differentiate, we make use of established derivatives called **standard forms**.

Standard Forms

To aid in the process of differentiation and to prevent laborious repetition of evaluating limits, use is made of a collection of standard forms. These are derivatives that have been previously derived from first principles. A few standard forms are listed here:

$f(x)$	$f'(x)$
a (constant)	0
ax^n	nax^{n-1}
$\sin x$	$\cos x$
$\cos x$	$-\sin x$
$\ln(x)$	$1/x$
e^x	e^x

Notice that this list of standard forms includes derivatives of the trigonometric functions. In the forms in which the derivatives are given it is essential that the angles be measured in radians.

EXAMPLES

1 Differentiate from first principles: $f(x) = 3x^2$

$$\begin{aligned} f'(x) &= \lim_{h\to 0} \frac{f(x+h) - f(x)}{h} \\ &= \lim_{h\to 0} \frac{3(x+h)^2 - 3x^2}{h} \\ &= \lim_{h\to 0} \frac{3x^2 + 6xh + 3h^2 - 3x^2}{h} \\ &= \lim_{h\to 0} \frac{6xh + 3h^2}{h} \\ &= \lim_{h\to 0} (6x + 3h) \\ &= 6x \end{aligned}$$

2 Find the gradient of the tangent to

$$f(x) = x^2 \text{ at } x = 1$$

Gradient of the tangent is $f'(x)$ where

$$\begin{aligned} f'(x) &= 2x^{2-1} \\ &= 2x \end{aligned}$$

At $x = 1$ this has the value

$$f'(1) = 2$$

3 Using the table of standard forms, write down the derivatives of each of the following:
(a) $f(x) = x^7$ **(b)** $f(x) = 5$ **(c)** $f(x) = \sin x$ **(d)** $f(x) = e^x$

(a) $f'(x) = 7x^{7-1}$
$= 7x^6$
(b) $f'(x) = 0$ 5 is a constant and constants do not change.
(c) $f'(x) = \cos x$
(d) $f'(x) = e^x$

EXERCISES

1 Differentiate from first principles: $f(x) = 2x^3$

$$f'(x) = \lim_{h \to 0} \frac{f(x+h) - f(x)}{h}$$

$$= \lim_{h \to 0} \frac{2(* + *)^3 - 2x^3}{h}$$

$$= \lim_{h \to 0} \frac{2(*^3 + 3*^2* + 3*(*)^2 + (*)^3) - 2x^3}{h}$$

$$= \lim_{h \to 0} \frac{6*^2* + 6*(*)^2 + 2(*)^3}{h}$$

$$= \lim_{h \to 0} (6* + 6* + 2(*)^2)$$

$$= *$$

2 Find the gradient of the tangent to

$$f(x) = -x^5 \text{ at } x = 3$$

Gradient of the tangent is $f'(x)$ where:

$$f'(x) = -*x^{*-*}$$

$$= -*x^*$$

At $x = 3$ this has the value

$$f'(3) = -**5$$

3 Using the table of standard forms, write down the derivatives of each of the following:

(a) $f(x) = x^{-1.5}$ **(b)** $f(x) = \frac{1}{3}$ **(c)** $f(x) = \cos x$ **(d)** $f(x) = \ln(x)$

(a) $f'(x) = -*x^{-*-*}$

$= -*x^{-*}$

(b) $f'(x) = *$ $\frac{1}{3}$ is a constant and constants do not *.

(c) $f'(x) = -*x$

(d) $f'(x) = */*$

Module 19: Further exercises

1 If a car travels 95 miles in 2 hours find its average speed over the journey.

2 A particle changes its velocity from 34 cm s^{-1} to 26 cm s^{-1} in 8 seconds. Find the average acceleration during that time.

3 A chord cuts the curve:

$$f(x) = -x^2$$

at (0, 0) and (3, −9). Find the gradient of the chord.

4 What is the rate of change of

$$f(x) = -5x^6$$

between $x = 5$ and $x = 10$?

5 Find the gradient of the tangent to:

$$f(x) = x^3 \text{ at } x = 4$$

6 Differentiate from first principles:

(a) $f(x) = 5x^3$ (b) $f(x) = (1 + 4x)^2$

7 Using the table of standard forms, write down the derivatives of each of the following:

(a) $f(x) = x^{-1/3}$ (b) $f(x) = \dfrac{15}{19}$ (c) $f(x) = \dfrac{-1}{x^{2/3}}$ (d) $f(x) = x^{12}$

Module 20: The Rules of Differentiation and Their Application

Aims: To develop the ability to differentiate expressions using the rules of differentiation, to demonstrate characteristic features of graphs and to demonstrate how to locate and identify stationary points.

Objectives: When you have read this module you will be able to:

▶▶ Use the rules of differentiation to differentiate compound expressions.

▶▶ Locate and identify various characteristic features of the graph of a function from a consideration of its first and second derivatives.

▶▶ Locate and identify the stationary points and the non-stationary points of inflexion of a function.

Unit 1 The Rules of Differentiation

Test yourself with the following:

Find the derivatives of each of the following:

(a) $f(x) = x - x^4$ (f) $f(x) = \cos(1 - 3x)$

(b) $f(x) = 12x^4$ (g) $f(x) = \exp(2x^3 - x^2)$

(c) $f(x) = x^3 \cos x$ (h) $f(x) = e^{4x}$

(d) $f(x) = \dfrac{\ln(x)}{x^4}$ (i) $f(x) = \cos 4x$

(e) $f(x) = \tan^2 x$

Derivatives of Combined Expressions

The rules of differentiation govern the differentiation of functions that have been combined under various operations. The rules are quoted as follows:

- **Sums** if

$$f(x) = g(x) + h(x)$$

then

$$f'(x) = g'(x) + h'(x)$$

For example, if $f(x) = x^2 + \sin x$ then $g(x) = x^2$ and $h(x) = \sin x$ so that

$$g'(x) = 2x, \; h'(x) = \cos x$$

Hence $f'(x) = 2x + \cos x$.

- **Products** If

$$f(x) = g(x)h(x)$$

then

$$f'(x) = g'(x)h(x) + g(x)h'(x)$$

For example, if $f(x) = x^2 \sin x$ then

$$f'(x) = 2x \sin x + x^2 \cos x$$

As a special case of the product rule, consider the function

$$f(x) = g(x)h(x) \text{ where } g(x) = a \text{ and where } a \text{ is a constant so that}$$

$$f(x) = ah(x)$$

In this case $g(x) = a$ so that $g'(x) = 0$, so

$$f'(x) = ah'(x)$$

For example, if $f(x) = 5x^2$ then here $a = 5$ and $h(x) = x^2$, hence

$$f'(x) = 5(2x) = 10x$$

- **Quotients** If

$$f(x) = \frac{g(x)}{h(x)}$$

then

$$f'(x) = \frac{g'(x)h(x) - g(x)h'(x)}{[h'(x)]^2}$$

For example, if $f(x) = x^2/\sin x$ then

$$f'(x) = \frac{2x \sin x - x^2 \cos x}{\sin^2 x}$$

Each of these rules is straightforward to prove. We shall prove the product rule and leave the others as exercises.

From the definition of a derivative:

$$\begin{aligned} f'(x) &= \lim_{k \to 0} \frac{f(x+k) - f(x)}{k} \\ &= \lim_{k \to 0} \frac{g(x+k)h(x+k) - g(x)h(x)}{k} \\ &= \lim_{k \to 0} \frac{g(x+k)h(x+k) - g(x)h(x+k) + g(x)h(x+k) - g(x)h(x)}{k} \end{aligned}$$

Here, adding and subtracting the same quantity $g(x)h(x+k)$ has the effect of adding zero but enables us to proceed by separating the limit into the sum of two limits:

$$f'(x) = \lim_{k\to 0} \frac{g(x+k)h(x+k) - g(x)h(x+k)}{k} + \lim_{k\to 0} \frac{g(x)h(x+k) - g(x)h(x)}{k}$$

Factorizing $h(x+k)$ in the first limit and $g(x)$ in the second, gives

$$f'(x) = \lim_{k\to 0} \frac{[g(x+k) - g(x)]h(x+k)}{k} + \lim_{k\to 0} \frac{g(x)[h(x+k) - h(x)]}{k}$$

$$= \lim_{k\to 0} \frac{g(x+k) - g(x)}{k} \cdot \lim h(x+k) + g(x) \lim \frac{h(x+k) - h(x)}{k}$$

$$= g'(x)h(x) + g(x)h'(x)$$

And so the rule is proved.

Compositions

The derivative of a function is the rate of change of output with respect to input. When functions are joined together in a composition, then there is a sequence of inputs and outputs and these must be taken into account.

If $f(x) = g[h(x)]$ then

$$f'(x) = g'[h(x)]h'(x)$$

In Figure 20.1 it is seen that the derivative of a function is found by differentiating the output with respect to the input. In a composition of functions, the output from the first function is the input for the second function. As a consequence, $g'[h(x)]$ represents the derivative of g with respect to the input $h(x)$.

For example, if $f(x) = (\sin x)^2$ then

$$f(x) = g[h(x)] \text{ where } h(x) = \sin x,\ g(x) = x^2$$

and, in particular,

$$g[h(x)] = h^2(x)$$

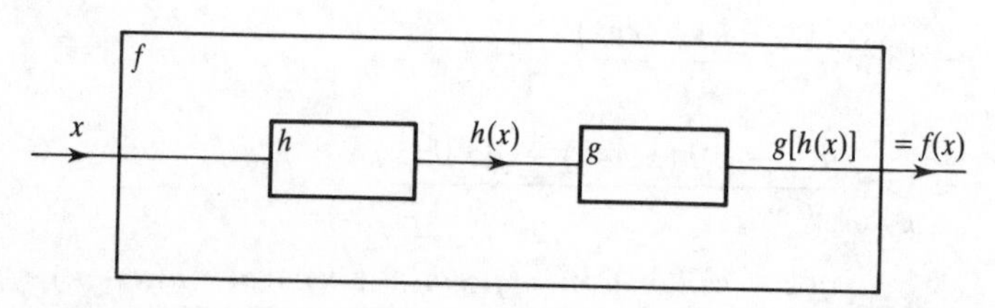

Figure 20.1

Thus:

$$g'[h(x)] = 2h(x)$$
$$= 2 \sin x$$

and

$$h'(x) = \cos x$$

so that

$$f'(x) = g'[h(x)]h'(x)$$
$$= 2 \sin x \cos x$$

As a special case of the composition rule, consider the function

$f(x) = g[h(x)]$ where $h(x) = ax$ and where a is a constant so that
$f(x) = g(ax)$

In this case $h'(x) = a$, so

$$f'(x) = ag'(ax)$$

For example, if $f(x) = \sin 2x$ then here $g(ax) = \sin ax$ and $a = 2$, so

$$f'(x) = 2 \cos 2x$$

EXAMPLES

Find the derivatives of each of the following:

(a) $f(x) = x^2 + x^3$
(b) $f(x) = 5x^6$
(c) $f(x) = x^2 e^x$
(d) $f(x) = \dfrac{x^2}{\cos x}$
(e) $f(x) = \tan x$
(f) $f(x) = \sin(2x + 1)$
(g) $f(x) = \exp(3x^2 + 1)$
(h) $f(x) = \ln(6x^2 - 1)$
(i) $f(x) = \sin 3x$

(a) $f'(x) = 2x + 3x^2$

(b) $f'(x) = 30x^5$

(c) $f'(x) = 2x\,e^x + x^2\,e^x$
$= x\,e^x(2 + x)$

(d) $f'(x) = \dfrac{2x \cos x - x^2(-\sin x)}{\cos^2 x}$
$= \dfrac{2x \cos x + x^2 \sin x}{\cos^2 x}$

(e) $\tan x = (\sin x)/(\cos x)$ therefore

$$\begin{aligned} f'(x) &= \frac{(\cos x)(\cos x) - (\sin x)(-\sin x)}{\cos^2 x} \\ &= \frac{\cos^2 x + \sin^2 x}{\cos^2 x} \\ &= \frac{1}{\cos^2 x} \\ &= \sec^2 x \end{aligned}$$

(f) Here we let $u(x) = 2x + 1$ and write

$$f(x) = \sin u(x)$$

then

$$\begin{aligned} f'(x) &= [\sin u(x)]' u'(x) \\ &= [\cos u(x)] . 2 \\ &= 2 \cos(2x + 1) \end{aligned}$$

(g) Here we let $u(x) = 3x^2 + 1$ and write

$$f(x) = \exp[u(x)]$$

then

$$\begin{aligned} f'(x) &= \{\exp[u(x)]\}' u'(x) \\ &= \{\exp[u(x)]\} . 6x \\ &= 6x \exp(3x^2 + 1) \end{aligned}$$

(h) Here we let $u(x) = 6x^2 - 1$ and write

$$f(x) = \ln u(x)$$

then

$$\begin{aligned} f'(x) &= [\ln u(x)]' u'(x) \\ &= \frac{1}{u(x)} . 12x \\ &= \frac{12x}{6x^2 - 1} \end{aligned}$$

(i) Here we let $u(x) = 3x$ and write

$$f(x) = \sin u(x)$$

then

$$\begin{aligned} f'(x) &= [\sin u(x)]' u'(x) \\ &= [\cos u(x)] . 3 \\ &= 3 \cos 3x \end{aligned}$$

EXERCISES

Find the derivatives of each of the following:

(a) $f(x) = x^{-3} + x^{-1}$

(b) $f(x) = \dfrac{2x^2}{3}$

(c) $f(x) = e^x \sin x$

(d) $f(x) = \dfrac{\ln(x)}{x}$

(e) $f(x) = \cot x$

(f) $f(x) = (1 + x^2)^3$

(g) $f(x) = \ln(2x^2 - x)$

(h) $f(x) = e^{3x}$

(i) $f(x) = \tan 4x$

(a) $f'(x) = -*x^{-3-*} + (-*)x^{-1-*}$

$= -*x^{-*} - x^{-*}$

(b) $f'(x) = \dfrac{2}{3} * x$

$= \dfrac{*x}{3}$

(c) $f'(x) = e^x * + * \cos x$

(d) $f'(x) = \dfrac{(*)x - [\ln(x)](*)}{x^2}$

$= \dfrac{* - * \ln(x)}{x^2}$

(e) Since $\cot x = (\cos x)/(\sin x)$:

$$f'(x) = \frac{(-*)(\sin x) - (*)(\cos x)}{*^2}$$

$$= \frac{* - *}{*^2}$$

$$= \frac{-1}{\sin^2 x}$$

$$= -*^2$$

(f) Here we let $u(x) = 1 + x^2$ and write

$$f(x) = *^3$$

then

$$f'(x) = (*^3)'u'(x)$$

$$= (3*^2)2x$$

$$= *(1 + x^2)^*$$

(g) Here we let $* = 2x^2 - x$ and write

$$f(x) = \ln(*)$$

then

$$f'(x) = [\ln(*)]'*'$$
$$= \left(\frac{1}{u(x)}\right)(4x - 1)$$
$$= \frac{4x - 1}{*}$$

(h) Here we let $u(x) = *$ and write

$$f(x) = e^*$$

then

$$f'(x) = (e^*)'u'(x)$$
$$= (e^*)*$$
$$= *e^*$$

(i) Here we let $* = *$ and write

$$f(x) = \tan *$$

then

$$f'(x) = (\tan *)'*'$$
$$= (\sec^2 *)*$$
$$= *$$

Unit 2 Graphical Characteristics

Test yourself with the following:

1 Without drawing the graph of the function, describe the characteristic features of each of the following:

(a) $f(x) = x^3 - x^2$ (b) $f(x) = 5(x - 2)^{1/5}$

2 Locate and identify the stationary points of

$$f(x) = x^3 - 2x^2 + x$$

3 Draw a sketch graph of the function defined by

$$f(x) = x^3 - 2x^2 + x$$

Graphical Features

When we look at the graph of a function, we notice that it possesses certain characteristic features. In the graph of Figure 20.2 we see that between points A and D the graph is rising, and that between points D and G it is falling – we always describe a graph of a function in the direction of increasing independent variable, that is, from left to right. Looking at the

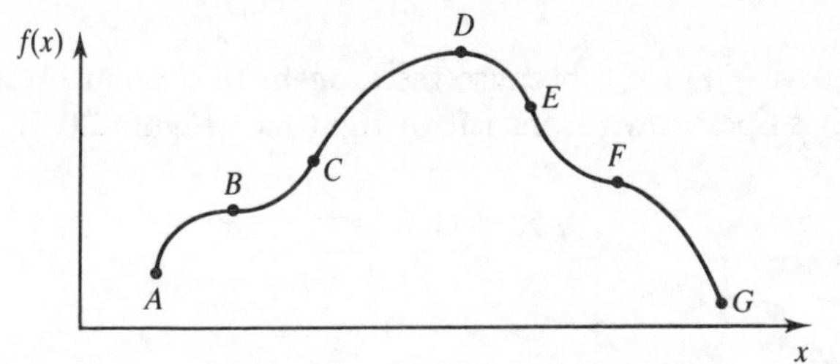

Figure 20.2

detail of the graph, we see that the way it rises between A and B and between C and D is different from the way it rises between B and C. Again, the way it falls between D and E and between F and G is different from the way it falls between E and F. At B it flattens out and at D it attains a maximum.

All of these features are easily seen by inspection. Now we wish to discover how these features relate to the algebraic behaviour of functions and their derivatives.

The Sign of the Derivative

The derivative of a function is a function. If a tangent can be constructed at a point on the graph of a function, the x-coordinate of this point forms the input to the derivative. The output is then equal to the gradient of the tangent and such gradients can be positive, negative or zero.

Positive Derivative

The tangent to any point on a rising curve has a positive gradient – it slopes up from left to right (see Figure 20.3). This means that $f'(x) > 0$ at any point on a rising curve. For example, the function f defined by

$$f(x) = x^3$$

has derivative

$$f'(x) = 3x^2$$

which is positive for all values of x except $x = 0$. The standard cubic is a rising curve except at the origin (see Figure 20.4).

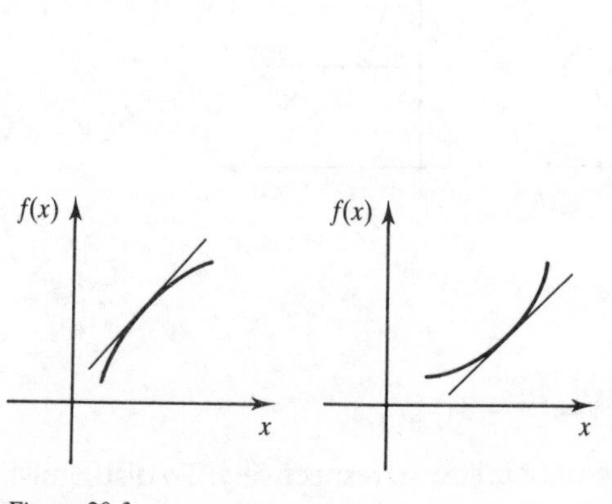

Figure 20.3

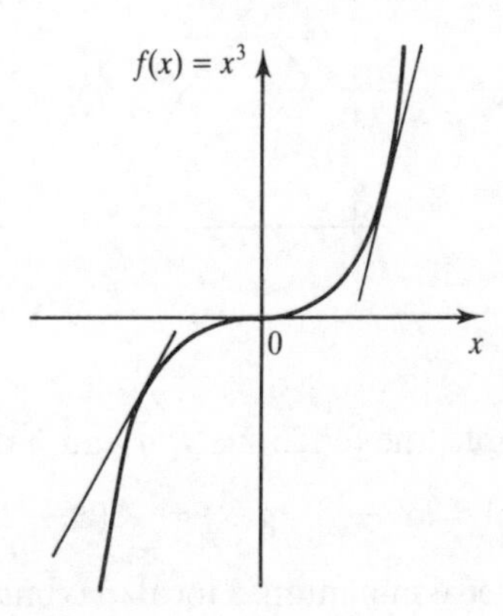

Figure 20.4

Negative Derivative

At all points on a falling curve $f'(x) < 0$ because the tangent to a point on a falling curve has a negative gradient – it slopes down from left to right (see Figure 20.5). For example, the function f defined by

$$f(x) = -x^3$$

has derivative

$$f'(x) = -3x^2$$

which is negative for all values of x except $x = 0$. This cubic is a falling curve except at the origin, as Figure 20.6 shows.

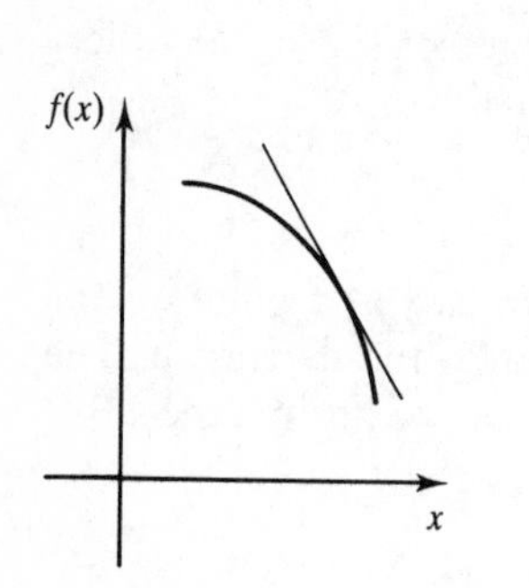

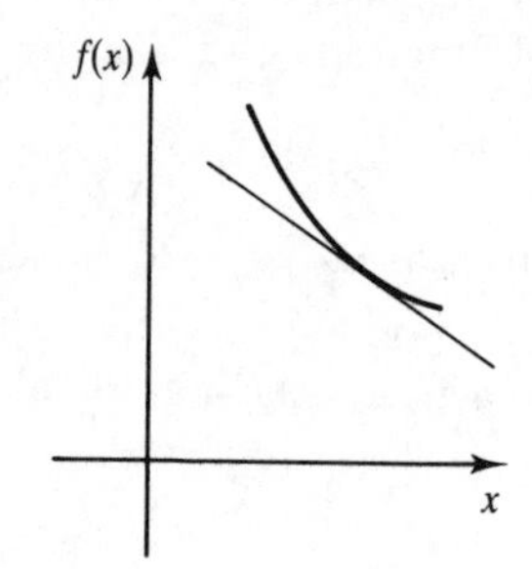

Figure 20.5

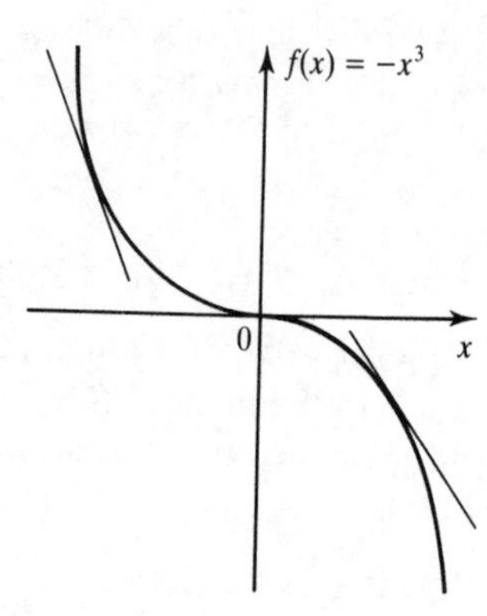

Figure 20.6

Zero Derivative

If $f'(x) = 0$ at a point on the graph of a function f then the tangent at that point has a zero gradient – it is parallel to the horizontal axis. Because the function itself is neither increasing nor decreasing at points where the derivative is zero, we call such points **stationary points.** A stationary point is either a local maximum, a local minimum or, sometimes, a point of inflexion.

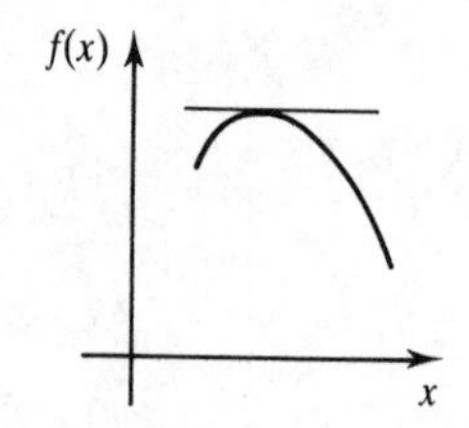

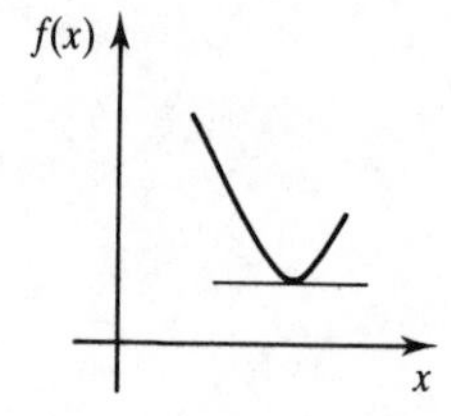

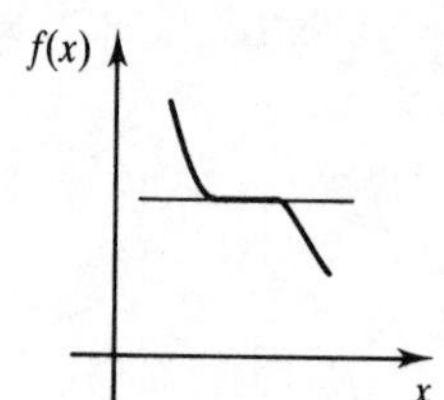

Figure 20.7

For example, the functions f, g and h defined by

$$f(x) = (x - 1)^2, \; g(x) = -(x - 2)^2 \text{ and } h(x) = (x - 3)^3$$

exhibit a local minimum, a local maximum and a point of inflexion respectively. To distinguish each of these features from the others, we consider the gradient of the tangent on either side of the point where $f'(x) = 0$ (see also Figure 20.8):

- **Local minimum** At a local minimum, the gradient of the tangent a little to the left is negative and a little to the right it is positive.
- **Local maximum** At a local maximum, the gradient of the tangent a little to the left is positive and a little to the right it is negative.
- **Point of inflexion** The sign of the gradient of the tangent is the same on both sides of a point of inflexion.

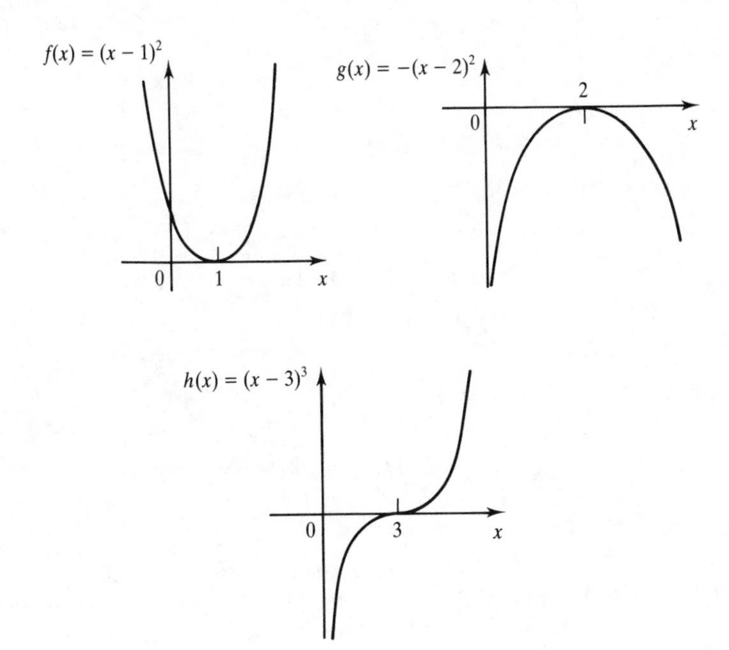

Figure 20.8

Undefined Derivative

The final case to consider is the location of points where the derivative is undefined. This can be for one of two reasons:

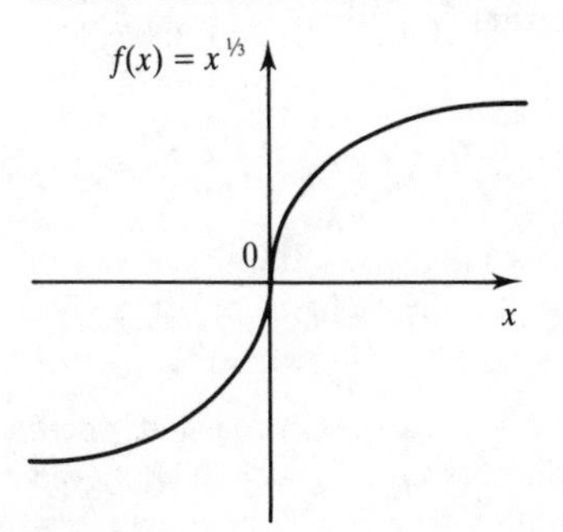

Figure 20.9

- **Vertical tangent** If a tangent to the graph of a function is vertical then its gradient cannot be defined. Hence, $f'(x)$ is not defined at that point. For example, the function defined by

$$f(x) = x^{1/3}$$

has derivative

$$f'(x) = \tfrac{1}{3}x^{-2/3}$$

which is undefined at the origin.

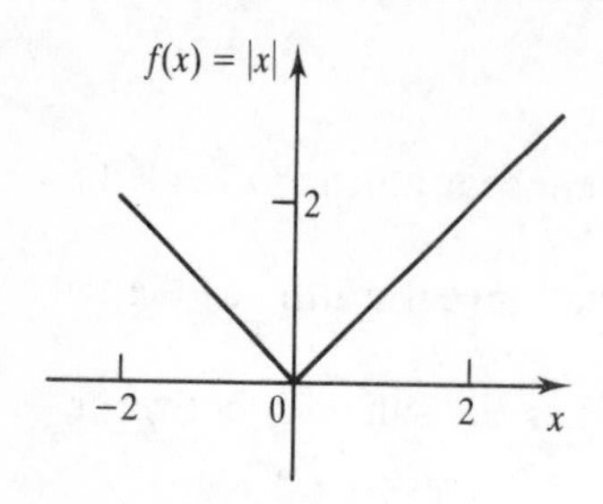

Figure 20.10

- **No tangent** It is not possible to define a tangent at points where the graph of a function is not a smooth curve. For example, the graph of

$$f(x) = |x|$$

does not have a tangent at $x = 0$ and so $f'(0)$ is undefined (note that $f'(x) = \pm 1$ for $x \neq 0$).

EXAMPLES

1 Without drawing the graph of the function, describe the characteristic features of each of the following:

$$\textbf{(a)}\ f(x) = \tfrac{1}{2}x^4 - x^2 + 3 \qquad \textbf{(b)}\ f(x) = 3(2x + 4)^{1/3}$$

(a) Many of the characteristic features of a function are deduced from a study of the derivative of the function. Differentiating $f(x)$ we find

$$f'(x) = 2x^3 - 2x$$

$\boldsymbol{f'(x) > 0}$ The derivative is positive when

$$2x^3 - 2x \equiv 2x(x^2 - 1)$$

is positive. This expression has two factors, $2x$ and $x^2 - 1$. If

$$x^2 - 1 > 0 \text{ then } x^2 > 1 \text{ so } x > +1 \text{ or } x < -1$$

Clearly, if $x > 1$ then because both factors are positive the derivative is positive. This means that for $x > 1$ the graph of the function is a rising curve. For $-1 < x < 0$ both factors are negative and so the derivative is positive. Again, over this interval the graph is a rising curve.

$\boldsymbol{f'(x) < 0}$ If $0 < x < 1$ then $x^2 - 1$ is negative and x is positive with the result that the derivative is negative for values of x in this interval. This means that for $0 < x < 1$ the graph of the function is a falling curve. If $x < -1$ then $x^2 - 1$ is positive but x is negative so the derivative remains negative and the graph is still a falling curve for $x < -1$.

$\boldsymbol{f'(x) = 0}$ The derivative is zero at three points – there are three stationary points:

(1) At $x = 1$ then $x^2 - 1 = 0$ so $f'(1) = 0$. The gradient of a tangent a little to the left of this point is negative and the gradient of a tangent a little to the right is positive. The stationary point at $x = 1$ is, accordingly, a minimum.

(2) At $x = 0$, $f'(0) = 0$. The gradient of a tangent a little to the left of this point is positive and the gradient of a tangent a little to the right is negative. The stationary point at $x = 0$ is, accordingly, a maximum.

(3) At $x = -1$ then $x^2 - 1 = 0$ so $f'(-1) = 0$. The gradient of a tangent a little to the left of this point is negative and the gradient of a tangent a little to the right is positive. The stationary point at $x = -1$ is, accordingly, another minimum.

In summary, from left to right the graph is a falling curve until the point $x = -1$ is reached where a local minimum occurs. From this point the curve rises to a local maximum at $x = 0$. Between $x = 0$ and $x = 1$ the curve falls until it achieves another local minimum at $x = 1$ after which it becomes a rising curve.

(b) $f'(x) = 2(2x + 4)^{-2/3}$

$f'(x) > 0$ The derivative is positive for all values of x because we can write

$$f'(x) = 2[(2x + 4)^{-1/3}]^2$$

This means that the graph of f is a rising curve for all values of x.

$f'(x)$ undefined The derivative is not defined when $2x + 4 = 0$, that is, when $x = -2$. At this point there is a vertical point of inflexion.

2 Locate and identify the stationary points of

$$f(x) = x^3 - 5x^2 + 8x - 4$$

The stationary points occur where $f'(x) = 0$:

$$f'(x) = 3x^2 - 10x + 8 \equiv (3x - 4)(x - 2)$$

So that $f'(x) = 0$ when $x = 4/3$ and when $x = 2$.

To identify the type of stationary point we look at the gradient of a tangent on either side of the stationary point:

(1) At $x = 4/3$, $f'(4/3) = 0$ and $f(4/3) = 4/27$. The gradient of a tangent a little to the left of this point where $x < 4/3$ is positive, and the gradient of a tangent a little to the right where $x > 4/3$ is negative. The point $x = 4/3$ is, accordingly, a maximum.

(2) At $x = 2$, $f'(2) = 0$ and $f(2) = 0$. The gradient of a tangent a little to the left of this point where $x < 2$ is negative, and the gradient of a tangent a little to the right where $x > 2$ is positive. The point $x = 2$ is, accordingly, a minimum.

In summary, the graph has a maximum point at $(4/3, 4/27)$ and a minimum point at $(2, 0)$.

3 Draw a sketch graph of the function defined by

$$f(x) = x^3 - 5x^2 + 8x - 4$$

In addition to the information contained in the solution to the previous question we note that:

- For $x < 4/3$, $f'(x) > 0$, so the graph is a rising curve for $x < 4/3$.
- For $4/3 < x < 2$, $f'(x) < 0$, so the graph is a falling curve during this interval.
- For $x > 2$, $f'(x) > 0$, so the graph is a rising curve for $x > 2$.
- Finally, $f(1) = f(2) = 0$, so by the factor theorem the curve meets the x-axis at $x = 1$ and $x = 2$.

A sketch of the graph can now be drawn. Do note that even though the question asks for a sketch graph it does not mean a thumbnail sketch. The sketch must be drawn on graph paper with a reasonable accuracy. Exact accuracy is not required but a reasonable representation of the graph's main features is required.

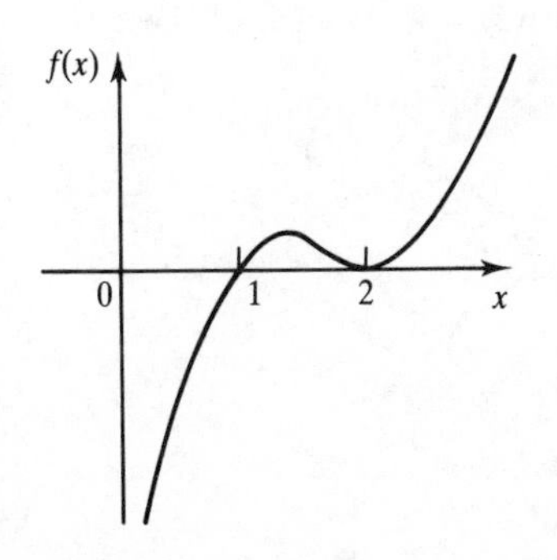

Figure 20.11

EXERCISES

1 Without drawing the graph of the function, describe the characteristic features of each of the following:

(a) $f(x) = 27x - x^3$ **(b)** $f(x) = (5x^2 - 20)^{1/5}$

(a) Many of the characteristic features of a function are deduced from a study of the derivative of the function. Differentiating $f(x)$ we find

$$f'(x) = * - *x^2$$

$\boldsymbol{f'(x) > 0}$ The derivative is positive when

$$* - *x^2 \equiv *(*)$$

is positive. This expression has the factor $*$. If

$$* > 0 \text{ then } * < * \text{ so } -* < x < +*$$

Clearly, if $-* < x < *$ the derivative is $*$tive. This means that for $-* < x < *$ the graph of the function is a $*$ing curve.

$\boldsymbol{f'(x) < 0}$ If $x > *$ then $* - *x^2$ is negative with the result that the derivative is $*$tive for values of $x > *$. This means that for $* > *$ the graph of the function is a $*$ing curve. If $* < -*$ then $* - *x^2$ is again $*$tive so the derivative is also $*$tive. This means that the graph is a $*$ing curve for $* < -*$.

$\boldsymbol{f'(x) = 0}$ The derivative is zero at two points:

(1) At $x = *$ then $* - *^2x = 0$ so $f'(*) = 0$. The gradient of a tangent a little to the left of this point is $*$tive and the gradient of a tangent a little to the right is $*$tive. The point $x = *$ is, accordingly, a m$*$imum.

(2) At $x = -*$ then $* - *x^2 = 0$ so $f'(-*) = 0$. The gradient of a tangent a little to the left of this point is $*$tive and the gradient of a tangent a little to the right is $*$tive. The point $x = -*$ is, accordingly, a m$*$imum.

In summary, from left to right the graph is a $*$ing curve until the point $x = -*$ is reached where a local m$*$imum occurs. From this point the curve $*$s to a m$*$imum at $x = *$. For values of x greater than $*$ the graph is a $*$ing curve.

(b) $f'(x) = 2(5x^2 - 20)^{-4/5}$

$\boldsymbol{f'(x) > 0}$ The derivative is $*$tive for all values of x because we can write

$$f'(x) = [2(5x^2 - 20)^{-1/5}]^4$$

This means that the graph of f is a $*$ing curve for $*$ values of x.

$\boldsymbol{f'(x)}$ **undefined** The derivative is not defined when $*$, that is, when $x = *$. At each of these two points there is a $*$.

2 Locate and identify the stationary points of

$$f(x) = 2x^3 + 9x^2 - 24x + 13$$

The stationary points occur where $* = 0$:

$$f'(x) = 6x^2 + 18x - 24 \equiv *(*x - *)(*x + *)$$

So that $* = 0$ when $x = *$ and when $x = *$.

To identify the type of stationary point we look at the gradient of a tangent on either side of the stationary point.

(1) At $x = *$, $f'(*) = *$ and $f(*) = *$. The gradient of a tangent a little to the left of this point where $x < *$ is *tive and the gradient of a tangent a little to the right where $x > *$ is *tive. The point $x = *$ is, accordingly, a m*imum.

(2) At $x = *$, $f'(*) = *$ and $f(*) = *$. The gradient of a tangent a little to the left of this point where $x < *$ is *tive and the gradient of a tangent a little to the right where $x > *$ is *tive. The point $x = *$ is, accordingly, a m*imum.

In summary, the graph *s to a m*imum point at $(*, *)$, *s to a m*imum point at $(*, *)$ and thereafter *s.

3 Draw a sketch graph of the function defined by

$$f(x) = 2x^3 + 9x^2 - 24x + 13$$

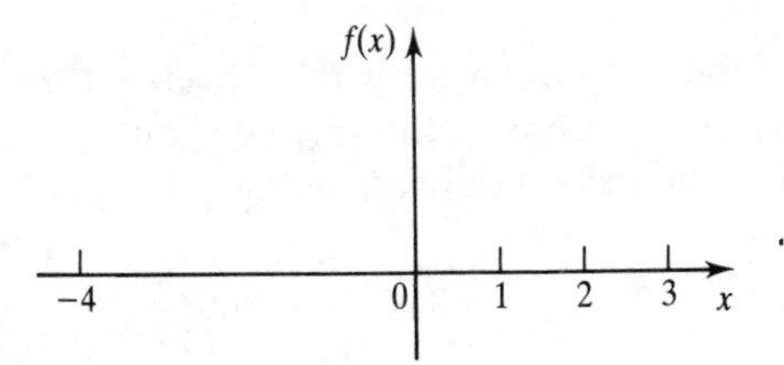

Figure 20.12

In addition to the information contained in the solution to the previous question we note that:

- For $x < *$, $* > 0$, so the graph is a *ing curve for $x < *$.
- For $* < x < *$, $* < 0$, so the graph is a *ing curve during this interval.
- For $x > *$, $* > 0$, so the graph is a *ing curve for $x > *$.

A sketch of the graph can now be drawn.

Unit 3 The Second Derivative

Test yourself with the following:

1 Find the second and third derivatives of each of the following:

(a) $f(x) = x^4$ (b) $f(x) = \sin x$ (c) $f(x) = e^{3x}$ (d) $f(x) = \ln x$

2 Describe the concavity of each of the following:

(a) $f(x) = x^3 + x^2$ (b) $(x - 1)^{2/3}$

3 Locate the points of inflexion of each of the following:

(a) $f(x) = x^3 + x^2$ (b) $(x - 1)^{2/3}$

The Derivative of the Derivative

The **second derivative** of a function is the derivative of the derivative:

$$f''(x) = \lim_{h \to 0} \frac{f'(x+h) - f'(x)}{h}$$

For this reason the derivative is often referred to as the **first derivative**. Just as the first derivative of a function is a function, so the second derivative is also a function. For example, if

$$f(x) = 3x^2 + 2x^4$$

then

$$f'(x) = 6x + 8x^3 \text{ and } f''(x) = 6 + 24x^2$$

Higher order derivatives are likewise defined. For example, the fourth derivative is the derivative of the third derivative. The third derivative being, naturally, the derivative of the second derivative.

The Sign of the Second Derivative

The second derivative represents the rate of change of the first derivative with respect to the independent variable x. Graphically, the second derivative represents the rate of change of the gradient of the tangent. Just like the first derivative, the second derivative at a point can be positive, negative or zero.

Second Derivative Positive

If, as the tangent moves around a curve from left to right, its gradient increases then the rate of change of the first derivative is positive. That is, the second derivative is positive. On a rising or falling curve, if $f''(x) > 0$, the curve is said to have **positive concavity**.

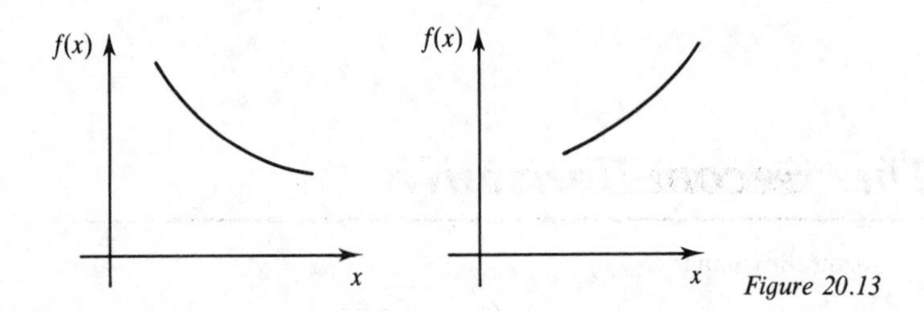

Figure 20.13

For example, the function defined by

$$f(x) = x^2$$

has first and second derivatives:

$$f'(x) = 2x \text{ and } f''(x) = 2$$

Here $f''(x) > 0$ for all values of x indicating that the parabolic graph with a minimum has positive concavity everywhere.

Second Derivative Negative

If, as the tangent moves around a curve from left to right, its gradient decreases so then is the second derivative negative. On a rising or falling curve, if $f''(x) < 0$, the curve is said to have **negative concavity**.

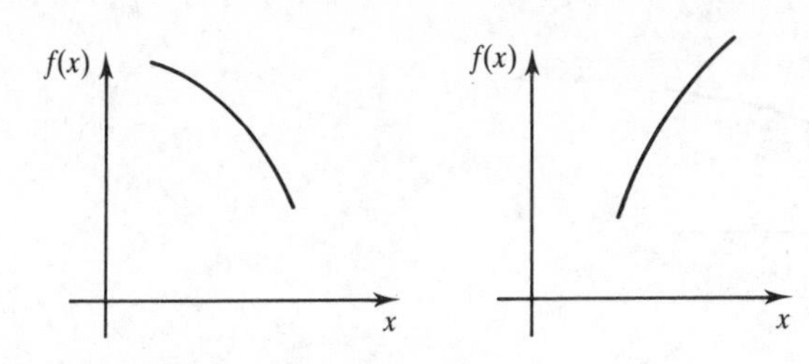

Figure 20.14

For example, the function defined by

$$f(x) = -x^2$$

has first and second derivatives:

$$f'(x) = -2x \text{ and } f''(x) = -2$$

Here $f''(x) < 0$ for all values of x indicating that the parabolic graph with a maximum has negative concavity everywhere.

Second Derivative Zero

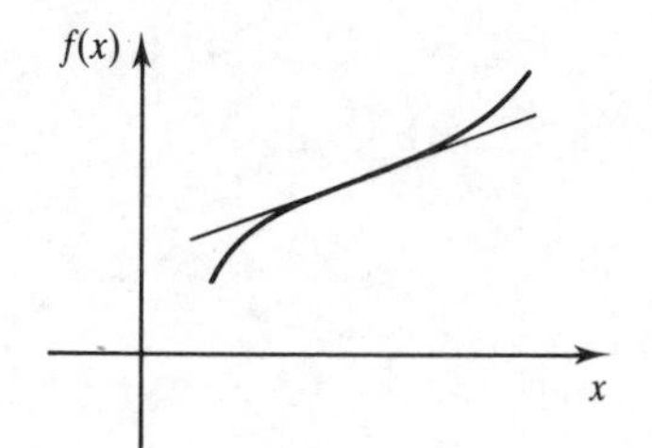

Figure 20.15

If the second derivative at a point on a curve is zero then the gradient of the tangent is neither increasing nor decreasing at that point. This can occur when, as the tangent traverses the point from left to right, it either stops increasing and starts to decrease or it stops decreasing and starts to increase. Such a point is called a **point of inflexion**, as shown in Figure 20.15.

For example, the function defined by

$$f(x) = \cos x$$

has first and second derivatives:

$$f'(x) = -\sin x \text{ and } f''(x) = -\cos x$$

Here $f''(x) = 0$ when $x = \pm\pi/2, \pm 3\pi/2, \pm 5\pi/2, \ldots$ which is at all those points where the graph of the cosine function crosses the x-axis. These points are all points of inflexion.

Second Derivative Undefined

If the second derivative at a point on a curve is undefined then the tangent at that point is either vertical or does not exist. If it is vertical then it crosses the graph at a point of inflexion. For example, the function defined by

$$f(x) = (x - 1)^{1/3}$$

has first and second derivatives:

$$f'(x) = \frac{1}{3}(x - 1)^{-2/3} \text{ and } f''(x) = \frac{-2}{9}(x - 1)^{-5/3}$$

Here, neither $f'(1)$ nor $f''(1)$ are defined because at $x = 1$ the tangent is vertical. As a consequence, at $x = 1, f(1) = 0$ the function has a point of inflexion, as shown in Figure 20.16.

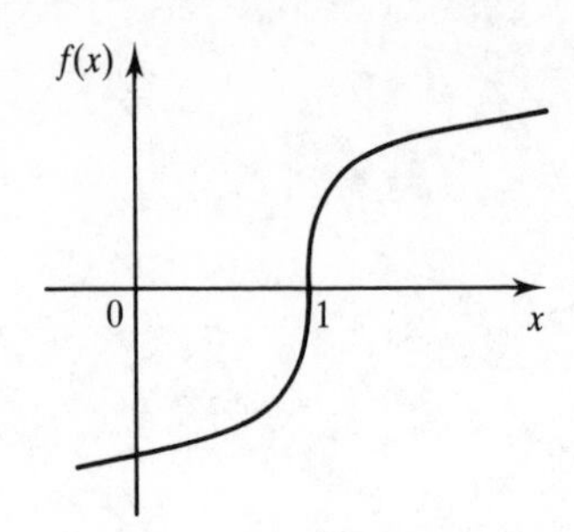

Figure 20.16

EXAMPLES

1 Find the second and third derivatives of each of the following:

(a) $f(x) = x^{-2}$ **(c)** $f(x) = xe^{-2x}$

(b) $f(x) = \tan x$ **(d)** $f(x) = \ln(x^2 - 1)$

(a) $f'(x) = -2x^{-3}$

$f''(x) = (-2)(-3)x^{-4} = 6x^{-4}$

$f'''(x) = -24x^{-5}$

(b) $f'(x) = \sec^2 x$

$f''(x) = 2\sec x \tan x \sec x = 2\sec^2 x \tan x$

$f'''(x) = 4\sec x \sec x \tan x \tan x + 2\sec^2 x \sec^2 x$

$= 4\sec^2 x \tan^2 x + 2\sec^4 x$

(c) $f'(x) = e^{-2x} + x(-2e^{-2x}) = (1 - 2x)e^{-2x}$

$f''(x) = -2e^{-2x} + (1 - 2x)(-2e^{-2x}) = 4(x - 1)e^{-2x}$

$f'''(x) = 4e^{-2x} + 4(x - 1)(-2e^{-2x}) = 4(3 - 2x)e^{-2x}$

(d) $f'(x) = \dfrac{2x}{x^2 - 1}$

$$f''(x) = \frac{2(x^2 - 1) - 2x(2x)}{(x^2 - 1)^2} = \frac{-2(1 + x^2)}{(x^2 - 1)^2}$$

$$f'''(x) = \frac{(-4x)(x^2 - 1)^2 + 2(1 + x^2)2(x^2 - 1)2x}{(x^2 - 1)^4} = \frac{4x(x^2 + 3)}{(x^2 - 1)^3}$$

2 Describe the concavity of each of the following:

(a) $f(x) = 4x^5 + 2x^4$ **(b)** $(3x + 6)^{1/3}$

(a) Concavity is determined by the sign of the second derivative:

$$f'(x) = 20x^4 + 8x^3$$

$$f''(x) = 80x^3 + 24x^2 = 8x^2(10x + 3)$$

$f''(x) = 0$ when $x = -3/10 = -0.3$, therefore

$$f''(x) < 0 \text{ when } x < -0.3 \text{ and } f''(x) > 0 \text{ when } x > -0.3$$

Consequently the graph of $f(x)$ has negative concavity for $x < -0.3$ and positive concavity for $x > -0.3$.

(b) Concavity is determined by the sign of the second derivative:

$$f'(x) = (3x + 6)^{-2/3}$$

$$f''(x) = -2(3x + 6)^{-5/3}$$

$f''(x) > 0$ when $3x + 6 < 0$, that is, when $x < -2$ and $f''(x) < 0$ when $3x + 6 > 0$, that is, when $x > -2$. Further, $f''(x)$ is not defined when $x = -2$.

Consequently, the graph of $f(x)$ has positive concavity for $x < -2$ and negative concavity for $x > -2$. When $x = -2$ the graph has an inflexion point at which the tangent is vertical.

3 Locate the points of inflexion of each of the following:

(a) $f(x) = 4x^5 + 2x^4$ **(b)** $(3x + 6)^{1/3}$

(a) From Example 2, $f'(x) = 4x^3(5x + 2)$ and so $f'(x) = 0$ when $x = -2/5$ and $x = 0$.

A little to the left of $x = -2/5$, $f'(x) > 0$ and a little to the right of $x = -2/5$, $f'(x) < 0$, so the stationary point is a maximum. A little to left of $x = 0$, $f'(x)$ is negative and a little to the right of $x = 0$, $f'(x)$ is positive. Therefore, $x = 0$, $f(0) = 0$ is a minimum. $f''(x) = 8x^2(10x + 3)$ and so $f''(-0.3) = 0$. A point of inflexion exists at $x = -0.3$.

(b) From Example 2, $f'(x) = (3x + 6)^{-2/3}$. For no value of x is $f'(x) = 0$ and so no stationary points exist on the graph of $f(x)$. We have seen, however, from Example 2 that there is a point of inflexion at $x = -2$. This point of inflexion is not a stationary point as the tangent is vertical there.

EXERCISES

1 Find the second and third derivatives of each of the following:

(a) $f(x) = 6x^3$ **(c)** $f(x) = (x^2 + 1)e^x$

(b) $f(x) = \cos^2 x$ **(d)** $f(x) = x \ln(x)$

(a) $f'(x) = 18x*$

$f''(x) = 18*x* = *x*$

$f'''(x) = *$

(b) $f'(x) = 2(\cos x)(*) = -2*$

$f''(x) = -2* + * = * \cos *$

$f'''(x) = *$

(c) $f'(x) = (2x)e^x + *(e^x) = (*)e^x$

$f''(x) = (*)e^x + *(e^x) = (*)e^x$

$f'''(x) = *e^x$

(d) $f'(x) = \ln(x) + x* = *$

$f''(x) = *$

$f'''(x) = *$

2 Describe the concavity of each of the following:

(a) $f(x) = 2x^2 - 8x^3$ **(b)** $(4 + 2x)^{1/3}$

(a) Concavity is determined by the sign of the $*$:

$$f'(x) = 4x* - *x*$$

$$f''(x) = *x* - *x* = 4(* - *x)$$

$f''(x) = 0$ when $x = *$, therefore

$$f''(x) < 0 \text{ when } x < * \text{ and } f''(x) > 0 \text{ when } x > *$$

Consequently, the graph of $f(x)$ has *tive concavity for $x < *$ and *tive concavity for $x > *$.

(b) Concavity is determined by the * of the *:

$$f'(x) = (4 + 2x)^{-*}$$

$$f''(x) = *(* + *x)^{-*}$$

$f''(x) < 0$ when $* + *x < *$, that is, when $x < *$, and $f''(x) > 0$ when $* + *x > *$, that is, when $x > *$. Further, $f''(x)$ is not defined when $x = *$.

Consequently, the graph of $f(x)$ has *tive concavity for $x < *$ and *tive concavity for $x > *$. When $x = *$ the graph has an * point at which the tangent is *.

3 Locate the points of inflexion of each of the following:

(a) $f(x) = 2x^2 - 8x^3$ **(b)** $(4 + 2x)^{1/3}$

(a) From Exercise 2, $f'(x) = *$ and $f'(x) = 0$ when $x = *$ and $x = *$.

A little to the left of $x = *$, $f'(x)$ is *tive and a little to the right of $x = *$, $f'(x)$ is also *tive. Therefore, $x = *$, $f(*) = *$ is a *.

A little to the left of $x = *$, $f'(x)$ is *tive and a little to the right of $x = *$, $f'(x)$ is also *tive. Therefore, $x = *$, $f(*) = *$ is a *.

(b) From Exercise 2, $f'(x) = *$.

For no value of x is $f'(x) = 0$ and so no stationary points exist on the graph of $f(x)$. We have seen, however, from Exercise 2 that there is a * at $x = *$. This * is not a stationary point as the * is * there.

Unit 4 Stationary Points and the Second Derivative

Test yourself with the following:

Locate and identify the stationary points in each of the following:

(a) $f(x) = x^6$ (b) $f(x) = \dfrac{1 - 2x}{6 + 3x}$ (c) $f(x) = x^3e^{-x}$

Local Maxima and Local Minima

A stationary point is located at a point on the graph of a function where the derivative is zero. This occurs at both a local maximum and a local minimum.

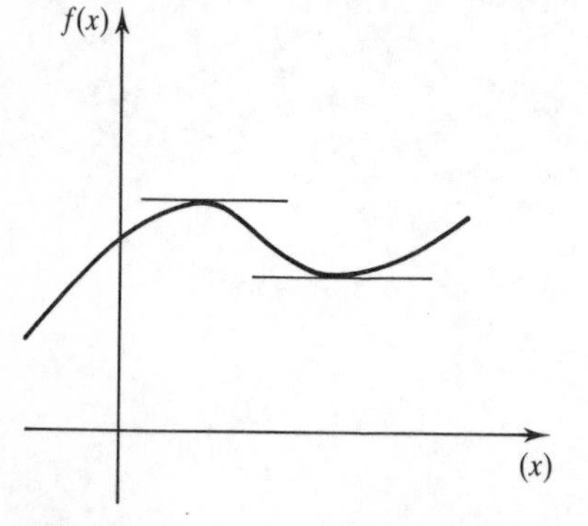

Figure 20.17

Locating Stationary Points

The stationary points of a function are located by solving the equation

$$f'(x) = 0$$

For example, the function defined by

$$f(x) = x^2 - 8x + 12$$

has the derivative

$$f'(x) = 2x - 8$$

Now $f'(x) = 0$ when $2x - 8 = 0$, that is, when $x = 4$. Consequently,

$$f'(4) = 0 \text{ and } f(4) = -4$$

so the point $(4, -4)$ is a stationary point.

Identifying Stationary Points

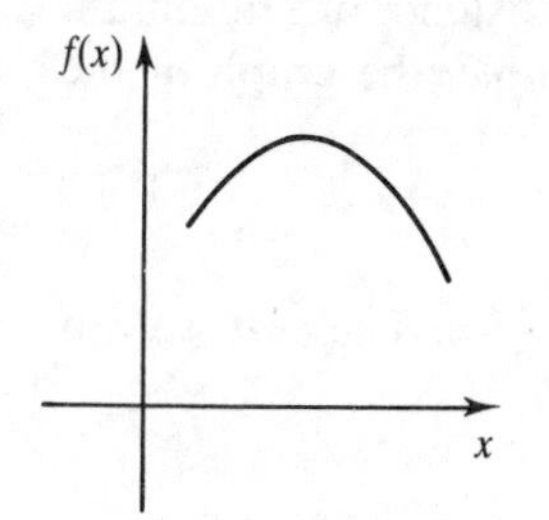

Figure 20.18

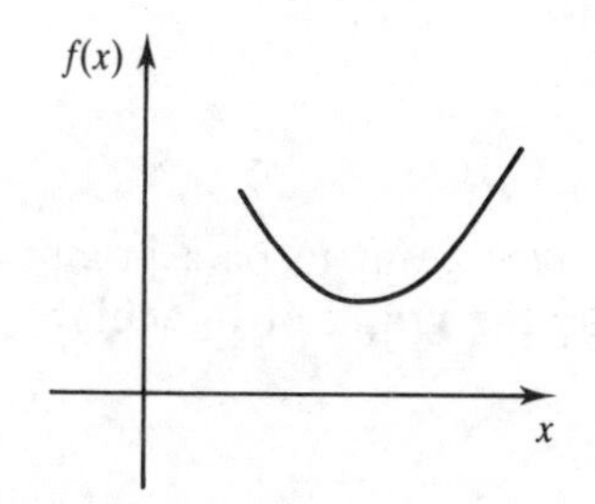

Figure 20.19

Having located a stationary point, we now wish to know what type of stationary point it is. To do this use is made of the following facts:

- If, at a stationary point, $f''(x) < 0$ then the graph has a negative concavity. This means that the stationary point is a *maximum* (Figure 20.18).
- If, at a stationary point, $f''(x) > 0$ then the graph has a positive concavity. This means that the stationary point is a *minimum* (Figure 20.19).

We can summarize this as follows:

- If, at $x = a$, $f'(a) = 0$ and $f''(a) < 0$ then $x = a$ is a local maximum.
- If, at $x = a$, $f'(a) = 0$ and $f''(a) > 0$ then $x = a$ is a local minimum.

For example, the function defined by

$$f(x) = -(x - 3)^2$$

has a first derivative

$$f'(x) = -2(x - 3) \text{ so that } f'(3) = 0$$

Further, the second derivative is

$$f''(x) = -2$$

which is negative not only at $x = 3$ but everywhere. Hence at the point $x = 3$, $f(3) = 0$, the function has a local maximum.

Be aware This argument does not work the other way. The existence of a maximum does necessarily mean that the second derivative is negative. For example, the graph of the function defined by

$$f(x) = -x^4$$

has a maximum at the origin and at the origin the second derivative is zero. The best we can say is that at a local maximum the second derivative is never positive.

As a further example, the function defined by

$$f(x) = (x + 2)^2$$

has a first derivative

$$f'(x) = 2(x + 2) \text{ so that } f'(-2) = 0$$

Further, the second derivative is

$$f''(x) = 2$$

which is positive not only at $x = -2$ but everywhere. Hence at $x = -2$, $f(-2) = 0$, the function has a local minimum.

Note, again, that the argument does not work the other way. The existence of a minimum does necessarily mean that the second derivative is positive. For example, the graph of the function defined by

$$f(x) = x^4$$

has a minimum at the origin and at the origin the second derivative is zero. The best we can say is that at a local minimum the second derivative is never negative.

Points of Inflexion

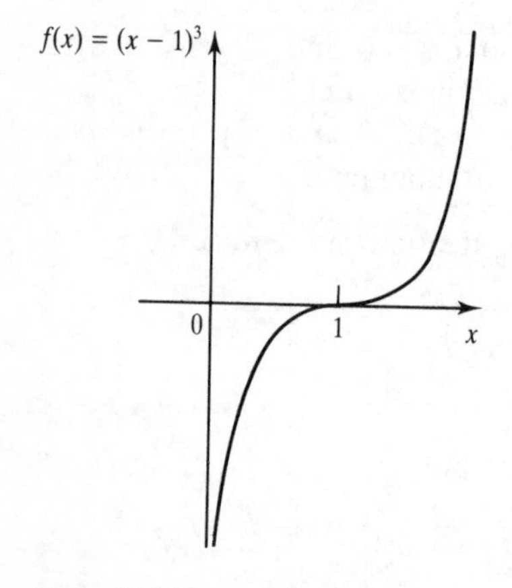

Figure 20.20

It is possible for a stationary point to be a point of inflexion. For example, the graph of the cubic defined by

$$f(x) = (x - 1)^3$$

has derivatives

$$f'(x) = 3(x - 1)^2 \text{ and } f''(x) = 6(x - 1)$$

Here we see that both $f'(1) = 0$ and $f''(1) = 0$. Testing the value of the gradient of the tangent a little to either side of the stationary point we see that it is positive on both sides. This means that the stationary point is a point of inflexion.

Consequently, a stationary point at which the second derivative is zero can be a local maximum, a local minimum or even a point of inflexion and its identification will only be revealed by considering the gradient of the tangent on either side of it.

EXAMPLES

1 Find the values of x for which $f'(x) = 0$ in each of the following:

(a) $f(x) = (x - 1)^2$ **(c)** $f(x) = \exp\left(\dfrac{x^3}{3} - x\right)$

(b) $f(x) = \sin(x + \pi)$

(a) $f'(x) = 2(x - 1)$ So $f'(x) = 0$ when $x - 1 = 0$, that is, when $x = 1$.
(b) $f'(x) = \cos(x + \pi)$ So $f'(x) = 0$ when $\cos(x + \pi) = 0$, that is, when

$$x + \pi = \pm\pi/2, \pm 3\pi/2, \pm 5\pi/2, \ldots$$

Consequently, $f'(x) = 0$ when

$$x = \pm\pi/2, \pm 3\pi/2, \pm 5\pi/2, \ldots$$

(c) $f'(x) = (x^2 - 1)\exp(x^3/3 - x)$ So $f'(x) = 0$ when $x^2 - 1 = 0$, that is, when $x = \pm 1$.

2 Locate and identify the stationary points in each of the following:

(a) $f(x) = x^2$ **(b)** $f(x) = \dfrac{1 - x^2}{1 + x^2}$ **(c)** $f(x) = x^2 e^x$

(a) Stationary points exist where $f'(x) = 0$.

$$f'(x) = 2x$$

and so $f'(x) = 0$ when $2x = 0$, that is, when $x = 0$. There is, therefore one stationary point at $x = 0$. To determine whether this is a maximum or a minimum, consider the value of $f''(0)$. $f''(x) = 2$ and so $f''(0) > 0$. Therefore, the stationary point is a minimum.

(b) Stationary points exist where $f'(x) = 0$.

$$f'(x) = \frac{(-2x)(1 + x^2) - (2x)(1 - x^2)}{(1 + x^2)^2}$$

$$= \frac{-4x}{(1 + x)^2}$$

and so $f'(x) = 0$ when $-4x = 0$, that is, when $x = 0$. There is, therefore, one stationary point at $x = 0$. To determine whether this is a maximum or a minimum, consider the value of $f''(0)$. $f''(x) = -4(1 + x)^2 + 8x(1 + x)$ and so $f''(0) < 0$. Therefore, the stationary point is a maximum.

(c) Stationary points exist where $f'(x) = 0$.

$$f'(x) = 2xe^x + x^2e^x$$
$$= xe^x(2 + x)$$

and so $f'(x) = 0$ when $x = 0$ and when $2 + x = 0$, that is, when $x = 0$ and when $x = -2$. There are, therefore, two stationary points at $x = 0$ and at $x = -2$. To determine whether these are maxima or minima consider the values of $f''(0)$ and $f''(-2)$. $f''(x) = e^x(x^2 + 4x + 2)$ and so $f''(0) > 0$ and $f''(-2) < 0$. Therefore, the stationary point at $x = 0$ is a minimum and the stationary point at $x = -2$ is a maximum.

EXERCISES

1 Find the values of x for which $f'(x) = 0$ in each of the following:

(a) $f(x) = x^2 - 4x$ **(c)** $f(x) = \ln(x^3 - 3x)$
(b) $f(x) = \cos^2(x - \pi/4)$

(a) $f'(x) = *x - *$ So $f'(x) = 0$ when $*x - * = 0$, that is, when $x = *$.
(b) $f'(x) = 2\cos(*)*(x - \pi)$ So $f'(x) = 0$ when $\cos(*) = 0$ or when $*(x - \pi) = 0$. $\cos(*) = 0$ when

$$* = \pm\pi/2, \pm\pi/3, \pm\pi/5, \ldots$$

that is, when

$$x = *$$

$*(x - \pi) = 0$ when

$$x - \pi = *$$

that is, when

$$x = *$$

Consequently, $f'(x) = 0$ when $x = \pm*\pi/*$.
(c) $f'(x) = (*)/(*)$ So $f'(x) = 0$ when $* = 0$, that is, when $x = \pm*$.

2 Locate and identify the stationary points in each of the following:

(a) $f(x) = -x^2$ **(b)** $f(x) = x\ln(x)$ **(c)** $f(x) = \dfrac{x^2 + x + 1}{x + 1}$

(a) Stationary points exist where $f'(x) = 0$.

$$f'(x) = *$$

and so $f'(x) = 0$ when $x = *$. There is, therefore, one stationary point at $x = *$. To determine whether this is a maximum or a minimum, consider the value of $f''(*)$. $f''(x) = *$ and so $f''(*) * 0$. Therefore, the stationary point at $x = 0$ is a m*imum.

(b) $f'(x) = *$ So $f'(x) = 0$ when $\ln(x) = -*$, that is, when $x = */*$. There is, therefore, one stationary point at $x = */*$. To determine whether this is a maximum or a minimum, consider the value of $f''(*)$. $f''(x) = *$ and so $f''(*) * 0$. Therefore, the stationary point at $x = */*$ is a m*imum.

$$\textbf{(c)}\ f'(x) = \frac{(*)(x+1) - (*)(x^2+x+1)}{(x+1)^2}$$

$$= \frac{*}{(x+1)^2}$$

and so $f'(x) = 0$ when $* = 0$, that is, when $x = *$ or $-*$. There are, therefore, two stationary points at $x = *$ and $x = -*$. To determine whether each is a maximum or a minimum consider the values of $f''(*)$ and $f''(*)$. $f''(x) = *$ and so $f''(*) * 0$ and $f''(*) * 0$. Therefore, the stationary point at $x = *$ is a m*imum and the stationary point at $x = -*$ is a m*imum.

Module 20: Further exercises

1 Find the derivatives of each of the following:

(a) $f(x) = x^{-2} + x^{-5}$ (d) $f(x) = \dfrac{x^3}{\sin x}$ (g) $f(x) = \ln(4x^3 - x^2)$

(b) $f(x) = \dfrac{9x^3}{5}$ (e) $f(x) = \cot^2 x$ (h) $f(x) = e^{-5x}$

(c) $f(x) = e^x \cos x$ (f) $f(x) = (1 - x^3)^2$ (i) $f(x) = \cot 6x$

2 Without drawing the graph of the function, describe the characteristic features of each of the following:

(a) $f(x) = (1/2)x^4 - x^2 + 3$ (b) $f(x) = 3(2x + 4)^{1/3}$

3 Locate and identify the stationary points of

$$f(x) = x^3 - 5x^2 - 8x - 4$$

4 Draw a sketch graph of the function defined by

$$f(x) = x^3 - 5x^2 - 8x - 4$$

5 Find the second and third derivatives of each of the following:

(a) $f(x) = x^{-2}$ (c) $f(x) = xe^{-2x}$

(b) $f(x) = \tan x$ (d) $f(x) = \ln(x^2 - 1)$

6 Describe the concavity of each of the following:

(a) $f(x) = 4x^5 + 2x^4$ (b) $(3x + 6)^{1/5}$

7 Locate the points of inflexion of each of the following:

(a) $f(x) = 4x^5 + 2x^4$ (b) $(3x + 6)^{1/5}$

8 Find the values of x for which $f'(x) = 0$ in each of the following:

(a) $f(x) = x^3 - 3x^2$ (c) $f(x) = \tan(x - \pi/2)$

(b) $f(x) = \cos^3 x$

9 Locate and identify the stationary points in each of the following:

(a) $f(x) = -x^4$ (c) $f(x) = \dfrac{x^2 - x - 1}{x + 1}$

(b) $f(x) = x^2 \ln(x)$

Module 21: The Differential and Its Uses

Aims: To examine the geometric significance of the differential and to apply it to the approximation of solutions to equations.

Objectives: When you have read this module you will be able to:

- ▶▶ Identify the differential from a graph.
- ▶▶ Recognize the alternative form for the derivative as a ratio of differentials.
- ▶▶ Use the differential to find the approximate change in one variable due to a change in another.

Unit 1 The Differential

Test yourself with the following:

1 Find the differential $df(x)$ of each of the following:

(a) $f(x) = x^2$ (d) $f(x) = e^{-2x}$
(b) $f(x) = \sin x$ (e) $f(x) = \ln(1/x)$
(c) $f(x) = \cot x$ (f) $f(x) = x^4 - x^2$

2 Find the primitive associated with each of the following:

(a) $\dfrac{df(x)}{dx} = x^3$ (d) $\dfrac{df(x)}{dx} = \cos 4x$

(b) $\dfrac{df(x)}{dx} = \dfrac{x^{-6}}{7}$ (e) $\dfrac{df(x)}{dx} = -9e^{-3x}$

(c) $\dfrac{df(x)}{dx} = x - 2x^4$ (f) $\dfrac{df(x)}{dx} = \dfrac{16}{x}$

Rising and Falling Tangents

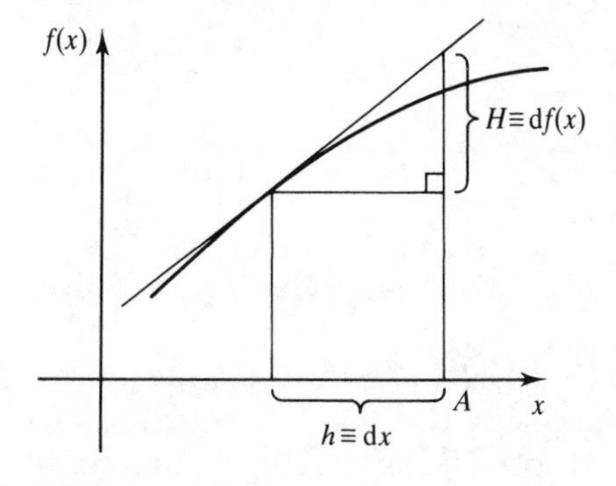

Figure 21.1

A tangent line touches a curve at a single point. On the x-axis, directly below the contact point, is the x-coordinate of the contact point. In Figure 21.1, point A on the x-axis is a distance h to the right of the x-coordinate of the contact point, and H is the amount the tangent line has risen above the contact point. Connected in this way, the quantities h and H are referred to as the **differentials** of x and $f(x)$ respectively and are represented by the symbols

$$h = dx \qquad \text{differential of } x$$

$$H = df(x) \qquad \text{differential of } f(x)$$

Ratios of Differentials

From the geometry of the differentials is it clear that their ratio is equal to the gradient of the tangent:

$$\text{Gradient of the tangent} = \frac{\mathrm{d}f(x)}{\mathrm{d}x}$$

$$= f'(x)$$

This permits us to relate the differential $\mathrm{d}f(x)$ to the differential $\mathrm{d}x$ as

$$\mathrm{d}f(x) = f'(x)\,\mathrm{d}x$$

It also allows the ratio of the differentials to be used as an alternative notation for the derivative, namely

$$\frac{\mathrm{d}f(x)}{\mathrm{d}x} \equiv f'(x)$$

Primitives

If, for a function f, we are given the ratio of the differentials as an expression in x then it is possible to find the type of function to which they relate by using the equation

$$\frac{\mathrm{d}f(x)}{\mathrm{d}x} = f'(x)$$

For example, if $\mathrm{d}f(x)/\mathrm{d}x = 0$ then

$$f'(x) = 0$$

and so

$$f(x) = c \qquad \text{(an arbitrary constant)}$$

All constant functions have a zero ratio of differentials. This conclusion can be verified by differentiating $f(x) = c$ to obtain $f'(x) = 0$.

As a further example, if $\mathrm{d}f(x)/\mathrm{d}x = a$ (a constant) then

$$f'(x) = a$$

and so

$$f(x) = ax + c \quad (c \text{ is, again, an arbitrary constant})$$

All linear functions have a constant ratio of differentials. Again, differentiating $f(x)$ gives $f'(x) = a$, which verifies the conclusion.

The type of function related to a given derivative is called a **primitive**, and the process whereby we obtain the primitive is the reverse process to that of differentiation. To 'un-differentiate' requires a deal of hindsight – we need to be familiar with the process of differentiation to enable us to 'see' the result. For example, to find the primitive associated with a linear ratio of differentials:

$$\frac{df(x)}{dx} = ax + b$$

we note that

$$f'(x) = ax + b$$

and that if $g(x) = x^2$ then $g'(x) = 2x$. Hence

$$\begin{aligned} f(x) &= a(x^2/2 + c) + b(x + c') \\ &= ax^2/2 + bx + (ac + bc') \\ &= ax^2/2 + bx + c'' \end{aligned}$$

All quadratic functions have a linear derivative. Again, differentiating $f(x)$ gives $f'(x)$ so verifying the conclusion.

Note that we have replaced $ac + bc'$ by the single constant c''. Because c and c' are arbitrary constants the value of $ac + bc'$ is still a single arbitrary constant – denoted by c''.

Standard Forms

To assist in the process of finding primitives a collection of standard forms is now given where, in each, c stands for an arbitrary constant:

$f'(x)$	$f(x)$	
a	$ax + c$	
x^n	$\frac{x^{n+1}}{(n+1)} + c$	$(n \neq -1)$
x^{-1}	$\ln(x) + c$	
$\sin x$	$-\cos x + c$	
$\cos x$	$\sin x + c$	
e^x	$e^x + c$	
$\sec^2 x$	$\tan x + c$	
$\text{cosec}^2 x$	$-\cot x + c$	

Each one of these standard forms can be verified by differentiating $f(x)$ to obtain the appropriate $f'(x)$.

EXAMPLES

1 Find the differential $df(x)$ of each of the following:

(a) $f(x) = x^3$ **(d)** $f(x) = e^{2x}$
(b) $f(x) = \cos x$ **(e)** $f(x) = \ln(x)$
(c) $f(x) = \tan x$ **(f)** $f(x) = x^3 + x^2 + x$

(a) $$\begin{aligned} df(x) &= f'(x)\,dx \\ &= 3x^2\,dx \end{aligned}$$

(b) $df(x) = f'(x)\,dx$
$= -\sin x\,dx$

(c) $df(x) = f'(x)\,dx$
$= \sec^2 x\,dx$

(d) $df(x) = f'(x)\,dx$
$= 2e^x\,dx$

(e) $df(x) = f'(x)\,dx$
$= \frac{1}{x}\,dx$

(f) $df(x) = f'(x)\,dx$
$= (3x^2 + 2x + 1)\,dx$

2 Find the primitive of each of the following:

(a) $\frac{df(x)}{dx} = x^4$ **(d)** $\frac{df(x)}{dx} = \sin 2x$

(b) $\frac{df(x)}{dx} = 5x^{-3}$ **(e)** $\frac{df(x)}{dx} = 2e^{3x}$

(c) $\frac{df(x)}{dx} = 3x^2 + 2x$ **(f)** $\frac{df(x)}{dx} = \frac{9}{x}$

(a) Using the table of standard forms:

$$f(x) = \frac{x^{4+1}}{(4+1)} + c$$
$$= \frac{x^5}{5} + c$$

(b) Using the table of standard forms:

$$f(x) = 5\left[\frac{x^{-3+1}}{-3+1} + c\right]$$
$$= \frac{5x^{-2}}{-2} + 5c$$
$$= \frac{5x^{-2}}{-2} + c'$$
$$= -\frac{5}{2}x^{-2} + c'$$

Here, the product $5c$ is still just an arbitrary constant.

(c) Here:

$$f(x) = 3\left(\frac{x^3}{3} + c\right) + 2\left(\frac{x^2}{2} + c'\right)$$
$$= x^3 + x^2 + 3c + 2c'$$
$$= x^3 + x^2 + c''$$

Here, the sum of the two arbitrary constants $3c + 2c'$ is still just a single arbitrary constant c''.

(d) Noting that the primitive of $\sin x$ is $-\cos x + c$ we see that

$$f(x) = \frac{-\cos 2x + c}{2}$$

$$= \frac{-\cos 2x}{2} + c'$$

(e) The primitive of e^x is $e^x + c$ and so

$$f(x) = 2\left(\frac{e^{3x}}{3} + c\right)$$

$$= \frac{2}{3}e^{3x} + 2c$$

$$= \frac{2}{3}e^{3x} + c'$$

(f) From the list of standard forms:

$$f(x) = 9[\ln(x) + c]$$

$$= 9\ln(x) + c'$$

EXERCISES

1 Find the differential $df(x)$ of each of the following:

(a) $f(x) = x^{-1}$ **(d)** $f(x) = e^{-x}$

(b) $f(x) = \sin 2x$ **(e)** $f(x) = \ln(x)^2$

(c) $f(x) = \cot 3x$ **(f)** $f(x) = x^2 + x^{-2}$

(a) $df(x) = f'(x)\,dx$
$= -x^{*}\,dx$

(b) $df(x) = f'(x)\,dx$
$= {*}\cos 2x\,dx$

(c) $df(x) = f'(x)\,dx$
$= 3{*}^2\,dx$

(d) $df(x) = f'(x)\,dx$
$= {*}\,dx$

(e) $df(x) = f'(x)\,dx$
$= ({*}/{*})\,dx$

(f) $df(x) = f'(x)\,dx$
$= ({*} - {*})\,dx$

2 Find the primitive of each of the following:

(a) $\frac{df(x)}{dx} = x^{-2}$ **(d)** $\frac{df(x)}{dx} = \sec^2 3x$

(b) $\frac{df(x)}{dx} = 4x^4$ **(e)** $\frac{df(x)}{dx} = 5e^{-2x}$

(c) $\frac{df(x)}{dx} = 5x^3 - x$ **(f)** $\frac{df(x)}{dx} = 12x^{-1}$

(a) Using the table of standard forms:

$$f(x) = \frac{x^{*+*}}{*+*} + c$$

$$= \frac{x^*}{*} + c$$

(b) Using the table of standard forms:

$$f(x) = 4\left(\frac{x^*}{*} + c\right)$$

$$= \frac{4x^*}{*} + 4c$$

$$= *x^* + *$$

(c) Using the table of standard forms:

$$f(x) = 5\left(\frac{x^*}{*} + c\right) - \left(\frac{x^*}{*} + c'\right)$$

$$= *x^* - *x^* + *$$

(d) Noting that the primitive of $\sec^2 x$ is $* + c$ we see that

$$f(x) = \frac{* + c}{3}$$

$$= \frac{*}{3} + *$$

(e) The primitive of e^x is $e^x + c$ and so

$$f(x) = 5\left(\frac{e^*}{*} + c\right)$$

$$= *e^* + *$$

(f) From the list of standard forms:

$$f(x) = 12(* + c)$$

$$= * + *$$

Unit 2 Numerical Approximations to Equations

Test yourself with the following:

1 Given that $x = 1.41$ is an approximate solution to the equation

$$x^3 = 3$$

find a better approximation.

2 Given that $x = 1.7$ is an approximate solution to the equation

$$x^2 = 3$$

find a better approximation.

Exact and Approximate Solutions to Equations

Let the value $x = a$ be an approximate solution to the equation

$$f(x) = 0$$

where $f(x)$ is some expression involving the variable x. Further, let the exact solution be $x = a - h$ so that

$$f(a - h) = 0$$

We can represent this graphically. If we draw the graph associated with the expression $f(x)$ then the exact solution $x = a - h$ will be the point where the graph meets the x-axis and the point $x = a$ will be some distance from it. If we draw the tangent to the curve at the point where the curve meets the x-axis then the length of the vertical line from $x = a$ to the tangent is the differential

$$\mathrm{d}f(x)$$

Provided that h is sufficiently small, we can say that the differential $\mathrm{d}f(x)$ is approximately equal to $f(a)$:

$$\mathrm{d}f(x) \approx f(a) \text{ and } \mathrm{d}x = h$$

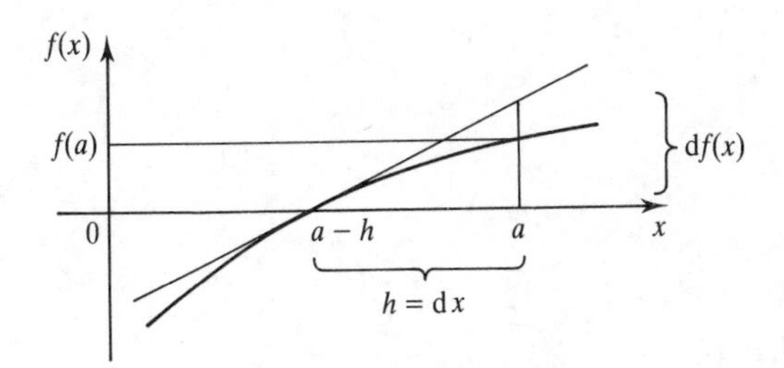

Figure 21.2

so that

$$\frac{\mathrm{d}f(x)}{\mathrm{d}x} \approx \frac{f(a)}{h}$$

The ratio of the differential $\mathrm{d}f(x)/\mathrm{d}x$ is equal to the derivative $f'(x)$ at $x = a - h$ and provided, again, that h is sufficiently small, this is approximately equal to the derivative at $x = a$. Thus

$$f'(a) \approx \frac{f(a)}{h}$$

or

$$h \approx \frac{f(a)}{f'(a)}$$

This means that

$$x = a - \frac{f(a)}{f'(a)}$$

is closer to the point where the graph meets the x-axis than the point $x = a$. In other words,

$$x = a - \frac{f(a)}{f'(a)}$$

is a better approximate solution to the equation $f(x) = 0$ than $x = a$.

For example, the equation

$$x^5 = 2$$

can be written as

$$x^5 - 2 = 0$$

Here $f(x) = x^5 - 2$. Assume that we have been given an approximate solution to this equation as $x = 1.2$ then

$$f(1.2) = 0.488\,32$$

and

$$f'(x) = 5x^4$$

so that

$$f'(1.2) = 10.368$$

This means that

$$\frac{f(1.2)}{f'(1.2)} = \frac{0.488}{10.368}$$

$$= 0.0471 \text{ to four dec. pl.}$$

Therefore

$$1.2 - \frac{f(1.2)}{f'(1.2)} = 1.2 - 0.0471$$

$$= 1.1529$$

and so $x = 1.1529$ is a better solution to the equation $x^5 = 2$. In fact, $f(1.1529) = 0.036\ldots$. It is not the exact solution, but it is a better solution than $x = 1.2$, having been interpolated between the original solution and the exact solution.

EXAMPLES

1 Given that $x = 2.2$ is an approximate solution to the equation

$$x^2 = 5$$

find a better approximation.

Here $f(x) = x^2 - 5$ and hence $f'(x) = 2x$. Therefore

$$f(2.2) = -0.16$$

and

$$f'(2.2) = 4.4$$

Therefore

$$\frac{f(2.2)}{f'(2.2)} = -0.036\,363\,6\ldots$$

so that

$$2.2 - \frac{f(2.2)}{f'(2.2)} = 2.2 + 0.04 \qquad \text{to two dec. pl.}$$

$$= 2.24$$

is a better approximation to the solution of the equation.

2 If $x = 1.1$ is taken as an approximate solution to the equation

$$x^2 + 2x - 4 = 0$$

find a better approximation.

Here $f(x) = x^2 + 2x - 4$ and hence $f'(x) = 2x + 2$. Therefore

$$f(1.1) = -0.59$$

and

$$f'(1.1) = 4.2$$

Therefore

$$\frac{f(1.1)}{f'(1.1)} = -0.140\,476\,1\ldots$$

so that

$$1.1 - \frac{f(1.1)}{f'(1.1)} = 1.1 + 0.14$$

$$= 1.24 \text{ to two dec. pl.}$$

is a better approximation to the solution of the equation.

EXERCISES

1 Given that $x = 1.3$ is an approximate solution to the equation

$$x^4 = 3$$

find a better approximation.

Here $f(x) = x^4 - 3$ and hence $f'(x) = *$. Therefore

$$f(1.3) = -*$$

and

$$f'(1.3) = 8.788$$

Therefore

$$\frac{f(1.3)}{f'(1.3)} = -0.0** **46\ldots$$

therefore

$$1.3 - \frac{f(1.3)}{f'(1.3)} = 1.3 + 0.0*$$

$$= 1.3* \text{ to two dec. pl.}$$

is a better approximation to the solution of the equation.

2 If $x = 1.4$ is taken as an approximate solution to the equation

$$x^3 = 2x$$

find a better approximation.

Here $f(x) = x^3 - 2x$ and hence $f'(x) = *$. Therefore

$$f(1.4) = -*$$

and

$$f'(1.4) = *$$

Therefore

$$\frac{f(1.4)}{f'(1.4)} = -0.0** *329\ldots$$

so that

$$1.4 - \frac{f(1.4)}{f'(1.4)} = 1.4 + 0.0*$$

$$= 1.4* \text{ to two dec. pl.}$$

is a better approximation to the solution of the equation.

Unit 3 The Newton–Raphson Method

Test yourself with the following:

1 Use the Newton–Raphson iterative method to solve the equation

$$x^3 = 3$$

to three decimal places given that 1.44 is a good first approximation.

2 Use the Newton–Raphson iterative method to solve the equation

$$x^2 = 3$$

to three decimal places given that 1.73 is a good first approximation.

Iterated Interpolations

The **Newton–Raphson method** of obtaining the numerical solution to an equation is an iterative procedure that repeatedly uses the expression

$$\frac{f(a)}{f'(a)}$$

to obtain a sequence of better, interpolated approximations.

Defining

$$h_1 = \frac{f(a)}{f'(a)}$$

makes $x = b = a - h_1$ an approximation to $a - h$. Similarly, defining

$$h_2 = \frac{f(b)}{f'(b)}$$

makes $x = c = b - h_2$ a better approximation to $a - h$. Defining

$$h_3 = \frac{f(c)}{f'(c)}$$

makes $x = d = c - h_3$ an even better approximation to $a - h$.

Proceeding with this iteration produces a solution that is as accurate as effort will permit. For example, for the equation

$$f(x) = x^5 - 2 = 0 \qquad f'(x) = 5x^4$$

we have found that $x = 1.15$ is a better approximation to the first approximation to the solution of $x = 1.2$. Defining

$$\begin{aligned} h_2 &= \frac{f(1.15)}{f'(1.15)} \\ &= \frac{0.011\,357\,1}{8.745\,031\,2} \\ &= 0.0013 \text{ to four dec. pl.} \end{aligned}$$

so that

$$\begin{aligned} 1.15 - h_2 &= 1.15 - 0.0013 \\ &= 1.149 \end{aligned}$$

is a better approximation. Repeating this process by defining

$$h_3 = \frac{f(1.149)}{f'(1.149)}$$

$$= \frac{0.002\,627\,35}{8.714\,653\,4}$$

$$= 0.000\,301\,486$$

$$= 0 \text{ to the third place of decimals}$$

Thus the solution to the equation $x^5 = 2$ is $x = 1.149$ to three decimal places.

EXAMPLES

1 Use the Newton–Raphson iterative method to solve the equation

$$x^2 = 5$$

to three decimal places given that 2.24 is a good first approximation.

$$f(x) = x^2 - 5 = 0 \qquad f'(x) = 2x$$

We have found that $x = 2.24$ is a better approximation to the first approximation to the solution of $x = 2.2$. Defining

$$h_2 = \frac{f(2.24)}{f'(2.24)}$$

$$= \frac{0.0176}{4.48}$$

$$= 0.0039 \text{ to four dec. pl.}$$

so that

$$2.24 - h_2 = 2.24 - 0.0039$$

$$= 2.236 \text{ to three dec. pl.}$$

is a better approximation. Repeating this process by defining

$$h_3 = \frac{f(2.236)}{f'(2.236)}$$

$$= \frac{-0.000\,304}{4.472}$$

$$= 0.000\,068$$

$$= 0 \text{ to the third place of decimals}$$

Thus the solution to the equation $x^2 = 5$ is $x = 2.236$ to three decimal places.

2 Use the Newton–Raphson iterative method to solve the equation

$$x^2 + 4x - 4 = 0$$

to three decimal places given that 0.9 is a good first approximation.

$$f(x) = x^2 + 4x - 4 = 0 \qquad f'(x) = 2x + 4$$

It is given that $x = 0.9$ is a good approximation to the equation. Defining

$$\begin{aligned} h_1 &= \frac{f(0.9)}{f'(0.9)} \\ &= \frac{0.41}{5.8} \\ &= 0.071 \text{ to three dec. pl.} \end{aligned}$$

so that

$$\begin{aligned} 0.9 - h_1 &= 0.9 - 0.071 \\ &= 0.83 \text{ to two dec. pl.} \end{aligned}$$

is a better approximation. Repeating this process by defining

$$\begin{aligned} h_2 &= \frac{f(0.83)}{f'(0.83)} \\ &= \frac{0.0089}{5.66} \\ &= 0.0016 \text{ to four dec. pl.} \end{aligned}$$

so that

$$\begin{aligned} 0.83 - h_2 &= 0.83 - 0.0016 \\ &= 0.828 \text{ to three dec. pl.} \end{aligned}$$

is a better approximation. Repeating this process by defining

$$\begin{aligned} h_3 &= \frac{f(0.828)}{f'(0.828)} \\ &= \frac{-0.002\,416}{5.656} \\ &= -0.0004 \\ &= 0 \text{ to the third place of decimals} \end{aligned}$$

Thus the solution to the equation $x^2 + 4x - 4$ is $x = 0.828$ to three decimal places.

EXERCISES

1 Use the Newton–Raphson iterative method to solve the equation

$$x^4 = 3$$

to three decimal places given that 1.32 is a good first approximation.

$$f(x) = x^4 - 3 = 0 \qquad f'(x) = 4x^3$$

It is given that $x = 1.32$ is a better approximation to the first approximation to the solution.

Defining

$$h_2 = \frac{f(1.32)}{f'(1.32)}$$

$$= \frac{*}{*}$$

$$= 0.00** \text{ to four dec. pl.}$$

so that

$$1.32 - h_2 = 1.32 - 0.00**$$

$$= 1.3** \text{ to three dec. pl.}$$

is a better approximation. Repeating this process by defining

$$h_3 = \frac{f(1.3**)}{f'(1.3**)}$$

$$= \frac{-0.000*7*7*}{*.1*6*9}$$

$$= *$$

$$= 0 \text{ to } * \text{ places of decimals}$$

Thus the solution to the equation $x^4 = 3$ is $x = 1.3**$ to three decimal places.

2 Use the Newton–Raphson iterative method to solve the equation

$$x^3 = 2x$$

to three decimal places given that 1.44 is a good first approximation.

$$f(x) = x^3 - 2x = 0 \qquad f'(x) = 3x^2 - 2$$

Defining

$$h_1 = \frac{f(1.44)}{f'(1.44)}$$

$$= \frac{*}{*}$$

$$= 0.0**1 \text{ to four dec. pl.}$$

so that

$$1.44 - h_1 = 1.44 - 0.0**1$$

$$= 1.4** \text{ to three dec. pl.}$$

is a better approximation. Repeating this process by defining

$$h_2 = \frac{f(1.4**)}{f'(1.4**)}$$

$$= \frac{*}{*}$$

$$= 0.000*9 \text{ to five dec. pl.}$$

so that

$$1.4** - h_2 = 1.4** - 0.00**$$
$$= 1.4*** \text{ to four dec. pl.}$$

is a better approximation. Repeating this process by defining

$$h_3 = \frac{f(1.4***)}{f'(1.4***)}$$
$$= \frac{-0.000\,054}{3.999\,885}$$
$$= *$$
$$= 0 \text{ to the third decimal place}$$

Thus the solution to the equation $x^3 - 2x$ is $x = 1.4**$ to three decimal places.

Module 21: Further exercises

1 Find the differential $df(x)$ of each of the following:

(a) $f(x) = x^{-4}$ (d) $f(x) = e^{3x}$
(b) $f(x) = \cos 4x$ (e) $f(x) = \ln(x^3)$
(c) $f(x) = \tan 6x$ (f) $f(x) = x - x^{-1}$

2 Find the primitive of each of the following:

(a) $\dfrac{df(x)}{dx} = x^{-5}$ (d) $\dfrac{df(x)}{dx} = \text{cosec}^2 5x$

(b) $\dfrac{df(x)}{dx} = \dfrac{2x^{-4}}{3}$ (e) $\dfrac{df(x)}{dx} = 12e^{4x}$

(c) $\dfrac{df(x)}{dx} = 7x - 9x^{-2}$ (f) $\dfrac{df(x)}{dx} = -3x^{-1}$

3 If $x = 2.0$ is taken as an approximate solution to the equation

$$x^6 = 72$$

find a better approximation.

4 If $x = 1.7$ is taken as an approximate solution to the equation

$$x^3 = 3x$$

find a better approximation.

5 Use the Newton–Raphson iterative method to solve the equation

$$x^6 = 72$$

to three decimal places given that 2.04 is a good first approximation.

6 Use the Newton–Raphson iterative method to solve the equation

$$x^3 = 3x$$

to three decimal places given that 1.73 is a good first approximation.

Chapter 5: Miscellaneous exercises

1 Plot the graph and differentiate each of the following:

$$\text{(a)}\ f(x) = \frac{x}{|x|} \qquad \text{(b)}\ f(x) = \frac{|x|}{x}$$

2 Is the function f defined by

$$f(x) = x^2/|x|$$

differentiable everywhere?

3 Plot the graph and differentiate the function f defined by

$$f(x) = |x^2 - 1|$$

4 Show that $f(x) = x + 1/x$ satisfies the equation

$$x^2 f''(x) + x f'(x) - f(x) = 0$$

5 Show that

$$f(x) = Ae^x + Be^{-x}$$

satisfies the equation

$$f''(x) - f(x) = 0$$

Find A and B if $f(0) = 0$ and $f'(0) = 2$.

6 If the equation

$$f'(x) + 2f(x) = 4x$$

is satisfied by

$$f(x) = ax^2 + bx + c$$

find a, b and c.

7 Show that the equation:

$$\frac{f''(x)f(x) - [f'(x)]^2}{[f(x)]^2} = \frac{-1 - \ln(x)}{(x \ln(x))^2}$$

is satisfied by $f(x) = \ln(x)$.

8 Functions f and g are defined by

$$f(x) = x^2 \text{ and } g(x) = \sin x$$

Show that, for these functions

$$(f \circ g)' \equiv g \circ f'$$

9 If, for functions f, g and h

$$h''(x) = 0 \text{ where } h \equiv f \circ g$$

show that

$$\frac{f''[g(x)]}{g''(x)} = \frac{-f'[g(x)]}{[g'(x)]^2}$$

10 Find the equation of the tangent to the graph of the function

$$f(x) = x^2$$

that touches the graph of f at $x = 1$.

11 The pressure of a gas at a constant temperature is inversely proportional to its volume. This is expressed by the equation

$$P = \frac{K}{V}$$

where P is the pressure and V is the volume of the gas. Show that

$$\frac{dP}{P} = -\frac{dV}{V}$$

If the pressure of the gas is increased by 5%, what is the resultant percentage change in the volume?

12 The periodic time of a simple pendulum is related to the length of the pendulum as follows.

$$T = 2\pi\sqrt{(kL)}$$

where T is the periodic time, L is the length and k is a constant. Show that a 2% increase in the length of the pendulum will result in an increase in the periodic time of 1%.

13 Show that if

$$f(x) = u(x)v(x)$$

then

$$df(x) = v(x)\,du(x) + u(x)\,dv(x)$$

and that a 2% increase in $u(x)$ coupled with a 3% decrease in $v(x)$ will result in a 1% decrease in $f(x)$.

14 If

$$f(x) = \frac{u(x)}{v(x)} \text{ for } v(x) \neq 0$$

find the percentage change in $f(x)$ due to a 5% increase in $u(x)$ and a 2% decrease in $v(x)$.

6
■ ■ ■ The Integral Calculus

The aims of this chapter are to:

1. Demonstrate the problems associated with defining the area of a shape enclosed by a curved boundary and solve the problem by defining the definite integral.
2. Demonstrate that the process of integration is the reverse process to differentiation and simplify the process of integration by permitting resort to tables of standard integrals.
3. Demonstrate a number of standard methods of converting integrals into standard form thereby permitting them to be evaluated by using the tables of standard integrals.

This chapter contains three modules:

Module 22: Area and the Definite Integral
The problem of defining the area contained within a curvilinear boundary is resolved by defining the definite integral.

Module 23: Integration
The process of evaluating an integral is shown to be the reverse process of differentiation. Also discussed are various properties of the definite integral.

Module 24: Methods of Integration
A number of methods whereby an integrand can be transformed into standard form are discussed. Once the integrand is in standard form integration is then a matter of using the table of standard forms.

Module 22: Area and the Definite Integral

Aims: To demonstrate the problems associated with defining the area of a shape enclosed by a curved boundary and to solve the problem by defining the definite integral.

Objectives: When you have read this module you will be able to:

- ▶▶ Appreciate the concept of area and recognize the problems of computing areas bounded by curves.
- ▶▶ Use the trapezoidal rule and Simpson's rule to calculate approximate values of areas under curves.
- ▶▶ Derive the area of a circle.
- ▶▶ Define the definite integral and evaluate some simple integrals.

Unit 1 Euclid's Definition of Area

Test yourself with the following:

1 Use the trapezoidal rule with six divisions of the x-axis to find the areas between the x-axis and the graph of each of the following:

(a) $f{:}f(x) = \cos x$ between $x = 0$ and $x = \pi/2$
(b) $f{:}f(x) = \ln(x)$ between $x = 1$ and $x = 3$
(c) $f{:}f(x) = (2x^2 + 1)^{1/2}$ between $x = 0$ and $x = 3$

2 Use Simpson's rule with six divisions of the x-axis to find the areas between the x-axis and the graph of each of the functions in Question 1.

The Rectangle and Non-overlapping Areas

The original definition of area was given by Euclid who said that:

(1)

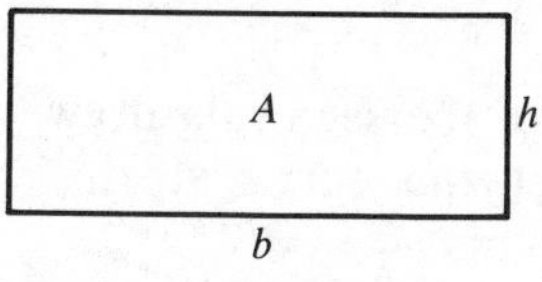

A h b

$A = b \times h$

Figure 22.1

The area of a rectangle is defined as the base length times the height – base × height.

(2) The combined area of a sum of non-overlapping areas is equal to the sum of the individual areas.

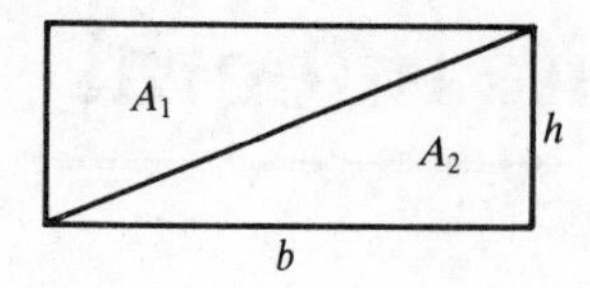

Figure 22.2

From this definition it can be seen that a triangle is equal in area to half a rectangle:

$$\tfrac{1}{2} \times \text{base} \times \text{height}$$

Rectilinear Boundaries

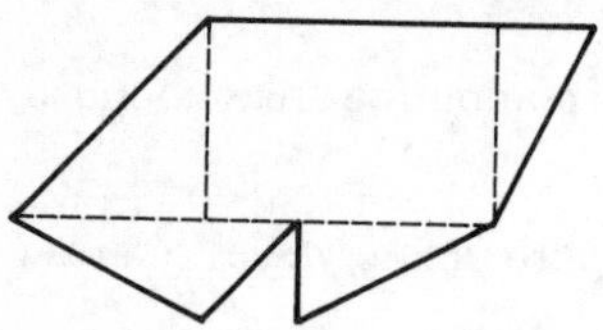

Figure 22.3

A plane figure bounded by rectilinear boundaries can have its area subdivided into a series of non-overlapping rectangles and triangles. The area of the whole is then the sum of the areas of its parts.

Curvilinear Boundaries

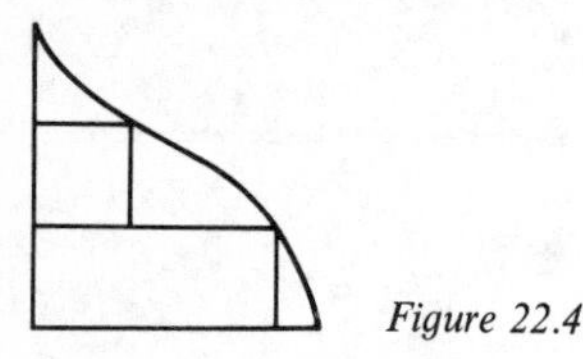

Figure 22.4

A plane figure bounded by curves cannot have its area completely subdivided into rectangles and triangles. As a consequence, the process of measuring its area cannot be performed accurately by adding up areas of triangles and rectangles. It is, however, possible to obtain its approximate area.

The Trapezoidal Rule

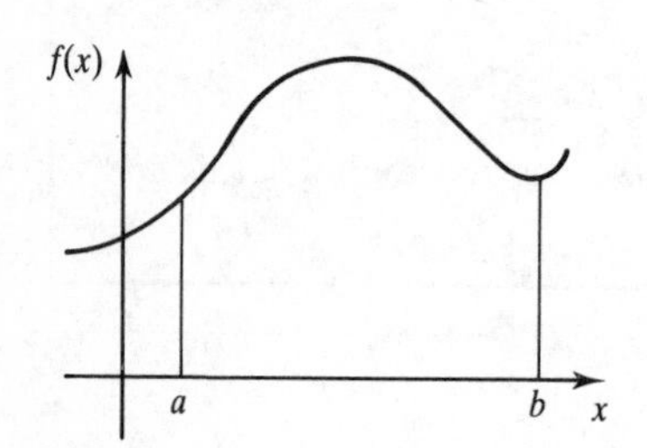

Figure 22.5

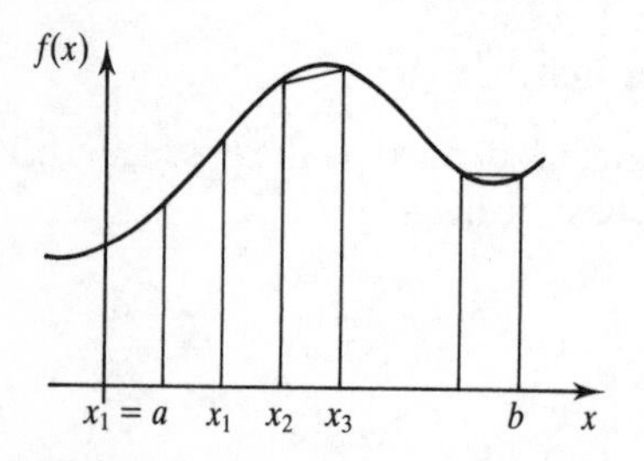

Figure 22.6

The **trapezoidal rule** is a method of approximating the area under a curve drawn against a pair of cartesian axes. A graph of the function f where $f(x) \geqslant 0$ for $a \leqslant x \leqslant b$ is shown in Figure 22.5. To find the area bounded by the curve, the x-axis and the vertical lines $x = a$ and $x = b$, we proceed as follows:

(1) Subdivide the interval of the x-axis between $x = a$ and $x = b$ into n equal segments each of width w. Label the end points of the segments as

$$x_1 = a, x_2, x_3, \ldots, b$$

(2) At the ends of the segments draw vertical lines to meet the curve at the vertical coordinate points

$$f(a), f(x_2), f(x_3), \ldots, f(b)$$

(3) Join the tops of adjacent vertical lines with a straight line.

The area in question has now been approximated by a series of trapezia. The area of the first trapezium on the left is

$$(w/2)[f(x_2) + f(a)]$$

and the area of the next trapezium is

$$(w/2)[f(x_3) + f(x_2)]$$

When the areas of all the trapezia have been added together the result is

$$A = (w/2)\{f(a) + 2[f(x_2) + f(x_3) + \ldots] + f(b)\}$$

where A is the approximation to the area under the curve. This is the trapezoidal rule for approximating the area under the curve.

Simpson's Rule

Simpson's rule is an improvement on the trapezoidal rule. The area beneath the curve is partitioned into an *even* number of strips and instead of joining the top of each strip with a straight line, each pair of adjacent strips have the three end points joined by a parabola. The resulting rule reads:

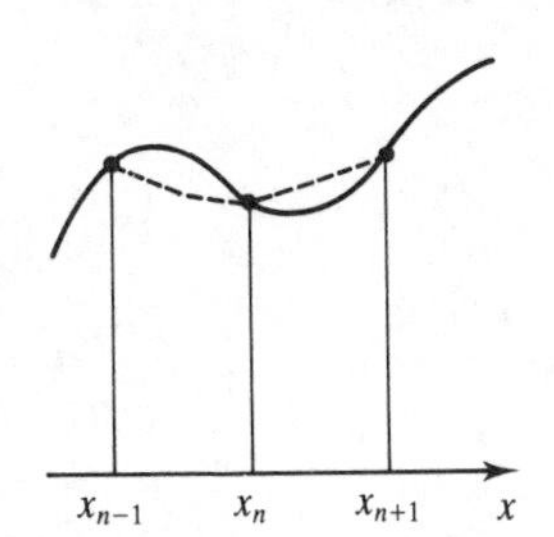

Figure 22.7

$$A = (w/3)[\text{sum of first and last ordinates} + 4(\text{sum of even ordinates}) + 2(\text{sum of remaining odd ordinates})]$$

$$A = (w/3)\{f(a) + f(b) + 4[f(x_2) + f(x_4) + \ldots] + 2[f(x_3) + f(x_5) + \ldots]\}$$

A being an approximation to the area beneath the curve.

EXAMPLES

1 Use the trapezoidal rule with six divisions of the x-axis to find the areas between the x-axis and the graph of each of the following:

(a) $f{:}f(x) = \sin x$ between $x = 0$ and $x = \pi$
(b) $f{:}f(x) = e^x$ between $x = -1$ and $x = 1$
(c) $f{:}f(x) = (x^2 - 1)^{-1/2}$ between $x = 4$ and $x = 16$

(a) The six divisions of the x-axis and the corresponding $f(x)$ values are, to three decimal places:

x	0	$\pi/6$	$\pi/3$	$\pi/2$	$2\pi/3$	$5\pi/6$	π
$f(x)$	0	0.5	0.866	1.0	0.866	0.5	0

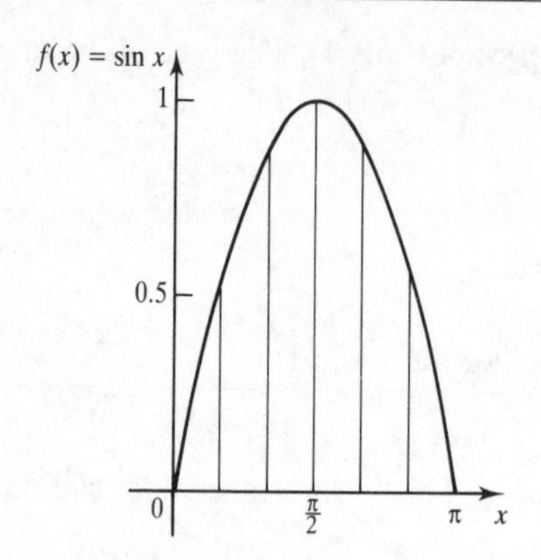

Here, $a = 0$ and $b = \pi$ with $w = (b - a)/6 = \pi/6$. Therefore, the area in question is found by applying the rule:

$$(w/2)\{f(a) + 2[f(x_2) + f(x_3) + \ldots] + f(b)\}$$
$$= (\pi/12)[0 + 2(0.5 + 0.866 + 1 + 0.866 + 0.5) + 0]$$
$$= 1.954 \text{ to three dec. pl.}$$

Figure 22.8

(b) The six divisions of the x-axis and the corresponding $f(x)$ values are, to three decimal places:

x	-1	$-2/3$	$-1/3$	0	$1/3$	$2/3$	1
$f(x)$	0.368	0.513	0.717	1.0	1.396	1.948	2.718

Here, $a = -1$ and $b = 1$ with $w = (b - a)/6 = 1/3$. Therefore, the area in question is found by applying the rule:

$$(w/2)\{f(a) + 2[f(x_2) + f(x_3) + \ldots] + f(b)\}$$
$$= (1/6)[0.368 + 2(0.513 + 0.717 + 1 + 1.396 + 1.948) + 2.718]$$
$$= 2.372 \text{ to three dec. pl.}$$

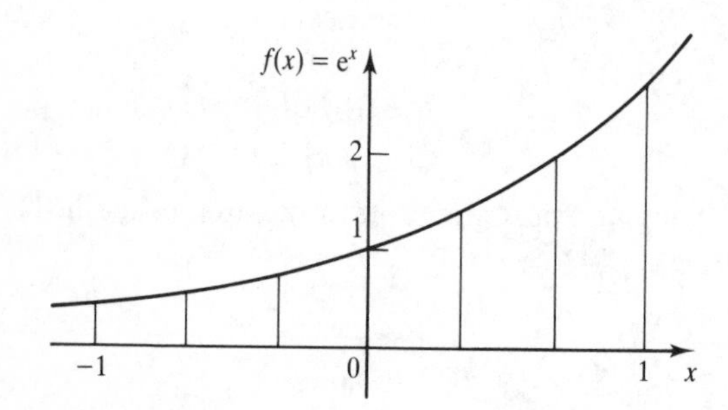

Figure 22.9

(c) The six divisions of the x-axis and the corresponding $f(x)$ values are, to three decimal places:

x	4	6	8	10	12	14	16
$f(x)$	0.258	0.169	0.126	0.101	0.084	0.072	0.063

Here, $a = 4$ and $b = 16$ with $w = (b - a)/6 = 2$. Therefore, the area in question is found by applying the rule:

$$(w/2)\{f(a) + 2[f(x_2) + f(x_3) + \ldots] + f(b)\}$$
$$= (1)[0.258 + 2(0.169 + 0.126 + 0.101 + 0.084 + 0.072) + 0.063]$$
$$= 1.425$$

2 Use Simpson's rule with six divisions of the x-axis to find the areas between the x-axis and the graph of each of the functions in Example 1.

(a) $f(x) = \sin x$ between $x = 0$ and $x = \pi$ The six divisions of the x-axis and the corresponding $f(x)$ value are, to three decimal places:

x	0	$\pi/6$	$\pi/3$	$\pi/2$	$2\pi/3$	$5\pi/6$	π
$f(x)$	0	0.5	0.866	1.0	0.866	0.5	0

Here, $a = 0$ and $b = \pi$ with $w = (b - a)/6 = \pi/6$. Therefore, the area in question is found by applying the rule:

$$(w/3)[\text{sum of first and last ordinate} + 4(\text{sum of even ordinates}) + 2(\text{sum of remaining ordinates})]$$

$$= (\pi/18)[0 + 0 + 4(0.5 + 1 + 0.5) + 2(0.866 + 0.866)]$$

$$= 2.001 \text{ to three dec. pl.}$$

(b) $f(x) = e^x$ between $x = -1$ and $x = 1$ The six divisions of the x-axis and the corresponding $f(x)$ values are, to three decimal places:

x	-1	$-2/3$	$-1/3$	0	$1/3$	$2/3$	1
$f(x)$	0.368	0.513	0.717	1.0	1.396	1.948	2.718

Here, $a = -1$ and $b = 1$ with $w = (b - a)/6 = 1/3$. Therefore, the area in question is found by applying the rule:

$$(w/3)[\text{sum of first and last ordinate} + 4(\text{sum of even ordinates}) + 2(\text{sum of remaining ordinates})]$$

$$= (1/9)[0.368 + 2.718 + 4(0.513 + 1 + 1.948) + 2(0.717 + 1.396)]$$

$$= 2.351 \text{ to three dec. pl.}$$

(c) $f(x) = (x^2 - 1)^{-1/2}$ between $x = 4$ and $x = 16$ The six divisions of the x-axis and the corresponding $f(x)$ values are, to three decimal places:

x	4	6	8	10	12	14	16
$f(x)$	0.258	0.169	0.126	0.101	0.084	0.072	0.063

Here, $a = 4$ and $b = 16$ with $w = (b - a)/6 = 2$. Therefore, the area in question is found by applying the rule:

$$(w/3)[\text{sum of first and last ordinate} + 4(\text{sum of even ordinates}) + 2(\text{sum of remaining ordinates})]$$

$$= (2/3)[0.258 + 0.063 + 4(0.169 + 0.101 + 0.072) + 2(0.126 + 0.084)]$$

$$= 1.406 \text{ to three dec. pl.}$$

EXERCISES

1 Use the trapezoidal rule with six divisions of the x-axis to find the areas between the x-axis and the graph of each of the following:

(a) $f{:}f(x) = \tan x$ between $x = 0$ and $x = \pi/4$
(b) $f{:}f(x) = xe^{-x}$ between $x = 0$ and $x = 1$
(c) $f{:}f(x) = (2 + x^3)^{-1}$ between $x = 0$ and $x = 5$

(a) The six divisions of the x-axis and the corresponding $f(x)$ values are, to three decimal places:

x	0	$\pi/24$	$\pi/12$	$\pi/8$	$\pi/6$	$5\pi/24$	$\pi/4$
$f(x)$	0	*	*	*	*	*	1.000

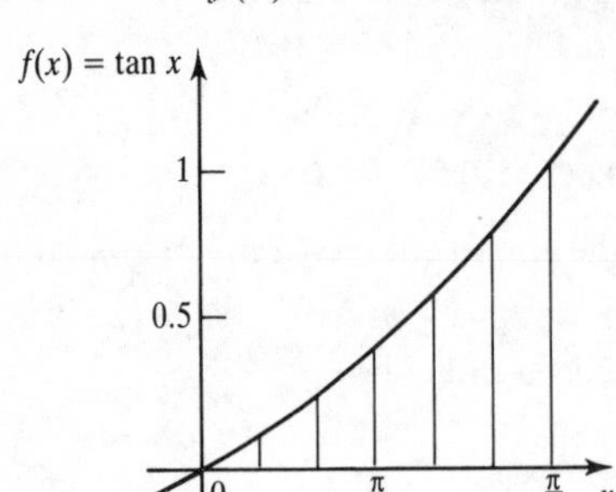

Here, $a = 0$ and $b = \pi/4$ with $w = (b - a)/6 = *$. Therefore, the area in question is found by applying the rule:

$$(w/2)\{f(a) + 2[f(x_2) + f(x_3) + \ldots] + f(b)\}$$
$$= (*)[0 + 2(* + * + * + * + *) + 1.000]$$
$$= *$$

Figure 22.10

(b) The six divisions of the x-axis and the corresponding $f(x)$ values are, to three decimal places:

x	0	*	*	*	*	*	1
$f(x)$	0	*	*	*	*	*	0.368

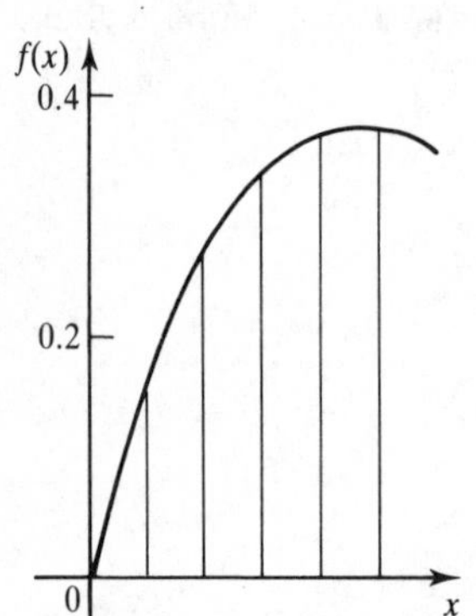

Here, $a = *$ and $b = *$ with $w = (b - a)/6 = *$. Therefore the area in question is found by applying the rule:

$$(w/2)\{f(a) + 2[f(x_2) + f(x_3) + \ldots] + f(b)\}$$
$$= (*)[0 + 2(*) + 0.368]$$
$$= *$$

Figure 22.11

(c) The six divisions of the x-axis and the corresponding $f(x)$ values are, to three decimal places:

x	*	*	*	*	*	*	*
$f(x)$	*	0.388	*	*	*	0.013	*

Here, $a = *$ and $b = *$ with $w = (b - a)/6 = *$. Therefore, the area in question is found by applying the rule:

$$(w/2)\{f(a) + 2[f(x_2) + f(x_3) + \ldots] + f(b)\}$$
$$= (*)[* + * + 2(0.388 + * + * + * + 0.013) + *]$$
$$= *$$

2 Use Simpson's rule with six divisions of the x-axis to find the areas between the x-axis and the graph of each of the functions in Exercise 1.

(a) $f(x) = \tan x$ between $x = 0$ and $x = \pi/4$ The six divisions of the x-axis and the corresponding $f(x)$ values are, to three decimal places:

x	0	$\pi/24$	$\pi/12$	$\pi/8$	$\pi/6$	$5\pi/24$	$\pi/4$
$f(x)$	0	*	*	*	*	*	1.000

Here, $a = 0$ and $b = \pi/4$ with $w = (b - a)/6 = *$. Therefore, the area in question is found by applying the rule:

$$(w/3)[\text{sum of first and last ordinate} + 4(\text{sum of even ordinates}) + 2(\text{sum of remaining ordinates})]$$
$$= (*)[0 + 1.000 + 4(* + * + *) + 2(* + *)]$$
$$= *$$

(b) $f(x) = xe^{-x}$ between $x = 0$ and $x = 1$ The six divisions of the x-axis and the corresponding $f(x)$ values are, to three decimal places:

x	0	*	*	*	*	*	1
$f(x)$	0	*	*	*	*	*	0.368

Here, $a = *$ and $b = *$ with $w = (b - a)/6 = *$. Therefore, the area in question is found by applying the rule:

$$(w/3)[\text{sum of first and last ordinate} + 4(\text{sum of even ordinates}) + 2(\text{sum of remaining ordinates})]$$
$$= (*)[0 + 0.368 + 4(*) + 2(*)]$$
$$= *$$

(c) $f(x) = (2 + x^3)^{-1}$ between $x = 0$ and $x = 5$. The six divisions of the x-axis and the corresponding $f(x)$ values are, to three decimal places:

x	*	*	*	*	*	*	*
$f(x)$	*	0.338	*	*	*	0.013	*

Here, $a = *$ and $b = *$ with $w = (b - a)/6 = *$. Therefore, the area in question is found by applying the rule:

$$(w/3)[\text{sum of first and last ordinate} + 4(\text{sum of even ordinates}) + 2(\text{sum of remaining ordinates})]$$
$$= (*)[* + * + 4(0.388 + * + 0.013) + 2(*)]$$
$$= *$$

Unit 2 The Area of a Circle and the Definite Integral

Test yourself with the following:

Find the value of each of the following integrals from first principles:

(a) $\int_{x=0}^{1} (x + 2)\,dx$ (b) $\int_{x=0}^{4} (x^2 - 2)\,dx$

Exact Areas Within Curvilinear Boundaries

The area of a circle is given as πr^2 and this area is exact, no approximation has been made. To show how this result is achieved, consider the circle with a regular polygon inscribed within it.

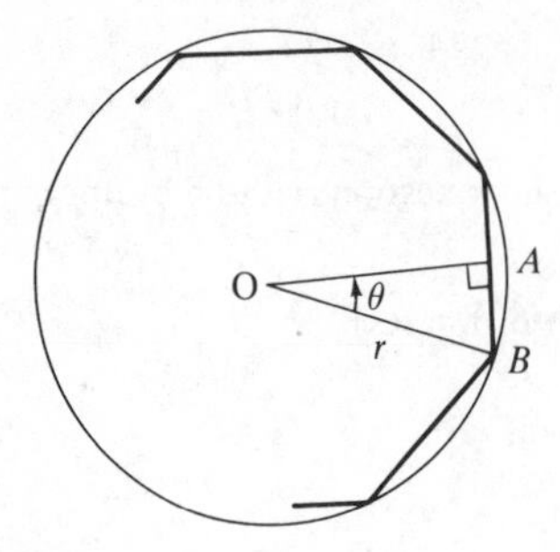

$OA = r\cos\theta \quad AB = r\sin\theta \quad \theta = \frac{\pi}{n}$ *Figure 22.12*

Inscribed Regular Polygons

The polygon has n equal sides. Each vertex of the polygon is joined to the centre of the circle O by a straight line, and in this way the interior of the polygon contains a series of n isosceles triangles – one for each side of the polygon. Each isosceles triangles is then bisected to form two right-angled triangles – $2n$ of them in total, each with a vertex angle π/n at the centre of the circle – the triangle OAB is typical. The base length AB is $r\sin(\pi/n)$ and the vertical height OA is $r\cos(\pi/n)$. The area of each right-angled triangle is then

$$\tfrac{1}{2} \times AB \times AO = \tfrac{1}{2}[r\sin(\pi/n)][r\cos(\pi/n)]$$

There are $2n$ such triangles. This means that the area of the inscribed polygon is

$$\begin{aligned} A &= 2n(1/2)[r\sin(\pi/n)][r\cos(\pi/n)] \\ &= n\,r\,sin(\pi/n)\,r\cos(\pi/n) \end{aligned}$$

The Limit

As the number of sides n of the polygon increases, so the area of the polygon approaches the area of the circle – the areas of the segments inside the circle but outside the polygon approach zero. The area of the circle is then defined to be equal to the **limit** of the area of the polygon as the number of sides n increases without bound:

$$\begin{aligned} \text{Area of circle} &= \lim_{n\to\infty} n\,r\sin(\pi/n)\,r\cos(\pi/n) \\ &= r^2 \lim_{n\to\infty} n\sin(\pi/n) \lim_{n\to\infty} \cos(\pi/n) \\ &= r^2 \lim_{n\to\infty} \frac{\sin(\pi/n)}{1/n} \end{aligned}$$

Let $\beta = 1/n$ to transform this limit to

$$\text{Area of circle} = r^2 \lim_{\beta\to 0} \frac{\sin\beta\pi}{\beta}$$

It can be shown by using methods that are outside the scope of this book that

$$\lim_{\beta \to 0} \frac{\sin \beta\pi}{\beta} = \pi$$

so that

$$\text{Area of circle} = \pi r^2$$

Hence the area of the circle is *defined* to be πr^2.

The Definite Integral

The principle of taking the limit of a sum of areas of rectilinear figures to define the area of a circle can be similarly employed to find the area of any figure bounded by curves. In particular, we are interested in the area enclosed between the curve of the graph of a function, the x-axis and the lines $x = a$ and $x = b$.

To find the required area we subdivide the interior into a collection of non-overlapping rectangles of uniform width. Note that in Figure 22.13 we do, in fact, have two sets of rectangles. Those with their tops above the curve and those with their tops below.

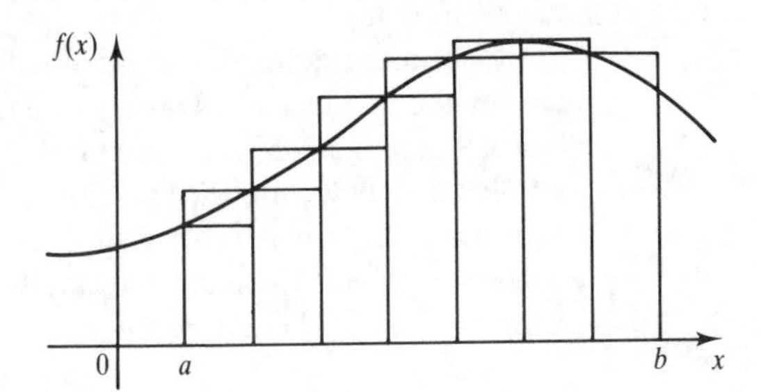

Figure 22.13

The area of each rectangle can be found by multiplying its height by its width. Adding up all the rectangles below the curve then gives an underestimated approximation to the area under the curve. Adding up all the rectangles with their tops above the curve gives an overestimated approximation to the area under the curve. If we decrease the width of each rectangle, we increase the number of rectangles. The sum of the areas of this greater collection then gives a better approximation to the area under the curve. The narrower each rectangle, the more rectangles there are, and the better the approximation made to the area under the curve. We define the area under the curve to be the limit as the width of the rectangles decreases to zero.

This process of defining area under a curve as the limit of a sum of rectangles in the limit as the number of such rectangles increases without bound is known as **integration**. The process is formalized as follows.

The area under the curve of the graph of the continuous function f where $f(x) \geqslant 0$ for $a \leqslant x \leqslant b$ and between the vertical lines $x = a$ and $x = b$ is obtained by escribing n rectangles each of width $\mathrm{d}x$, adding the areas of the rectangles and taking the limit as the number of such strips increases without bound – that is, their width approaches zero.

The height of the rth rectangle is $f(a + r\,\mathrm{dx})$ and so the area of the rth rectangle is

$$f(a + r\,\mathrm{dx})\,\mathrm{dx}$$

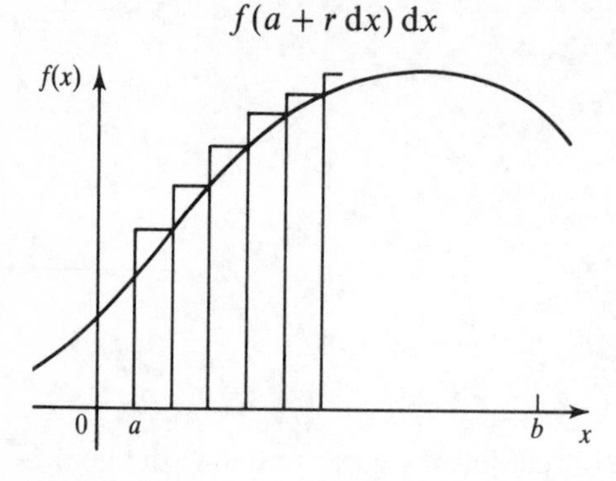

Figure 22.14

where $\mathrm{dx} = (b - a)/n$. The area of all such rectangles is

$$\sum_{r=1}^{n} [f(a + r\,\mathrm{dx})\,\mathrm{dx}]$$

The area in question is then defined as the definite integral:

$$\int_{x=a}^{b} f(x)\,\mathrm{dx} = \lim_{n\to\infty} \left\{ \sum_{r=1}^{n} [f(a + r\,\mathrm{dx})\,\mathrm{dx}] \right\}$$

The symbol $\int$ is an old English S and it denotes the **integral** – the limit of a sum. The expression $f(x)$ is referred to as the **integrand** and the numbers a and b are referred to as the **limits** of the integral. The quantity **dx** can be considered as the differential of x where x is referred to as the **variable of integration**. Note that in this derivation we have only considered those rectangles with their tops above the curve. A similar result would be obtained if those with their tops below the curve were to be considered instead.

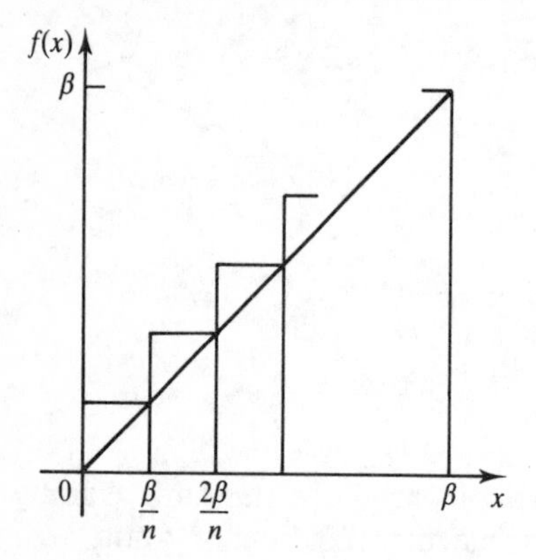

Figure 22.15

The area under the line defined by $f(x) = x$ between the x-values $x = 0$ and $x = \beta$ is a triangular area of size $\beta^2/2$. This result can also be obtained as follows (see Figure 22.15). Divide the interval from $x = 0$ to $x = \beta$ into n subintervals each of width $\mathrm{dx} = \beta/n$. Construct the escribed rectangles of width β/n and height $f(a + r\,\mathrm{dx})$, where

$$f(a + r\,\mathrm{dx}) = f\left(0 + \frac{r\beta}{n}\right)$$

$$= \frac{\beta r}{n} \text{ (since } f(x) = x\text{)}$$

so that

$$\int_{x=0}^{\beta} x\,\mathrm{dx} = \lim_{n\to\infty} \left\{ \sum_{r=1}^{n} [f(a + r\,\mathrm{dx})]\,\mathrm{dx} \right\}$$

$$= \lim_{n\to\infty} \sum_{r=1}^{n} \frac{\beta r}{n} \cdot \frac{\beta}{n}$$

$$= \lim_{n\to\infty} \left(\frac{\beta^2}{n^2} \sum_{r=1}^{n} r \right)$$

since

$$\sum_{r=1}^{n} r = \frac{1}{2} n(n+1) \text{ (the } n\text{th term of an arithmetic series)}$$

then

$$\lim_{n\to\infty}\left(\frac{\beta^2}{n^2}\cdot\sum_{r=1}^{n} r\right) = \lim_{n\to\infty}\frac{\beta^2(\frac{1}{2}n)(n+1)}{n^2}$$

$$= \frac{\beta^2}{2}\lim_{n\to\infty}\left(\frac{n}{n}\cdot\frac{n+1}{n}\right)$$

$$= \frac{\beta^2}{2}\lim_{n\to\infty}\left(1+\frac{1}{n}\right)$$

$$= \frac{\beta^2}{2}$$

which agrees with the area of the triangle of base length β and height β.

Rules of Sums

To assist in the evaluation of such definite integrals the following rules of sums are needed:

$$\sum_{n=j}^{k}\{f(n)+g(n)\} = \sum_{n=j}^{k} f(n) + \sum_{n=j}^{k} g(n)$$

$$\sum_{n=j}^{k} af(n) = a\sum_{n=j}^{k} f(n)$$

In particular for $f(n) = 1$

$$\sum_{n=1}^{N} a = aN \text{ (a is a constant)}$$

EXAMPLES

Find the value of each of the following integrals:

$$\textbf{(a)} \int_{x=0}^{\beta} (x+1)\,dx \qquad \textbf{(b)} \int_{x=0}^{\beta} x^2\,dx$$

(a) The integrand $x + 1$ is the sum of two functions:

$$g(x) = x \text{ and } h(x) = 1$$

If we plot both of these functions on the same cartesian graph, we can see that the area beneath the function defined by

$$f(x) = x + 1$$

is the same as the area beneath the sum of $g(x)$ and $h(x)$. The area beneath $g(x)$ is the triangular area and the area beneath $h(x)$ is the rectangle of width β and height 1. That is

$$\int_{x=0}^{\beta} (x+1)\,dx = \int_{x=0}^{\beta} x\,dx + \int_{x=0}^{\beta} dx = \frac{\beta^2}{2} + \beta$$

(b) We shall evaluate this integral from first principles making note that $f(x) = x^2$, $a = 0$, $b = \beta$ so that $\mathrm{d}x = \beta/n$. Hence

$$\int_{x=0}^{\beta} x^2\,\mathrm{d}x = \operatorname*{Lim}_{n\to\infty} \sum_{r=1}^{n} [f(a + r\,\mathrm{d}x)]\,\mathrm{d}x$$

$$= \operatorname*{Lim}_{n\to\infty} \sum_{r=1}^{n} \left(\frac{\beta r}{n}\right)^2 \frac{\beta}{n}$$

$$= \operatorname*{Lim}_{n\to\infty} \left(\frac{\beta}{n}\right)^3 \sum_{r=1}^{n} r^2$$

To evaluate this sum we need to know the value of Σr^2. Consider

$$\sum_{r=1}^{n} [r^3 - (r-1)^3] = \sum_{r=1}^{n} (r^3 - r^3 + 3r^2 - 3r + 1)$$

$$= 3\sum_{r=1}^{n} r^2 - 3\sum_{r=1}^{n} r + \sum_{r=1}^{n} 1$$

$$= 3\sum_{r=1}^{n} r^2 - 3(1/2)n(n+1) + n$$

The sum on the left is a telescoping series – all but the smallest and largest terms cancel out.

$$\sum_{r=1}^{n} [r^3 - (r-1)^3] = (1^3 - 0^3) + (2^3 - 1^3) + (3^3 - 2^3)$$

$$+ \ldots + [n^3 - (n-1)^3]$$

$$= -0^3 + (1^3 - 1^3) + (2^3 - 2^3)$$

$$+ \ldots + [(n-1)^3 - (n-1)^3] + n^3$$

$$= n^3$$

Therefore

$$n^3 = 3\sum_{r=1}^{n} r^2 - 3.\frac{1}{2}n(n+1) + n$$

so that

$$3\sum_{r=1}^{n} r^2 = n^3 + \frac{3}{2}n^2 + \frac{3}{2}n - n$$

$$= \frac{1}{2}n(n+1)(2n+1)$$

so that

$$\sum_{r=1}^{n} r^2 = \frac{1}{6}n(n+1)(2n+1)$$

Substituting this result into the integral of x^2 we find that

$$\int_{x=0}^{\beta} x^2\,dx = \lim_{n\to\infty}\left(\frac{\beta}{n}\right)^3 \frac{1}{6}n(n+1)(2n+1)$$

$$= \lim_{n\to\infty} \beta^3 . \frac{1}{6}\left(1+\frac{1}{n}\right)\left(2+\frac{1}{n}\right)$$

$$= \frac{\beta^3}{3}$$

EXERCISES

Find the value of each of the following integrals:

(a) $\displaystyle\int_{x=0}^{5} (2x+6)\,dx$ **(b)** $\displaystyle\int_{x=0}^{\beta} x^3\,dx$

(a) The integrand * is the sum of two functions:

$g(x) = *$ and $h(x) = *$

If we plot both of these functions on the same cartesian graph, we can see that the area beneath the function defined by:

$f(x) = 2x + 6$

is the same as the area beneath the sum of $g(x)$ and $h(x)$. The area beneath $g(x)$ is the * area and the area beneath $h(x)$ is the * of width β and height *. That is

$$\int_{x=0}^{\beta} (2x+6)\,dx = \int_{x=0}^{\beta} *\,dx + \int_{x=0}^{\beta} *\,dx$$

$$= * + *$$

(b) We shall evaluate this integral from first principles making note that $f(x) = x^3$, $a = 0$, $b = \beta$, so that $dx = \beta/n$ and

$$\int_{x=0}^{\beta} x^3\,dx = \lim_{n\to\infty}\sum_{r=1}^{n} [f(a + r\,dx)]\,dx$$

$$= \lim_{n\to\infty}\sum_{r=1}^{n}\left(\frac{\beta r}{n}\right)^3 \frac{\beta}{n}$$

$$= \lim_{n\to\infty}\left(\frac{\beta}{n}\right)^4 \sum_{r=1}^{n} r^3$$

To evaluate this sum we need to know the value of Σr^3. Consider

$$\sum_{r=1}^{n} [r^4 - (r-1)^4] = \sum_{r=1}^{n} (r^4 - r^4 + 4r^3 - 6r^2 + 4r - 1)$$

$$= 4\sum_{r=1}^{n} r^3 - * \sum_{r=1}^{n} r^2 + * \sum_{r=1}^{n} r - \sum_{r=1}^{n} 1$$

$$= 4\sum_{r=1}^{n} r^3 - *(1/6)n(n+1)(2n+1) + *(1/2)n(n+1) - n$$

The sum on the left is a telescoping series and its value is n^4. Therefore:

$$n^4 = 4\sum_{r=1}^{n} r^3 - n(*) + 2n(*) - n$$

so that

$$4\sum_{r=1}^{n} r^3 = n^4 + n^2(*)$$

$$= [n(n+1)]^2$$

so that

$$\sum_{r=1}^{n} r^3 = \left[\frac{1}{2}n(n+1)\right]^2$$

Substituting this result into the integral of x^3 we find that:

$$\int_{x=0}^{\beta} x^3 \, dx = \lim_{n\to\infty} \left(\frac{\beta}{n}\right)^4 \left[\frac{1}{2}n(n+1)\right]^2$$

$$= \lim_{n\to\infty} \beta^4 (*)^2$$

$$= \frac{\beta^4}{4}$$

Note the interesting result that the sum of the first n integers cubed is equal to the square of the sum of the first n integers.

Module 22: Further exercises

1 Use the trapezoidal rule with six divisions of the x-axis to find the areas between the x-axis and the graph of each of the following:

(a) $f{:}f(x) = \cot x$ between $x = \pi/4$ and $x = \pi/2$
(b) $f{:}f(x) = x \ln(x)$ between $x = 1$ and $x = 7$
(c) $f{:}f(x) = (5 - x^2)^{-1}$ between $x = -2$ and $x = +2$

2 Use Simpson's rule with six divisions of the x-axis to find the areas between the x-axis and the graph of each of the functions in Exercise 1.

3 Find the value of each of the following integrals from first principles:

(a) $\int_{x=0}^{2} (3x+5) \, dx$ (b) $\int_{x=0}^{8} x^4 \, dx$

Module 23: Integration

Aims: To demonstrate that the process of integration is the reverse process to differentiation and to simplify the process of integration by permitting resort to tables of standard integrals.

Objectives: When you have read this module you will be able to:

▶▶ Recognize that integration is the reverse process of differentiation.

▶▶ Use standard derivatives to solve simple integrals.

Unit 1 Integration as the Reverse of Differentiation

Test yourself with the following:

Use the fundamental theorem of calculus to evaluate the definite integrals of each of the following:

(a) $f(x) = x^4$ between $x = -1$ and $x = 2$
(b) $f(x) = \sec^2 x$ between $x = 0$ and $x = \pi/4$
(c) $f(x) = e^{-x}$ between $x = 0$ and $x = 1$
(d) $f(x) = -\cos x$ between $x = \pi/2$ and $x = \pi$

Properties of Integrals

The definite integral has been introduced as a measure of area, and we shall continue this theme to demonstrate various properties of areas that are reflected in the properties of integrals.

Notation

The simplest integral is the one with an integrand of 1 whose value is equal to the area of the rectangle of height 1 and base length $(b - a)$:

$$\int_{x=a}^{b} dx = (b - a)$$

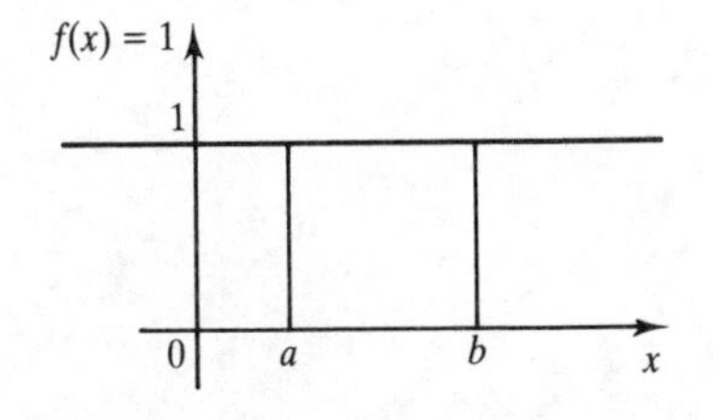

Figure 23.1

We shall interpose an intermediate step here to introduce a new notation by writing

$$\int_{x=a}^{b} dx = [x]_{x=a}^{b}$$

$$= (b - a)$$

The contents of the square brackets are evaluated by giving x the value of the upper limit and subtracting from it the value of x at the lower limit. For example, we have seen that

$$\int_{x=0}^{\beta} x\,\mathrm{d}x = \frac{\beta^2}{2}$$

Using the square bracket notation this becomes

$$\int_{x=0}^{\beta} x\,\mathrm{d}x = \left[\frac{x^2}{2}\right]_{x=0}^{\beta}$$

$$= \left(\frac{\beta^2}{2} - \frac{0^2}{2}\right)$$

$$= \frac{\beta^2}{2}$$

Sums of Non-overlapping Areas

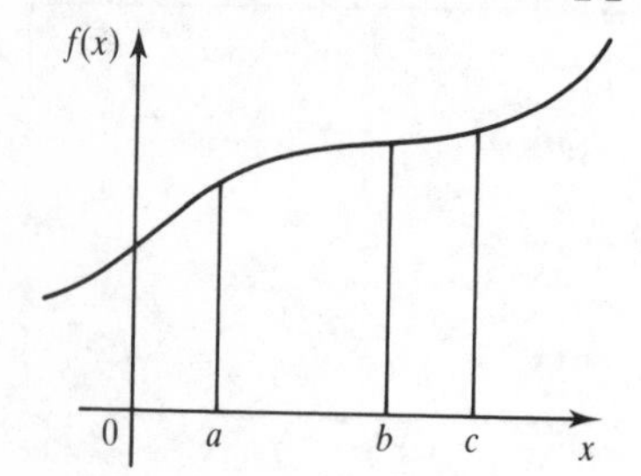

Figure 23.2

In Figure 23.2 the area beneath the curve between $x = a$ and $x = c$ is equal to the sum of the area beneath the curve between a and b and the area beneath the curve between b and c.

This is expressed in integral format as

$$\int_{x=a}^{c} f(x)\,\mathrm{d}x = \int_{x=a}^{b} f(x)\,\mathrm{d}x + \int_{x=b}^{c} f(x)\,\mathrm{d}x$$

In particular:

$$\int_{x=b}^{c} f(x)\,\mathrm{d}x = \int_{x=a}^{c} f(x)\,\mathrm{d}x - \int_{x=a}^{b} f(x)\,\mathrm{d}x$$

For example, we have already seen that

$$\int_{x=0}^{p} x\,\mathrm{d}x = \left[\frac{x^2}{2}\right]_{x=0}^{p}$$

$$= \frac{p^2}{2}$$

so that

$$\int_{x=0}^{q} x\,\mathrm{d}x = \left[\frac{x^2}{2}\right]_{x=0}^{q}$$

$$= \frac{q^2}{2}$$

Hence

$$\int_{x=p}^{q} x\,dx = \int_{x=0}^{q} x\,dx - \int_{x=0}^{p} x\,dx$$
$$= \frac{q^2}{2} - \frac{p^2}{2}$$
$$= \left[\frac{x^2}{2}\right]_{x=p}^{q}$$

You may have noticed that when evaluating integrals in this unit the quantity in the square brackets was similar to the primitive of the integrand. For example,

$$\int_{x=a}^{b} dx = [x]_{x=a}^{b} \qquad \int_{x=a}^{b} x\,dx = \left[\frac{x^2}{2}\right]_{x=a}^{b}$$

This is no accident, the quantity in the square brackets is indeed the primitive, though each has been written here without the arbitrary constant. To demonstrate the validity of this general conclusion, we consider the fundamental theorem of calculus, but before we do so we must look at another theorem, the **mean value theorem**.

The Mean Value Theorem

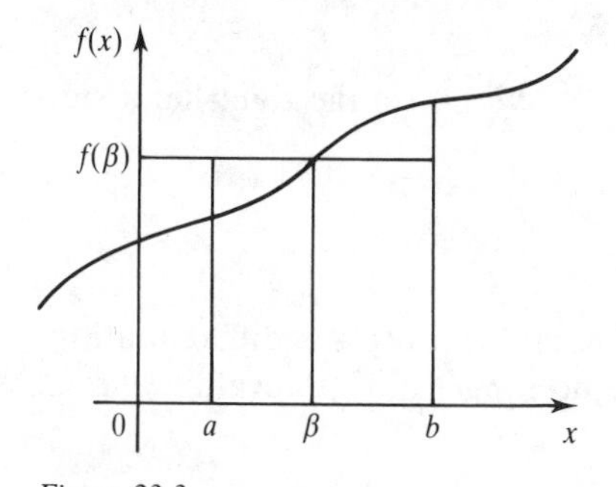

Figure 23.3

If a function f, defined by $f(x)$, is continuous over the interval $a \leqslant x \leqslant b$ then it is possible to find a rectangle with base length $(b - a)$ whose area is equal to the area beneath the curve between the values $x = a$ and $x = b$. Figure 23.3 illustrates this theorem which states that it is possible to find a value of x, say $x = \beta$ where

$$\int_{x=a}^{b} f(x)\,dx = (b-a)f(\beta)$$

The value of $f(\beta)$ is an average or mean value of $f(x)$ in the interval $f(a)$ to $f(b)$ that produces the rectangular equivalent to the area under the curve. We are now ready to consider the fundamental theorem.

The Fundamental Theorem of Calculus

From the mean value theorem, it is possible to find a value for x, say $x = \beta$, such that the area beneath the curve of the continuous function defined by $f(x)$ between the values $x = a$ and $x = a + h$ is given as

$$\int_{x=a}^{a+h} f(x)\,dx = [(a+h) - a]f(\beta)$$
$$= hf(\beta)$$

If we now define $F(x)$ to be such that

$$\int_{x=a}^{b} f(x)\,\mathrm{d}x = [F(x)]_{x=a}^{b}$$

then

$$\int_{x=o}^{a+h} f(x)\,\mathrm{d}x = [F(x)]_{x=0}^{a+h}$$

$$= F(a+h) - F(a)$$
$$= hf(\beta)$$

That is

$$hf(\beta) = F(a+h) - F(a)$$

so, dividing through by h:

$$f(\beta) = \frac{F(a+h) - F(a)}{h}$$

As h becomes small so the point at $x = \beta$ approaches $x = a$. In the limit:

$$\lim_{h\to 0} f(\beta) = \lim_{h\to 0}\left[\frac{F(a+h) - F(a)}{h}\right]$$

The limit on the left-hand side of this equation is $f(a)$ and the limit on the right-hand side is the derivative of $F(x)$ at $x = a$. That is

$$f(a) = F'(a)$$

so that $F(x)$ is the primitive of $f(x)$.

This result means that the process of integration is the reverse process of differentiation, which makes integration much less of a tedious task – we no longer have to evaluate limits of sums. For example, to evaluate

$$\int_{x=1}^{3} x^3\,\mathrm{d}x$$

we merely note that the primitive of $f(x) = x^3$ is $F(x) = x^4/4 + c$, so that

$$\int_{x=1}^{3} x^3\,\mathrm{d}x = \left[\frac{x^4}{4} + c\right]_{x=1}^{3}$$

$$= \left[\left(\frac{81}{4} + c\right) - \left(\frac{1}{4} + c\right)\right]$$

$$= 20$$

Because the arbitrary constant in the primitive is always subtracted out on the right-hand side of this equation we shall, in future, exclude it from the square brackets.

The Indefinite Integral

We have investigated the *definite* integral a great deal. There is another type of integral, called the **indefinite integral** which is an alternative name for the primitive. The integral is written with a single upper limit in the form of a variable. For example,

$$\int^{t} f(x)\,\mathrm{d}x = F(t) + c$$

Note that here the arbitrary constant in the primitive is retained – it is called the **constant of integration**. Indefinite integrals are used when solving differential equations, and as these are outside the scope of this book we shall not consider indefinite integrals any further.

Standard Derivatives and Standard Integrals

To assist with the process of integration, we resort to tables of standard integrals. A short table, which has been generated by using the table of standard derivatives, is listed here:

$f(x) = F'(x)$	$F(x)$	
x^n	$\dfrac{x^{n+1}}{n+1} + c$	$n \neq -1$
x^{-1}	$\ln(x) + c$	
$\cos x$	$\sin x + c$	
$\sin x$	$-\cos x + c$	
e^x	$e^x + c$	

These can be checked to be correct by differentiating the expressions in the right-hand column to obtain the corresponding expressions in the left-hand column.

EXAMPLES

Use the fundamental theorem of calculus to evaluate the definite integrals of each of the following:

(a) $f(x) = x^2$ between $x = -2$ and $x = 2$
(b) $f(x) = \sin x$ between $x = 0$ and $x = \pi$
(c) $f(x) = e^x$ between $x = 0$ and $x = 1$

(a) $$\int_{a=-2}^{2} x^2\,\mathrm{d}x = \left[\frac{x^3}{3}\right]_{x=-2}^{2} \qquad \left(\text{the primitive is } \frac{x^3}{3} + c\right)$$

$$= \frac{8}{3} - \left(\frac{-8}{3}\right)$$

$$= \frac{16}{3}$$

(b) $$\int_{x=0}^{\pi} \sin x \, dx = [-\cos x]_{x=0}^{\pi} \qquad \text{(the primitive is } -\cos x + c)$$

$$= -\cos \pi - (-\cos 0)$$

$$= 1 - (-1)$$

$$= 2$$

(c) $$\int_{x=0}^{1} e^x \, dx = [e^x]_{x=0}^{1} \qquad \text{(the primitive is } e^x + c)$$

$$= e - 1$$

EXERCISES

Use the fundamental theorem of calculus to evaluate the definite integrals of each of the following:

(a) $f(x) = x^{-1}$ between $x = 1$ and $x = 2$
(b) $f(x) = \cos x$ between $x = 0$ and $x = \pi/2$
(c) $f(x) = 3$ between $x = 1$ and $x = 4$

(a) $$\int_{x=1}^{2} x^{-1} \, dx = [\ln(x)]_{x=1}^{2} \qquad \text{(the primitive is } *)$$

$$= \ln(*) - \ln(*)$$
$$= \ln(2)$$

(b) $$\int_{x=0}^{\pi/2} \cos x \, dx = [*]_{x=0}^{\pi/2} \qquad \text{(the primitive is } *)$$

$$= * - *$$
$$= 1$$

(c) $$\int_{x=1}^{4} 3 \, dx = [*]_{x=1}^{4} \qquad \text{(the primitive is } *)$$

$$= * - *$$
$$= 9$$

Unit 2 Properties of the Definite Integral

Test yourself with the following:

Find the value of the definite integral of each of the following:

(a) $f(x) = x^5 + \cos x$ between $x = \pi/6$ and $x = \pi/3$
(b) $f(x) = 7x^4$ between $x = -2$ and $x = 5$
(c) $f(x) = \begin{cases} \cos x & x < \pi/2 \\ x - \pi/2 & x \geqslant \pi/2 \end{cases}$ between $x = 0$ and $x = 3\pi/4$

Four Further Properties

In addition to the properties of the definite integral discussed in the previous unit, we need to consider four further properties to enable us to be more efficient in our integration.

Integrand Multiplied by a Constant

An extension of the sum of non-overlapping areas concerns the integral of an expression multiplied by a constant. The constant can be taken outside the integral. For example,

$$\int_{x=a}^{b} \beta f(x)\,dx = \beta \int_{x=a}^{b} f(x)\,dx$$

so that

$$\begin{aligned}\int_{x=1}^{2} 5x\,dx &= 5 \int_{x=1}^{2} x\,dx \\ &= 5\left[\frac{x^2}{2}\right]_{x=1}^{2} \\ &= 5\left(2-\frac{1}{2}\right) \\ &= \frac{15}{2}\end{aligned}$$

The Integral of a Sum

The integral of a sum is a sum of integrals:

$$\int_{x=a}^{b} [f(x)+g(x)]\,dx = \int_{x=a}^{b} f(x)\,dx + \int_{x=a}^{b} g(x)\,dx$$

For example,

$$\begin{aligned}\int_{x=0}^{\pi/6} (x-\cos x)\,dx &= \int_{x=0}^{\pi/6} x\,dx - \int_{x=0}^{\pi/6} \cos x\,dx \\ &= \left[\frac{x^2}{2}\right]_{x=0}^{\pi/6} - [\sin x]_{x=0}^{\pi/6} \\ &= \frac{\pi^2}{72} - (\sin \pi/6 - \sin 0) \\ &= \frac{\pi^2}{72} - \frac{1}{2}\end{aligned}$$

Negative Values

So far, when we have considered the area beneath the graph of a function, we have tacitly assumed that $f(x) \geqslant 0$ for all values of x in the interval of the x-axis in question. There is good reason for this. If one constructs rectangles between a curve and the x-axis for $f(x)$ values that are negative then according to our formalism the heights of the rectangles are negative. Since negative areas are not defined it is at this point that the concept of an integral departs from the concept of area. However, the use of the integral is far more wide-ranging than its use as a measure of area, and for this reason negative values of the definite integral are permitted.

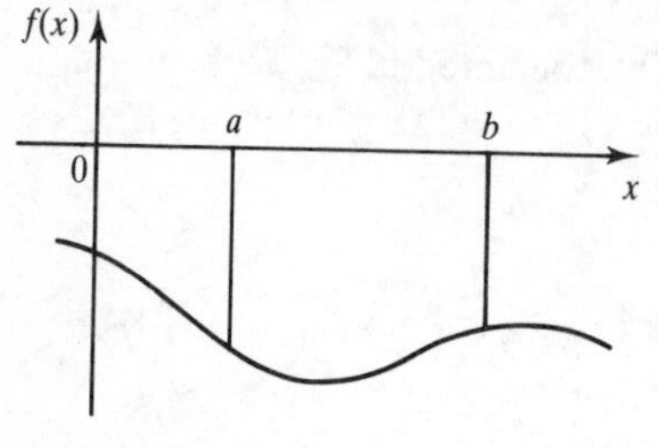

Figure 23.4

If the area in question is beneath the x-axis then the value of the definite integral is negative:

$$\int_{x=a}^{b} f(x)\,\mathrm{d}x < 0$$

Interchanging Limits

Interchanging the limits changes the sign of the integral:

$$\int_{x=a}^{b} f(x)\,\mathrm{d}x = -\int_{x=b}^{a} f(x)\,\mathrm{d}x$$

EXAMPLES

Find the value of the definite integral of each of the following:

(a) $f(x) = x^2 + \sin x$ between $x = 0$ and $x = \pi/2$

(b) $f(x) = 5x^3$ between $x = 2$ and $x = 4$

(c) $f(x) = \begin{cases} 2x/\pi & x < \pi/2 \\ \sin x & x \geqslant \pi/2 \end{cases}$ between $x = 0$ and π

(d) $f(x) = -x$ between $x = 0$ and $x = 3$

(a)
$$\int_{x=0}^{\pi/2} (x^2 + \sin x)\,\mathrm{d}x = \int_{x=0}^{\pi/2} x^2\,\mathrm{d}x + \int_{x=0}^{\pi/2} \sin x\,\mathrm{d}x$$

$$= \left[\frac{x^3}{3}\right]_{x=0}^{\pi/2} + [-\cos x]_{x=0}^{\pi/2}$$

$$= \frac{\pi^3}{24} + [-\cos \pi/2 - (-\cos 0)]$$

$$= \frac{\pi^3}{24} + 1$$

(b) $$\int_{x=2}^{4} 5x^3\,dx = 5\left[\frac{x^4}{4}\right]_{x=2}^{4}$$

$$= 5(2^6 - 2^2)$$

$$= 5(60)$$

$$= 300$$

(c) $f(x) = \begin{cases} 2x/\pi & x < \pi/2 \\ \sin x & x \geqslant \pi/2 \end{cases}$ between $x = 0$ and $x = \pi$

Here

$$\int_{x=0}^{\pi} f(x)\,dx = \int_{x=0}^{\pi/2} f(x)\,dx + \int_{x=\pi/2}^{\pi} f(x)\,dx$$

$$= \int_{x=0}^{\pi/2} \left(\frac{2x}{\pi}\right) dx + \int_{x=\pi/2}^{\pi} \sin x\,dx$$

$$= \frac{2}{\pi}\left[\frac{x^2}{2}\right]_{x=0}^{\pi/2} + [-\cos x]_{x=\pi/2}^{\pi}$$

$$= \frac{2}{\pi}\frac{\pi^2}{8} + [-\cos\pi - (-\cos\pi/2)]$$

$$= \frac{\pi}{4} + 1$$

(d) $$\int_{x=0}^{3} (-x)\,dx = -\int_{x=0}^{3} x\,dx$$

$$= \int_{x=3}^{0} x\,dx$$

$$= \left[\frac{x^2}{2}\right]_{x=3}^{0}$$

$$= \frac{0}{2} - \frac{9}{2}$$

$$= \frac{-9}{2}$$

EXERCISES

Find the value of the definite integral for each of the following:

(a) $f(x) = x^4 - \sec^2 x$ between $x = 0$ and $x = \pi/3$

(b) $f(x) = 8x^{-2}$ between $x = 1$ and $x = 3$

(c) $f(x) = \begin{cases} e^x & x < 1 \\ e(2 - x) & x \geqslant 1 \end{cases}$ between $x = 0$ and $x = 3$

(a) $$\int_{x=0}^{\pi/3} (x^4 - \sec^2 x)\,dx = \int_{x=0}^{\pi/3} *\,dx - \int_{x=0}^{\pi/3} *\,dx$$
$$= [*]_{x=0}^{\pi/3} - [\tan x]_{x=0}^{\pi/3}$$
$$= (* - *) - (* - *)$$
$$= *$$

(b) $$\int_{x=1}^{3} 8x^{-2}\,dx = 8[*]_{x=1}^{3}$$
$$= 8(* - 8*)$$
$$= \frac{*}{3}$$

(c) $$f(x) = \begin{cases} e^x & x < 1 \\ e(2-x) & x \geqslant 1 \end{cases} \qquad \text{between } x = 0 \text{ and } x = 3$$

Here

$$\int_{x=0}^{3} f(x)\,dx = \int_{x=0}^{*} f(x)\,dx + \int_{x=*}^{*} f(x)\,dx$$
$$= \int_{x=0}^{*} *\,dx + \int_{x=*}^{*} *\,dx$$
$$= \int_{x=0}^{1} *\,dx + * \int_{x=1}^{3} dx - e\int_{x=1}^{3} *\,dx$$
$$= [*]_{x=0}^{1} + 2e[*]_{x=1}^{3} - e[*]_{x=1}^{3}$$
$$= (* - *) + 2e(* - *) - e(* - *)$$
$$= -*$$

Module 23: Further exercises

1 Use the fundamental theorem of calculus to evaluate the definite integrals of each of the following:

(a) $f(x) = x^{-2}$ between $x = 1$ and $x = 3$

(b) $f(x) = \text{cosec}^2\, x$ between $x = \pi/4$ and $x = \pi/3$

(c) $f(x) = 5$ between $x = 3$ and $x = 6$

2 Find the value of the definite integrals of each of the following:

(a) $f(x) = x^{-6} - \sec^2 x$ between $x = \pi/6$ and $x = \pi/4$

(b) $f(x) = 9x^7$ between $x = 3$ and $x = 8$

(c) $f(x) = \begin{cases} x - 1 & x \leqslant 1 \\ 1 - 1/x & x > 1 \end{cases}$ between $x = 1/2$ and $x = 3/2$

Module 24: Methods of Integration

Aims: To demonstrate a number of standard methods of converting integrals into standard form, thereby permitting them to be evaluated by using the tables of standard integrals.

Objectives: When you have read this module you will be able to evaluate integrals:

- ▶▶ By changing the variable of integration
- ▶▶ With integrands of the form $f'(x)/f(x)$
- ▶▶ By partial fractions
- ▶▶ By parts

Unit 1 Change of Variable Using an Algebraic Substitution

Test yourself with the following:

Integrate each of the following:

(a) $f(x) = (x+1)^4$ between $x = 1$ and $x = 2$
(b) $f(x) = \cos(3x - 4\pi)$ between $x = 0$ and $x = 4\pi$
(c) $f(x) = \exp(2x + 3)$ between $x = -2$ and $x = 0$
(d) $f(x) = x(x^2 - 1)^2$ between $x = 3$ and $x = 4$
(e) $f(x) = 3x \sin(x^2)$ between $x = 0$ and $x = \sqrt{\pi}$

Methods of Integration

Many expressions within an integral are not in a standard form and this means that the process of integration is more than simply looking up the answer in a table of standard integrals. The various **methods of integration** are concerned with manipulating the integrand in such a manner that it is converted into a standard form, thereby making it a simple matter to complete the integration. The first of these methods concerns the changing of the variable of integration by using an algebraic substitution. For example, the integral

$$I = \int_{x=a}^{b} \sin(2x+3)\,dx$$

is not in standard form. The problem is the $2x + 3$ as the argument of the sine. Were the argument just the single symbol x then it would be in the standard form.

We are going to change the variable of integration so that the argument of the sine is a single symbol. We shall define a new variable $u(x)$ where

$$u(x) = 2x + 3$$

The derivative of $u(x)$ is then

$$u'(x) = 2$$

so that the differential $du(x)$ is given as

$$\begin{aligned} du(x) &= u'(x)\,dx \\ &= 2\,dx \end{aligned}$$

making $dx = du(x)/2$. Also

when $x = a$, $u(x) = 2a + 3$ and when $x = b$, $u(x) = 2b + 3$

Writing u for $u(x)$, integral I can now be rewritten in terms of the new variable u as

$$I = \int_{u=2a+3}^{2b+3} \sin u \frac{du}{2}$$

Note that the limits have been changed as the variable of integration has been changed from x to u. When $x = a$ then $u = 2a + 3$, and when $x = b$ then $u = 2b + 3$. The $\frac{1}{2}$ can now come outside the integral to give

$$I = \frac{1}{2} \int_{u=2a+3}^{2b+3} \sin u\, du$$

This integral is now in standard form and the solution is obtained from the table as

$$\begin{aligned} I &= \tfrac{1}{2}[-\cos u]_{u=2a+3}^{2b+3} \\ &= \tfrac{1}{2}\{-\cos(2b + 3) - [-\cos(2a + 3)]\} \\ &= \tfrac{1}{2}[\cos(2a + 3) - \cos(2b + 3)] \end{aligned}$$

EXAMPLES

Integrate each of the following:

(a) $f(x) = (x - 4)^5$ between $x = -2$ and $x = 2$
(b) $f(x) = \sin(5x + 3\pi)$ between $x = 0$ and $x = 3\pi/5$
(c) $f(x) = x(x^2 + 5)^3$ between $x = 1$ and $x = 2$
(d) $f(x) = 2x \exp(x^2)$ between $x = 0$ and $x = 1$

(a) $$\int_{x=-2}^{2} (x - 4)^5\, dx$$

Here we shall change the variable x to $u(x)$ where

$$u(x) = x - 4$$

so that:

$$u'(x) = 1 \text{ and } du(x) = dx$$

When $x = -2$ then $u(x) = -6$ and when $x = 2$ then $u(x) = -2$. Writing u for $u(x)$ and substituting into the integral we find that

$$\int_{u=-6}^{-2} u^5 \, du = \left[\frac{u^6}{6}\right]_{u=-6}^{-2}$$

$$= \{(-2)^6 - (-6)^6\}16$$

$$= -\frac{46\,592}{6}$$

(b) $\displaystyle\int_{x=0}^{3\pi/5} \sin(5x + 3\pi)\, dx$

Here we shall change the variable x to $u(x)$ where

$$u(x) = 5x + 3\pi \text{ so that } du(x) = 5\, dx$$

When $x = 0$ then $u(x) = 3\pi$ and when $x = 3\pi/5$ then $u(x) = 6\pi$. Writing u for $u(x)$ and substituting into the integral we find

$$\int_{u=3\pi}^{6\pi} \sin u \left(\frac{du}{5}\right) = \frac{1}{5}[-\cos u]_{u=3\pi}^{6\pi}$$

$$= \frac{1}{5}[-\cos 6\pi - (-\cos 3\pi)]$$

$$= \frac{1}{5}(-1 - 1)$$

$$= \frac{-2}{5}$$

(c) $\displaystyle\int_{x=1}^{2} x(x^2 + 5)^3\, dx$

In this integral we note that x is half of the derivative of $x^2 + 5$ and so we shall change the variable x to $u(x)$ where

$$u(x) = x^2 + 5 \text{ so that } du(x) = 2x\, dx$$

When $x = 1$ then $u(x) = 6$ and when $x = 2$ then $u(x) = 9$. Writing u for $u(x)$ and substituting into the integral we find

$$\int_{u=6}^{9} u^3 \left(\frac{du}{2}\right) = \frac{1}{2}\left[\frac{u^4}{4}\right]_{u=6}^{9}$$

$$= \frac{1}{2}\left(\frac{9^4}{4} - \frac{6^4}{4}\right)$$

$$= \frac{1}{2}\left(\frac{6561}{4} - \frac{1296}{4}\right)$$

$$= \frac{5265}{8}$$

(d) $\displaystyle\int_{x=0}^{1} 2x \exp(x^2)\, dx$

Here we notice that $2x$ is the derivative of x^2 and so we shall change the variable x to $u(x)$ where

$$u(x) = x^2 \text{ so that } du(x) = 2x\, dx$$

When $x = 0$ then $u(x) = 0$ and when $x = 1$ then $u(x) = 1$. Writing u for $u(x)$ and substituting into the integral we find

$$\int_{u=0}^{1} \exp(u)\, du = [\exp(u)]_{u=0}^{1} = e - 1$$

EXERCISES

Integrate each of the following:

(a) $f(x) = (1 - x)^{-3}$ between $x = -1$ and $x = 0$
(b) $f(x) = \sec^2(2\pi - 5x)$ between $x = 0$ and $x = \pi/5$
(c) $f(x) = x \exp(x^2 - 2)$ between $x = 0$ and $x = 1$
(d) $f(x) = 2x \cos(x^2)$ between $x = \sqrt{(\pi/4)}$ and $x = \sqrt{(\pi/2)}$

(a) $\displaystyle\int_{x=-1}^{0} (1 - x)^{-3}\, dx$

Here we shall change the variable x to $u(x)$ where

$$u(x) = 1 - x \text{ so that } du(x) = -dx$$

When $x = -1$ then $u(x) = 2$ and when $x = 0$ then $u(x) = 1$. Writing u for $u(x)$ and substituting into the integral we find

$$\int_{u=2}^{1} *(-du) = -\int_{u=2}^{1} *\, du$$

$$= \int_{u=1}^{2} *\, du$$

$$= \left[\frac{*}{-*}\right]_{u=1}^{2}$$

$$= (* - *)$$

$$= \frac{3}{8}$$

(b) $\displaystyle\int_{x=0}^{\pi/5} \sec^2(2\pi - 5x)\, dx$

Here we shall change the variable x to $u(x)$ where

$$u(x) = 2\pi - 5x \text{ so that } du(x) = *\, dx$$

When $x = 0$ then $u(x) = 2\pi$ and when $x = \pi/5$ then $u(x) = \pi$. Writing u for $u(x)$ and substituting into the integral we find

$$\int_{u=2\pi}^{\pi} \sec^2 u(* \, du) = *[\tan u]_{u=\pi}^{\text{limit to } \pi}$$

$$= *(* - *)$$

$$= 0$$

(c) $\displaystyle\int_{x=0}^{1} x \exp(x^2 - 2)\, dx$

Here we notice that the derivative of $x^2 - 2$ is $*$ and so we shall change the variable x to $u(x)$ where

$$u(x) = * \text{ so that } du(x) = * \, dx$$

When $x = 0$ then $u(x) = -2$ and when $x = 1$, $u(x) = -1$. Writing u for $u(x)$ and substituting into the integral we find

$$\int_{u=-2}^{-1} * \left(\frac{du}{*}\right) = *[*]_{u=-2}^{-1}$$

$$= *(* - *)$$

$$= *$$

(d) $\displaystyle\int_{x=\sqrt{(\pi/4)}}^{\sqrt{(\pi/2)}} 2x \cos(x^2)\, dx$

Again we notice that $*$ is the derivative of $*$ and so we shall change the variable x to $u(x)$ where

$$u(x) = * \text{ so that } du(x) = * \, dx$$

When $x = \sqrt{(\pi/4)}$ then $u(x) = *$ and when $x = \sqrt{(\pi/2)}$, $u(x) = *$. Writing u for $u(x)$ and substituting into the integral we find

$$\int_{u=*}^{*} * \, du = [*]_{u=*}^{*}$$

$$= (* - *)$$

$$= *$$

Unit 2 Change of Variable Using a Trigonometric Substitution

Test yourself with the following:

Integrate each of the following:

(a) $f(x) = 3 \sin^2 x$ between $x = 0$ and $x = \pi$
(b) $f(x) = 4(1 + \tan^2 x)$ between $x = \pi/6$ and $x = \pi/3$
(c) $f(x) = \sin 2x \cos 2x$ between $x = 0$ and $x = \pi/4$

Using Trigonometric Identities

A change of variable can sometimes be effected by using the trigonometric identities. For example, the integral

$$I = \int_{x=0}^{\pi/4} \cos^2 x \, dx$$

is not a standard integral, but can be converted into a standard form by using the identity

$$\cos 2x = 2\cos^2 x - 1$$

This identity can be transposed to read

$$\cos^2 x = \tfrac{1}{2}(\cos 2x + 1)$$

Substituting this identity into the integral gives

$$I = \int_{x=0}^{\pi/4} \frac{1}{2}(\cos 2x + 1)\, dx$$

$$= \frac{1}{2}\int_{x=0}^{\pi/4} \cos 2x \, dx + \frac{1}{2}\int_{x=0}^{\pi/4} dx$$

$$= \frac{1}{2}\left[\frac{\sin 2x}{2}\right]_{x=0}^{\pi/4} + \frac{1}{2}[x]_{x=0}^{\pi/4}$$

$$= \frac{1}{2}\left[\left(\frac{\sin \pi/2}{2}\right) - 0\right] + \frac{1}{2}\left(\frac{\pi}{4} - 0\right)$$

$$= \frac{1}{4} + \frac{\pi}{8}$$

EXAMPLES

Integrate each of the following:

(a) $f(x) = 4\sin^2 x$ between $x = 0$ and $x = \pi$

(b) $f(x) = 5(1 + \tan^2 x)$ between $x = \pi/6$ and $x = \pi/3$

(c) $f(x) = \sin 4x \cos 2x$ between $x = 0$ and $x = \pi/2$

(a) $$\int_{x=0}^{\pi} 4\sin^2 x \, dx$$

Here we make the observation that

$$\cos 2x = 1 - 2\sin^2 x$$

so that

$$\sin^2 x = \tfrac{1}{2}(1 - \cos 2x)$$

and this forms the trigonometric substitution. The integral becomes

$$\int_{x=0}^{\pi} 4\frac{1}{2}(1 - \cos 2x)\,dx = 2\left[x - \frac{\sin 2x}{2}\right]_{x=0}^{\pi}$$

$$= 2\pi$$

(b) $\displaystyle\int_{x=\pi/6}^{\pi/3} 5(1 + \tan^2 x)\,dx$

Here we make the observation that

$$1 + \tan^2 x = \sec^2 x$$

and this forms the trigonometric substitution. The integral becomes

$$\int_{x=\pi/6}^{\pi/3} 5\sec^2 x\,dx = 5[\tan x]_{x=\pi/6}^{\pi/3}$$

$$= 5(\sqrt{3} - 1/\sqrt{3})$$

(c) $\displaystyle\int_{x=0}^{\pi/2} (\sin 4x \cos 2x)\,dx$

Here we make the observation that

$$\sin 6x + \sin 2x = 2 \sin 4x \cos 2x$$

so that

$$\sin 4x \cos 2x = \tfrac{1}{2}(\sin 6x + \sin 2x)$$

and this forms the trigonometric substitution. The integral becomes

$$\int_{x=0}^{\pi/2} \frac{1}{2}(\sin 6x + \sin 2x)\,dx = \frac{1}{2}\left[\frac{-\cos 6x}{6} + \frac{-\cos 2x}{2}\right]_{x=0}^{\pi/2}$$

$$= \frac{1}{2}\left[\left(\frac{-\cos 3\pi}{6} + \frac{-\cos \pi}{2}\right) - \left(\frac{-\cos 0}{6} + \frac{-\cos 0}{2}\right)\right]$$

$$= \frac{1}{2}\left[\left(\frac{1}{6} + \frac{1}{2}\right) - \left(\frac{-1}{6} - \frac{1}{2}\right)\right]$$

$$= \frac{2}{3}$$

EXERCISES

1 Integrate each of the following:

(a) $f(x) = 2(1 + \cot^2 x)$ between $x = \pi/6$ and $x = \pi/4$
(b) $f(x) = \tan^2 x \sec^2 x$ between $x = \pi/6$ and $x = \pi/3$
(c) $f(x) = \sin 2x \sin 6x$ between $x = 0$ and $x = \pi/12$

(a) $$\int_{x=\pi/6}^{\pi/4} 2(1+\cot^2 x)\,\mathrm{d}x$$

Here we make the observation that

$$1+\cot^2 x = \operatorname{cosec}^2 x$$

and this forms the trigonometric substitution. The integral becomes

$$\int_{x=\pi/6}^{\pi/4} * \,\mathrm{d}x = 2[-*]_{x=\pi/6}^{\pi/4}$$

$$= *(* + \sqrt{3})$$

(b) $$\int_{x=\pi/6}^{\pi/3} \tan^2 x \sec^2 x\,\mathrm{d}x$$

Here we make the observation that if $u(x) = \tan x$ then

$$u'(x) = * \text{ and so } \mathrm{d}u(x) = * \,\mathrm{d}x$$

and this forms the trigonometric substitution. The integral becomes

$$\int_{u=*}^{*} u^2\,\mathrm{d}u = \left[\frac{*^3}{*}\right]_{u=*}^{*}$$

$$= \sqrt{3} - *$$

(c) $$\int_{x=0}^{\pi/12} (\sin 2x \sin 6x)\,\mathrm{d}x$$

Here we make the observation that

$$\cos 8x - \cos 4x = *$$

so that

$$\sin 2x \sin 6x = *$$

and this forms the trigonometric substitution. The integral becomes

$$\int_{x=0}^{\pi/12} * \,\mathrm{d}x = *[1]_{x=0}^{\pi/12}$$

$$= \frac{\sqrt{3}}{32}$$

Unit 3 Integrals with Integrand of the Form f′(x)/f(x)

Test yourself with the following:

Integrate each of the following:

(a) $\dfrac{2x+1}{x^2+x+2}$ between $x = 1$ and $x = 2$

(b) $\cot x$ between $x = \pi/6$ and $x = \pi/3$

(c) $\dfrac{1/x}{1+\ln(x)}$ between $x = 1$ and $x = 2$

Logarithms

Consider the integral of the form

$$\int_{x=a}^{b} \frac{f'(x)}{f(x)}\,\mathrm{d}x$$

Since $f'(x)\,\mathrm{d}x = \mathrm{d}f(x)$ the ratio $[f'(x)/f(x)]\,\mathrm{d}x$ can be rewritten as

$$\frac{\mathrm{d}f(x)}{f(x)}$$

Writing f for $f(x)$ and substituting into the integral, we find the integral becomes

$$\int_{f=f(a)}^{f(b)} \frac{\mathrm{d}f}{f} = [\ln f]_{f=f(a)}^{f(b)}$$

$$= [\ln f(b)] - [\ln f(a)]$$

$$= \ln\frac{f(b)}{f(a)}$$

If the numerator is the derivative of the denominator then the integral is the natural logarithm of the denominator. For example, the integral

$$I = \int_{x=2}^{3} \frac{2x}{x^2-1}\,\mathrm{d}x$$

has an integrand of the form $f'(x)/f(x)$ where $f(x) = x^2 - 1$ so $f'(x) = 2x$. The integral is then evaluated as

$$I = [\ln(x^2-1)]_{x=2}^{3}$$

$$= \ln(8) - \ln(3)$$

$$= \ln(\tfrac{8}{3})$$

EXAMPLES

Integrate each of the following:

(a) $\dfrac{4x-3}{2x^2-3x+5}$ between $x = 1$ and $x = 4$

(b) $\dfrac{\sec^2 x}{\tan x}$ between $x = \pi/4$ and $x = \pi/3$

(c) $\dfrac{2e^{2x}}{5+e^{2x}}$ between $x = 2$ and $x = 4$

(a) $$\int_{x=1}^{4} \frac{4x-3}{2x^2-3x+5}\,dx$$

Here it is noted that if $f(x) = 2x^2 - 3x + 5$ then $f'(x) = 4x - 3$, and so the numerator is the derivative of the denominator. Hence

$$\int_{x=1}^{4} \frac{4x-3}{2x^2-3x+5}\,dx = [\ln(2x^2-3x+5)]_{x=1}^{4}$$

$$= \ln(25) - \ln(4)$$

$$= \ln(\tfrac{25}{4})$$

(b) $$\int_{x=\pi/4}^{\pi/3} \frac{\sec^2 x}{\tan x}\,dx$$

Here it is noted that the numerator is the derivative of the denominator. Hence

$$\int_{x=\pi/4}^{\pi/3} \frac{\sec^2 x}{\tan x}\,dx = [\ln(\tan x)]_{x=\pi/4}^{\pi/3}$$

$$= \ln(\surd 3) - \ln(1)$$

$$= \tfrac{1}{2}\ln(3)$$

(c) $$\int_{x=2}^{4} \frac{2e^{2x}}{5+e^{2x}}\,dx$$

Here it is noted that the numerator is the derivative of the denominator. Hence

$$\int_{x=2}^{4} \frac{2e^{2x}}{5+e^{2x}}\,dx = [\ln(5+e^{2x})]_{x=2}^{4}$$

$$= \ln(5+e^8) - \ln(5+e^4)$$

$$= \ln\left(\frac{5+e^8}{5+e^4}\right)$$

EXERCISES

Integrate each of the following:

(a) $(x+1)^{-1}$ between $x = 0$ and $x = 1$

(b) $\tan x$ between $x = 0$ and $x = \pi/4$

(c) $\dfrac{e^x}{e^x + 1}$ between $x = -1$ and $x = 1$

(a) $\displaystyle\int_{x=0}^{1} \frac{1}{x+1}\,dx$

Here it is noted that if $f(x) = x + 1$ then $f'(x) = *$ so that the $*$ is the derivative of the $*$. Hence

$$\int_{x=0}^{1} \frac{1}{x+1}\,dx = [\ln(*)]_{x=0}^{1}$$
$$= \ln(2)$$

(b) $\displaystyle\int_{x=0}^{\pi/4} \tan x\,dx = \int_{x=0}^{\pi/4} \frac{*}{*}\,dx$

Here it is noted that $\tan x = \sin x/\cos x$ and that the derivative of $\sin x$ is $*$. Thus, in this integral the numerator is the negative of the derivative of the denominator. Hence

$$\int_{x=0}^{\pi/4} \tan x\,dx = -[\ln(*)]_{x=0}^{\pi/4}$$
$$= \ln(*)$$

(c) $\displaystyle\int_{x=-1}^{1} \frac{e^x}{e^x+1}\,dx$

Here it is noted that the $*$ is the derivative of the $*$. Hence

$$\int_{x=-1}^{1} \frac{e^x}{e^x+1} = [*]_{x=-1}^{1}$$
$$= *$$

Unit 4 Integration by Partial Fractions

Test yourself with the following:

Integrate each of the following:

(a) $f(x) = \dfrac{2}{x^2 - 1}$ between $x = 2$ and $x = 4$

(b) $f(x) = \dfrac{4x^2 - 6x + 8}{x(x-4)}$ between $x = 5$ and $x = 10$

(c) $f(x) = \dfrac{1 - x^2}{x(x-6)^2}$ between $x = 7$ and $x = 8$

Rational Algebraic Expressions

If the integrand is a rational expression of polynomials then it may be possible to convert the integral into a standard form by using partial fractions. For example, the integral

$$I = \int_{x=3}^{4} \frac{5x - 2}{x^2 + x - 2}\,dx$$

has an integrand that can be split into partial fractions:

$$\frac{5x - 2}{x^2 + x - 2} = \frac{1}{x - 1} + \frac{4}{x + 2}$$

The integral can thus be written as

$$I = \int_{x=3}^{4} \left(\frac{1}{x - 1} + \frac{4}{x + 2}\right) dx$$

$$= [\ln(x - 1)]_{x=3}^{4} + 4[\ln(x + 2)]_{x=3}^{4}$$

$$= \ln(\tfrac{3}{2}) + 4\ln(\tfrac{6}{5})$$

$$= \ln(\tfrac{1944}{625})$$

EXAMPLES

Integrate each of the following:

(a) $f(x) = \dfrac{3x - 1}{x^2 - 2x - 3}$ between $x = 4$ and $x = 5$

(b) $f(x) = \dfrac{2 + 3x - x^2}{x^2(x + 1)}$ between $x = 1$ and $x = 2$

(c) $f(x) = \dfrac{x^3 - 15x}{x^2 - 4x + 1}$ between $x = 6 - \sqrt{3}$ and $x = 6 + \sqrt{3}$

(a) The algebraic fraction

$$\frac{3x - 1}{x^2 - 2x - 3}$$

can be written as a sum of partial fractions in the form

$$\frac{1}{x + 1} + \frac{2}{x - 3}$$

So that:

$$\int_{x=4}^{5} \frac{3x - 1}{x^2 - 2x - 3}\,dx = \int_{x=4}^{5} \frac{1}{x + 1}\,dx + \int_{x=4}^{5} \frac{2}{x - 3}\,dx$$

$$= [\ln(x + 1)]_{x=4}^{5} + [2\ln(x - 3)]_{x=4}^{5}$$

$$= [\ln(6) - \ln(5)] + [2\ln(2) - 2\ln(1)]$$

$$= \ln(\tfrac{24}{5})$$

(b) The algebraic fraction

$$\frac{2 + 3x - x^2}{x^2(x + 1)}$$

can be written as a sum of partial fractions in the form

$$\frac{2}{x^2} + \frac{1}{x} - \frac{2}{x + 1}$$

So that

$$\int_{x=1}^{2} \frac{2 + 3x - x^2}{x^2(x + 1)}\,dx = \int_{x=1}^{2} \frac{2}{x^2}\,dx + \int_{x=1}^{2} \frac{1}{x}\,dx - \int_{x=1}^{2} \frac{2}{x + 1}\,dx$$

$$= 2\left[\frac{x^{-1}}{-1}\right]_{x=1}^{2} + [\ln(x)]_{x=1}^{2} - 2[\ln(x + 1)]_{x=1}^{2}$$

$$= 2(-1/2 + 1) + (\ln(2) - \ln(1)) - 2(\ln(3) - \ln(2))$$

$$= \ln(\tfrac{8}{9}) + 1$$

(c) The algebraic fraction

$$\frac{x^3 - 15x}{x^2 - 4x + 1}$$

has a numerator polynomial of degree higher than the denominator polynomial and therefore we must perform the division first. It is found that

$$\frac{x^3 - 15x}{x^2 - 4x + 1} = (x + 4) - \frac{4}{x^2 - 4x + 1}$$

The latter algebraic fraction cannot be written as a sum of partial fractions at the moment. Instead we complete the squares to yield:

$$\frac{x^3 - 15x}{x^2 - 4x + 1} = (x + 4) - \frac{4}{(x - 2)^2 - 3}$$

$$= (x + 4) - \frac{4}{(x - 2)^2 - (\sqrt{3})^2}$$

So that

$$\int_{x=6-\sqrt{3}}^{6+\sqrt{3}} \frac{x^3 - 15x}{x^2 - 4x + 1}\,dx = \int_{x=6-\sqrt{3}}^{6+\sqrt{3}} (x + 4)\,dx - \int_{x=6-\sqrt{3}}^{6+\sqrt{3}} \frac{4}{(x - 2)^2 - (\sqrt{3})^2}\,dx$$

In the second integral we change the variable to $u(x) = x - 2$ so that, writing u for $u(x)$ we find

$$\frac{4}{(x - 2)^2 - (\sqrt{3})^2} = \frac{4}{u^2 - (\sqrt{3})^2}$$

$$= \frac{4}{(u + \sqrt{3})(u - \sqrt{3})}$$

$$= \frac{2}{\sqrt{3}}\left(\frac{1}{u - \sqrt{3}} - \frac{1}{u + \sqrt{3}}\right)$$

The integral now becomes

$$I = \int_{x=6-\sqrt{3}}^{6+\sqrt{3}} (x+4)\,dx - \int_{u=4-\sqrt{3}}^{4+\sqrt{3}} \frac{2/\sqrt{3}}{u-\sqrt{3}}\,du + \int_{u=4-\sqrt{3}}^{4+\sqrt{3}} \frac{2/\sqrt{3}}{u+\sqrt{3}}\,du$$

$$= [x^2/2 + 4x]_{x=6-\sqrt{3}}^{6+\sqrt{3}} - \frac{2}{\sqrt{3}}[\ln(u-\sqrt{3})]_{u=4-\sqrt{3}}^{4+\sqrt{3}} + \frac{2}{\sqrt{3}}[\ln(u+\sqrt{3})]_{u=4-\sqrt{3}}^{4+\sqrt{3}}$$

$$= 20\sqrt{3} - \frac{2}{\sqrt{3}}\left(\ln\frac{4}{4-2\sqrt{3}} - \ln\frac{4+2\sqrt{3}}{4}\right)$$

$$= \frac{4}{\sqrt{3}}[15 - \ln(2)]$$

EXERCISES

Integrate each of the following:

(a) $f(x) = \dfrac{5x+17}{2x^2+x-10}$ between $x = 3$ and $x = 6$

(b) $f(x) = \dfrac{8x^2-10x+8}{x^2(x-2)}$ between $x = 4$ and $x = 6$

(c) $f(x) = \dfrac{x^4-1}{x^2-4}$ between $x = 3$ and $x = 4$

(a) The algebraic fraction

$$\frac{5x+17}{2x^2+x-10}$$

can be written as a sum of partial fractions in the form

$$\frac{*}{x-2} - \frac{*}{2x+5}$$

So that:

$$\int_{x=3}^{6} \frac{5x+17}{2x^2+x-10}\,dx = \int_{x=3}^{6} \frac{*}{x-2}\,dx - \int_{x=3}^{6} \frac{*}{2x+5}\,dx$$

$$= *[\ln(x-2)]_{x=3}^{6} - *[\ln(2x+5)]_{x=3}^{6}$$

$$= \ln(64)\frac{\sqrt{11}}{\sqrt{17}}$$

(b) The algebraic fraction

$$\frac{8x^2-10x+8}{x^2(x-2)}$$

can be written as a sum of partial fractions in the form

$$\frac{*}{x-2}+\frac{*}{x^2}+\frac{*}{x}$$

So that

$$\int_{x=4}^{6} \frac{8x^2-10x+8}{x^2(x-2)}\,dx = \int_{x=4}^{6} \frac{*}{x-2}\,dx + \int_{x=4}^{6} \frac{*}{x^2}\,dx + \int_{x=4}^{6} \frac{*}{x}\,dx$$

$$= *[*]_{x=4}^{6} + *[x^*/*]_{x=4}^{6} + *[\ln(*)]_{x=4}^{6}$$

$$= \ln(108) - \tfrac{1}{3}$$

(c) The algebraic fraction

$$\frac{x^4-1}{x^2-4}$$

has a numerator polynomial of degree higher than the denominator polynomial and therefore we must perform the division first. It is found that

$$\frac{x^4-1}{x^2-4} = * + \frac{*}{x^2-4}$$

The latter algebraic fraction can be written as a sum of partial fractions to yield

$$\frac{x^4-1}{x^2-4} = * + *\left(\frac{1}{x-2} * \frac{1}{x+2}\right)$$

So that

$$\int_{x=3}^{4} \frac{x^4-1}{x^2-4}\,dx = \int_{x=3}^{4} *\,dx + * \int_{x=3}^{4} \frac{1}{x-2}\,dx * * \int_{x=3}^{4} \frac{1}{x+2}\,dx$$

$$= [*]_{x=3}^{4} + *[\ln(x-2)]_{x=3}^{4} * *[\ln(x+2)]_{x=3}^{4}$$

$$= \tfrac{15}{4}\ln(\tfrac{5}{3}) + \tfrac{49}{3}$$

Unit 5 Integration by Parts

Test yourself with the following:

Integrate each of the following:

(a) $f(x) = x\sec^2 x$ between $x = \pi/6$ and $x = \pi/3$
(b) $f(x) = x^2\ln(x)$ between $x = 2$ and $x = 4$
(c) $f(x) = x\,\mathrm{cosec}^2 x$ between $x = \pi/4$ and $x = 3\pi/4$

Products of Dissimilar Types of Expression

If the integrand is a product of two dissimilar types of expression then integration by parts may be employed. For example, the integral

$$\int_{x=0}^{\pi/2} x \sin x \, dx$$

has an integrand that is the product of x and $\sin x$ – two dissimilar types of expression – x being an algebraic expression and $\sin x$ being a trigonometric expression. The method of integration by parts is derived from the product rule for differentiation:

$$[u(x)v(x)]' = u'(x)v(x) + u(x)v'(x)$$

Integrating this equation gives

$$\int_{x=a}^{b} [u(x)v(x)]' \, dx = \int_{x=a}^{b} u'(x)v(x) \, dx + \int_{x=a}^{b} u(x)v'(x) \, dx$$

that is

$$\int_{x=a}^{b} d[u(x)v(x)] = \int_{x=a}^{b} v(x) \, du(x) + \int_{x=a}^{b} u(x) \, dv(x)$$

$$= [u(x)v(x)]_{x=a}^{b}$$

So that

$$\int_{x=a}^{b} u(x) \, dv(x) = [u(x)v(x)]_{x=a}^{b} - \int_{x=a}^{b} v(x) \, du(x)$$

This is the formula for integration by parts. For example, in the integral

$$\int_{x=0}^{\pi/2} x \sin x \, dx$$

we shall choose $u(x) = x$ because in the parts formula we have to differentiate $u(x)$ and doing this will reduce x to 1. That is,

$$du(x) = u'(x) \, dx = dx$$

Having chosen $u(x)$ to be x we are left with

$$dv(x) = \sin x \, dx$$

and by integration $v(x) = -\cos x$. Substituting into the parts formula we have

$$\int_{x=0}^{\pi/2} x \sin x \, dx = \int_{x=0}^{\pi/2} u(x) \, dv(x)$$

$$= [u(x)v(x)]_{x=0}^{\pi/2} - \int_{x=0}^{\pi/2} v(x) \, du(x)$$

$$= [-x \cos x]_{x=0}^{\pi/2} - \int_{x=0}^{\pi/2} (-\cos x) \, dx$$

$$= 0 + [\sin x]_{x=0}^{\pi/2}$$

$$= 1$$

EXAMPLES

Integrate each of the following:

(a) $f(x) = x \cos x$ between $x = \pi/6$ and $x = \pi/3$
(b) $f(x) = \ln(x)$ between $x = 1$ and $x = 4$
(c) $f(x) = x \sec^2 x$ between $x = 0$ and $x = \pi/4$

(a) $$\int_{x=\pi/6}^{\pi/3} x \cos x \, dx = \int_{x=\pi/6}^{\pi/3} u \, dv$$

$$= [u \,.\, v]_{x=\pi/6}^{\pi/3} - \int_{x=\pi/6}^{\pi/3} v \, du$$

We choose $u(x)$ to be x so that when we differentiate $u(x)$ in the parts formula we reduce the x to 1. Having chosen $u(x) = x$ we are left with $dv(x) = \cos x \, dx$. This means that $du = dx$ and $v = \sin x$.

Substitution into the parts formula then yields

$$\int_{x=\pi/6}^{\pi/3} x \cos x \, dx = [x(\sin x)]_{x=\pi/6}^{\pi/3} - \int_{x=\pi/6}^{\pi/3} \sin x \, dx$$

$$= [x \sin x + \cos x]_{x=\pi/6}^{\pi/3}$$

$$= [(\pi/3) \sin \pi/3 + \cos \pi/3] - [(\pi/6) \sin \pi/6 + \cos \pi/6]$$

$$= (\pi/2\sqrt{3} + 1/2) - (\pi/12 + \sqrt{3}/2)$$

$$= \pi(2\sqrt{3} - 1)/12 + (1 - \sqrt{3})/2$$

(b) $$\int_{x=1}^{4} \ln(x) \, dx = \int_{x=1}^{4} u \, dv$$

$$= [u \,.\, v]_{x=1}^{4} - \int_{x=1}^{4} v \, du$$

Here we choose $dv = dx$ because if we chose $dv = \ln(x)\,dx$ we would have to integrate $\ln(x)$ and that would be begging the question. Choosing $dv = dx$ giving $v = x$ means that $u = \ln(x)$ and $du = dx/x$. Substituting yields

$$\int_{x=1}^{4} \ln(x)\,dx = [x\ln(x)]_{x=1}^{4} - \int_{x=1}^{4} x\left(\frac{1}{x}\right)dx$$

$$= [x\ln(x)]_{x=1}^{4} - \int_{x=1}^{4} dx$$

$$= [x\ln(x) - x]_{x=1}^{4}$$

$$= (4\ln(4) - 4) - (1\ln(1) - 1)$$

$$= 4\ln(4) - 3$$

(c) $$\int_{x=0}^{\pi/4} x\sec^2 x\,dx = \int_{x=0}^{\pi/4} u\,dv$$

$$= [u.v]_{x=0}^{\pi/4} - \int_{x=0}^{\pi/4} v\,du$$ where $u = x$ so $du = dx$, and $dv = \sec^2 x\,dx$ so $v = \tan x$

$$= [x\tan x]_{x=0}^{\pi/4} - \int_{x=0}^{\pi/4} \tan x\,dx$$

$$= [x\tan x + \ln(\cos x)]_{x=0}^{\pi/4}$$

$$= [(\pi/4)\tan\pi/4 + \ln(\cos\pi/4)] - [(0)\tan 0 + \ln(\cos 0)]$$

$$= [\pi/4 + \ln(1/\sqrt{2})] - [0 + \ln(1)]$$

$$= \pi/4 - \ln(\sqrt{2})$$

EXERCISES

Integrate each of the following:

(a) $f(x) = x\exp x$ between $x = 0$ and $x = 1$
(b) $f(x) = x\ln(x)$ between $x = 1$ and $x = 2$

(a) $$\int_{x=0}^{1} x\exp x\,dx = \int_{x=0}^{1} u\,dv$$

$$= [u.v]_{x=0}^{1} - \int_{x=0}^{1} v\,du$$ where $u = *$ so $du = *\,dx$, and $dv = *\,dx$ so $v = *$

$$= [*]_{x=0}^{1} - \int_{x=0}^{1} *\,dx$$

$$= [* - *]_{x=0}^{1}$$

$$= 1$$

(b) $$\int_{x=1}^{2} x\ln(x)\,dx = \int_{x=1}^{2} u\,dv$$

$$= [u.v]_{x=1}^{2} - \int_{x=1}^{2} v\,du$$ where $u = \ln(x)$ so $du = * \,dx$, and $dv = x\,dx$ so $v = *$

$$= [*\ln(x)]_{x=1}^{2} - \int_{x=1}^{2} *\,dx$$

$$= [*\ln(x) - *]_{x=1}^{2}$$

$$= *$$

Module 24: Further exercises

1 Integrate each of the following by using an algebraic substitution:
(a) $f(x) = (x+9)^2$ between $x = 0$ and $x = 9$
(b) $f(x) = \operatorname{cosec}^2(3x-5)\pi/4$ between $x = 2$ and $x = 8/3$
(c) $f(x) = 5x\sec^2(x^2)$ between $x = 0$ and $x = \sqrt{(\pi/6)}$

2 Integrate each of the following by using a trigonometric substitution:
(a) $f(x) = 5 - 2\sin^2 6x$ between $x = \pi/6$ and $x = 5\pi/6$
(b) $f(x) = \cos^2 3x$ between $x = 0$ and $x = \pi/3$
(c) $f(x) = \cos 4x\cos 2x$ between $x = \pi/6$ and $x = \pi/4$

3 Integrate each of the following:
(a) $\dfrac{5x^2}{x^3-1}$ between $x = 2$ and $x = 3$
(b) $\dfrac{\sin 2x}{\sin^2 x}$ between $x = \pi/4$ and $x = 3\pi/4$
(c) $\dfrac{1+\ln(x)}{x\ln(x)}$ between $x = \mathrm{e}$ and $x = 2\mathrm{e}$

4 Integrate each of the following:
(a) $f(x) = \dfrac{x-4}{2x^2-x-1}$ between $x = 2$ and $x = 3$
(b) $f(x) = \dfrac{8-6x+3x^2}{2x^2-6x}$ between $x = 4$ and $x = 6$
(c) $f(x) = \dfrac{1}{x(3x+6)^2}$ between $x = 1$ and $x = 2$
(d) $f(x) = \dfrac{1}{x^2-x-1}$ between $x = \frac{1}{2} + \sqrt{5}$ to $\frac{1}{2} + 2\sqrt{5}$

5 Integrate each of the following by parts:
(a) $f(x) = 2x\cos 3x$ between $x = 0$ and $x = \pi/4$
(b) $f(x) = x\exp(4x)$ between $x = 0$ and $x = 1/4$
(c) $f(x) = 3x\ln(5x)$ between $x = 1/5$ and $x = \mathrm{e}/10$

Chapter 6: Miscellaneous exercises

1 Evaluate each of the following integrals:

$$\text{(a)} \int_{x=-1}^{1} |x|\, dx \qquad \text{(b)} \left| \int_{x=-1}^{1} x\, dx \right|$$

2 Evaluate

$$\int_{x=-2}^{2} |x^2 - 1^2|\, dx$$

3 Show that

$$\int_{x=a}^{b} \frac{f''(x)}{f'(x)}\, dx = [\ln f'(x)]_{x=a}^{b}$$

provided $f'(x) \neq 0$ for $a \leqslant x \leqslant b$.

4 Show that

$$\int_{x=a}^{b} d\left[\frac{u(x)}{v(x)}\right] = \int_{x=a}^{b} \frac{1}{v(x)}\, du(x) - \int_{x=a}^{b} \frac{u(x)}{v^2(x)}\, dv(x)$$

provided $v(x) \neq 0$ for $a \leqslant x \leqslant b$.

5 Show that if f is an even function then

$$\int_{x=-a}^{a} f(x)\, dx = 2 \int_{x=0}^{a} f(x)\, dx$$

and that if g is an odd function then

$$\int_{x=-a}^{a} g(x)\, dx = 0$$

6 Evaluate

$$\int_{x=-\pi/2}^{\pi/2} \sin^5 x\, dx$$

7 If we define a function F by

$$F(s) = \lim_{k \to \infty} \int_{t=0}^{k} f(t)\, e^{-st}\, dt$$

provided

$$\lim_{k \to \infty} [f(t)\, e^{-st}] = 0$$

show that

(a) if $f(t) = 1$ then $F(s) = 1/s$
(b) if $f(t) = e^{-t}$ then $F(s) = 1/(s+1)$

(*Hint* Evaluate each integral by parts and then take the limit.)

8 Show that if

$$F(s) = \lim_{k \to \infty} \int_{t=0}^{k} f(t)\, e^{-st}\, dt$$

provided

$$\lim_{t \to \infty} [f(t)\, e^{-st}] = 0$$

then

$$sF(s) - f(0) = \lim_{k \to \infty} \int_{t=0}^{k} f'(t)\, e^{-st}\, dt$$

9 Given the equation

$$f'(x) + f(x) = 1 \text{ where } f(0) = 0$$

use the results of Questions 8 and 7 to show that

$$F(s) = \frac{1}{s} - \frac{1}{s+1}$$

10 Using the results of Question 7 find the expression $f(t)$ that corresponds to

$$F(s) = \frac{1}{s} - \frac{1}{s+1}$$

and show that it satisfies the equation

$$f'(x) + f(x) = 1$$

11 Evaluate

(a) $\displaystyle\int_{x=0}^{k} x\, e^{-x^2}\, dx$ (c) $\displaystyle\int_{x=0}^{k} x^5\, e^{-x^2}\, dx$

(b) $\displaystyle\int_{x=0}^{k} x^3\, e^{-x^2}\, dx$

12 From the results of Question 11, can you deduce the value of the integral

$$\int_{x=0}^{k} x^{2n+1}\, e^{-x^2}\, dx$$

where n is a positive integer?

7
■ ■ ■ Sets and Probability

The aims of this chapter are to:

1. Introduce the notion of a random experiment and describe such experiments using the language of sets.
2. Describe events as subsets of a sample space and discuss their associated probabilities within the ambit of a random experiment.
3. Develop the idea of a probability distribution and discuss specific probability distributions, both continuous and discrete.

This chapter contains three modules:

Module 25: Sets and Random Experiments
Sets are introduced using the notion of events associated with a random experiment where the elements of the set are the mutually exclusive outcomes of the experiment.

Module 26: Events and Probabilities
Probability is defined as a measure of the chance of the occurrence of a simple event, probabilities being assigned to events either *a priori* or by statistical regularity. Probabilities of a union of events are discussed, as are probabilities of events formed from sequences of random experiments, the latter demonstrated by using a probability tree.

Module 27: Probability Distributions
The binomial probability distribution is derived from a consideration of sequences of Bernoulli trials. The Poisson probability distribution is considered and is used as an approximation to the binomial probability distribution for rare events. The concept of a continuous probability distribution is introduced via the piecewise continuous extension of a binomial probability distribution bar graph into a histogram. This is followed by an extensive discourse on the use of tables to evaluate the probabilities associated with a normal (or Gaussian) probability distribution.

Module 25: Sets and Random Experiments

Aim: To introduce sets and set manipulation techniques via the concepts of random experiments and their associated events.

Objectives: When you have read this module you will be able to:

- ▶▶ List the outcomes from a random experiment.
- ▶▶ Create events from a sample space of outcomes.
- ▶▶ Construct Venn diagrams.
- ▶▶ Manipulate sets using the operations of union, intersection and complement coupled with their properties.

Unit 1 Sets and Random Experiments

Test yourself with the following:

1 Write down the sample space of:
 (a) selecting a sock from a drawer containing black, blue and grey socks
 (b) selecting a page from a book containing 200 pages

2 Describe each of the following sets in an alternative way:
 (a) $\{1, 1.5, 2, 2.5, 3, 3.5\}$
 (b) $\{x\text{: } x \text{ is a day of the week}\}$

3 Which of the following are true statements:
 (a) $1 \in \{1, 2, 3\}$ (b) $\{3\} \in \{1, 2, 3\}$ (c) $\mathrm{D} \in \{\mathrm{A}, \mathrm{B}, \mathrm{C}\}$

4 List the subsets of the set $\{\mathrm{A}, \mathrm{B}, 1, 2, 3\}$

5 Which of the following are true statements:
 (a) $\{x\text{: } x \text{ is an integer and } -2 < x \leqslant 3\} \subseteq \{-1, 0, 1, 2, 3, 4\}$
 (b) $\{\mathrm{A}, \mathrm{B}, \mathrm{C}\} \subset \{\mathrm{A}, \mathrm{B}, \mathrm{C}\}$

6 Define suitable universal sets for each of the following random experiments:
 (a) selecting a day of the week (b) picking a colour of the rainbow.

Random Experiments

An experiment is any action or sequence of actions performed to find the value or values of some unknown quantity or quantities. For example, an experiment could be conducted to find the average height of a group of people. The experimental method would involve measuring the heights of everyone in the group and from these measurements, calculating the average height. The sequence of actions in this experiment consists of measuring and calculating; the value of the unknown quantity found being the average height. The unknown quantity whose value we are trying to find is called the **result** of the experiment and its value, found from an experiment, is called an **outcome** of the result. By the very nature of an experiment an outcome is unknown before the experiment is performed; a result can be anticipated but its actual value – the outcome – is unknown until the experiment is completed.

Take, for example, the simple experiment of tossing a coin. Before the coin is tossed it is not known whether it will show a head or a tail but it is known that it will show one or the other – it will produce a result. We can, therefore, list the **possible outcomes** of the result as:

Head, Tail or H, T for short

If an ordinary six-sided die were to be rolled the result will be a numbered face showing uppermost. The possible outcomes of that result are then:

One, Two, Three, Four, Five, Six or 1,2,3,4,5,6 for short

Any experiment that has a result with more than one possible outcome is referred to as a **random experiment** and the complete list of possible outcomes is called the **sample space** of the experiment. The only requirement that is made of the outcomes of a random experiment is that they be **mutually exclusive**, that is, that only one of the possible outcomes can actually occur when the experiment is performed – the result can have only one value.

Sets

A **set** is a collection of objects such as a set of tools or a set of numbers or a set of outcomes of the result from a random experiment. The notation used to denote a set consists of a pair of curly brackets $\{\ldots\}$ where the individual objects that comprise the set are described inside the brackets. For example, the set consisting of the whole numbers 1 to 5 can be written as:

$\{1,2,3,4,5\}$

where the contents are **listed** individually. Alternatively, the set could be written as:

$\{x\text{: } x \text{ is a whole number and } 1 \leqslant x \leqslant 5\}$

This is read as:

The set of x values, where the value of x is a whole number and x is greater than or equal to one and less than or equal to five.

The colon (:) stands for the word 'where'. In this description of the set the contents are **prescribed** by describing the properties that they hold in common. Notice that sets:

$\{1, 1, 2, 2, 2, 3, 4, 5\}$ and $\{1, 2, 3, 4, 5\}$

are the same set; *sets do not contain repeated elements.*

It is often more convenient to refer to a set without continually describing its contents, in which case we use capital letters. For example:

$A = \{\text{Head, Tail}\}$

Here the set consisting of the possible outcomes from tossing a coin has been called set A and can now be referred to in future by simply using the letter A.

The individual objects that are contained within a set are called *elements* of the set and the symbol $\in$ is used to denote membership of a set. For example, if:

$$A = \{a, b, c\} \text{ and } B = \{c, d, e\}$$

then $a \in A$ and $d \in B$. That is, *a is an element of set A* and *b is an element of set B*. Note that $a \notin B$ and $d \notin A$ – the slash through the symbol negates the symbol so that $\notin$ means **not a member**. Hence, *a is not an element of B* and *d is not an element of A*.

Set notation is used to describe the sample space of possible outcomes of the result of a random experiment. For example, if a random experiment were performed to select a coloured ball from a bag containing red, blue, yellow and green balls then the sample space (denoted by set *S*) of this experiment would be:

$$S = \{\text{Red, Blue, Yellow, Green}\}$$

Here the result of the experiment is a ball drawn from the bag and its possible values are its possible colours – the four possible outcomes.

Subsets

Additional sets can be formed by using the elements of a given set. For example, given the set:

$$A = \{a, b, c\}$$

then by using just the elements of set *A* the following additional sets could be formed:

$$\{a\}, \{b\}, \{c\}, \{a, b\}, \{a, c\}, \{b, c\}$$

Each of these sets is called a **subset** of *A* and the notation used to denote a subset is $\subseteq$ or $\supseteq$. For example:

$$\{a, b\} \subseteq \{a, b, c\} \text{ or, alternatively, } \{a, b, c\} \supseteq \{a, b\}$$

which means that the set $\{a, b\}$ is a subset of the set $\{a, b, c\}$.

If two sets *A* and *B* are such that $A \subseteq B$ and $B \subseteq A$ then $A = B$. Two sets are equal if they both contain the same elements.

If $A \subseteq B$ and $A \neq B$ then *A* is called a **proper subset** of *B*. The symbol to denote a proper subset is $\subset$ or $\supset$. That is, $\{a, b\} \subset \{a, b, c\}$ or $\{a, b, c\} \supset \{a, b\}$.

The Empty Set

The **empty set** is defined as the set that contains no elements at all. It is denoted by $\{ \}$ or, more commonly, by ϕ. ϕ is a Greek lower-case f, pronounced 'fi'. The empty set may seem to be an odd set to define but it is necessary to complete the rules governing the manipulation of sets. The empty set is unique and every set has the empty set as a subset.

Any pair of subsets that have no elements in common are said to be **mutually exclusive** subsets so that $\{a\}$, $\{b\}$ and $\{c\}$ are all mutually exclusive, as are $\{a\}$ and $\{b, c\}$, because they have no elements in common. However, sets $\{a\}$ and $\{a, b\}$ are not mutually exclusive, because they both contain the element *a*.

The Universal Set

The universal set – denoted by U – is the set that contains all the elements under discussion. For example, if

$A = \{x\text{: } x \text{ is a person with blue eyes}\}$ and $B = \{x\text{: } x \text{ is a person with green eyes}\}$

then the universal set could be:

$\mathrm{U} = \{x\text{: } x \text{ is a person with blue or green eyes}\}$

Notice the words 'could be'. The universal set is not unique. In the example just given the universal set could have been defined as:

$\mathrm{U} = \{x\text{: } x \text{ is a person}\}$

The definition of a universal set depends upon the context of the problem. The overriding factor is that all the sets that are going to be discussed *must be subsets of the universal set.* For a random experiment the universal set is the sample space:

$\mathrm{U} = S$

EXAMPLES

1 Write down the sample space of:

(a) measuring lengths up to 1 cm using a measuring scale graded to 1 mm
(b) answering a Male/Female question on a questionnaire

(a) {1 mm, 2 mm, 3 mm, 4 mm, 5 mm, 6 mm, 7 mm, 8 mm, 9 mm, 10 mm}
(b) {Males, Female}

2 Describe each of the following sets in an alternative way:

(a) $\{2, 3, 5, 7, 11, 13\}$ **(b)** $\{n\text{: } n^2 - 1 = 0\}$

(a) $\{x\text{: } x \text{ is a prime number} \leqslant 13\}$
(b) $\{-1, 1\}$

3 Which of the following are true statements?

(a) $\phi \in \{-1, 0, 1\}$ **(b)** $\{\phi\} \in \{-1, 0, 1\}$ **(c)** $\pi \in \{x\text{: } x \text{ is rational}\}$

(a) The empty set is a subset of every set so this is true.
(b) The set containing the empty set is not a member of the set $\{-1, 0, 1\}$.
(c) π is irrational, and not rational, so the statement is false.

4 List all the subsets of the set $\{1/2, -1/2, -3/2, -5/2\}$.

ϕ $\{1/2\}$, $\{-1/2\}$, $\{-3/2\}$, $\{-5/2\}$, $\{1/2, -1/2\}$, $\{1/2, -3/2\}$, $\{1/2, -5/2\}$, $\{-1/2, -3/2\}$, $\{-1/2, -5/2\}$, $\{-3/2, -5/2\}$, $\{1/2, -1/2, -3/2\}$, $\{1/2, -1/2, -5/2\}$, $\{1/2, -3/2, -5/2\}$, $\{-1/2, -3/2, -5/2\}$, $\{1/2, -1/2, -3/2, -5/2\}$.

5 Which of the following are true statements?

(a) $\{x\text{: } x \text{ is a natural number less than } 5\} \supseteq \{1, 2, 3, 4, 5\}$
(b) $\{P, Q, R\} \supset \{P, Q, R, S\}$

(a) Not true because 5 is not less than 5.
(b) Not true. The set on the right has more elements than the set on the left.

6 Define suitable universal sets for each of the following random experiments:

(a) choosing a whole number between 1 and 7
(b) selecting a letter from the first eight letters of the alphabet

(a) $\{1, 2, 3, 4, 5, 6, 7\}$
(b) $\{a, b, c, d, e, f, g, h\}$

EXERCISES

1 Write down the sample space of:

(a) choosing a suit from a deck of playing cards
(b) selecting a month of the year

(a) $\{*, *, *, *\}$
(b) $\{*, *, *, *, *, *, *, *, *, *, *, *\}$

2 Describe each of the following sets in an alternative way:

(a) $\{-4, -2, 0, 2, 4\}$
(b) $\{x\text{: } x \text{ is a colour of the rainbow}\}$

(a) $\{x\text{: } x \text{ is even and } *\}$
(b) $\{*, *, *, *, *, *, *\}$

3 Which of the following are true statements?

(a) $\phi \in \{\{\phi\}\}$
(b) $\{\phi\} \in \{\{\phi\}, \phi\}$
(c) $2/3 \in \{x\text{: } x \text{ is irrational}\}$

(a) ϕ is a * of every * so this statement is *.
(b) $\{\phi\}$ is the * which is * in the right-hand set, in which case this statement is *.
(c) 2/3 is/is not irrational so the statement is *.

4 List the subsets of the set $\{1, \{1\}\}$.

$*, \{1\}, \{*\}, \{*\}$.

5 Which of the following are true statements?

(a) $\{x\text{: } x^2 - 4 = 0\} \subseteq \{y\text{: } y^3 - 8 = 0\}$
(b) $\{a, b, c, d\} \subset \{a, b, c, d, e\}$

(a) the solution of $y^3 - 8 = 0$ is $y = *$ and the solution of $x^2 - 4 = 0$ is * so the statement is *.
(b) all the * of the set on the left are also * of the set on the right and so the statement is *.

6 Define suitable universal sets for each of the following random experiments:

(a) selecting a working or defective component from an assembly line
(b) selecting a month of the summer

(a) $\{*, *\}$
(b) $\{*, *, *\}$

Unit 2 Outcomes and Events

Test yourself with the following:

1 Construct all possible simple events from each of the following sample spaces:

(a) $\{1, 2, 3\}$
(b) {king, spade, diamond, club}

2 Construct all possible simple events from each of the following random experiments:
   (a) selecting a colour of the rainbow
   (b) selecting a letter from the last six letters of the alphabet

3 Draw an outcome tree for each of the following random experiments:
   (a) selecting a vowel
   (b) choosing a colour from the sample space {red, orange, green, blue, white}

4 Construct all possible simple events from the following sequence of random experiments:
   (a) A bowl contains twelve salt and vinegar crisps and a single cheese and onion crisp.
      (i) a crisp is taken out of the bowl, its flavour recorded and then it is replaced
      (ii) a second crisp is selected and its flavour recorded.
   (b) What is the result if the first crisp selected is eaten before the second crisp is selected?

Range of Outcomes

While the result of a random experiment will be a single outcome we may not be interested in the specific outcome but in whether the outcome lies within a range of possible outcomes. For example, if we roll a six-sided die we may be interested to know not the actual value but whether the value showing on the uppermost face is less than 4. Here the range of possible outcomes is:

1, 2 or 3

Events

To cater for ranges of possible outcomes we define an **event**. An event is a subset of the sample space. In the example just considered we would define event E as:

$$E = \{1, 2, 3\}$$

An event is a subset of the sample space and different events may or may not be mutually exclusive. An event containing a single outcome is called a **simple event**. For example, for the random experiment of tossing a coin, the sample space is:

{H, T}

and the collection of mutually exclusive subsets

{H} and {T}

are the only two possible simple events associated with the random experiment. All random experiments have a unique collection of simple events. However, they are not always the only events that can be defined. In the random experiment of throwing a die we could define the events:

$E1$ An even number
$E2$ A number less than four.

These events are then described by the sets

$$E1 = \{2, 4, 6\} \text{ and } E2 = \{1, 2, 3\}$$

which are not mutually exclusive because the number 2 is common to both.

Sequences of Random Experiments

When two or more random experiments are performed one after the other, the final outcomes of the sequence of experiments will consist of combinations of the outcomes of the individual experiments. For example, consider the two random experiments of tossing a silver coin followed by tossing a copper coin. We shall denote the sample space of the first by

{SH, ST}

where S stands for silver, and of the second by

{CH, CT}

where C stands for copper. We can describe this sequence of experiments by using an **outcome tree**.

Experiment 1, represented in Figure 25.1 by the square containing the number 1, is performed first and the two possible outcomes of its result are represented by the two branches labelled SH and ST respectively. Experiment 2, represented by the square containing the number 2, is then performed and the two possible outcomes of its result are represented by the two branches labelled CH and CT. Notice that there are two sets of two branches for experiment 2 because of the two possible outcomes from experiment 1.

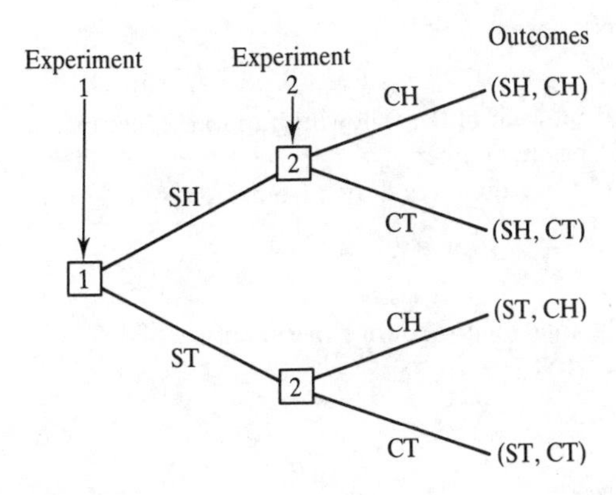

Figure 25.1

Each experiment has two possible outcomes so the sequence of two experiments has four possible combined outcomes and each combined outcome is represented by a specific route through the tree. For example, if the first experiment produced an outcome SH and the second an outcome CH then the final, combined outcome is SH and CH, represented by the topmost route through the tree. Each route is labelled with the two combined outcomes along that route.

From the outcome tree it can be seen that the sample space of the random experiment consisting of these two experiments performed in sequence is:

{(SH, CH), (SH, CT), (ST, CH), (ST, CT)}

Combining Outcomes into Events

Events are not unique and are created from the set of outcomes according to the needs of the particular problem at hand. From the previously described sequence of experiments we shall define the following events:

$E1$ One silver head
$E2$ Two heads
$E3$ At least one tail

Considering the sample space we find that:

$$E1 = \{(\text{SH}, \text{CH}), (\text{SH}, \text{CT})\}$$
$$E2 = \{(\text{SH}, \text{CH})\}$$

and

$$E3 = \{(\text{SH}, \text{CT}), (\text{ST}, \text{CH}), (\text{ST}, \text{CT})\}$$

Notice that $E2 \subset E1$ – the one element in $E2$ is also an element of $E1$.

EXAMPLES

1 Construct all possible simple events from each of the following sample spaces:

(a) {sock, shirts, trousers, skirts}
(b) $\{-2, -1, 0, 1, 2\}$

(a) {sock}, {shirts}, {trousers}, {skirts}
(b) $\{-2\}, \{-1\}, \{0\}, \{1\}, \{2\}$

2 Construct all possible simple events from each of the following random experiments:

(a) selecting a single-digit prime number
(b) finding the averages of three numbers drawn from the set $\{1, 2, 3, 4\}$

(a) $\{2\}, \{3\}, \{5\}, \{7\}$
(b) $\{2\}, \{2.333\ldots\}, \{2.666\ldots\}, \{3\}$

3 Draw an outcomes tree for each of the following random experiments:

(a) choosing a day of the week

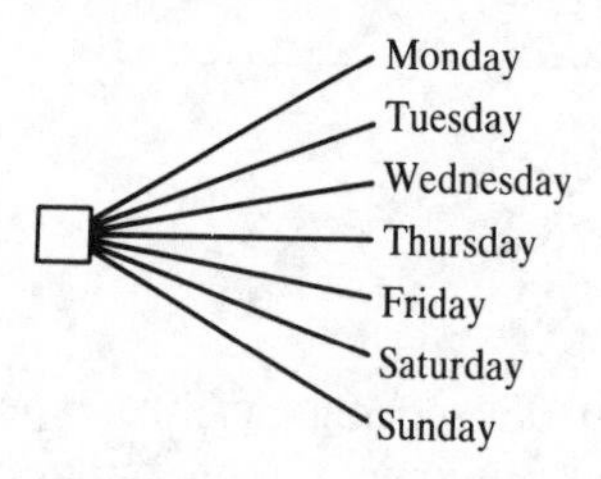

(b) answering a Yes/No question on a questionnaire

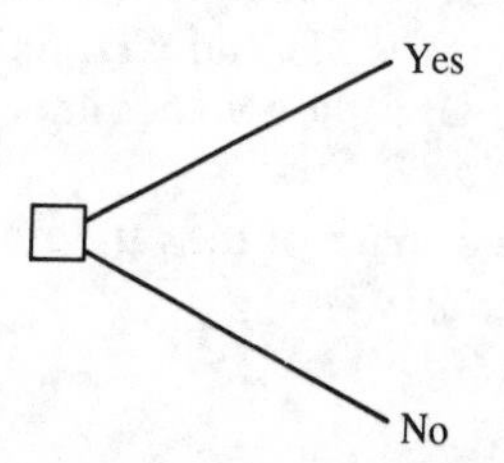

4 Construct all possible simple events from the following sequence of random experiments:
From a bag containing 3 red balls and 1 blue ball

(i) take out one ball, record its colour and then replace it
(ii) take out another ball and record its colour

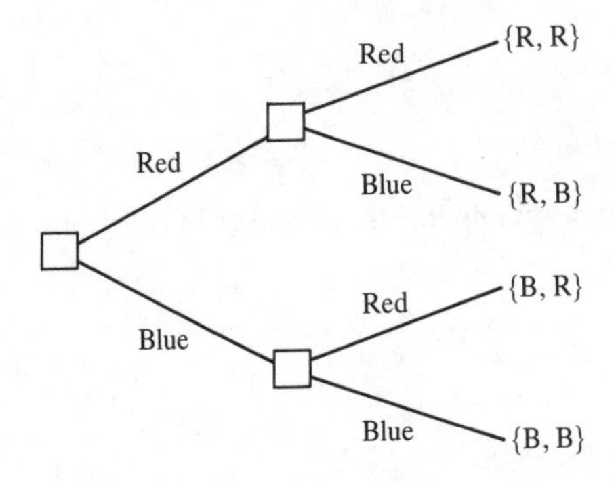

{R, R}, {R, B}, {B, B}. (Note that {R, B} and {B, R} are the same event as order is not specified.)

(iii) what is the effect of not replacing the first ball before taking the second?

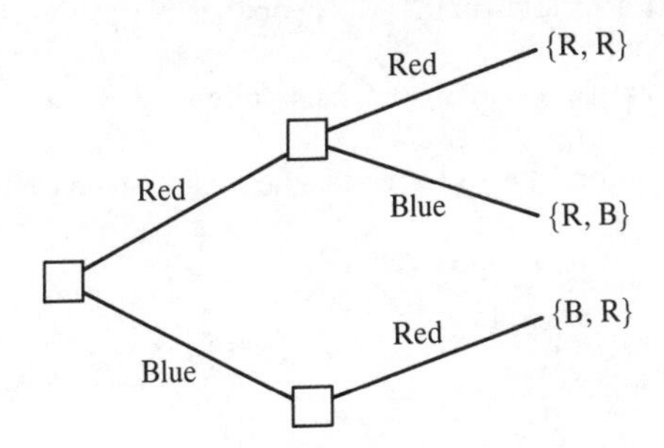

{R, R}, {R, B} – if a blue ball is taken out of the bag in the first selection and not replaced then it is not possible to take out a second blue ball.

EXERCISES

1 Construct all possible simple events from each of the following sample spaces:

(a) $\{a, b, c\}$
(b) {red, blue, yellow, green}

(a) $\{*\}$, $\{b\}$, $\{*\}$
(b) {red}, {*}, {*}, {*}

2 Construct all possible simple events from each of the following random experiments:

(a) selecting an even number between 3 and 12
(b) choosing a vowel

(a) {*}, {6}, {*}, {*}
(b) {*}, {*}, {i}, {*}, {*}

3 Draw an outcomes tree for each of the following random experiments:

(a) selecting a positive, even number less than 10

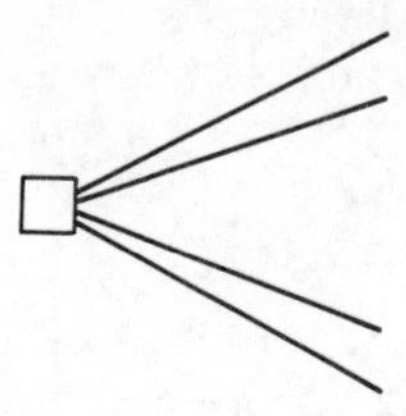

(b) choosing an item from an assembly line that is either in-order or defective

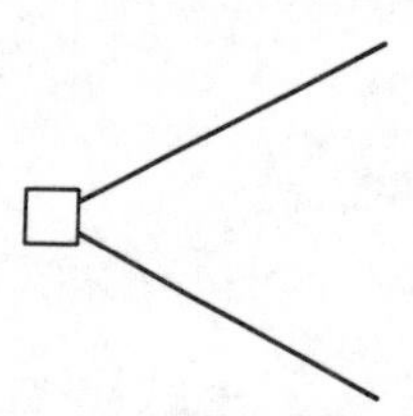

4 There are 5 black taxis and 1 white taxi at a taxi rank. Two people arrive at the same time and book a taxi one immediately after the other.

(a) Construct all possible simple events from this following sequence of random experiments.

(b) What is the result if the second person arrived after the first taxi had returned to the taxi rank?

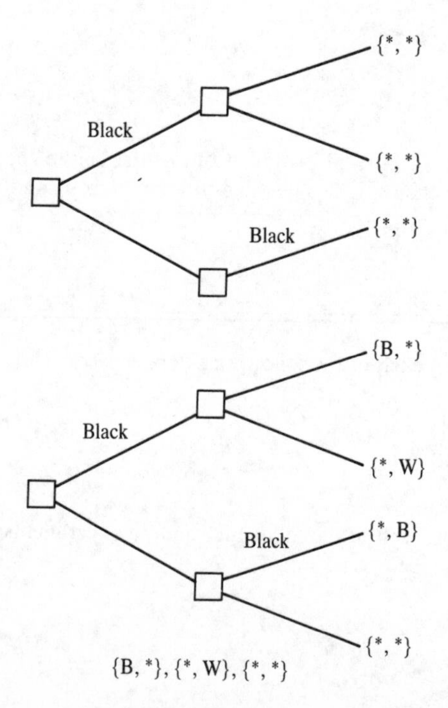

Unit 3 Combining Events

Test yourself with the following:

1 Find the union, intersection and complement of the following pairs of sets:

$\{1, 2, 3, 4\}$ and $\{-1, 0, 1\}$ where the universal set is the set of integers.

2 Draw the Venn diagram to represent each of the following:

(a) $(A \cup B) \cap C$

(b) $A' \cap B'$

3 From the random experiment of rolling a die find the union, intersection and complement of the following events:

$E1 = \{\text{odd number} \geqslant 3\}$ and $E2 = \{\text{even number} < 4\}$

4 Construct events according to the following requirements:

A die is rolled and it is required to know if it displays a 1 or an even number on its uppermost face.

Union, Intersection and Complement

Events can be combined using *or* and *and*. For example, in the previous sequence of experiments of tossing a silver and a copper coin we could define the events:

*E*4: One silver head *or* At least one tail

in which case:

$$E4 = \{(SH, CH), (SH, CT), (ST, CH), (ST, CT)\}$$

Here *E*4 consists of all the elements of *E*1 and all the elements of *E*3 – with no repetition. We say that *E*4 is the **union** of *E*1 and *E*3 and symbolically write this as:

$$\begin{aligned} E4 &= E1 \cup E3 \\ &= \{(SH, CH), (SH, CT)\} \cup \{(SH, CT), (ST, CH), (ST, CT)\} \\ &= \{(SH, CH), (SH, CT), (ST, CH), (ST, CT)\} \end{aligned}$$

where the symbol $\cup$ stands for the union of the two sets.

Notice that *E*4 is the sample space of the random experiment. We could further define the event *E*5 where:

*E*5: A silver head *and* A copper tail

in which case *E*5 consists of just the single element:

$$E5 = \{(SH, CT)\}$$

Here *E*5 consists of that element of *E*1 that is also in *E*3. We say that *E*5 is the **intersection** of *E*1 and *E*3 and symbolically write this as:

$$\begin{aligned} E5 &= E1 \cap E3 \\ &= \{(SH, CH), (SH, CT)\} \cap \{(SH, CT), (ST, CH), (ST, CT)\} \\ &= \{(SH, CT)\} \end{aligned}$$

where the symbol $\cap$ stands for the intersection of the two sets – the one element $E1$ and $E3$ have in common. Finally, we define the event $E6$ where

$E6$: No silver head

in which case:

$E6 = \{(\text{ST, CH}), (\text{ST, CT})\}$

Here $E6$ consists of all the elements of the sample space that are *not* elements of $E1$. We say that $E6$ is the **complement** of $E1$ and symbolically write this as:

$E6 = E1'$

where the symbol $'$ stands for the complement of the set $E1$.

Union, intersection and complement are **operations** and though they have been applied here between events of a sample space they are in fact operations between any sets. For example, if

$A = \{a, b, c, d\}$ and $B = \{c, d, e, f\}$

where the universal set is:

$\text{U} = \{a, b, c, d, e, f, g, h\}$

then the union of A and B is:

$A \cup B = \{a, b, c, d, e, f\}$

the set of elements in A *or* B. The intersection of A and B is:

$A \cap B = \{c, d\}$

the set of those elements in A *and* B. The complement of A is:

$A' = \{e, f, g, h\}$

the set of those elements in the universal set that are *not* in A.

Venn Diagrams

There is a very useful diagrammatic aid to manipulating sets called the Venn diagram. The universal set – or sample space – is represented by a rectangle and any subset – or event – is represented by a circle enclosed within the rectangle. In the Venn diagram in Figure 25.2 the sample space

$\{(\text{SH, CH}), (\text{SH, CT}), (\text{ST, CH}), (\text{ST, CT})\}$

is represented by the rectangle and the event $E1$ is represented by the circle.

Figures 25.3–25.5 represent $E4 = E1 \cup E3$, $E5 = E1 \cap E3$ and $\text{E}6 = E1'$, respectively.

$E4 = E1 \cup E3 = \{(\text{SH, CH}), (\text{SH, CT}), (\text{ST, CH}), (\text{ST, CT})\}$. The union of $E1$ and $E3$ consists of all the elements of $E1$ and $E3$ combined together. Notice that the element (SH, CT) is common to both but when the union is listed it is not repeated.

$E5 = E1 \cap E3 = \{(\text{SH, CT})\}$. The intersection of $E1$ and $E3$ consists of the element that is in both $E1$ and $E3$.

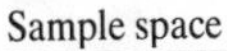

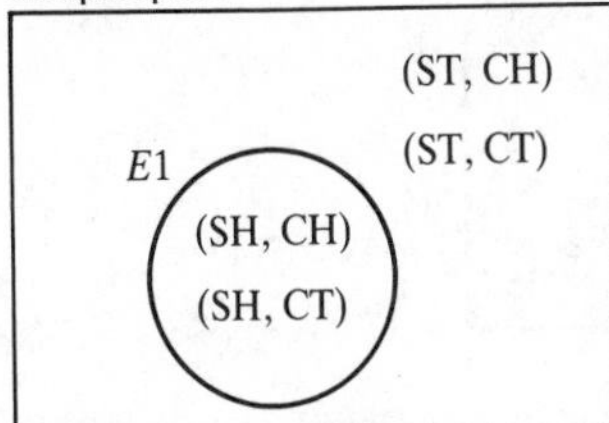

Figure 25.2

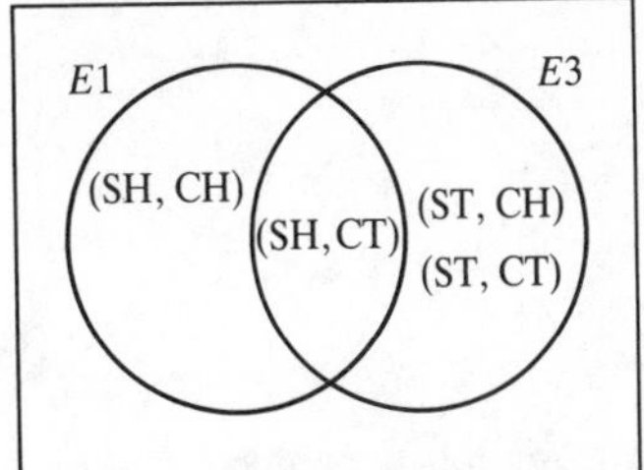

Notice: $E1 \cup E3$ = sample space

Figure 25.3

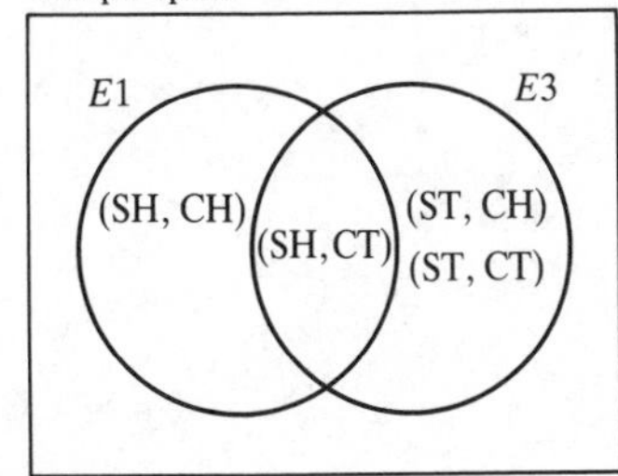

Notice: $E1 \cap E3 = \{(\text{SH, CT})\}$

Figure 25.4

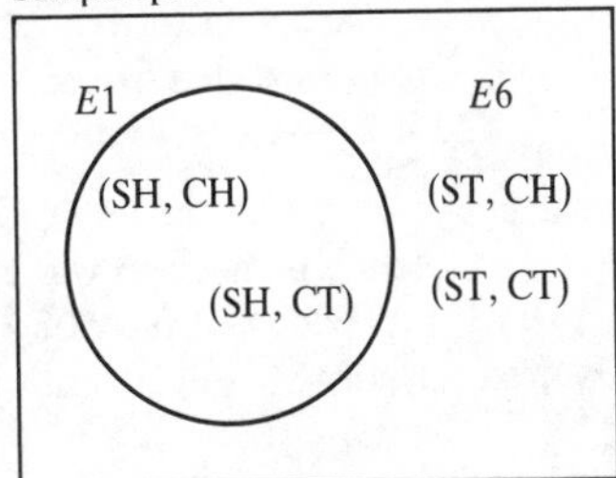

Notice: $E6 = E1' = \{(\text{ST, CH}), (\text{ST, CT})\}$

Figure 25.5

$E6 = E1' = \{(\text{ST, CH}), (\text{ST, CT})\}$. The complement of $E1$ consists of those elements of the sample space that are not in $E1$. In this case we have not represented $E6$ by a circle in order to stress the fact that $E6$ contains those elements of the sample space *not in* $E1$.

EXAMPLES

1 Find the union, intersection and complement of the following pair of sets:

{red, orange, green} and {blue, yellow, indigo} where the universal set is the set of colours of the rainbow.

{red, orange, green} ∪ {blue, yellow, indigo} = {red, orange, green, blue, yellow, indigo}
{red, orange, green} ∩ {blue, yellow, indigo} = ϕ
{red, orange, green}′ = {blue, yellow, indigo, violet}
{blue, yellow, indigo}′ = {red, orange, green, violet}

2 Draw the Venn diagram to represent each of the following:

(a) $(A \cap B) \cup C'$

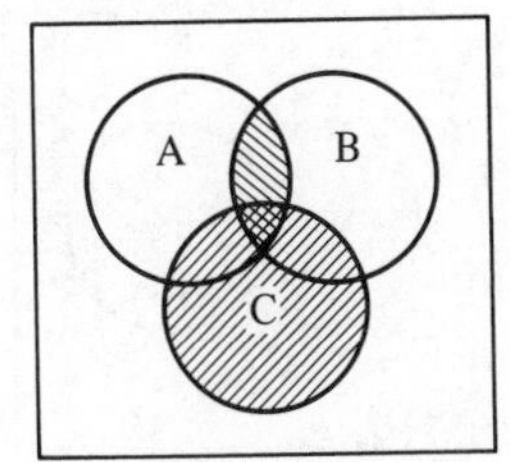

(b) $(A \cup B)'$

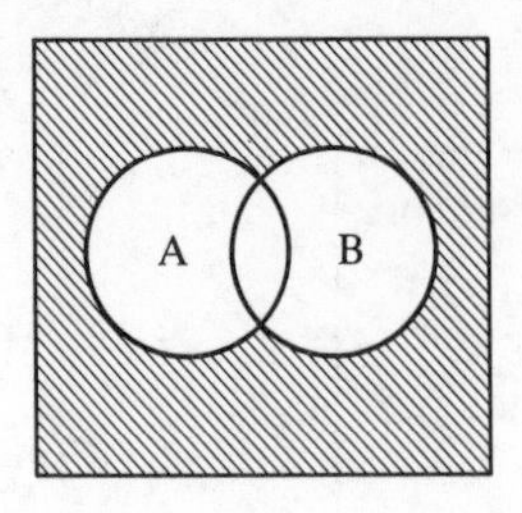

3 From the random experiment of selecting a card from a deck of playing cards find union, intersection and complement of the events:

$E1 = \{\text{court card}\}$ and $E2 = \{\text{red card}\}$

$E1 \cup E2 = \{\text{court card or a red card}\}$, $E1 \cap E2 = \{\text{red court card}\}$, $E1' = \{\text{numbered card}\}$, $E2' = \{\text{black card}\}$

4 Construct events according to the following requirements:

Items are selected from a supermarket shelf according to the criteria that they are made in the UK and cost from 50p to 99p.

Universal set $\text{U} = \{x\text{: } x \text{ is an item on a supermarket shelf}\}$
$E1 = \{x\text{: } x \in \text{U and } x \text{ is made in the UK}\}$
$E2 = \{x\text{: } x \in \text{U and } x \text{ costs from 50p to 99p}\}$
The required selection is an element of $E1 \cap E2$.

EXERCISES

1 Find the union, intersection and complement of the following pairs of sets:

{a, c, e, f, h} and {d, e, f, g} where the universal set contains the first ten letters of the alphabet.

$\{a, c, e, f, h\} \cap \{d, e, f, g\} = \{*, *\}$
$\{a, c, e, f, h\} \cup \{d, e, f, g\} = \{*, *, *, *, *, *, *\}$
$\{a, c, e, f, h\}' = \{*, *, *, *, *\}$
$\{d, e, f, g\}' = \{*, *, *, *, *, *\}$

2 Draw the Venn diagram to represent each of the following:

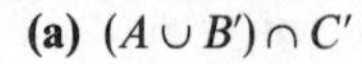
(a) $(A \cup B') \cap C'$

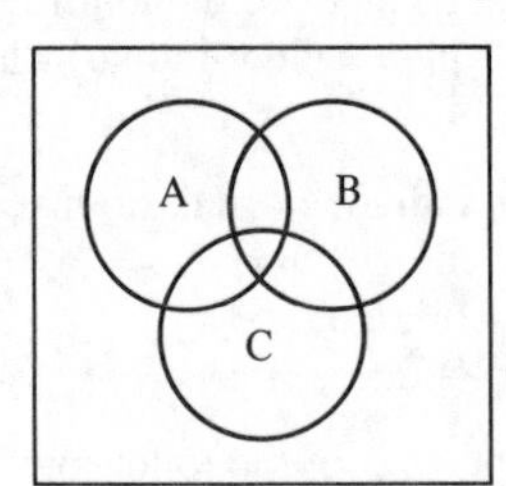

(b) $A' \cup B'$

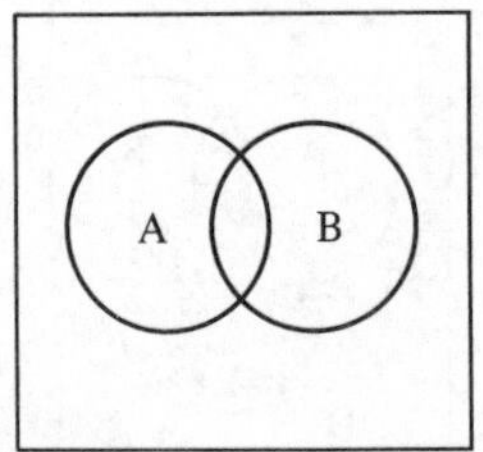

3 From the random experiment involving picking a ball from a bag containing 6 red balls numbered with a 1, 4 blue balls numbered with a 2 and 2 green balls numbered with a 3 find union, intersection and complement of the events:

$$E1 = \{\text{red ball}\} \text{ and } E2 = \{\text{number} \leqslant 2\}$$

$E1 \cup E2 = \{* \text{ or } *\}$, $E1 \cap E2 = \{\text{red number } *\}$, $E1' = \{* \text{ or } * \text{ ball}\}$,
$E2' = \{\text{number } *\}$

4 Construct events according to the following requirements:

A car is to be purchased from among {Volvo, Toyota, Ford} whose colour is from {red, white, blue}, and what is required is either a Toyota or a white Ford.

Universal set $U = \{x: x \text{ is a } *\}$
$E1 = \{x: x \in U \text{ and } x \text{ is a } *\}$, $E2 = \{x: x \in U \text{ and } x \text{ is a } *\}$, $E3 = \{x: x \in U \text{ and } x \text{ is a } *\}$
$E4 = \{x: x \in U \text{ and the colour of } x \text{ is } *\}$, $E5 = \{x: x \in U \text{ and the colour of } x \text{ is } *\}$
$E6 = \{x: x \in U \text{ and the colour of } x \text{ is } *\}$
Selection is an element of $(E*)*[(E*)*(E*)]$

Unit 4 Set Properties

Test yourself with the following:

1 Use a Venn diagram to demonstrate the validity of $(A \cap B) \cup A' = B \cup A$

2 Show that $(A' \cup B) \cap A = B \cap A$ for

$A = \{a, b, c, d, e\}$, $B = \{d, e, f, g, h\}$ and $C = \{h, j\}$ where the universal set
$U = \{a, b, c, d, e, f, g, h, j, k, l\}$

3 100 hundred people applied for a job in a computer service bureau. Of the 100, 15 were left-handed, 48 were male and 63 had previous experience in information technology. Of the 15 left-handers, 3 were women with past experience in IT, 6 were men without past experience in IT and 2 were men with past experience of IT. Of the women a total of 30 had past experience of IT. How many women were there that were right-handed with no previous experience of information technology?

Set Algebra

Sets combine together using the operations of union and intersection according to a collection of rules known as the **rules of set algebra**. These are described as follows:

Commutative Rule

It does not matter whether set A is unioned with B or vice versa, the resulting set is the same. The same applies to intersection:

$$A \cup B = B \cup A \text{ and } A \cap B = B \cap A$$

Associative Rule

The order in which sets are unioned does not matter; the resulting set is the same. Again, the same applies to intersection:

$$A \cup (B \cup C) = (A \cup B) \cup C \text{ and } A \cap (B \cap C) = (A \cap B) \cap C$$

To demonstrate the associativity of union consider the sets:

$A = \{1, 2, 3\}$, $B = \{2, 3, 4, 5\}$ and $C = \{1, 3, 5, 7\}$

then:

$$\begin{aligned} A \cup (B \cup C) &= \{1, 2, 3\} \cup (\{2, 3, 4, 5\} \cup \{1, 3, 5, 7\}) \\ &= \{1, 2, 3\} \cup \{1, 2, 3, 4, 5, 7\} \\ &= \{1, 2, 3, 4, 5, 7\} \end{aligned}$$

Similarly:

$$\begin{aligned} (A \cup B) \cup C &= (\{1, 2, 3\} \cup \{2, 3, 4, 5\}) \cup \{1, 3, 5, 7\} \\ &= \{1, 2, 3, 4, 5\} \cup \{1, 3, 5, 7\} \\ &= \{1, 2, 3, 4, 5, 7\} \\ &= A \cup (B \cup C) \end{aligned}$$

Distributivity Rule

Union is distributed over intersection and intersection is distributed over union:

$A \cup (B \cap C) = (A \cup B) \cap (A \cup C)$ and $A \cap (B \cup C) = (A \cap B) \cup (A \cap C)$

The latter can be seen from Figure 25.6.

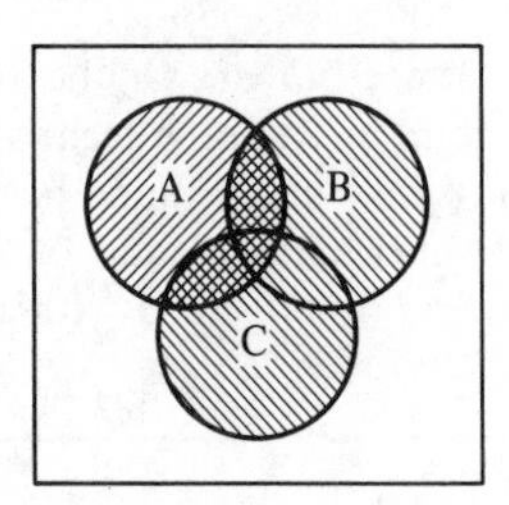

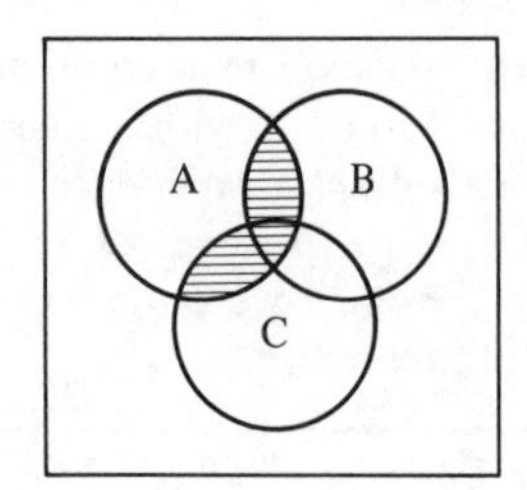

Figure 25.6

De Morgans Rules

The complement of a union of sets is the intersection of the complements of each set

$(A \cup B)' = A' \cap B'$

The complement of an intersection of sets is the union of their complements .

$(A \cap B)' = A' \cup B'$

To demonstrate the first of the De Morgans rules let:

$A = \{\text{Red, Yellow, Violet}\}$ and $B = \{\text{Orange, Yellow, Blue, Indigo}\}$

where the universal set is:

$U = \{\text{Red, Orange, Yellow, Green, Blue, Indigo, Violet}\}$

$$(A \cup B)' = (\{\text{Red, Yellow, Violet}\} \cup \{\text{Orange, Yellow, Blue, Indigo}\})'$$
$$= \{\text{Red, Orange, Yellow, Blue, Indigo, Violet}\}'$$
$$= \{\text{Green}\}$$

and:

$$A' \cap B' = \{\text{Red, Yellow, Violet}\}' \cap \{\text{Orange, Yellow, Blue, Indigo}\}'$$
$$= \{\text{Orange, Green, Blue, Indigo}\} \cap \{\text{Red, Green, Violet}\}$$
$$= \{\text{Green}\}$$
$$= (A \cup B)'$$

A number of additional rules are self evident:

The union of set A with the empty set is A and the intersection of set A with the empty set is the empty set.

$$A \cup \phi = A \text{ and } A \cap \phi = \phi$$

The union of set A with the universal set is the universal set and the intersection of set A with the universal set is set A

$$A \cup \mathrm{U} = \mathrm{U} \text{ and } A \cap \mathrm{U} = A$$

The union of A with its complement is the universal set and the intersection of set A with its complement is the empty set.

$$A \cup A' = \mathrm{U} \text{ and } A \cap A' = \phi$$

The union or intersection of set A with itself is itself

$$A \cup A = A \text{ and } A \cap A = A$$

The double complement rule: the complement of the complement of A is A

$$(A')' = A$$

EXAMPLES

1 Use a Venn diagram to demonstrate the validity of $(A \cap B) \cup A' = B \cup A'$.

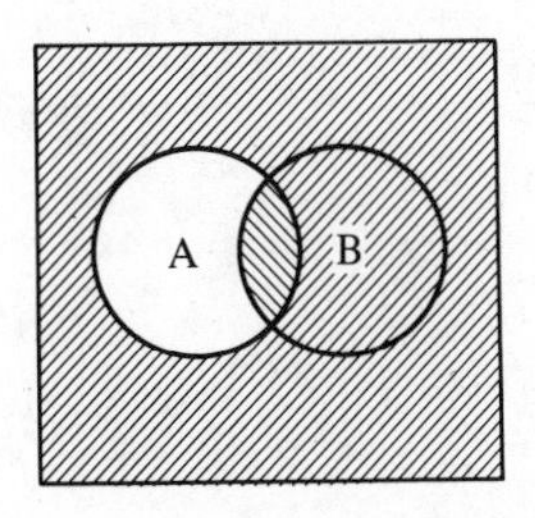

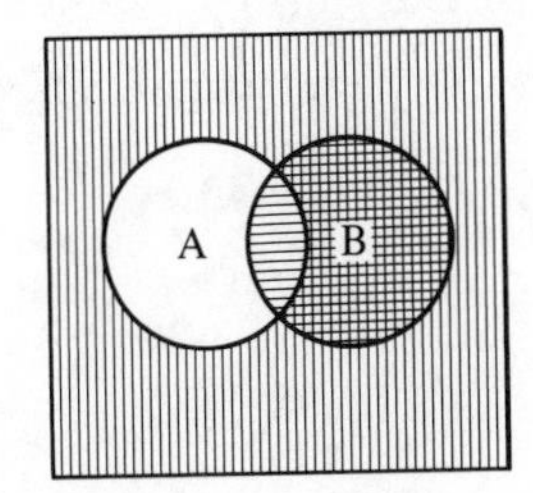

2 Show that $(A \cup B') \cap (A' \cup B) = (A \cap B) \cup (B \cup A)'$ for

$A = \{a, b, c, d, e\}$ and $B = \{d, e, f, g, h\}$
where the universal set $\mathrm{U} = \{a, b, c, d, e, f, g, h, k, l\}$

$A' = \{f, g, h, k, l\}$ and $B' = \{a, b, c, k, l\}$ so that:
$(A \cup B') = \{a, b, c, d, e\} \cup \{a, b, c, k, l\} = \{a, b, c, d, e, k, l\}$ and
$(A' \cup B) = \{f, g, h, k, l\} \cup \{d, e, f, g, h\} = \{d, e, f, g, h, k, l\}$ so that
$(A \cup B') \cap (A' \cup B) = \{d, e, k, l\}$
Also:
$(A \cap B) = \{a, b, c, d, e\} \cap \{d, e, f, g, h\} = \{d, e\}$ and
$(B \cup A)' = (\{d, e, f, g, h\} \cup \{a, b, c, d, e\})' = \{a, b, c, d, e, f, g, h\}' = \{k, l\}$ so that
$(A \cap B) \cup (B \cup A') = \{d, e\} \cup \{k, l\} = \{d, e, k, l\} = (A \cup B') \cap (A' \cup B)$

3 Of 1500 students in the faculty of engineering, mathematics and science of a university, 500 studied mathematics and science but no engineering, 450 studied mathematics and engineering but no science and 230 studied all three subjects. If mathematics was compulsory for engineering and science students how many students studied mathematics by itself?

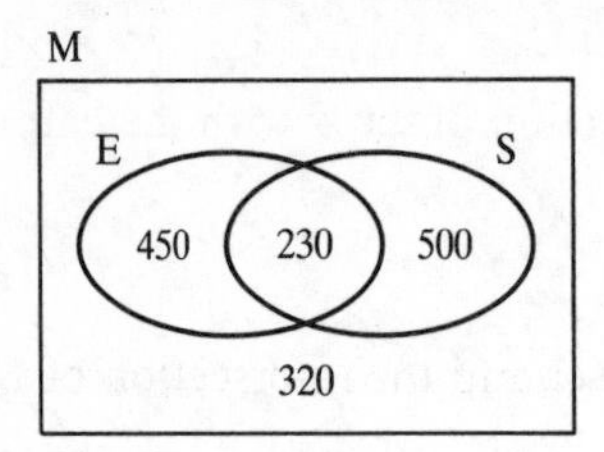

EXERCISES

1 Use a Venn diagram to demonstrate the validity of $(A \cap B') \cup B = A \cup B$.

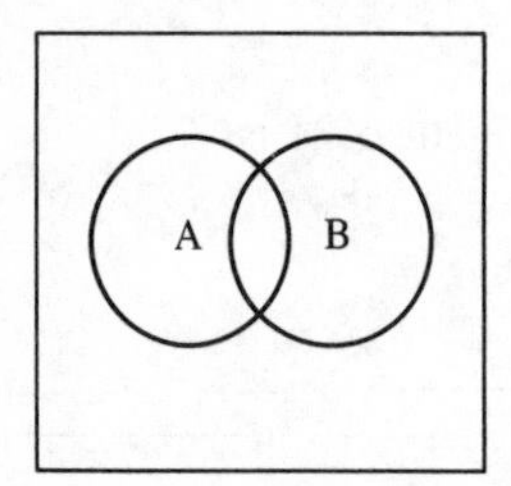

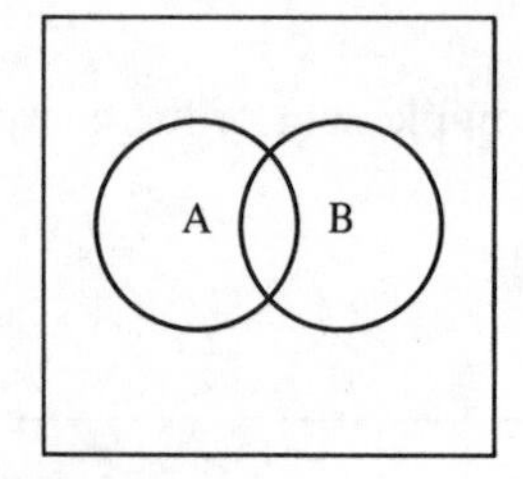

2 Show that: $(A \cap B') \cup (A' \cap B) = (A \cup B) \cap (A' \cup B')$ for

$A = \{a, b, c, d, e\}, B = \{d, e, f, g, h\}$ and $C = \{h, j\}$ where the universal set $U = \{a, b, c, d, e, f, g, h, j, k, l\}$

$A' = \{*, *, *, *, *, *\}$ and $B' = \{*, *, *, *, *, *\}$ so that:
$(A \cap B') = *$ and
$(A' \cap B) = *$ so that
$(A \cap B') \cup (A' \cap B) = *$
Also:
$(A \cup B) = * \cap * = *$ and
$(A' \cup B')' = * \cup *' = *' = *$ so that
$(A \cup B) \cap (A' \cup B')' = * \cup * = * = (A \cap B') \cup (A' \cap B)$

3 On one summer Sunday a theme park had 150 000 visitors. Of the 150 000, 94 000 were under 15 years old and were admitted at half price, 86 500 went on the thrill rides and 84 000 ate a meal

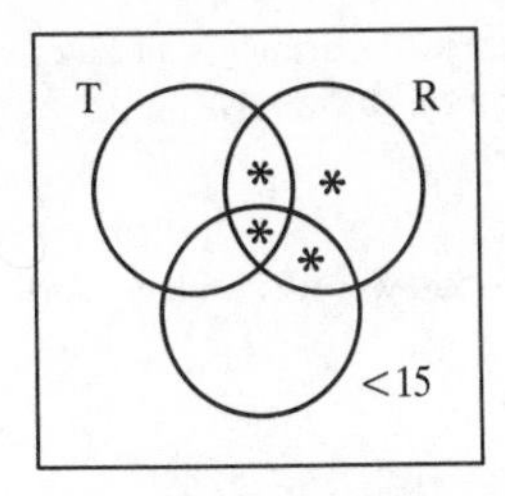

in one of the twelve restaurants. Of the under 15s, 28 000 went on the thrill rides and ate in the restaurants and of all those who ate in the restaurants 43 500 went on the thrill rides. Of the over 15s, 12 000 ate in a restaurant only and 24 000 went on a thrill ride only. How many of the over 15s neither ate in a restaurant nor went on a thrill ride?

Module 25: Further exercises

1 Write down the sample space of:
(a) buying a Volvo, Toyota or a Ford car (b) paying a bill by cash or by cheque

2 Describe each of the following sets in an alternative way:
(a) $\{0, 1\}$ (b) $\{x: x^2 - 5x + 6 = 0\}$

3 Which of the following are true statements?
(a) $\{1\} \in \{1, \{1\}, \{\{1\}\}\}$ (b) bus $\in \{x: x \text{ is a mode of transport}\}$ (c) $3 \in \{x: x \text{ is a complex number}\}$

4 List all the subsets of the set $\{\phi, \{\phi\}\}$.

5 Which of the following are true statements?
(a) $\{\phi\} \subseteq \{\phi, \{\phi\}\}$ (b) $\{1\} \subset \{1, \{1\}\}$

6 Define suitable universal sets for each of the following random experiments:
(a) setting three on/off switches (b) choosing a chapter from a book containing 6 chapters

7 Construct all possible simple events from each of the following sample spaces:
(a) $\{W, X, Y, Z\}$ (b) {Monday, Wednesday, Friday, Sunday}

8 Construct all possible simple events from each of the following random experiments:
(a) selecting a colour from the colours in the Union Jack (b) selecting a positive integer less than 9

9 Draw an outcomes tree for each of the following random experiments:
(a) rolling a six-sided die (b) selecting a court card from a deck of playing cards

10 Construct all possible simple events from the following sequence of random experiments:
(a) From a bowl containing 3 blue chips displaying the letter A, 4 red chips displaying the letter B and 1 red chip displaying the letter C
(i) take out one chip, record its colour and then replace it
(ii) take out another chip and then record the letter displayed
(b) Repeat without replacement after (i).

11 Find the union, intersection and complement of the following pairs of sets:
$\{x: x \text{ is a colour of the rainbow}\}$ and $\{x: x \text{ is a colour of the Union Jack}\}$ where the universal set is the set of all colours.

12 Draw the Venn diagram to represent each of the following:
(a) $(A' \cap B) \cup C'$ (b) $(A \cap B)'$

13 From the random experiment, selecting an employee from among the three departments of sales, management and finance find union, intersection and complement in words from the events:

$E1 = \{\text{male}\}$ and $E2 = \{\text{management}\}$

14 Construct events according to the following requirements:

A card is selected from a deck of playing cards and it is required to know if it is a black card or a court card.

15 Use a Venn diagram to demonstrate the validity of:

$(A \cap B) \cup (A' \cap B') = (A' \cup B) \cap (A \cup B')$

16 Show that $(A \cup B) \cap C = (A \cap C) \cup (B \cup C)$ for

$A = \{a, b, c, d, e\}$, $B = \{d, e, f, g, h\}$ and $C = \{h, j\}$ where the universal set $\text{U} = \{a, b, c, d, e, f, g, h, j, k, l\}$

17 In a school of 1200 pupils, 650 were boys, 400 stayed for lunch and 380 were juniors. Of the juniors, 75 were boys who stayed for lunch and 85 were girls who did not stay for lunch. 405 of the pupils were senior boys of whom 150 stayed for lunch. How many senior girls did not stay for lunch?

Module 26: Events and Probabilities

Aim: To appreciate the concept of probability and to be able to assign probabilities to simple and compound events that arise from a random experiment.

Objectives: When you have read this module you will be able to:

- Define the probabilities of simple events.
- Understand the use of the connective 'OR' and the concept of the mutual exclusivity of simple outcomes from a random experiment.
- Understand the use of the connective 'AND' and the concept of the interdependence of the events of successive random experiments.
- Construct and interpret probability trees.

Unit 1 Probabilities of Simple Events

Test yourself with the following:

1 Every card in a deck of 52 has an equal chance of being selected. Find the probability of selecting:
(a) an even numbered card (b) a red Six (c) the Ace of clubs

2 What is the *a priori* probability of:
(a) a given name being drawn from a hat containing 60 names if every name has the same chance of being chosen
(b) selecting a date from the month of August
(c) picking a word at random from a dictionary

3 What is the probability that:
(a) an opened yogurt will prove to be inedible if, from past experience, 2 out of every 13 bought are inedible
(b) your new car will break down within 10 000 miles if AA statistics claim that 16 out of every 132 of your model breaks down within the first 10 000 miles
(c) selecting a plumber that is immediately available from a list of 25 plumbers if at any time it is estimated that only 5% are immediately available

4 A bag contains 7 black beads, 6 white beads and 11 grey beads. A bead is selected and every bead has an equal probability of being chosen. What are the probabilities of each of the following events:
(a) $E1 = \{\text{black bead}\}$ (b) $E2 = \{\text{white or grey bead}\}$
(c) $E3 = \{\text{not a white bead}\}$

5 An experienced typist strikes, at random, three letter keys from the top row of letters on a typewriter keyboard. What is the probability that they print out TOP?

Probability

Murphy's law states that buttered toast always lands butter-side down. As an expression of a law of nature it is a pretty poor one but as an expression of luck and the law of chance it is a good attempt. Chance is defined in the Oxford dictionary as the *absence of design or discoverable cause* and many events that we observe seem to happen by chance. Indeed, some people attempt to measure chance in their daily lives; they refuse to walk under a ladder, they throw salt over their left shoulder, they bet money on a horse to win a race. In all of these activities there is an in-built measure of likelihood or **probability**. If you walk under a ladder you feel that you are likely to be unlucky. If you throw salt over your left shoulder you feel that you are likely to avoid bad luck. If you bet on a horse winning a race what is the probability of it winning? Who is to know? There are so many factors involved it is impossible to say with certainty. All you can do is to **weight the odds** – compare the probability of your horse winning against the probability of any other horse winning.

Consider the random experiment of tossing a coin that has a head on both sides – a double-headed coin. If you toss this coin then it is **certain** to fall down heads. The event $\{H\}$ is a certain event and we define the probability of a certain event as unity:

$$\Pr(\{H\}) = \Pr(\text{certainty}) = 1$$

Similarly, it is impossible for it to fall down tails – the event $\{T\}$ is an impossible event. We define the probability of an impossible event as zero:

$$\Pr(\{T\}) = \Pr(\text{impossibility}) = 0$$

If the double-headed coin is now replaced by a normal coin possessing a tail as well as a head then the event $\{H\}$ is no longer certain and the event $\{T\}$ is no longer impossible. In each case their probabilities lie between zero and unity:

$$0 < \Pr(\{H\}) < 1 \text{ and } 0 < \Pr(\{T\}) < 1$$

Assigning correct numbers to these probabilities is not simple but what we do say is that because when a normal coin is tossed it is certain to show either a head or a tail then the two probabilities add up to the probability of certainty, that is unity:

$$\Pr(\{H\}) + \Pr(\{T\}) = \Pr(\text{certainty}) = 1$$

Assigning Probabilities

Before a random experiment is performed there are a number of possible outcomes to the result. Because the actual outcome that occurs when the experiment is performed does so according to the rules of chance we can assign probabilities to outcomes or, more precisely, we can assign probabilities to events.

Probabilities can be assigned to events either *a priori* – beforehand – or by **statistical regularity** – afterwards. For example, if we make the assumption that a coin is fair then we have tacitly assigned probabilities beforehand because the notion of fairness means that there is an equal chance of the tossed coin falling down heads or tails. That is:

$$\Pr(\{H\}) + \Pr(\{T\}) = 1 \text{ and } \Pr(\{H\}) = \Pr(\{T\})$$

so that:

$$\Pr(\{H\}) = 1/2 = 0.50 \text{ and}$$

$$\Pr(\{T\}) = 1/2 = 0.50$$

If instead of assuming the coin to be fair we tossed it 1000 times and found that we ended up with 510 heads and 490 tails then we would say that, based on statistical regularity, the probabilities were defined as:

$$\frac{\text{number of times a simple event occurs in a sequence of trials}}{\text{total number of trials}}$$

Consequently,

$$\text{probability of a head, } \Pr(\{H\}) = 510/1000 = 0.51$$

$$\text{probability of a tail, } \Pr(\{T\}) = 490/1000 = 0.49$$

Remember that the probabilities must add up to 1.

As a further example, consider the random experiment of picking a ball out of a bag in which there are 1 white, 4 red and 5 blue balls – 10 balls in total. We shall define the possible outcomes as W, R and B so the sample space is given as:

$$\{W, R, B\}$$

We can define three simple events:

$$E1 = \{W\}$$

$$E2 = \{R\}$$

$$E3 = \{B\}$$

In order to assign probabilities to these events we picked a ball out of the bag, noted its colour and then replaced it. This process was repeated 1000 times and we found that the colours occurred with the following frequencies:

Colour	Frequency	Relative frequency
W	110	110/1000 = 0.11
R	380	380/1000 = 0.38
B	510	510/1000 = 0.51
Total	1000	1.00

The three relative frequencies are now used to define the probabilities of the three events:

$$p(E1) = 0.11$$

$$p(E2) = 0.38$$

$$p(E3) = 0.51$$

These three probabilities collectively form a **probability distribution** for the random experiment. We shall say more about probability distributions in Module 27, where we discuss how probabilities are assigned to combined events given the probabilities of the simple events related to a random experiment.

Summary

To every simple event of a random experiment we can assign a measure of the chance of that event occurring. This measure is called the probability of that event occurring.

The probability of a certain event is 1 and the probability of an impossible event is 0. Any other event that is neither certain nor impossible will have a probability between 0 and 1.

If the probabilities of all the simple events of a random experiment are added together their sum is 1.

Probabilities can be assigned to events *a priori* – beforehand – by weighing the odds and using the concept of fairness. Alternatively, probabilities can be assigned by statistical regularity by using the ratio:

$$\frac{\text{number of times a simple event occurs in a sequence of trials}}{\text{total number of trials}}$$

EXAMPLES

1 Every card in a deck of 52 has an equal chance of being selected. Find the probability of selecting:

(a) a red card **(b)** a black Ace **(c)** the Jack of diamonds

(a) Pr({red card)} = 26/52 = 0.5, as 26 of the 52 cards are red.
(b) Pr({black ace}) = 2/52. There are 4 aces and two of them are black.
(c) Pr({Jack of diamonds}) = 1/52, as there is only 1 jack of diamonds.

2 What is the *a priori* probability of:

(a) a given number showing on a roll of a fair, six-sided die.
(b) selecting a male from 100 people where 36 of them are female if everyone has the same chance of being chosen.
(c) picking a raffle ticket out of a hat if every ticket has an equal chance of being drawn.

(a) 1/6 as all sides have an equal chance of showing.
(b) 64/100 as there are 64 males.
(c) 1/(total number of tickets)

3 What is the probability that:

(a) a defective article is selected from an assembly line if, during the previous week, 32 defective items had been selected from a total of 10 000 items.
(b) tossing a head with a coin that regularly lands tail up 10 times every 25 tosses.
(c) selecting a hatted pedestrian from a shopping precinct if, on average, 12% of pedestrians have been recorded as wearing hats.

(a) 32/10 000 by statistical regularity.
(b) 10/25 by statistical regularity.
(c) 12/100 by statistical regularity.

4 A bag contains 3 red beads, 5 blue beads and 6 green beads. A bead is selected and every bead has an equal probability of being chosen. What are the probabilities of each of the following events?

(a) $E1$ = {red bead} $E2$ = {blue or green bead} **(c)** $E3$ = {not a green bead}

(a) Pr($E1$) = 3/14. There are 14 beads in total, of which 3 are red.
(b) Pr($E2$) = 11/14. 11 are either blue or green.
(c) Pr($E3$) = 8/14. There are only 8 that are not green.

5 3 balls numbered 1, 2 and 3 are placed at random in three slots, numbered 1, 2, 3. What is the probability that the numbers on the balls match the numbers on the slots?

The total number of different arrangements of the three balls in the three slots is:

$$3! = 6$$

of which only one is the matching arrangement. Thus the probability of a match is 1/6, assuming each arrangement is equally probable.

EXERCISES

1 Every card in a deck of 52 has an equal chance of being selected. Find the probability of selecting:
(a) a court card **(b)** a red Ten **(c)** the Queen of hearts

(a) Pr({court card}) = */52 = * as there are * of the 52 cards which are court cards.
(b) Pr({red Ten}) = */* = *. There are * Tens and * of them are red.
(c) Pr({Queen of hearts}) = * as there is *.

2 What is the *a priori* probability of:
(a) a given number showing on a ball selected from a bag containing 10 balls numbered 1 to 10.
(b) selecting a poem from a book of 120 poems if every poem has the same chance of being chosen.
(c) picking a compact disc from a cabinet of discs if every disc has an equal chance of being drawn.

(a) * as all * have an equal chance of being selected.
(b) */* as there are *.
(c) 1/*.

3 What is the probability that:
(a) a bus will arrive at a given bus stop within the next five minutes if one has just left and it has been recorded that 8 buses per hour leave from the stop.
(b) rolling 4 with a die that regularly rolls up an even number 6 times every 20 rolls.
(c) selecting a car that has overstayed its limit in a cark park if 17% of cars regularly overstay their time.

(a) * by statistical regularity.
(b) the probability of the number being even is * by statistical regularity. If even numbers have an equal chance then the probability of a 4 is *.
(c) * by statistical regularity.

4 A bag contains 8 purple beads, 2 yellow beads and 9 grey beads. A bead is selected and every bead has an equal probability of being chosen. What are the probabilities of each of the following events:
(a) $E1 = \{\text{grey bead}\}$
(b) $E2 = \{\text{yellow or grey bead}\}$
(c) $E3 = \{\text{not a yellow bead}\}$

(a) $\Pr(E1)$ = */*. There are * beads in total of which * are *.
(b) $\Pr(E2)$ = *. * are either * or *.
(c) $\Pr(E3)$ = *. There are only * that are *.

5 Five cards displaying, on both sides, the letters, B, R, A, C, E are thrown in the air and they all

land apart from each other. What is the probability that they spell the word BRACE when read from left to right.

The total number of different arrangements of the * is:

*! = *

of which * is the correct arrangement. Thus the probability of BRACE is *, assuming *.

Unit 2 The Connective 'OR'

Test yourself with the following:

1 A bead is drawn from a bag containing 6 red beads, 4 green beads, 2 yellow beads and 3 blue beads. What is the probability that a bead drawn at random is:
   (a) either a red bead or a blue bead
   (b) not a yellow bead and not a red bead

2 A card is drawn at random from a deck of 52 playing cards. What is the probability that it is:
   (a) either a red card or an ace
   (b) not a heart and not a Jack
   (c) either a black card or a King but not both

3 10 men and 6 women sit around a table. Of the men 6 have brown eyes, 3 have green eyes and one has blue eyes. Of the women, 5 have green eyes and 1 has blue eyes. An individual is selected at random. Find the probability that the person selected:
   (a) is male or has green eyes
   (b) is female or has brown eyes
   (c) is not male and does not have green eyes
   (d) is either female or has blue eyes.

OR and Union

Let $E1$ and $E2$ be two events associated with a random experiment. These two events can be connected via the set operation of **union** to form the event $E3$:

$$E3 = E1 \cup E2$$

Event $E3$ can be described in English as:

event $E1$ OR event $E2$

In other words:

either event $E1$ occurs *or* event $E2$ occurs or *both* occur

This is an **inclusive or** because it permits both events to occur simultaneously.

The probability of $E3$ occurring depends upon whether $E1$ and $E2$ are:

- Mutually exclusive events
- Not mutually exclusive events

Mutually exclusive events

If events $E1$ and $E2$ are mutually exclusive they contain no outcomes in common.

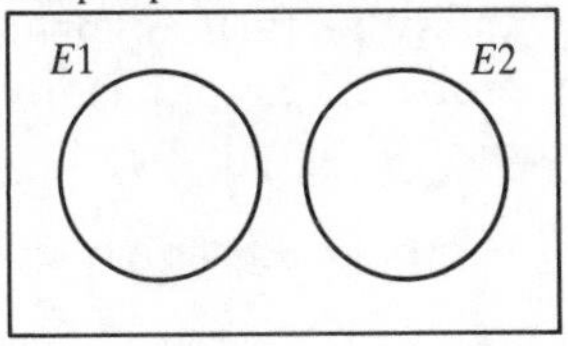

Sample space

$E1$ $E2$

Figure 26.1

In this case the probability of the union of $E1$ and $E2$ is equal to the sum of each probability:

$$\Pr(E1 \cup E2) = \Pr(E1) + \Pr(E2)$$

For example, consider the random experiment of drawing a card from a deck of 52 where the result is the suit of the card drawn. The sample space of this random experiment is:

$$\{\text{Heart, Spade, Club, Diamond}\}$$

which is the union of the four, mutually exclusive, simple events:

$$\{\text{Heart}\} \cup \{\text{Spade}\} \cup \{\text{Club}\} \cup \{\text{Diamond}\}$$

Because each suit has an equal chance of being selected the probability of each simple event is 0.25. That is:

$$\begin{aligned}
\Pr(\{\text{Heart}\}) &= 0.25 \\
\Pr(\{\text{Spade}\}) &= 0.25 \\
\Pr(\{\text{Club}\}) &= 0.25 \\
\Pr(\{\text{Diamond}\}) &= 0.25
\end{aligned}$$

The probability that the card drawn is *either* a Heart *or* a Diamond is then:

$$\begin{aligned}
\Pr(\{\text{Heart}) \cup \{\text{Diamond}\} &= \Pr(\{\text{Heart}\}) + \Pr(\{\text{Diamond}\}) \\
&= 0.25 + 0.25 \\
&= 0.5
\end{aligned}$$

Notice that the probability of the sample space is 1.

$$\begin{aligned}
\Pr(\{\text{Heart, Spade, Club, Diamond}\}) &= \Pr(\{\text{Heart}\} \cup \{\text{Spade}\} \cup \{\text{Club}\} \cup \{\text{Diamond}\}) \\
&= \Pr(\{\text{Heart}\}) + \Pr(\{\text{Spade}\}) + \Pr(\{\text{Club}\}) + \Pr(\{\text{Diamond}\}) \\
&= 0.25 + 0.25 + 0.25 + 0.25 \\
&= 1
\end{aligned}$$

This is a mathematical statement that the result is bound to occur – the sample space is certain to occur. By the same reasoning the probability of the empty set is zero:

$$\Pr(\phi) = 0$$

The empty set represents a null result, which is impossible.

Non-mutually Exclusive Events

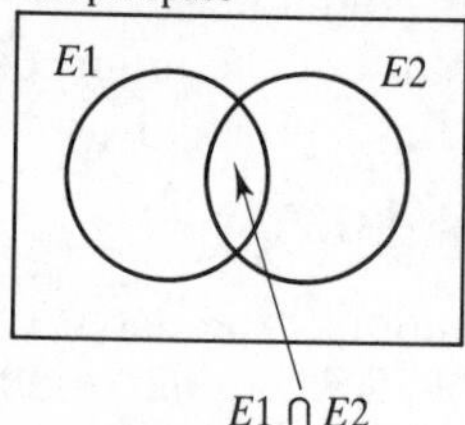

Figure 26.2

If events $E1$ and $E2$ do have outcomes in common then they are not mutually exclusive.

In this case the probability of their union is equal to the sum of the probabilities minus the probability of their intersection:

$$\Pr(E1 \cup E2) = \Pr(E1) + \Pr(E2) - \Pr(E1 \cap E2)$$

This also holds true when the events are mutually exclusive because the intersection of two mutually exclusive events is the empty set and the probability of the empty set is zero.

For example, consider the random experiment of rolling a six-sided die where the result is the number showing uppermost. The sample space is then:

$$\{1, 2, 3, 4, 5, 6\}$$

If we define the two events:

$E1$ = the number showing is less than 5

$E2$ = the number showing is even

then

$$E1 = \{1, 2, 3, 4\} \text{ and } E2 = \{2, 4, 6\}$$

and

$$\Pr(E1) = 4/6 \text{ and } \Pr(E2) = 3/6$$

The event that the number showing is *either* less than five *or* is an even number is $E3$ where:

$$\begin{aligned} E3 &= E1 \cup E2 \\ &= \{1, 2, 3, 4\} \cup \{2, 4, 6\} \\ &= \{1, 2, 3, 4, 6\} \end{aligned}$$

and so:

$$\Pr(E1 \cup E2) = 5/6$$

but

$$\Pr(E1) + \Pr(E2) = 7/6 > 1$$

and this is not allowed.

However:

$$\begin{aligned} E1 \cap E2 &= \{1, 2, 3, 4\} \cap \{2, 4, 6\} \\ &= \{2, 3\} \end{aligned}$$

and so:

$$Pr(E1 \cap E2) = 2/6$$

so that:

$$Pr(E1) + Pr(E2) - Pr(E1 \cap E2) = 7/6 - 2/6 = 5/6 = Pr(E1 \cup E2)$$

EXAMPLES

1 A red die and a blue die are thrown. What is the probability that:

(a) either a red 3 or a blue 5 is displayed
(b) a red 1 is not displayed and neither is a blue 2

(a) Pr({red 3}) = 1/6, Pr({blue 5}) = 1/6. Both events are mutually exclusive so that:
Pr({red 3}OR{blue 5}) = Pr({red 3}) + Pr({blue 5}) = 1/3
(b) Pr({red 1}′AND{blue 2}′) = Pr([{red 1}OR{blue 2}]′) and
Pr([{red 1}OR{blue 2}]′) = 1 − Pr({red 1}OR{blue 2}) = 1 − 1/3 = 2/3

2 A car park contains 10 red Ford cars, 8 white Ford cars, 12 red Honda cars and 3 white Honda cars. A car is selected at random. What is the probability that it is:

(a) either a red car or a Ford
(b) not a white car and not a Honda
(c) either a white car or a Honda but not both

(a) Pr({red car}) = 22/33, Pr({Ford}) = 18/33. Both events are not mutually exclusive so that:

Pr({red car}OR{Ford}) = Pr({red car}) + Pr({Ford}) − Pr({red Ford})

= 22/33 + 18/33 − 10/33 = 30/33

(b) Pr({white car}′AND{Honda}′) = Pr([{white car}OR{Honda}]′)
and
Pr([{white car}OR{Honda}]′) = 1 − Pr({white car}OR{Honda}) = 1 − (11/33 + 15/33 − 3/33)
= 10/33

(c) There are 3 white Hondas and Pr({white car}OR{Honda}) = 23/33 so that Pr{either a white car or a Honda but not both} = 23/33 − 3/33 = 20/33

3 A library shelf contains 25 hardback and 48 softback books. Of the hardback books, 10 are novels, 5 are detective fiction and the remainder are text books. Of the softback books, 15 are text books, 21 are detective fiction and the remainder are novels. A book is selected at random. Find the probability that the book selected is:

(a) a hardback or a novel
(b) not a softback and is not detective fiction
(c) either hardback or a novel or a text book
(d) a novel or a softback but not both.

(a) Pr({hardback}) = 25/73, Pr({novel}) = 22/73, Pr({hardback} $\cap$ {novel}) = 10/73
Therefore, Pr({hardback}OR{novel}) = 25/73 + 22/73 − 10/73 = 37/73
(b) Pr({softback}′) = 25/73, Pr({detective fiction}′) = 47/73 and

Pr({softback}′)AND({detective fiction}′) = Pr([{softback})OR({detective fiction}]′)

= 1 − Pr([{softback})OR({detective fiction}])

= 1 − (48/73 + 26/73 − 21/73)

= 20/73

(c) Pr({hardback}) = 25/73, Pr({novel}) = 22/73, Pr({text book}) = 25/73
Therefore Pr({hardback}OR{novel}OR{textbook})
= Pr({hardback}) + Pr({novel}) + Pr({text book})
− Pr({hardback} ∩ {novel}) − Pr({hardback} ∩ {text book})
= 25/73 + 22/73 + 25/73 − 10/73 − 10/73
= 57/73

(d) Pr({novel or a softback but not both})
= Pr({hardback novel or a softback which is not a novel})
= 10/73 + 36/73
= 46/73

EXERCISES

1 A computer lab contains 5 PCs and 6 Apple Macs. Of the PCs 3 have a colour monitor and of the Apple Macs only 1 has a colour monitor. A computer is selected at random. What is the probability that the computer is:

(a) either a colour PC or a monochrome Apple Mac
(b) not a monochrome PC and not a colour Apple Mac

(a) Pr({colour PC}) = *, Pr({monochrome Apple Mac}) = *.
Therefore:
Pr({colour PC}OR{monochrome Apple Mac})
= Pr({colour PC}) * Pr({monochrome Apple Mac})
= *

(b) Pr({monochrome PC}′) = *, Pr({*}′) = *.
Also:
Pr({*}′AND{*}′) = Pr([{*}OR{*}]′) so that
Pr({monochrome PC}′AND{*}′) = Pr([{*}OR{*}]′)
= 1 − Pr([{*}OR{*}])
= *

2 Two dice are thrown and their score recorded. What is the probability that the score is:

(a) either an odd number or less than 6
(b) not an even number and not greater than 8
(c) either is made up of two even numbers or amounts to 10 but not both

(a) Pr({odd number}) = *, Pr({less than 6}) = *, Pr({odd number} ∩ {less than 6}) = *
Therefore:
Pr({odd number}OR{less than 6}) = * + * − * = *

(b) Pr({*}′) = *, Pr({*}′) = *.
Also:
Pr({*}′AND{*}′) = Pr([{*}OR{*}]′) so that
Pr({*}′AND{*}′) = Pr([{*}OR{*}]′)
= 1 − Pr([{*}OR{*}])
= *

(c) Pr({two even numbers}) = *, Pr({amounts to 10}) = *
Therefore
Pr({two even numbers or amounts to 10 but not both}) = *

3 A company has three departments, sales, personnel and management. Sales employs 5 women and 3 men, personnel employs 6 men and 4 women and management employs 2 women and 3 men. An individual is selected at random. Find the probability that the person selected:

(a) is male or works in management
(b) is female or works in sales
(c) is not male and does not work in personnel
(d) is either female or works in management but not both

	Sales	Personnel	Management
Men	3	6	3
Women	5	4	2

(a) $\Pr(\{\text{male}\}) = *$, $\Pr(\{\text{works in management}\}) = *$,
$\Pr(\{\text{male}\} \ll \{\text{works in management}\}) = *$
Therefore:
$\Pr(\{\text{male}\}\text{OR}\{\text{works in management}\}) = * + * - * = *$

(b) $\Pr(\{\text{female}\}) = *$, $\Pr(\{\text{works in sales}\}) = *$,
$\Pr(\{*\} \ll \{*\}) = *$. Therefore:
$\Pr(\{\text{female}\}\text{OR}\{\text{works in sales}\}) = * + * - * = *$

(c) $\Pr(\{*\}') = *$, $\Pr(\{*\}') = *$.
Also:
$\Pr(\{*\}'\text{AND}\{*\}') = \Pr([\{*\}\text{OR}\{*\}]')$ so that
$\Pr(\{*\}'\text{AND}\{*\}') = \Pr([\{*\}\text{OR}\{*\}]')$
$= 1 - \Pr([\{*\}\text{OR}\{*\}])$
$= *$

(d) $\Pr(\{\text{female}\}) = *$, $\Pr(\{\text{works in management}\}) = *$
Therefore
$\Pr(\{* \text{ or } * \text{ but not both}\}) = *$

Unit 3 The Connective 'AND'

Test yourself with the following:

1 A bead is drawn from a bag containing 6 red beads, 4 green beads, 2 yellow beads and 3 blue beads. Its colour is recorded and then it is replaced. A second bead is withdrawn and its colour again recorded. What is the probability that:

(a) a red bead and a blue bead were drawn
(b) neither a yellow bead nor a red bead were drawn
(c) both beads were green

2 A card is drawn at random from a deck of 52 playing cards and without it being replaced a second card is drawn. What is the probability that:

(a) a red numbered card and an ace were drawn
(b) both cards were black
(c) how would replacement of the first card have affected the answers to (a) and (b)?

3 10 men and 6 women sit around a table. Of the men 6 have brown eyes, 3 have green eyes and one has blue eyes. Of the women, 5 have green eyes and 1 has blue eyes. Two individuals are selected at random. Find the probability that the two people selected:

(a) are of different sexes
(b) are a man with brown eyes and a woman with green eyes
(c) are neither a man with green eyes nor a woman with brown eyes
(d) are both blue eyed

AND and Intersection

Two random experiments are performed in sequence. Let $E1$ be an event associated with a first random experiment and $E2$ be an event associated with a second random experiment. These two events can be connected via the set operation of **intersection** to form the event $E3$:

$$E3 = E1 \cap E2$$

Event E3 can be described in English as:

event $E1$ *and* event $E2$

In other words:

both event $E1$ occurs *and* event $E2$ occurs

The probability of $E3$ occurring depends upon whether $E1$ and $E2$ are:

- Dependent events
- Independent events

Dependent Events

If two random experiments are performed in sequence, one after the other, then it may be possible for the outcome of the first experiment to affect the outcome of the second. If this is the case the outcomes are **dependent** upon each other and the probabilities change after the first experiment has been performed.

For example, a drawer contains one blue sock, one red sock and one white sock. Let experiment 1 be selecting a sock out of our drawer, and experiment 2 be selecting another sock out of our drawer.

Before experiment 1 is performed the probabilities, assigned *a priori*, associated with the experiment's simple events are:

$$\Pr(\{\text{selection of a blue sock}\}) = 1/3$$

$$\Pr(\{\text{selection of a red sock}\}) = 1/3$$

$$\Pr(\{\text{selection of a white sock}\}) = 1/3$$

Let the result of experiment 1 be the event {selection of a blue sock} which has probability:

$$\Pr(\{\text{selection of a blue sock}\}) = 1/3$$

This now affects the probabilities for our second selection as there are only *two* socks left in the drawer. For experiment 2

$$\Pr(\{\text{selection of a red sock}\}) = 1/2$$

$$\Pr(\{\text{selection of a white sock}\}) = 1/2$$

$$\Pr(\{\text{selection of a blue sock}\}) = 0$$

The probabilities have now changed. Notice that this last probability is zero as there are no blue socks in the drawer after the first selection. This outcome is **impossible**.

Independent Events

If the outcome of the first experiment does not affect the outcome of the second experiment then the outcomes are **independent** of each other and the probabilities will not change after the first experiment has been performed. If we had replaced the blue sock before experiment 2 then we should have had **independent events**. There would have been three socks in the drawer ready for experiment 2 with

$$\Pr(\{\text{selection of a red sock}\}) = 1/3$$

$$\Pr(\{\text{selection of a white sock}\}) = 1/3$$

$$\Pr(\{\text{selection of a blue sock}\}) = 1/3$$

The probabilities have not changed.

In either case, whether the events are independent or dependent:

$$\Pr(E1 \cap E2) = \Pr(E1)\Pr(E2)$$

For example, the probability of selecting a blue sock and then drawing a white sock out of our drawer *without replacing the blue sock first* is:

$$\begin{aligned}
&\Pr(\{\text{selection of a blue sock}\}\ \textit{and}\ \{\text{selection of a white sock}\})\\
&\quad = \Pr(\{\text{selection of a blue sock}\})\Pr(\{\text{selection of a white sock}\})\\
&\quad = (1/3) \times (1/2)\\
&\quad = 1/6
\end{aligned}$$

In comparison, the probability of selecting a blue sock and then drawing a white sock out of our drawer *after* replacing the blue sock is:

$$\begin{aligned}
&\Pr(\{\text{selection of a blue sock}\}\ \textit{and}\ \{\text{selection of a white sock}\})\\
&\quad = \Pr(\{\text{selection of a blue sock}\})\Pr(\{\text{selection of a white sock}\})\\
&\quad = (1/3) \times (1/3)\\
&\quad = 1/9
\end{aligned}$$

Probability Trees

We are already familiar with the idea of a sequence of random experiments and the outcome tree that results from it. Now, instead of dealing with outcomes we are dealing with events and their associated probabilities and as a result the outcome tree becomes a probability tree. A probability tree is completely analogous to an outcome tree where, instead of outcomes listed against the branches of the tree, the probabilities are given. The probabilities of combined events are then products of probabilities, as shown in Figure 26.3.

For example, in a factory, items pass through two processes, namely cleaning and painting. The probability that an item has a cleaning fault is 0.2 and the probability that an item has a painting fault is 0.3. Cleaning and painting faults occur independently of one

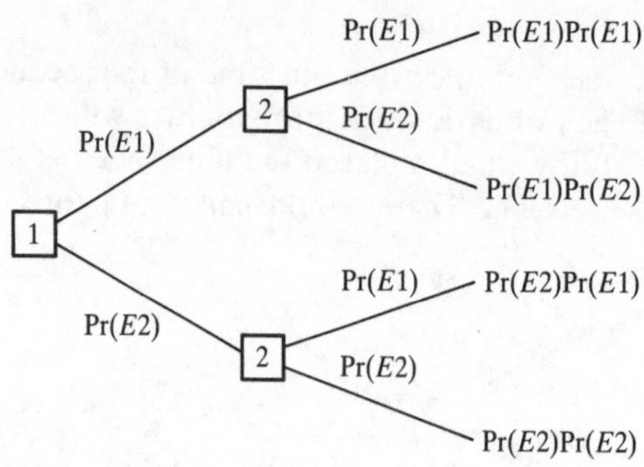

Figure 26.3

another. Calculate the probability that a selected item

(a) Has both cleaning and painting faults.
(b) Has neither cleaning nor painting faults.
(c) Has a cleaning fault but does not have a painting fault.

To solve this problem we first define the random experiments and the associated events that are involved.

Cleaning and painting are independent of each other. Let

Cleaning be experiment 1 with sample space {cleaning fault, no cleaning fault}
Painting be experiment 2 with sample space {painting fault, no painting fault}

In experiment 1, we are given:

$$\Pr(\{\text{cleaning fault}\}) = 0.2$$

hence

$$\Pr(\{\text{no cleaning fault}\}) = 0.8$$

since any item either does or does not have a cleaning fault and the sum of the probabilities must equal 1. In experiment 2 we are given:

$$\Pr(\{\text{painting fault}\}) = 0.3$$

hence

$$\Pr(\{\text{no painting fault}\}) = 0.7$$

(a) $$\begin{aligned}\Pr(\{\text{cleaning fault}\}\ \textit{and}\ \{\text{painting fault}\}) &= \Pr(\{\text{cleaning fault}\})\Pr(\{\text{painting fault}\})\\ &= (0.2)(0.3)\\ &= 0.06\end{aligned}$$

(b) $$\begin{aligned}&\Pr(\{\text{no cleaning fault}\}\ \textit{and}\ \{\text{no painting fault}\})\\ &\qquad = \Pr(\{\text{no cleaning fault}\})\Pr(\{\text{no painting fault}\})\\ &\qquad = (0.8)(0.7)\\ &\qquad = 0.56\end{aligned}$$

(c) $\Pr(\{\text{cleaning fault}\}$ *and* $\{\text{no painting fault}\})$

$$= \Pr(\{\text{cleaning fault}\})\Pr(\{\text{no painting fault}\})$$
$$= (0.2)(0.78)$$
$$= 0.14$$

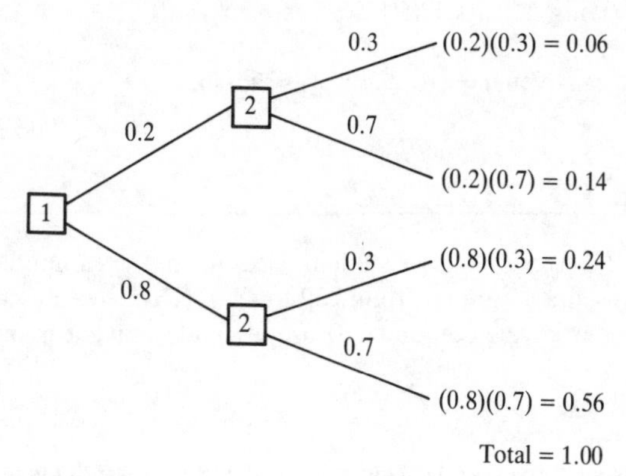

Figure 26.4

EXAMPLES

1 A red die and a blue die are thrown. What is the probability that:

(a) a red 3 and a blue 5 is displayed
(b) neither a red 1 nor a blue 2 is displayed

(a) $\Pr(\{\text{red } 3\}) = 1/6$, $\Pr(\{\text{blue } 5\}) = 1/6$. Both events are independent so that:
$\Pr(\{\text{red } 3\}\text{AND}\{\text{blue } 5\}) = \Pr(\{\text{red } 3\})\Pr(\{\text{blue } 5\}) = 1/9$

(b) $\Pr(\{\text{red } 1\}'\text{AND}\{\text{blue } 2\}') = \Pr(\{\text{red } 1\}')\Pr(\{\text{blue } 2\}') = (5/6)(5/6) = 25/36$

2 A car park contains 10 Ford cars and 12 Honda cars. Two cars are stolen at different times, being selected at random. What is the probability that the stolen cars are:

(a) a Honda and a Ford
(b) both Ford cars

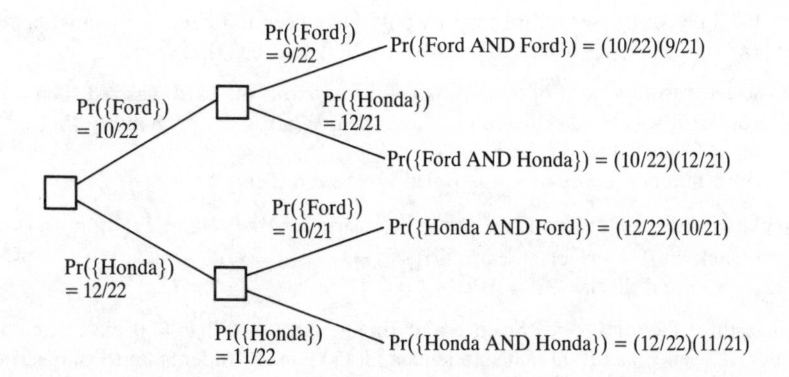

(a) The two events are dependent because after one car has been stolen, it is removed, thereby changing the probabilities of selecting from the remainder.
Pr({Honda is stolen}) = 12/22 leaving 21 cars. Pr({Ford is stolen}) = 10/21.
Therefore:
Pr({Honda is stolen}AND{Ford is stolen}) = (12/22)(10/21) = 120/462.

(b) The two events are again dependent.
Pr({Ford is stolen}) = 10/22 leaving 21 cars. Pr({Ford is stolen}) = 9/21.
Therefore:
Pr({Ford is stolen}AND{Ford is stolen}) = (10/22)(9/21) = 90/462.

EXERCISES

1 A computer lab contains 5 PCs and 6 Apple Macs. A student used one of the computers from noon to 2:00 p.m. and a second student used a computer from 3:00 to 4:00. No one else used a computer during that time. If each student selected the computer to use at random, what is the probability that the computers were:

(a) a PC and an Apple Mac
(b) both PCs
(c) how would your answers to (a) and (b) have differed if the two students had been in the laboratory at the same time?

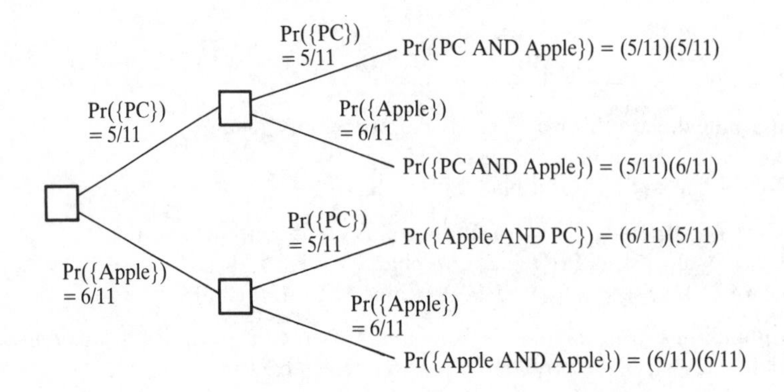

(a) (5/11)(*) + (*)(*) = */121
(b) (*)(*) = *
(c) The probabilities of the second selection would have been different so the answer to (a) would be (5/11)(*) + (*)(*) = */110, and the answer to (b) would be (*)(*) = *).

2 A card was selected from a deck of 52 and its colour noted. If the card was red then a red die was rolled and if the card was black a black die was rolled. What is the probability that:

(a) a 6 was rolled on the black die
(b) a number less than 4 was rolled on the red die

These two experiments are independent. Also, Pr({black card}) = * and Pr({number}) = *
(a) Pr({6 on black die}) = Pr({black card})Pr({*}) = (*)(1/*) = */12
(b) Pr({1, 2 or 3 on red die}) = Pr({*})Pr({*}) = *

3 A college has three faculties, Arts, Science and Engineering. Arts has 800 male students and 750 female students, Science has 1200 male students and 450 female students and Engineering has 1800

male students and 200 female students. Two individuals are selected at random. Find the probability that the people selected are:

(a) male and study Arts
(b) neither female nor study Engineering
(c) female and study Arts or Science

(a) There are * men who study Arts so the probability of selecting two people who are men and study Arts is:

Pr({male Arts1})Pr({male Arts2}) = (500/*)(499/*) = *.

(b) Pr({female}'AND{study Engineering}') = Pr({male}AND[{study Arts}OR{study Science}])

= (*)[(*) + (*)]

= *

(c) Pr({female}AND[{study Arts}OR{study Science}]) = *

Module 26: Further exercises

1 Every card in a deck of 52 has an equal chance of being selected. Find the probability of selecting:
(a) an odd numbered red card
(b) a black Ace
(c) the Queen of diamonds

2 What is the *a priori* probability of:
(a) a given number being drawn from a drum containing 150 numbers if every name has the same chance of being chosen
(b) selecting a day from the week
(c) picking a particular hair from a person's head

3 What is the probability that:
(a) a purchased item will prove to be defective if, from past experience, 7 out of every 1000 made are defective
(b) contacting your friend at home on the telephone if he is out of the house for 39% of the day

4 A box contains 3 black buttons, 2 white buttons and 9 grey buttons. A button is selected and every button has an equal probability of being chosen. What are the probabilities of each of the following events?
(a) $E1$ = {black button}
(b) $E2$ = {white or grey button}
(c) $E3$ = {not a white button}

5 A dartboard is numbered from 1 to 20. If three darts are thrown at random so that every number has an equal chance of being hit, what is the probability of the three darts hitting 1, 2 and 3 in sequence?

6 A junior school has a choir comprised of all the students in the first and second year. In the first year there are 10 boys and 12 girls and in the second year there are 8 boys and 14 girls. A chorister is selected at random. What is the probability that the chorister is:
(a) either a first year boy or a second year girl
(b) not a boy and not a first year girl

7 Find the probability that the chorister selected in the previous question is:
(a) either a boy or a second year student
(b) not a girl and not a first year student
(c) either a boy or a first year student but not both

8 A jury consists of 7 men and 5 women. Of the men 5 think the defendant is guilty, 1 is undecided and 1 thinks the defendant not guilty. Of the women 2 think the defendant is guilty, 2 are undecided and the remaining 1 thinks the defendant not guilty. A juror is selected at random. Find the probability that the juror selected:

(a) is male or thinks the defendant guilty
(b) is female or is undecided
(c) is not male and does not think the defendant guilty
(d) is either female or is undecided but not both

9 A lamp is taken at random from a production line and tested. If the bulb does not light up two possible causes of the fault are either a faulty switch or a faulty bulb. The switch is likely to be faulty with a probability of 0.1 and the bulb is likely to be faulty with a probability of 0.3. Switch and bulb faults are independent of each other.

Calculate the probability that a defective lamp

(a) has both a switch fault and a bulb fault
(b) has neither a switch fault nor a bulb fault
(c) has a switch fault but does not have a bulb fault
(d) has either a switch fault or a bulb fault but not both

10 A junior school has a choir comprised of all the students in the first and second year. In the first year there are 10 boys and 12 girls and in the second year there are 8 boys and 14 girls. Two different choristers are selected at random, each to read a lesson. What is the probability that the choristers are:

(a) a first year boy and a second year girl
(b) neither a boy nor a first year girl

11 A jury consisted of 7 men and 5 women. Of the men 5 think the defendant is guilty, 1 is undecided and 1 thinks the defendant not guilty. Of the women 2 think the defendant is guilty, 2 are undecided and the remaining 1 thinks the defendant not guilty. Two jurors are selected at random and removed from the jury room. Find the probability that the jurors selected are:

(a) male and think the defendant guilty
(b) female and are undecided
(c) neither male nor think the defendant guilty

Module 27: Probability Distributions

Aim: To introduce the concept of a probability distribution and to distinguish between discrete and continuous probability distributions.

Objectives: When you have read this module you will be able to:

- ▶▶ Define a random variable and assign values to it by coding events.
- ▶▶ Obtain the mean and variance of a random variable by using the concept of expectation.
- ▶▶ Understand the meaning of a Bernoulli trial.
- ▶▶ Manipulate the binomial probability distribution and use the Poisson probability distribution for rare events.
- ▶▶ Appreciate the role of the standard normal probability distribution and be able to use tables to solve problems involving the normal probability distribution.

Unit 1 Probability Distributions

Test yourself with the following:

1 A bag contains 4 red beads, 3 blue beads, 2 yellow beads and 1 green bead. A bead is drawn from the bag and its colour noted. If each bead has an equal chance of being drawn code the simple events of this random experiment and construct a probability distribution.

2 Find the mean and variance of the probability distribution of Question 1.

3 In a game, cards are selected at random from a deck of 52 playing cards. If a heart is selected you score 3, if a spade is selected you score 2. If either a club or a diamond is selected you score 1.

(a) Construct a probability distribution for the random variable x that represents the score.

(b) Find the mean and variance of this random variable.

(c) What is the expected score for any selection?

4 On any day, the probability of a bus arriving at a certain bus stop within a 2 minute wait is 0.4. What is the probability that:

(a) there will only be two days out of five when you will have to wait for longer than 2 minutes?

(b) you will have to wait more than 2 minutes for every one of the five days?

(c) you will have to wait more than two minutes for no more than 3 days out of the five?

Random Variable

Every random experiment gives rise to a collection of mutually exclusive simple events, each with an associated probability of that event occurring. This collection of probabilities is called the **probability distribution** of the random experiment and we have seen how a probability distribution can be divined either from a relative frequency distribution using the notion of statistical regularity or from *a priori* considerations.

Divining the probability distribution using statistical regularity is reasonably straightforward – the random experiment is repeated a number of times sufficient to do the job. Divining the probability distribution from *a priori* considerations is not always quite so straightforward. The process can, however, be greatly assisted when a formalized method is available but to create such a formalized method requires the notion of a **random variable**.

Every random experiment has a result and associated with that result is a set of possible simple events. Because all the simple events are mutually exclusive it is possible to code each simple event with its own unique number. For example, the random experiment of tossing a coin gives rise to the two events $\{H\}$ and $\{T\}$, which can be coded as:

$\{H\}$ coded as 1 and $\{T\}$ coded as 2

Next, the result of the random experiment, namely that *a face will show*, will be represented by the variable x where 1 and 2 are the only two allowed values of the variable. If the coin is **fair** then:

$$\Pr(\{H\}) = \Pr(x = 1) = \Pr(1) = 0.5$$

and

$$\Pr(\{T\}) = \Pr(x = 2) = \Pr(2) = 0.5$$

Notice that the coding of the simple events is not unique; we could have coded as:

$\{H\}$ coded as -1 and $\{T\}$ coded as 1

in which case:

$$\Pr(-1) = 0.5 \text{ and } \Pr(1) = 0.5$$

By coding in this manner a relationship is defined between the values of the variable x and their associated probabilities. This relationship is expressed as a probability distribution. Because the variable x is associated with the result of a random experiment it is referred to as a random variable.

Expectation

Every permitted value of a random variable x associated with a random experiment has a probability $\Pr(x)$ of being realized.

In complete analogy with defining the average values of a collection of data as the sum of the products of the various datum values with their relative frequencies, so we can define an **average** value of a random variable as the sum of the products of its values with their respective probabilities.

$$\mu = \Sigma x \Pr(x)$$

(the symbol μ is a lower-case Greek m, pronounced 'myou'). Here, the *average* random variable, μ, is called the **expectation of** x, denoted by $E(x)$. That is:

$$\mu = E(x), \text{ the expectation of } x$$

Note that the use of the word 'expectation' can be misleading. It does not mean that the most likely value to be expected is μ, because the most likely value to be expected is that with the highest probability. For example, in the case of the random experiment of tossing a *fair* coin the expectation is given as:

$$E(x) = 1 \cdot \Pr(1) + 2 \cdot \Pr(2) = 1(0.5) + 2(0.5) = 1.5$$

Because the coin is fair neither simple event $\{H\}$ or $\{T\}$ is more likely. However, the expectation of the random variable x is 1.5 – halfway between the two values 1 and 2. Notice also that the expectation is not necessarily one of the possible values of the random variable, just as the arithmetic mean of a collection of data may not be one of the data.

In a similar manner, the variance σ^2 of the random variable is given as:

$$\begin{aligned}\sigma^2 &= E([x - \mu]^2)\\ &= \Sigma(x - \mu)^2 \Pr(x)\end{aligned}$$

In the previous example of the toss of a fair coin

$$\begin{aligned}\sigma^2 &= \Sigma(x - \mu)^2 \Pr(x)\\ &= (1 - 1.5)^2(0.5) + (2 - 1.5)^2(0.5)\\ &= (0.25)(0.5) + (0.25)(0.5)\\ &= 0.25\\ &= E([x - \mu]^2)\end{aligned}$$

The symbol used for the variance of a random variable is σ^2, where σ is a lower-case Greek s, pronounced 'sigma'. Notice that σ is the standard deviation of the random variable.

Bernoulli trials

A **Bernoulli trial** is any random experiment whose result gives rise to only two possible simple events which we shall call $\{\text{success}\}$ with probability p and $\{\text{failure}\}$ with probability q:

$$\Pr(\{\text{success}\}) = p \text{ and } \Pr(\{\text{failure}\}) = q$$

where, naturally:

$$p + q = 1.$$

A typical Bernoulli trial is the tossing of a coin where a head could be considered as success and a tail as failure. The actual connotations of success and failure are unimportant. What is important is that the random experiment has *only two possible outcomes.*

In what follows we shall be concerned not with just a single Bernoulli trial but rather with a succession of such trials where each successive trial is independent of the previous trials. For example, consider the five successive tossings of a single coin – the fivefold repetition of a single random experiment where the result of each repetition is independent of the result of its predecessor. One such sequence of outcomes could be:

H,H,T,T,T

Here there are 2 heads and 3 tails.

The event associated with this outcome is:

$$E = \{H\} \cap \{H\} \cap \{T\} \cap \{T\} \cap \{T\}$$

If we assume the coin to be possibly unfair with a probability of a head being p and the probability of a tail being q then the probability of event E is:

$$p \cdot p \cdot q \cdot q \cdot q = p^2 q^3$$

The combination of two heads and three tails given here is not unique. For example, the following is another such sequence:

H,T,T,H,T

and this has the same probability as the other sequence, namely:

$$p \cdot q \cdot q \cdot p \cdot q = p^2 q^3$$

The next question to ask is:

How many such combinations of 2 heads and 3 tails are possible?

We have already seen the answer to this question in Unit 2 of Module 6 where Pascal's triangle and combinations were discussed. There it was found that the number of different combinations of 2 identical items among 5 locations is given as:

$$\begin{aligned} {}^5C_2 &= 5!/[(5-2)!2!] \\ &= 5!/[3!2!] \\ &= (5 \cdot 4 \cdot 3 \cdot 2 \cdot 1)/[(3 \cdot 2 \cdot 1)(2 \cdot 1)] \\ &= (5 \cdot 4)/(2 \cdot 1) \\ &= 10 \end{aligned}$$

As a consequence there are 10 different combinations of 2 heads and 3 tails, each combination having the same probability:

$$p^2 q^3$$

This gives the probability of tossing two heads in five tosses of a single coin as:

$$10p^2 q^3$$

This probability we shall denote by the symbol Pr(2:5). In general, we can say that the probability of 2 successes in 5 Bernoulli trials is:

$$\Pr(2:5) = {}^5C_2 p^2 q^{5-2} = 10p^2 q^3$$

This result can be generalized to the probability of r successes in n trials, Pr(r:n). The result is given in the form:

$$\Pr(r:n) = {}^nC_r p^r q^{n-r}$$

where:

$$^nC_r = n!/[(n-r)!r!]$$

Notice that here the random variable is r which can take on $n+1$ values $0, 1, 2, \ldots, n$.

Expectation for a Bernoulli trial

For a single Bernoulli trial there are two simple events, namely {success} and {failure}. The probabilities of these two events are:

$$\Pr(0{:}1) = q, \text{ the probability of \{failure\}}$$

which occurs when the random variable $r = 0$.

$$\Pr(1{:}1) = p, \text{ the probability of \{success\}}$$

which occurs when the random variable $r = 1$. Consequently, the expectation of r for a single Bernoulli trial is:

$$E(r) = \mu = \Sigma r\Pr(r) = 0 \cdot q + 1 \cdot p = p$$

Similarly, the variance σ^2 is $E((r-\mu)^2)$ where:

$$\begin{aligned} E((r-\mu)^2) &= \Sigma(r-\mu)^2\Pr(r) \\ &= (0-p)^2 q + (1-p)^2 p \\ &= p^2 q + q^2 p \\ &= pq(p+q) \\ &= pq \end{aligned}$$

The mean μ of a single Bernoulli trial is p and the variance σ^2 is pq.

EXAMPLES

1 A sample of 30 packets of cereal were taken from a supermarket shelf and weighed. They were all supposed to contain 200 g of cereals but the weights were as follows:

2 weighed 190 g
4 weighed 195 g
15 weighed 200 g
5 weighed 205 g
3 weighed 210 g and
1 weighed 215 g

From this random experiment construct a probability distribution.

Weight (x)	Number	Probability $\Pr(x)$
190	2	2/30 = 0.0666
195	4	4/30 = 0.1333
200	15	15/30 = 0.5000
205	5	5/30 = 0.1666
210	3	3/30 = 0.1000
215	1	1/30 = 0.0333
Total	30	1.0000

Notice the similarity between this table and the frequency distribution tables of Module 2. Indeed, the very process of assigning probability by statistical regularity is identical to the computation of relative frequencies where the relative frequency of a datum is the frequency of the datum divided by the total number of data.

2 Find the mean and variance of the probability distribution of Question 1.

Weight (x)	Number	Probability $\Pr(x)$	$x\Pr(x)$	$x-\mu$	$(x-\mu)^2$	$(x-\mu)^2\Pr(x)$
190	2	2/30 = 0.067	12.730	−10.995	120.890	8.100
195	4	4/30 = 0.133	25.935	−5.995	35.940	4.780
200	15	15/30 = 0.500	100.000	−0.995	0.990	0.495
205	5	5/30 = 0.167	34.235	4.005	16.040	2.679
210	3	3/30 = 0.100	21.000	9.005	81.090	8.109
215	1	1/30 = 0.033	7.095	14.005	196.140	6.473
Total	30		$\mu = 200.995$			$\sigma^2 = 30.636$ $\sigma = 5.535$

3 Two dice are thrown. If you score a 7 or an 11 then you win, otherwise you lose. Define a random variable x whose values denote a win or a lose. Find both the mean and standard deviation of this random variable.

When two dice are thrown there are $6 \times 6 = 36$ possible scores, not all of them different. A score of 7 can be achieved in 6 ways, by throwing:

1 and 6, 2 and 5, 3 and 4, 4 and 3, 5 and 2 and finally, 6 and 1

A score of 11 can be achieved in two ways, by rolling:

6 and 5 or 5 and 6

As a consequence there are 8 ways of winning out of the 36 scores, so the probability of a win is:

$$p = 8/36 = 0.2222$$

and the probability of a lose is:

$$q = 28/36 = 0.7778$$

(both to 4 dec. pl.) Define a random variable to be $x = 0$ for a lose and $x = 1$ for a win. Then the mean is given by:

$$\mu = (0)(0.7778) + (1)(0.2222) = 0.2222 = p$$

and the variance is given as:

$$\sigma^2 = (0 - 0.222)^2(0.7778) + (1 - 0.2222)^2(0.2222)$$
$$= 0.1728$$

so that $\sigma = 0.5543$. Notice that:

$$pq = 0.1728$$

which confirms the fact that $\sigma^2 = pq$

4 A company has found that on average 20% of its 50 accounts are outstanding at the end of a month. What is the probability that at the end of a particular month:

(a) ten accounts are outstanding
(b) fifty accounts are outstanding

(a) The probability of an outstanding account is $p = 0.2$. The probability that ten accounts are outstanding is:

$$\begin{aligned}\Pr(10{:}60) &= {}^{60}C_{10}(0.2)^{10}(0.8)^{50}\\ &= (60!/[50!10!])(0.2)^{10}(0.8)^{50}\\ &= 7.5394.10^{10}(1.024.10^{-7})(1.4272.10^{-5})\\ &= 0.1102 \text{ to four dec. pl.}\end{aligned}$$

(b) The probability that fifty accounts are outstanding is:

$$\begin{aligned}\Pr(50{:}60) &= {}^{60}C_{50}(0.2)^{50}(0.8)^{10}\\ &= (60!/[50!10!])(0.2)^{50}(0.8)^{10}\\ &= 1.0272.10^{10}(1.1259.10^{-35})(0.1074)\\ &= 0.0000 \text{ to four dec. pl.}\end{aligned}$$

EXERCISES

1 In a random experiment a person is selected from a room containing 32 people. Of the people in the room 5 were less than 15 years old, 7 were aged between 15 and 30, 12 were aged between 30 and 45 and 8 were aged between 45 and 60. If each person has an equal chance of being selected and their age is taken to be the midpoint of their age range, construct a probability distribution for this random experiment.

Age range	Midpoint (x)	Number	Probability $\Pr(x)$
0.00–14.99	7.495	5	5/32 = *
15.00–29.99	*	*	* = *
30.00–44.99	*	*	* = *
45.00–60.00	52.50	*	* = *
	Total	*	= 1.0000

2 Find the mean and variance of the probability distribution of Question 1.

Age (x)	Number	Probability $\Pr(x)$	$x\Pr(x)$	$x-\mu$	$(x-\mu)^2$	$(x-\mu)^2\Pr(x)$
7.495	5	5/32 = *	*	−*	*	*
*	*	* = *	*	−*	*	*
*	*	* = *	*	*	*	*
*	*	* = *	*	*	*	*
Total	32		$\mu = 33.2775$			$\sigma^2 = 228.344$
						$\sigma = 15.11$

3 A die is made in such a manner that the probability of a given number being rolled is proportional to the number rolled. Find the probabilities associated with the various scores and find the mean and variance of the probability distribution. Find also the probability that the score is:

(a) even

(b) neither 2 nor 5

x	$\Pr(x)$	$x\Pr(x)$	$x - \mu$	$(x - \mu)^2$	$(x - \mu)^2\Pr(x)$
1	$1/21 = *$	*	*	*	*
2	*	*	*	*	*
3	*	*	*	*	*
4	*	*	*	*	*
5	*	*	*	*	*
6	*	*	*	*	*
Total 21	1.00	$\mu = *$			$\sigma^2 = *$
					$\sigma = *$

4 Of 835 birds captured during an ornithological expedition in the Hebrides it was found that 146 of them had been previously ringed. If a sample of 12 birds were taken from the 835 captured what is the probability that:

(a) 5 birds had been previously ringed

(b) 7 birds had not been previously ringed

(c) at most 2 birds had been previously ringed

(a) ${}^{12}C_5 p^5 q^7 = *$ where $p = 146/835 = 0.175$ and $q = 689/835 = 0.825$ (both to 3 dec. pl.)

(b) ${}^{12}C_* p^* q^* = *$

(c) ${}^{12}C_* p^* q^* + {}^{12}C_* p^* q^* + {}^{2}C_* p^* q^* = *$

Unit 2 The Binomial Probability Distribution

Test yourself with the following:

A certain public examination has a consistent past record of a 25% failure rate. For a group of 6 people sitting the examination this year.

(a) Find the probabilities that:

(i) 1, 2 or 3 people of the 6 pass the examination

(ii) at least 4 people pass the examination

(b) Draw the bar graph of this probability distribution.

(c) From the bar graph of this probability distribution construct the histogram that represents the piecewise continuous extension of the binomial probability distribution.

(d) From the histogram find the probability that more than 2 but less than 4 people from the sample of 6 pass the examination.

The Binomial Distribution

The binomial distribution concerns itself with the probability of r successes in n Bernoulli trials and is given by:

$$\Pr(r{:}n) = {}^nC_r(p)^r(q)^{n-r} \qquad \text{where } {}^nC_r = n!/[(n-r)!r!]$$

where p is the probability of success and q is the probability of failure in each independent trial and where $p + q = 1$.

For example, a production line produced items of which 1% are defective. If a sample of 10 is taken from the line what is the probability that:

(a) one item is defective
(b) all items are perfect
(c) no more than two items are defective

To solve this problem we designate the selection of a defective component as a success and because 1% are defective the probability of success is:

$$\Pr(\{\text{success}\}) = p = 0.01$$

Hence the probability of failure, $q = 0.99$ where failure is the selection of a non-defective item.

(a) The probability of one defective item out of 10 is the probability of 1 success out of ten Bernoulli trials and is given by:

$$\begin{aligned}\Pr(1{:}10) &= {}^{10}C_1(p)^1(q)^{10-1}\\ &= {}^{10}C_1(0.01)^1(0.99)^9\\ &= 10!/[9!1!](0.01)^1(0.99)^9\\ &= 10(0.01)(0.91)\\ &= 0.091\end{aligned}$$

(b) The probability that all items are perfect is the probability of no defective items out of the 10 selected – the probability of 0 successes out of ten Bernoulli trials and is given by:

$$\begin{aligned}\Pr(0{:}10) &= {}^{10}C_0(p)^0(q)^{10-0}\\ &= {}^{10}C_0(0.01)^0(0.99)^{10}\\ &= 10!/[10!0!](0.01)^0(0.99)^{10}\end{aligned}$$

This gives the probability of no successes in ten trials as:

$$\Pr(0{:}10) = 1(1)(0.99)^{10}$$

Notice that $(0.01)^0 = 1$ (any number raised to the power zero is defined to be 1). Hence:

$$\Pr(0{:}10) = 0.904$$

(c) If no more than one item is defective then it is possible that no items are defective as well as it being possible that one item is defective. This event is the union of the two mutually exclusive events:

$$\{0 \text{ successes}\} \cup \{1 \text{ success}\}$$

So that the probability of no more than one item out of the 10 being defective is the sum:

$$\begin{aligned}\Pr(\{0 \text{ successes}\} \cup \{1 \text{ success}\}) &= \Pr(\{0 \text{ defective items}\}) + \Pr(\{1 \text{ defective item}\})\\ &= 0.904 + 0.091\\ &= 0.995\end{aligned}$$

Expectation

Each simple event associated with a binomial probability distribution is a sequence of Bernoulli trials and the random variable of a binomial distribution is r – the number of successes in a sequence of n Bernoulli trials. The expectation of each Bernoulli trial is p – the probability of success. The expectation of a sequence of n Bernoulli trials is the sum of the expectations of each trial:

$$\sum_{1}^{n} p = np$$

Consequently, the expectation of the random variable r of a binomial probability distribution, the *mean* μ of the binomial probability distribution, is:

$$\begin{aligned}\mu &= E(r)\\ &= \sum_{r=1}^{n} p\\ &= np\end{aligned}$$

Similarly, the variance is

$$\begin{aligned}\sigma^2 &= E([r-\mu]^2)\\ &= \sum_{r=1}^{n} pq\\ &= npq\end{aligned}$$

where pq is the variance of a single Bernoulli trial.

Graph of the Binomial Distribution

The graph of the binomial distribution can be displayed as a bar chart. For example, if, in a Bernoulli trial the probability of {success} p is 0.8 and the probability of {failure} q is 0.2 and we consider each simple event as a sequence of three Bernoulli trials then the resulting binomial probability distribution is:

$$\begin{aligned}\Pr(0{:}3) &= {}^3C_0 p^0 q^3\\ &= 1(0.8)^0(0.2)^3\\ &= 0.008 \qquad \text{the probability of no successes}\end{aligned}$$

$$\begin{aligned}\Pr(1{:}3) &= {}^3C_1 p^1 q^2\\ &= 3(0.8)^1(0.2)^2\\ &= 0.096 \qquad \text{the probability of 1 success}\end{aligned}$$

$$\begin{aligned}\Pr(2{:}3) &= {}^3C_2 p^2 q^1\\ &= 3(0.8)^2(0.2)^2\\ &= 0.384 \qquad \text{the probability of 2 successes}\end{aligned}$$

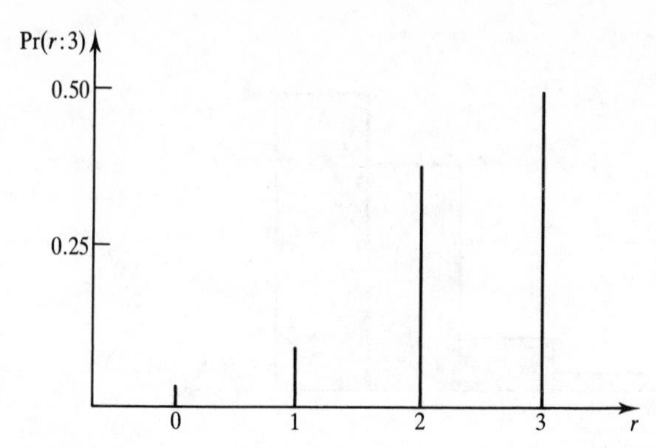

Figure 27.1

$$\begin{aligned}\Pr(3{:}3) &= {}^3C_3p^3q^0 \\ &= 1(0.8)^3(0.2)^0 \\ &= 0.512 \qquad \text{the probability of 3 successes}\end{aligned}$$

We can now plot r against Pr(r:3) to obtain the bar chart shown in Figure 27.1. Note that the actual plot of this binomial probability distribution consists of just the four points in the plane; the vertical bars are drawn in to give a visual aid to the location of the points. Note also that the probabilities are *equal to the lengths of the bars* – hence the sum of the lengths of the bars is 1. In addition, the vertical axis is not drawn through the origin. It has been shifted to the left to prevent it merging with the plot through $r = 0$.

From Discrete to Continuous

A **discrete** random variable is a variable that can only assume isolated values, whereas a **continuous** random variable can assume values that are arbitrarily close to one another. For example, the days of the month are listed by discrete numbers but the time of the day can be given to any desired precision on a scale of continuous numbers.

The graph of the binomial probability distribution is a graph of **discrete points** – each plotted point is isolated from any other plotted point because each point relates to a value of the discrete random variable r. However, just as we did in Unit 2 of Module 1, we can convert this discrete graph into a piecewise continuous graph by constructing the histogram shown in Figure 27.2.

In Figure 27.2 each discrete point has been replaced by a straight line extending between the midpoints of two adjacent values of the random variable r. Now, because the width of each rectangle in the histogram is equal to 1, the probability of each value of the random variable r is equal to the magnitude of the area of the appropriate rectangle.

What has happened here is that each discrete value of the random variable r has been replaced by a continuous set of values ranging from mid-way between adjacent r-values. In doing so, the original identification of the probability associated with a given value of r being

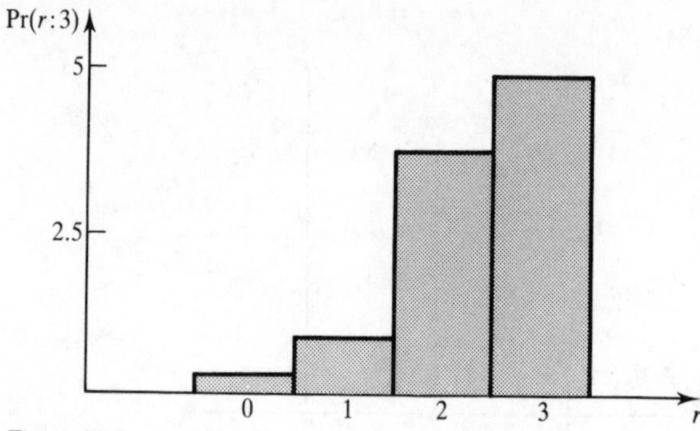

Figure 27.2

the height of the respective bar has been changed. Now the probability associated with a given value of r has become the probability that the value of r lies between the midpoints of adjacent r-values and is identified as the magnitude of the area of the appropriate rectangle. Hence the total area beneath the graph of the binomial distribution is unity.

This difference between the geometric interpretations of the probabilities associated with a discrete variable and the probabilities associated with a continuous variable is very important and will be referred to in the last unit of this module.

EXAMPLES

An assembly line has a history of producing one defective component in every 50 being produced. For a sample of three components taken at random from the assembly line:

(a) Find the probabilities that:
  (i) at least 2 components are defective
  (ii) no components are defective

(b) Draw the bar graph of this probability distribution

(c) From the bar graph of this probability distribution construct the histogram that represents the piecewise continuous extension of the binomial probability distribution.

(d) From the histogram find the probability that more than 0 but less than 3 components are defective.

(a) The probability of a defective component is $p = 1/50 = 0.02$. The probability of a non-defective component is, therefore, $q = 0.98$. From a sample of 3 components selected at random the probability that r of them are defective is:

$${}^3C_r(0.02)^r(0.98)^{3-r}$$

(i) ${}^3C_2(0.02)^2(0.98)^1 = \{3!/[1!2!]\}(0.02)^2(0.98) = 0.001176$

${}^3C_3(0.02)^3(0.98)^0 = \{3!/[0!3!]\}(0.02)^3 = 0.000008$

The probability that at least 2 are defective is, therefore $0.001176 + 0.000008 = 0.001184$

(iii) ${}^3C_0(0.02)^0(0.98)^3 = \{3!/[3!0!]\}(0.02)^0(0.98)^3 = 0.941192$

(b) To complete the probability distribution we require:

${}^3C_1(0.02)^1(0.98)^2 = \{3!/[2!1!]\}(0.02)^1(0.98)^2 = 0.057624$

the graph is then as follows:

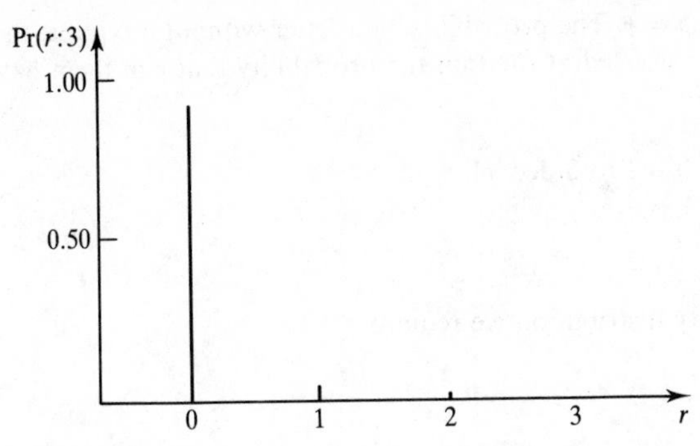

Figure 27.3

(c) The histogram is as shown:

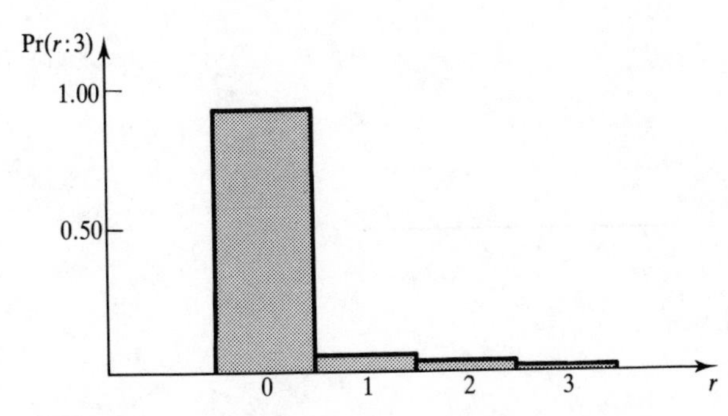

Figure 27.4

(d) From the histogram the probability that the random variable has a value greater than 0 and less than 3 is the area beneath the histogram between the two values $r = 0.5$ and $r = 2.5$ where r now refers not to the discrete random variable but to the continuous extension of the discrete random variable. This area is equal to the sum of the two middle rectangles, each being of width 1. The required probability is $(0.057624) \times (1) + (0.001176) \times (1) = 0.0588$.

EXERCISES

1 A typist finds that one in five letters typed has at least one typo. For a sample of 6 typed letters selected at random:

(a) Find the probabilities that:

(i) 0, 3 or 5 letters had a typo

(ii) at least 4 letters had a typo

(b) Draw the bar graph of this probability distribution

(c) From the bar graph of this probability distribution construct the histogram that represents the piecewise continuous extension of the binomial probability distribution.

(d) From the histogram find the probability that more than 1 but less than 3 letters had a typo.

(a) The probability of a typo $p = *$. The probability of a letter without a typo is, therefore, $q = *$. From a sample of 6 letters selected at random the probability that r of them have a typo is:

$$^6C_*(*)^*(*)^{6-*}$$

(i) $^6C_*(*)^*(*)^* = 0.262$ to 3 dec. pl.
$\quad {}^*C_3(*)^3(*)^* = *$
$\quad {}^*C_3(*)^*(*)^* = *$

(ii) $* + * + * + * = *$

(b) To complete the probability distribution we require:

$$^6C_6(*)^6(*)^0 = *$$

the graph is then as follows:

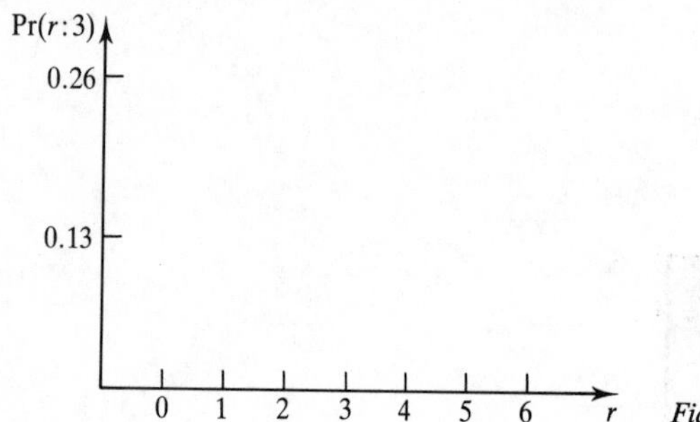

Figure 27.5

(c) The histogram is as shown:

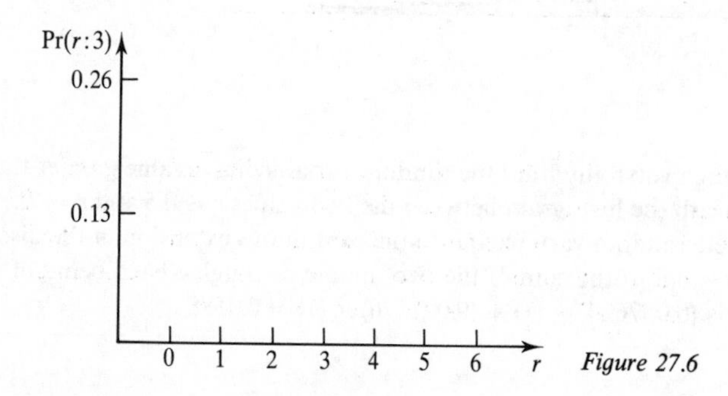

Figure 27.6

(d) From the histogram the probability that the random variable has a value between 1 and 3 is the area beneath the histogram between those two values, that is the area of the * rectangles, each having a width of *. This is *.

Unit 3 Rare Events and the Poisson Distribution

Test yourself with the following:

1 Use a calculator to evaluate $\Pr(r) = \lambda^r e^{-\lambda}/r!$ where:

(a) $r = 2$ and $\lambda = 0.1$
(b) $r = 3$ and $\lambda = 1.5$

2 The probability of selecting a defective item from an assembly line is 0.001. Using a Poisson probability distribution, find the probability of finding, from a sample of 150 items selected off the assembly line:

(a) 2 defective items

(b) more than 2 defective items

Check your answers using a binomial probability distribution.

Rare events

A **rare event** is one whose probability is very small. For example, a roulette wheel displays 37 numbers from 0 to 36 and against each number is a slot for a rolling ball to fall into. The wheel is spun and the ball is rolled. When the wheel stops the ball comes to rest alongside a number and if you had placed a bet against this number you would have won. Assuming that in a single spin of the wheel each number has an equal chance of being selected by the ball then the probability of {success} – the probability that your selection comes up – is:

$$p = 1/37 = 0.027$$

which is sufficiently small to be considered a rare event.

For a player of roulette each spin of the wheel is a Bernoulli trial and because each spin is independent of the previous spin the probabilities of winning and losing during a succession of spins are governed by the binomial probability distribution. For example, the probability of winning once in 10 plays is given by:

$$\begin{aligned}\Pr(1{:}10) &= {}^{10}C_1(0.027)^1(1 - 0.027)^{10-1}\\ &= 10(0.027)(0.973)^9\\ &= 0.211\end{aligned}$$

and the probability of winning twice in 10 plays is:

$$\begin{aligned}\Pr(2{:}10) &= {}^{10}C_2(0.27)^2(1 - 0.27)^{10-2}\\ &= 45(0.027)^2(0.973)^8\\ &= 0.026\end{aligned}$$

These two calculations are reasonably straightforward to execute. However, for the confirmed gambler it may be required to know what the chances are of winning 100 times in 1000 spins of the wheel. This is given by the binomial probability distribution as:

$$\Pr(100{:}1000) = {}^{1000}C_{100}(0.027)^{100}(1 - 0.027)^{900}$$

To evaluate this probability would take a computational power beyond even the most ambitious hand calculator – it would have to be able to evaluate 1000!

Fortunately, for rare events it is found that the Poisson probability distribution is a reasonable approximation to the binomial probability distribution and much easier to evaluate. The Poisson probability distribution is given as:

$$\Pr(r) = \lambda^r e^{-\lambda}/r!$$

where $\lambda = np$ – the mean of the binomial probability distribution. To maintain a reasonable approximation to the binomial distribution we must ensure that $\lambda < 2$.

For example, to compute the chances of winning once in 10 plays of roulette where the probability of {success} p is 0.027 we see that:

$$n = 10,\ p = 0.27 \text{ so } \lambda = (10)(0.27) = 0.27 < 2$$

As a consequence the probability of winning once in 10 plays is given to a good approximation by the Poisson distribution as:

$$\begin{aligned}\Pr(1) &= \lambda^1 e^{-\lambda}/1! \\ &= (0.27)e^{-0.27} \\ &= 0.206\end{aligned}$$

and the probability of winning twice in 10 plays is:

$$\begin{aligned}\Pr(2) &= \lambda^2 e^{-\lambda}/2! \\ &= (0.27)^2 e^{-0.27}/2 \\ &= 0.028\end{aligned}$$

which shows reasonable agreement with the binomial probability distribution to two decimal places.

EXAMPLES

1 Use a calculator to evaluate $\Pr(r) = \lambda^r e^{-\lambda}/r!$ where:

(a) $r = 2$ and $\lambda = 0.1$
(b) $r = 3$ and $\lambda = 1.5$

(a) $\Pr(2) = (0.1)^2 e^{-0.1}/2! = (0.01)(0.9048\ldots)/(2) = 0.0045$ to four dec. pl.
(b) 0.1255 to four dec. pl.

2 A weekly raffle is held between 100 members of a club. There is only one winner each week and the draw is fair. Find the probability of one individual member winning:

(a) once in 52 weeks
(b) more than 2 times in 52 weeks

(a) The probability of an individual member winning a given weekly raffle is $1/100 = 0.01$. For the Poisson distribution:

$$\lambda = (52)(0.01) = 0.52$$

hence the probability of a member winning just once in 52 weeks is given as:

$$\Pr(1) = (0.52)^1 e^{-0.52}/1! = 0.3092 \text{ to four dec. pl.}$$

(b) $\Pr(0) = (0.52)^0 e^{-0.52}/0! = 0.5945$ to four dec. pl. and $\Pr(2) = (0.52)^2 e^{-0.52}/2! = 0.0804$ to four dec. pl. Consequently the probability of winning *two times or less* is:

$$\Pr(0) + \Pr(1) + \Pr(2) = 0.5945 + 0.3092 + 0.0804 = 0.9841$$

Hence the probability of winning more than twice is:

$$1 - 0.9841 = 0.0159 \text{ to four dec. pl.}$$

EXERCISES

1 Use a calculator to evaluate $\Pr(r) = \lambda^r e^{-\lambda}/r!$ where:

(a) $r = 4$ and $\lambda = 0.8$
(b) $r = 6$ and $\lambda = 2.1$

(a) $\Pr(4) = (0.8)^4 e^{-0.8}/4! = (*)(*)/(*) = *$ to four dec. pl.
(b) * to four dec. pl.

2 A disease is endemic in a certain population with incidence of 1 in 10 000. If 1000 people are selected from the population, what is the probability that:

(a) one person has the disease
(b) more than one has the disease

(a) The probability of a individual member of the population having the disease is *. For the Poisson distribution, with a sample of *:

$$\lambda = (*)(*) = *$$

hence the probability of a member of the sample having the disease is given as:

$$\Pr(1) = (*)^1 e^{-*}/1! = * \text{ to four dec. pl.}$$

(b) $\Pr(0) = (*)^0 e^{-*}/0! = *$ to four dec. pl. Consequently the probability of *one person or no person* having the disease in the sample is:

*

Hence the probability of more than one person having the disease is:

*

Unit 4 The Normal Distribution

Test yourself with the following:

1 It is given that variable x is a normally distributed random variable with a Gaussian probability curve enclosing a total area of 1. Inspection of the curve shows that the area:

(i) to the left of $x = 15$ is 0.346 and
(ii) to the right of $x = 29$ is 0.012

Draw a sketch of the curve, indicating these two areas and find the probability that a randomly selected value of x is such that:

(a) $15 \leqslant x < 29$
(b) $x \geqslant 15$
(c) $x < 29$

2 A normally distributed random variable x has a mean $\mu = 34$ and a standard deviation $\sigma = 2.5$. Find the z-values of the standard normal distribution that correspond to:

(a) $x = 37$
(b) $x = 30$
(c) $29 \leqslant x < 36$

3 Find the area beneath the standard normal distribution between z_1 and z_2, where:

(a) $z_1 = 0$ and $z_2 = 1.21$
(b) $z_1 = 0.51$ and $z_2 = 0.82$
(c) $z_1 = 0.94$ and $z_2 = \infty$
(d) $z_1 = -1.63$ and $z_2 = 0$
(e) $z_1 = -1.84$ and $z_2 = -0.96$

(f) $z_1 = -\infty$ and $z_2 = -0.54$
(g) $z_1 = -2.41$ and $z_2 = 1.36$

4 The heights of 250 people were measured and recorded. It was found that the mean height was 167.64 cm with a standard deviation of 8.89 cm. Assuming the heights to be normally distributed:

(a) what is the probability of selecting one person from among the 250 whose height is:
(i) less than 170 cm
(ii) more than 165 cm
(iii) between 164 cm and 173 cm

(b) of the 250 people, how many can you expect to have a height:
(i) less than 170 cm
(ii) more than 165 cm
(iii) between 164 cm and 173 cm

Continuous Probability Distributions

The binomial and the Poisson probability distributions are examples of **discrete** probability distributions – they give the probabilities associated with the values of a discrete random variable r.

While many random experiments such as tossing coins or rolling dice can be described using a discrete random variable, many others cannot. For example, if the random experiment involved measuring the height of a person the result of the experiment would be a measured height. This result would then have to be described using a **continuous** variable because height is a continuous quantity – no individual can grow from 1 m to 2 m in height without attaining a continuous scale of heights in between. It does not matter that the recorded measure might be restricted to a given number of decimal places so that in reality the recorded measure would be discrete, the value of the result itself must be considered as being available on a continuous scale of values.

For a continuous random variable to have with it associated probabilities, the probability distribution itself must necessarily be continuous. The most useful and most common of these continuous probability distributions is the **normal** or **Gaussian probability distribution**, the graph of which is characterized by the bell-shaped curve of Figure 27.7, where x represents the random variable and $P(x)$ the continuous probability distribution:

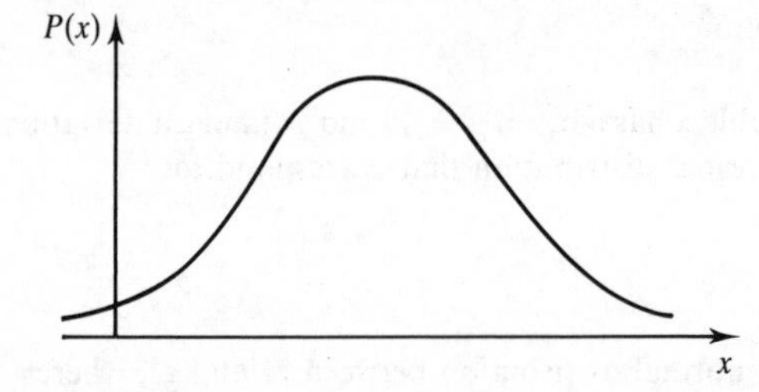

Figure 27.7

In the case of a discrete probability distribution each value of the random variable has an associated probability and the sum of all such probabilities is unity. For a continuous probability distribution we cannot associate a probability to a given value of the random variable because there is an infinity of such values and hence there would be an infinity of

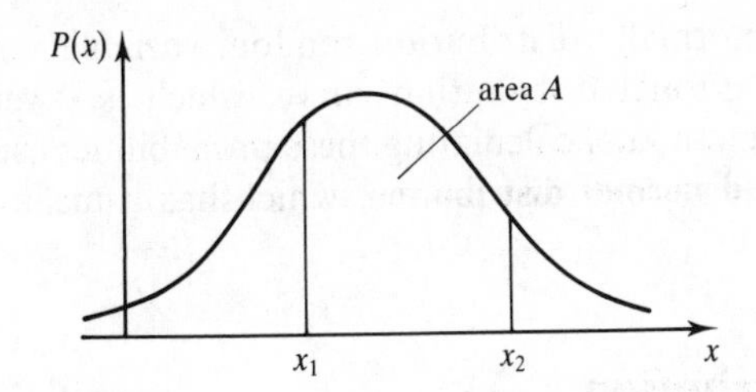

Figure 27.8

associated probabilities whose sum would exceed unity. Instead we associate probabilities to **ranges of values** of the random variable, where the probabilities are equal to the area beneath the graph of the probability distribution within that range. For example, in the graph of the normal probability distribution of the random variable x shown in Figure 27.8 the probability that x is found to have a value less than x_2 but greater than x_1 is equal to the area A:

$$P(x_1 < x < x_2) = A$$

Notice that this probability is the same as:

$$P(x_1 \leqslant x \leqslant x_2),\ P(x_1 < x \leqslant x_2) \text{ and } P(x_1 \leqslant x \leqslant x_2) = A$$

The addition of single points at the ends of the range does not affect the probability.

The Normal or Gaussian Probability Distribution

Just as with discrete probability distributions where the sum of probabilities is unity, with a continuous probability distribution, the *area* under the curve is unity. Thus the probability that a variable has a value lying within a given range of values is given by the area under the curve between the two extremes of the range of values. In Figure 27.8, the probability that $x_1 \leqslant x \leqslant x_2$ is equal to the area A. As can be seen from the graph, the bell-shaped curve, rising from small $P(x)$ to a maximum and then falling to small $P(x)$ again, is symmetric about a vertical line through the maximum. Where this vertical line meets the x axis is the mean μ of the normal distribution (Figure 27.9).

As the curve falls from its maximum it undergoes a change in curvature. Initially as it falls it is concave and then it switches to convex. The horizontal distance between the vertical line of symmetry and the point at which the curve switches curvature is the standard deviation of the normal distribution, as shown in Figure 27.10.

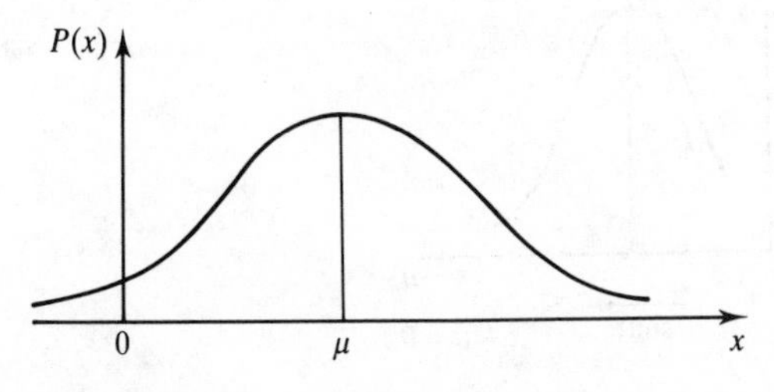

Figure 27.9

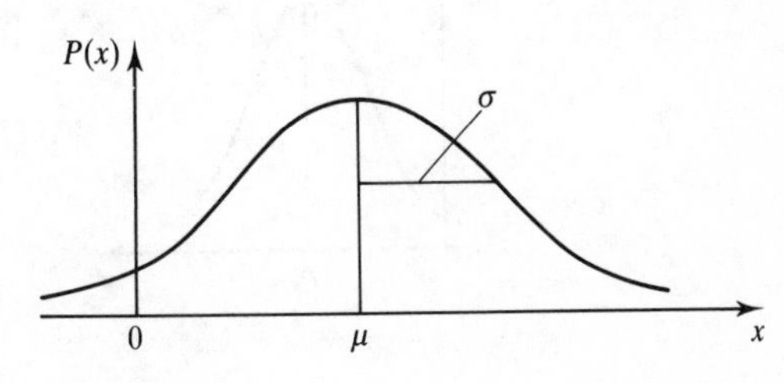

Figure 27.10

To calculate probabilities of a normally distribution random variable x requires the calculation of areas under a general normal distribution curve, which is a very involved process. Fortunately there is a simpler method of calculating these probabilities using a special normal distribution called the **standard normal distribution**, which has a mean of 0 and a standard deviation of 1.

The Standard Normal Distribution

The standard normal distribution, shown in Figure 27.11, is a normal distribution with a mean of 0 and a standard deviation of 1, and the random variable along the horizontal axis is no longer x but z for reasons that will soon become clear.

Any normal distribution can be transformed into the standard normal distribution by first shifting the vertical axis across so that it coincides with the mean value μ, thus changing the horizontal values from x-values to $(x - \mu)$-values, as shown in Figure 27.12.

Having shifted the vertical axis so that it coincides with the axis of symmetry the distribution can now be either stretched or squeezed to convert it into a normal distribution with a standard deviation of 1 (Figure 27.13).

If it has to be squeezed it is because the original normal distribution had a standard deviation $\sigma > 1$ and if it has to be stretched, the original normal distribution had a standard deviation that was less than 1. In either case the shape of the curve is dictated by the fact that the area beneath it should remain at 1. The result of this is that:

$(x - \mu)$-values become $[(x - \mu)/\sigma]$-values

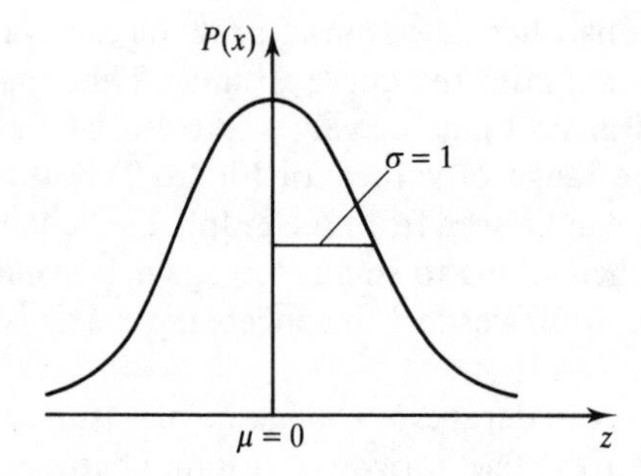

Figure 27.11

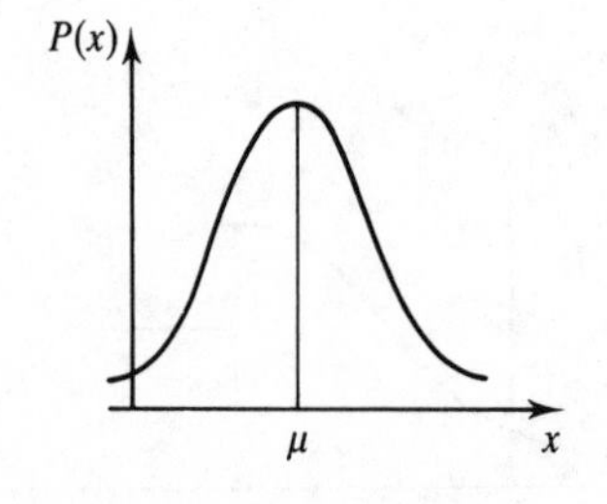

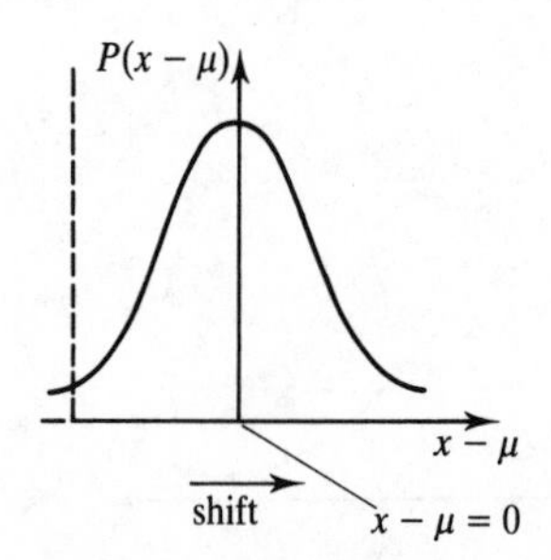

Figure 27.12

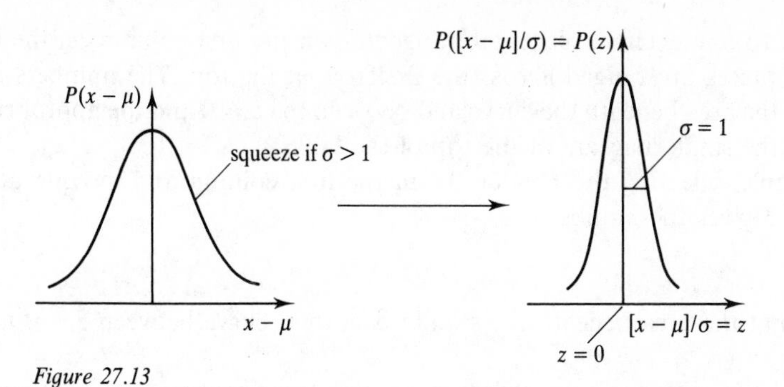

Figure 27.13

In conclusion, to convert a normal distribution with mean μ and standard deviation σ into a standard normal distribution the x-values of the original normal distribution are converted into z-values for the standard normal distribution by the transformation:

$$z = (x - \mu)/\sigma$$

For example, the heights of 100 people were measured and found to have a mean of $\mu = 1.72$ m with a standard deviation of $\sigma = 0.2$ m. It is assumed that the height, represented by the random variable x, is normally distributed with the measured mean and standard deviation. Find the standard normal distribution z-value corresponding to each of the following heights:

(a) 1.95 m
(b) 1.72 m
(c) 1.57 m

(a) $x = 1.95$, therefore $z = (1.95 - 1.72)/0.2 = 1.15$
(b) $x = 1.72$, therefore $z = (1.72 - 1.72)/0.2 = 0$
(c) $x = 1.57$, therefore $z = (1.57 - 1.72)/0.2 = -0.75$

Areas Beneath the Standard Normal Distribution

Areas beneath the standard normal distribution have been calculated and are contained in the Statistical Table on p. 553. A section of this table is reproduced below.

Area beneath the standard normal curve

z	**0**	**1**	**2**	**3**	**4**	**5**	**6**	**7**	**8**	**9**
0.0	0.000	0.004	0.008	0.012	0.016	0.020	0.024	0.028	0.032	0.036
0.1	0.040	0.044	0.048	0.052	0.056	0.060	0.064	0.068	0.071	0.075
0.2	0.080	0.083	0.087	0.091	0.095	0.099	0.103	0.106	0.110	0.114
0.3	0.118	0.122	0.126	0.129	0.133	0.137	0.141	0.144	0.148	0.152
0.4	0.155	0.159	0.163	0.166	0.177	0.174	0.177	0.181	0.184	0.188
0.5	0.192	0.195	0.199	0.202	0.205	0.209	0.212	0.216	0.219	0.222
0.6	0.226	0.229	0.232	0.236	0.239	0.242	0.245	0.249	0.252	0.255
0.7	0.258	0.261	0.264	0.267	0.270	0.273	0.276	0.279	0.282	0.285

The values of z, to one decimal place, are ranged down the first column on the left and the second decimal places are ranged across the first row at the top. The numbers in the body of the table give the area beneath the curve and between the $z = 0$ and the appropriate z-value, as indicated by the small diagram at the top of the table.

For example, selecting the z value 0.4 in the first column and moving across to the column headed 3 gives the number:

0.166

This tells you that the area beneath the standard normal curve between $z = 0$ and $z = 0.43$ is 0.166.

Because the area beneath the standard normal curve between a pair of z-values is the same as the area beneath the normal curve for equivalent x-values we now have a means of calculating probabilities for *any* normal distribution.

For example, if the ages of a group of students are normally distributed with a mean of 18 years and a standard deviation of 1.6 years we can use the standard normal distribution to find the probability of selecting a student whose age lies between any two ages. As an instance we can find the probability of selecting a student who is aged between 18 and 19 by finding the equivalent z-values for these two ages. Remember that:

$z = (x - \mu)/\sigma$ where $\mu = 18$ and $\sigma = 1.6$

so that:

$$z_{18} = (18 - 18)/1.6 = 0$$

$$z_{19} = (19 - 18)/1.6 = 0.625 = 0.63 \text{ to 2 decimal places}$$

From the tables we find that the area beneath the standard normal curve between $z = 0$ and $z = 0.63$ is:

0.236

This means that the area beneath the normal distributions of student ages between $x = 18$ and $x = 19$ is also:

0.236

Hence the probability of selecting a student aged between 18 and 19 is 0.236. In other words, 23.6% of the students are aged between 18 and 19.

Not all problems are quite as straightforward as that. Other problems are likely to require a greater knowledge of how to use the tables.

Using Tables

Looking closely at the tables, you will see that they only give areas to the right of $z = 0$. By the symmetry, areas to the left of $z = 0$ are the same as equivalent areas to the right so these can be easily read from the table. There are a number of cases of areas between two z-values z_1 and z_2 that we should consider. Refer to the table on page 553.

Case 1: $z_1 = 1$ and $z_2 = \infty$

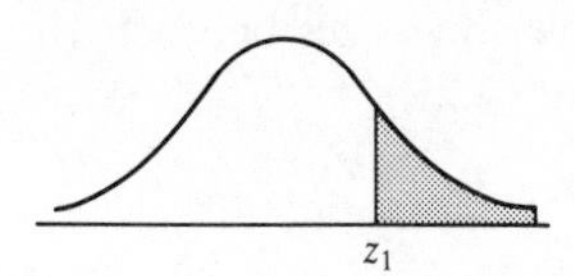

From the table, the area between $z = 0$ and $z = 1$ is given as 0.341. From the symmetry of the curve the area between $z = 0$ and $z = \infty$ is half the total area. Thus the area between $z = 1$ and $z = \infty$ is:

$$0.500 - 0.341 = 0.159$$

Case 2: $z_1 = -1.5$ and $z_2 = 0$

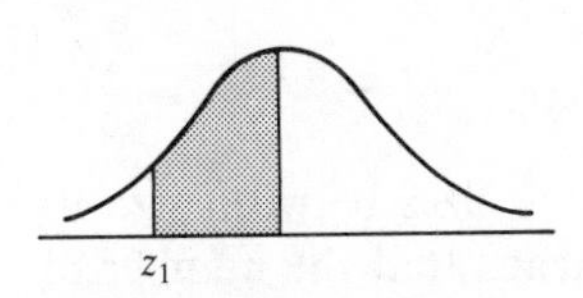

Areas to the left of $z = 0$ match areas to the right. Consequently, the area between $z = -1.5$ and $z = 0$ is the same as the area between $z = 0$ and $z = +1.5$, which is 0.433.

Case 3: $z_1 = -\infty$ and $z_2 = -1.5$

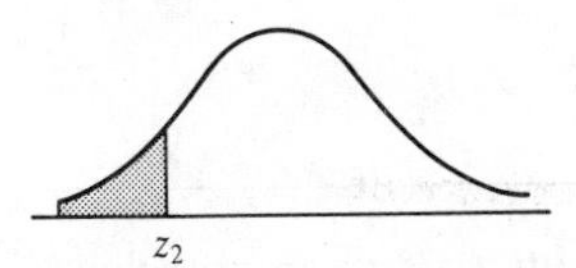

Again, this is the same area as the area between $z = +1.5$ and $z = +\infty$, which is:

$$0.500 - 0.433 = 0.067$$

Case 4: $z_1 = -\infty$ and $z_2 = 1$

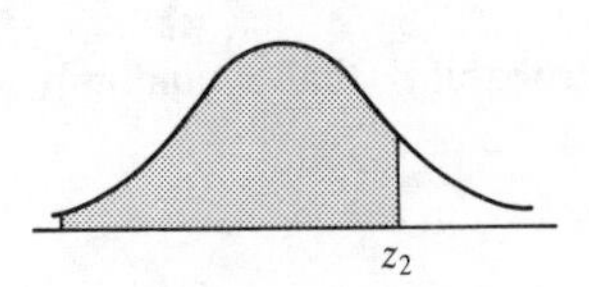

The area between $z = -\infty$ and $z = 0$ is half of the total area and the area between $z = 0$ and $z = 1$ is 0.341. As a result, the area between $z = -\infty$ and $z = 1$ is:

$$0.500 + 0.341 = 0.841$$

Case 5: $z_1 = -1.5$ and $z_2 = \infty$

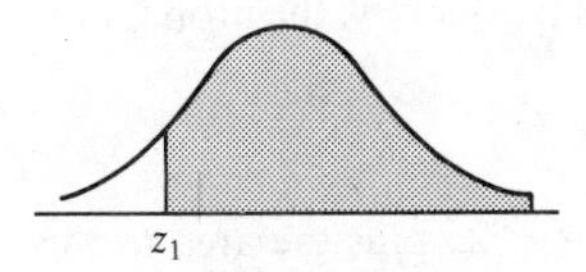

By the symmetry of the curve this area is the same as the area between $z = 0$ and $z = +1.5$ plus half the area beneath the curve:

$$0.433 + 0.500 = 0.933$$

Case 6: $z_1 = -1.5$ and $z_2 = 1$

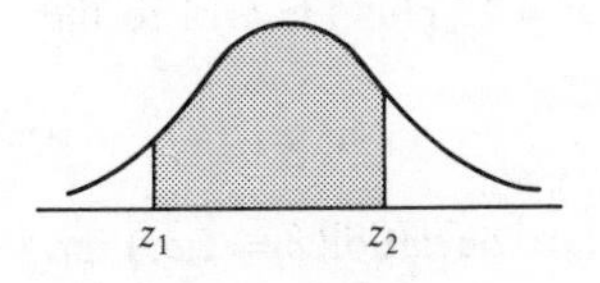

The area between $z = -1.5$ and $z = 0$ is 0.433. The area between $z = 0$ and $z = 1$ is 0.341. Consequently, the area between $z = -1.5$ and $z = 1$ is:

$$0.433 + 0.341 = 0.774$$

Case 7: $z_1 = 1$ and $z_2 = 1.5$

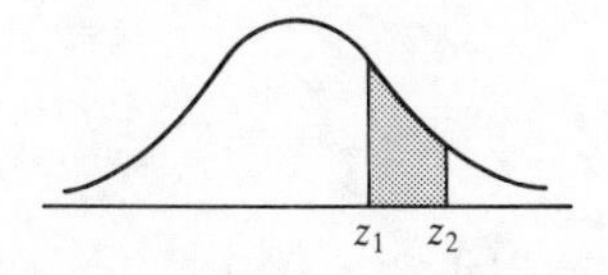

The area between $z = 0$ and $z = 1.5$ is 0.433. The area between $z = 0$ and 1 is 0.341. As a result, the area between $z = 1$ and $z = 1.5$ is:

$$0.433 - 0.341 = 0.092$$

For example, the shoe sizes of 2000 people are normally distributed with a mean size of 9.5 and a standard deviation of 1.25 sizes. How many people have shoe sizes between 10 and 11? Here, $\mu = 9.5$ and $\sigma = 1.25$, hence:

$$z_{10} = (10 - 9.5)/1.25 = 0.40$$

$$z_{11} = (11 - 9.5)/1.25 = 1.20$$

From the table the area between $z = 0$ and $z = 0.40$ is 0.155 and the area between $z = 0$ and $z = 1.20$ is 0.385. Consequently, the area beneath the standard normal distribution between $z = 0.40$ and $z = 1.20$ is:

$$0.385 - 0.155 = 0.230$$

This also means that the area beneath the normal distribution of shoe sizes between size 10 and size 11 is 0.230. Since the total area under the curve is 1 this means that the number of people with shoe sizes between 10 and 11 is:

$$2000(0.230) = 460$$

EXAMPLES

1 It is given that variable x is a normally distributed random variable with a Gaussian probability curve enclosing a total area of 1. Inspection of the curve shows that the area:

(i) to the left of $x = 12$ is 0.421 and
(ii) to the right of $x = 25$ is 0.104

Draw a sketch of the curve, indicating these two areas and find the probability that a randomly selected value of x is such that:

(a) $12 \leqslant x < 25$
(b) $x \geqslant 12$
(c) $x < 25$

(a) Area to the left of $x = 12$ is 0.421, area to the right of $x = 25$ is 0.104, the total area is 1 so that the area between $x = 12$ and $x = 25$ is $1 - (0.421 + 0.104) = 0.475$. Consequently, the probability that a randomly selected value of x is such that:

$$12 \leqslant x < 25 \text{ is } 0.475$$

(b) Area to the right of $x = 12$ is equal to the area between $x = 12$ and $x = 25$ plus the area to the right of $x = 25$. That is:

$$\Pr(\{x \geqslant 12\}) = 0.475 + 0.104 = 0.579$$

(c) Area to the left of $x = 25$ is equal to the area between $x = 12$ and $x = 25$ plus the area to the left of $x = 12$. That is:

$$\Pr(\{x < 25\}) = 0.475 + 0.421 = 0.896$$

2 A normally distributed random variable x has a mean $\mu = 34$ and a standard deviation $\sigma = 2.5$. Find the z-values of the standard normal distribution that correspond to:

(a) $x = 37$
(b) $x = 30$
(c) $29 \leqslant x < 36$

(a) $z = [37 - 34]/2.5 = 1.20$
(b) $z = [30 - 34]/2.5 = -1.60$
(c) $z_1 = [29 - 34]/2.5 = -2.00$ and $z_2 = [36 - 34]/2.5 = 0.80$ therefore $-2.00 \leqslant z < 0.80$

3 Find the area beneath the standard normal distribution between z_1 and z_2, where:

(a) $z_1 = 0$ and $z_2 = 1.54$
(b) $z_1 = 0.31$ and $z_2 = 0.72$
(c) $z_1 = 0.86$ and $z_2 = \infty$
(d) $z_1 = -1.49$ and $z_2 = 0$
(e) $z_1 = -1.76$ and $z_2 = -0.72$
(f) $z_1 = -\infty$ and $z_2 = -0.49$
(g) $z_1 = -2.23$ and $z_2 = 2.04$

Reading from the tables, taking into account that a tabular value represents an area to the left of the z-value and the origin and that the curve is symmetric about the origin:

(a) 0.438
(b) $0.264 - 0.122 = 0.142$
(c) $0.500 - 0.305 = 0.195$
(d) 0.432 (same as the area between 0 and $+1.49$)
(e) $0.461 - 0.264 = 0.197$ (same as the area between 0.72 and $+1.76$)
(f) $0.500 - 0.188 = 0.312$ (same as the area between $+0.49$ and ∞)
(g) $0.487 + 0.479 = 0.966$

4 2000 parcels being delivered by QuickPost Ltd were weighed and were found to have a mean weight of 62.55 kg with a standard deviation of 12.34 kg:

(a) what is the probability of selecting one parcel from the 2000 whose weight is:
(i) less than 69 kg
(ii) more than 45 kg
(iii) between 42 kg and 71 kg

(b) of the 2000 parcels, how many can you expect to have a weight:
(i) less than 59 kg
(ii) more than 45 kg
(iii) between 42 kg and 71 kg

(a) Conversion from x-values to z-values is according to $z = [x - 62.55]/12.43$

(i) $x = 69$ so that the corresponding standard normal variable value is

$$z = [69 - 62.55]/12.43 = 0.52 \text{ to 2 dec. pl.}$$

Hence, $\Pr(\{x < 69\}) = \Pr(\{z < 0.52\}) = 0.500 + 0.199 = 0.699$

(ii) $x = 45$ so that the corresponding normal variable value is

$$z = [45 - 62.55]/12.43 = -1.41 \text{ to 2 dec. pl.}$$

Hence, $\Pr(\{x > 45\}) = \Pr(\{z > -1.41\}) = 0.500 + 0.421 = 0.921$

(iii) when $x = 42$, $z = -1.65$ and when $x = 71$, $z = 0.68$ so that:

$$\Pr(\{42 < x < 71\}) = \Pr(\{-1.65z < 0.68\}) = 0.451 + 0.252 = 0.703$$

(b) (i) The number of parcels less than 69 kg is expected as $(2000)(0.699) = 1398$
(ii) The number of parcels more than 45 kg is expected as $(2000)(0.921) = 1842$
(iii) The number of parcels between 42 kg and 71 kg is expected as $(2000)(0.703) = 1406$

EXERCISES

1 It is given that variable x is a normally distributed random variable with a Gaussian probability curve enclosing a total area of 1. Inspection of the curve shows that the area:

(i) to the left of $x = 9$ is 0.304 and
(ii) to the right of $x = 17$ is 0.005

Draw a sketch of the curve, indicating these two areas and find the probability that a randomly selected value of x is such that:

(a) $9 \leqslant x < 17$
(b) $x \geqslant 9$
(c) $x < 17$

(a) Area to the left of $x = 9$ is *, area to the right of $x = 17$ is *, the total area is * so that the area between $x = 9$ and $x = 17$ is $* - (* + *) = *$. Consequently, the probability that a randomly selected value of x is such that:

$9 \leqslant x < 17$ is *

(b) Area to the right of $x = 9$ is equal to the area between $x = *$ and $x = *$ plus the area to the right of $x = *$. That is:

$\Pr(\{x \geqslant 9\}) = * + * = *$

(c) Area to the left of $x = *$ is equal to the area between $x = 9$ and $x = 17$ plus the area to the left of $x = 9$. That is:

$\Pr(\{x < 17\}) = *$

2 A normally distributed random variable x has a mean $\mu = 165$ and a standard deviation $\sigma = 24.8$. Find the z-values of the standard normal distribution that correspond to:

(a) $x = 176$
(b) $x = 158$
(c) $140 \leqslant x < 180$

(a) $z = [176 - 165]/24.8 = *$ to 2 dec. pl.
(b) $z = [158 - 165]/24.8 = *$ to 2 dec. pl.
(c) $z_1 = [140 - 165]/24.8 = *$ and $z_2 = [180 - 165]/24.8 = *$ therefore $* \leqslant z < *$

3 Find the area beneath the standard normal distribution between z_1 and z_2, where:

(a) $z_1 = 0$ and $z_2 = 2.22$
(b) $z_1 = 0.18$ and $z_2 = 0.69$
(c) $z_1 = 0.54$ and $z_2 = \infty$
(d) $z_1 = -0.35$ and $z_2 = 0$
(e) $z_1 = -2.01$ and $z_2 = -0.46$
(f) $z_1 = -\infty$ and $z_2 = -0.11$
(g) $z_1 = -1.59$ and $z_2 = 1.66$

Reading from the tables, taking into account that a tabular value represents an area to the left of the z-value and the origin and that the curve is symmetric about the origin:

(a) *
(b) $0.255 - * = *$
(c) $0.500 - * = *$
(d) * (same as the area between 0 and +*)
(e) $* - * = *$ (same as the area between * and *)
(f) * (same as the area between * and ∞)
(g) $* + * = *$

4 135 radioactive substances had their half-lives measured and it was found that the mean half-life was 137 ms with a standard deviation of 15 ms:

(a) what is the probability of selecting one radioactive substance from among the 135 whose half-life is:

(i) less than 145 ms
(ii) more than 129 ms
(iii) between 130 ms and 140 ms

(b) of the 135 radioactive substances, how many can you expect to have a half-life:
(i) less than 145 ms
(ii) more than 129 ms
(iii) between 130 ms and 140 ms

(a) Conversion from x-values to z-values is according to $z = [x - *]/*$
(i) $x = 145$ so that the corresponding standard normal variable is

$$z = [* - *]/* = * \text{ to 2 dec. pl.}$$

Hence, $\Pr(\{x < *\}) = \Pr(\{z < *\}) = 0.500 + * = *$.
(ii) $x = *$ so that the corresponding standard normal variable value is

$$z = [* - *]/* = * \text{ to 2 dec. pl.}$$

Hence, $\Pr(\{x > *\}) = \Pr(\{z > *\}) = *$

(iii) when $x = 130$, $z = *$ and when $x = *$, $z = *$, so that:

$$\Pr(\{* < x < *\}) = \Pr(\{*z < *\}) = * + * = *$$

(b) (i) The number of radioactive substances with a half-life less than $*$ is expected as $(135)(*) = *$.
(ii) The number of radioactive substances with a half-life more than $*$ is expected as $(*)(*) = *$.
(iii) The number of radioactive substances with a half-life between $*$ and $*$ is expected as $*$.

Module 27: Further Exercises

1 A library shelf contains 10 works of fiction, 24 historical novels, 4 reference works and 12 text books. A book is selected at random and its type noted. If each book has an equal chance of being drawn code the simple events of this random experiment and construct a probability distribution.

2 Find the mean and variance of the probability distribution of Question 1.

3 In a game, a coin is tossed 4 times. If you toss four heads you score 3, if you toss 3 heads you score 2 and if you toss anything else you score 0.
(a) Construct a probability distribution for the random variable x that represents the score.
(b) Find the mean and variance of this random variable.

4 Each day of the week you drive through a certain traffic light controlled intersection. Over a long period of time you have observed that you can drive through the light on green without stopping first only 15 times out of 64. What is the probability that, in any one week:
(a) there will only be two days out of five when you will have to wait for the lights to turn green.
(b) you will have to wait for the lights to turn green for every one of the five days
(c) you will have to wait for the lights to turn green for no more than three days out of the five.

5 The probability that a piece of buttered toast will fall to the floor buttered side down has been estimated as 0.6. For a recorded sample of 8 pieces of buttered toast that have fallen to the floor:
(a) Find the probabilities that:
(i) 2, 5 or 7 fell buttered side down
(ii) at least 6 fell buttered side down
(b) Draw the bar graph of this probability distribution.
(c) From the bar graph of this probability distribution construct the histogram that represents the piecewise continuous extension of the binomial probability distribution.
(d) From the histogram find the probability that more than 1 but less than 4 pieces fell to the ground buttered side down.

6 Use a calculator to evaluate $\Pr(r) = \lambda^r e^{-\lambda}/r!$ where:
(a) $r = 4$ and $\lambda = 0.7$
(b) $r = 1$ and $\lambda = 1.2$
(c) $r = 3$ and $\lambda = 0.2$

7 The chances of scoring 100% in a national examination are rated at 1 in 5000. From a sample of 200 who sat the examination what is the probability of finding:

(a) one 100% score

(b) more than 2 100% scores

8 It is given that variable x is a normally distributed random variable with a Gaussian probability curve enclosing a total area of 1. Inspection of the curve shows that the area:

(i) to the left of $x = 253$ is 0.482 and

(ii) to the right of $x = 304$ is 0.223

Draw a sketch of the curve, indicating these two areas and find the probability that a randomly selected value of x is such that:

(a) $253 \leqslant x < 304$

(b) $x \geqslant 253$

(c) $x < 304$

9 A normally distributed random variable x has a mean $\mu = 96$ and a standard deviation $\sigma = 14$. Find the z-values of the standard normal distribution that correspond to:

(a) $x = 105$

(b) $x = 82$

(c) $80 \leqslant x < 110$

10 Find the area beneath the standard normal distribution between z_1 and z_2, where:

(a) $z_1 = 0$ and $z_2 = 2.03$

(b) $z_1 = 0.11$ and $z_2 = 0.76$

(c) $z_1 = 1.23$ and $z_2 = \infty$

(d) $z_1 = -0.95$ and $z_2 = 0$

(e) $z_1 = -2.22$ and $z_2 = -1.22$

(f) $z_1 = -\infty$ and $z_2 = -0.23$

(g) $z_1 = -0.99$ and $z_2 = 1.67$

11 Traffic density on a motorway was measured in vehicles per minute passing beneath a certain bridge. The survey lasted 24 hours and the mean density was found to be 87 vehicles per minute with a standard deviation of 6 vehicles per minute. Assuming the traffic density to be normally distributed:

(a) what is the probability of counting the number of vehicles passing under the bridge in a randomly selected minute and finding that the traffic density was:

(i) less than 90 vehicles per minute

(ii) more than 75 vehicles per minute

(iii) between 80 vehicles per minute and 95 vehicles per minute

(b) how many minutes in the 24 hours have a traffic density of:

(i) less than 90 vehicles per minute

(ii) more than 75 vehicles per minute

(iii) between 80 vehicles per minute and 95 vehicles per minute

Chapter 7: Miscellaneous Exercises

1 List the possible outcomes of the result of each of the following random experiments:

(a) selecting a sock from a drawer containing blue and grey socks.

(b) selecting a page from a book containing 200 pages.

(c) measuring lengths up to 1 cm using a measuring scale graded to 1 mm.

(d) answering a Male/Female question on a questionnaire.

2 Write down in symbolic form, the following sets:

(a) the set of whole numbers between 5 and 9.

(b) the set of numbers between 5 and 9.

(c) the set of dogs.

(d) the set of vowels.

3 Find all possible subsets of the set A where:

(a) $A = \{1, 2, 3\}$
(b) $A = x$: x is a vowel$\}$
(c) $A = \{$red, white, blue$\}$

4 Which of the following pairs of sets, A and B, are mutually exclusive:

(a) $A = \{1, 2, 3, 4\}$
$B = \{x\colon x \geqslant 4\}$
(b) $A = \{$red, white, blue$\}$
$B = \{x\colon x$ is a colour in the German national flag$\}$
(c) $A = \{x\colon x$ is a vowel$\}$
$B = \{x\colon x$ is a consonant$\}$

5 List all the simple events associated with each of the random experiments in Question 1.

6 Find the sample space for each of the following pairs of random experiments performed in the sequence in which they are written:

(a) selecting a sock from a drawer containing blue and grey socks.
tossing a coin.
(b) tossing a coin.
selecting a sock from a drawer containing blue and grey socks.

7 A card is drawn from a deck of 52 and the suit is noted, after which a coin is tossed. Draw an outcome tree for this sequence of experiments.

8 Use a Venn diagram to find the union and intersection of each of the following pairs of sets:

(a) $A = \{1, 3, 5\}$ and $B = \{2, 3, 4, 5\}$
(b) $A = \{x\colon -1 \leqslant x < 2\}$ and $B = \{x\colon 0 \leqslant x \leqslant 3\}$
(c) $A = \{$Head, Tail$\}$ and $B = \{$Club, Heart, Spade, Diamond$\}$

9 Write down the sample space of each of the following experiments and the complement of the given event:

(a) tossing a coin; $E = \{$Head$\}$
(b) rolling a six-sided die; $E = \{1, 2\}$
(c) drawing a card from a deck of 52; $E = \{$Hearts, Clubs, King of Spades$\}$
(d) selecting a component from an assembly line; $E = \{x\colon x$ is a defective component$\}$
(e) asking a person their age; $E = \{x\colon x$ is an age and $x \leqslant 25\}$
(f) measuring the height of a tree; $E = \{x\colon x$ is the height of a tree and $x < 8$ m$\}$

10 A box contains lead pencils of three different grades, B, HB and H. The pencils are painted either brown or yellow. The following table shows the quantity of each in the box.

	B	HB	H
Brown	20	15	5
Yellow	25	20	15

A pencil is selected at random. Calculate the probability that it is:

(a) Painted yellow
(b) Not a B grade
(c) A brown HB
(d) Not a yellow H
(e) Either a brown HB or a red B

11 From a deck of 52 playing cards one card is selected at random. What are the probabilities that the card is:

(a) An Ace?
(b) A red card
(c) The Jack of clubs?
(d) Not a black seven?
(e) A red four or a black two?
(f) The Queen of diamonds or the Ace of hearts?

12 If the card in Question 11 is replaced in the deck, the deck shuffled and a second card drawn, what are the probabilities that:

(a) Both cards were an Ace?
(b) Both cards were red?
(c) The first card was not a Four and the second was not a Two?
(d) At least one of the two cards was a King?
(e) If the first card drawn had not been replaced would the probabilities of (a) to (d) have been different?

13 In an experiment to study the learning process 20 rats were introduced into a maze, at the centre of which is a reward. The following table records the time taken by the rats to achieve the reward.

Time (min)	1	2	3	4
Number of rats	1	5	10	4

What is the probability that a rat chosen at random achieved the reward:

(a) in 2 minutes
(b) in less than 3 minutes
(c) either in 2 minutes or in more than 1 minute

The same collection of rats are sent into the maze a second time and the following frequency table results:

Time (min)	1	2	3	4
Number of rats	1	9	8	2

What are the probabilities that a rat chosen at random achieved the reward in its second attempt:

(d) in 2 minutes
(e) in less than 3 minutes

What is the probability that a rat chosen at random achieved the reward both times:

(f) in 2 minutes
(g) in less than 3 minutes.

14 A market research team find that only 1% of its approaches to potential customers result in a sale. What is the probability that out of 150 approaches:

(a) 3 sales result
(b) no more than 2 sales result
(c) more than 2 sales result
(d) no sales result.

15 The chances of contracting malaria during a visit to a mosquito infested country is 0.001, provided that anti-malarial drugs are consumed. What are the chances of catching malaria under these conditions in:

(a) one visit
(b) ten visits
(c) one hundred visits.

16 The percentage attendance at a night class is normally distributed with a mean of 80% and a standard deviation of 12%. What is the probability that on any given night the attendance is:

(a) greater than 75%
(b) less than 85%
(c) between 75% and 85%

17 Computer floppy disk failure occurs after a disk has been used a number of times. The non-failure usage rates are normally distributed with a mean of 1020 uses and a standard deviation of 22 uses. What is the probability that a disk will fail after:

(a) 1020 uses
(b) 2000 uses
(c) 900 uses
(d) between 850 and 1150 uses

Area beneath the standard normal curve

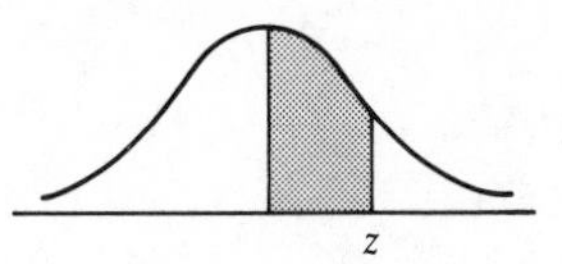

z	0	1	2	3	4	5	6	7	8	9
0.0	0.000	0.004	0.008	0.012	0.016	0.020	0.024	0.028	0.032	0.036
0.1	0.040	0.044	0.048	0.052	0.056	0.060	0.064	0.068	0.071	0.075
0.2	0.080	0.083	0.087	0.091	0.095	0.099	0.103	0.106	0.110	0.114
0.3	0.118	0.122	0.126	0.129	0.133	0.137	0.141	0.144	0.148	0.152
0.4	0.155	0.159	0.163	0.166	0.177	0.174	0.177	0.181	0.184	0.188
0.5	0.192	0.195	0.199	0.202	0.205	0.209	0.212	0.216	0.219	0.222
0.6	0.226	0.229	0.232	0.236	0.239	0.242	0.245	0.249	0.252	0.255
0.7	0.258	0.261	0.264	0.267	0.270	0.273	0.276	0.279	0.282	0.285
0.8	0.288	0.291	0.294	0.297	0.300	0.302	0.305	0.308	0.311	0.313
0.9	0.316	0.319	0.321	0.324	0.326	0.329	0.332	0.334	0.337	0.339
1.0	0.341	0.344	0.346	0.349	0.351	0.353	0.355	0.358	0.360	0.362
1.1	0.364	0.367	0.369	0.371	0.372	0.375	0.377	0.379	0.381	0.383
1.2	0.385	0.387	0.389	0.391	0.393	0.394	0.396	0.398	0.400	0.402
1.3	0.403	0.405	0.407	0.408	0.410	0.412	0.413	0.415	0.416	0.418
1.4	0.420	0.421	0.422	0.424	0.425	0.427	0.428	0.429	0.431	0.432
1.5	0.433	0.435	0.436	0.437	0.438	0.439	0.441	0.442	0.443	0.444
1.6	0.445	0.446	0.447	0.448	0.450	0.451	0.452	0.453	0.454	0.455
1.7	0.455	0.456	0.457	0.458	0.459	0.460	0.461	0.462	0.463	0.463
1.8	0.464	0.465	0.466	0.466	0.467	0.468	0.469	0.469	0.470	0.471
1.9	0.471	0.472	0.473	0.473	0,474	0.474	0.475	0.476	0.476	0.477
2.0	0.477	0.478	0.478	0.479	0.479	0.480	0.480	0.481	0.481	0.482
2.1	0.482	0.483	0.483	0.483	0.484	0.484	0.485	0.485	0.485	0.486
2.2	0.486	0.486	0.487	0.487	0.488	0.488	0.488	0.488	0.489	0.489
2.3	0.489	0.490	0.490	0.490	0.490	0.491	0.491	0.491	0.491	0.492
2.4	0.492	0.492	0.492	0.493	0.493	0.493	0.493	0.493	0.493	0.494
2.5	0.494	0.494	0.494	0.494	0.495	0.495	0.494	0.495	0.495	0.495
2.6	0.495	0.496	0.496	0.496	0.496	0.496	0.496	0.496	0.496	0.496
2.7	0.497	0.497	0.497	0.497	0.497	0.497	0.497	0.497	0.497	0.497
2.8	0.497	0.498	0.498	0.498	0.498	0.498	0.498	0.498	0.498	0.498
2.9	0.498	0.498	0.498	0.498	0.498	0.498	0.499	0.499	0.499	0.499
3.0	0.499	0.499	0.499	0.499	0.499	0.499	0.499	0.499	0.499	0.499
3.1	0.499	0.499	0.499	0.499	0.499	0.499	0.499	0.499	0.499	0.499
3.2	0.499	0.499	0.499	0.499	0.499	0.499	0.499	0.499	0.499	0.499

▪ ▪ ▪ Solutions

CHAPTER 1

Module 1

Unit 1: Test

1 (a) -5 (b) 44 (c) 3 (d) -13
2 (a) 164: 2, 41: $2 \times 2 \times 41$; 1025: 5, 41: $5 \times 5 \times 41$
(b) 164: 1, 2, 4, 41, 82, 164; 1025: 1, 5, 25, 41, 205, 1025
(c) HCF: 41, LCM: 4100
3 (a) 120, 100, 0
(b) 550, 500, 1000
(c) -950, -900, -1000
4 Estimate: 295; exact: 313.7 to 1 dec. pl.

Unit 1: Exercises

1 (c) -51
2 (a) 66: 2, 3, 11: $2 \times 3 \times 11$
144: 2, 3: $2 \times 2 \times 2 \times 2 \times 3 \times 3$
(b) 66: 1, 2, 3, 6, 11, 22, 33, 66
144: 1, 2, 3, 4, 6, 8, 9, 12, 16, 18, 24, 36, 48, 72, 144
(c) HCF: 6
3 (a) 1040, 1000, 1000
(b) 820, 800, 1000
(c) -550, -500, -1000

Unit 2: Test

1 (a) -10.53 (b) 79.27 (c) -233.38 to two dec. pl.
2 (a) 13.5, 13.45 (b) -7.06, -7.06 (c) 20.0, 20.00
3 (a) 5/8 (b) 5.11 to two dec. pl.
4 (a) 10.01, 100.1, 1001: 0.1001, 0.010 01, 0.001 001
(b) 0.502, 5.02, 50.2: 0.005 02, 0.000 502, 0.000 050 2
(c) -1931.124, -19311.24, $-193\,112.4$: $-19.311\,24$, $-1.931\,124$, $-0.193\,112\,4$
5 (a) 1.1523×10^2 (b) 1.23×10^{-3}
6 Estimate: 0.015, exact: 0.0165 to three sig. fig.

Unit 2: Exercises

2 (a) 13.7, 13.66 (b) -3.16, -3.16 (c) 0.0915, 0.09
3 (a) 0.1101, 1.101: 0.000 110 1, 0.000 011 01
(b) 60.234, 6023.4: 0.602 34, 0.006 023 4
(c) -2501.023, $-250\,10.23$: $-2.501\,023$, $-0.250\,102\,3$
5 (a) 2.74312×10^3 (b) 1.0×10^{-5}
6 -63.22 to four sig. fig.

Unit 3: Test

1 (a) 1/5 (b) 2/3 (c) 7/3
2 (a) 11/15 (b) $-1/20$ (c) 3/2 (d) 5/3
(e) 14/15 (f) 1/8 (g) 2 (h) 1/8
3 (a) $0.\dot{2}$ (b) $0.\dot{5}7142\dot{8}$ (c) $4.\dot{8}\dot{1}$
4 2:1:3
5 (a) 20 (b) 25%

Unit 4: Test

4 (a) $\pm 3j$ (b) 1 (c) $\pm 9j$ (d) j
(e) $\pm j/(2\sqrt{2})$

Unit 4: Exercises

1 (a) rational (b) rational (c) irrational
2 (a) $\pm 9j$ (b) $\pm j/(4\sqrt{2})$ (c) $-j$ (d) 243j

Module 1: Further exercises

1 (a) 2 (b) 166 (c) -13 (d) -34
2 (a) 255: 1, 3, 5, 15, 17, 51, 85, 255: 3, 5, 17
1020: 1, 2, 3, 4, 5, 6, 10, 12, 15, 17, 20, 30, 34, 51, 60, 68, 85, 102, 170, 204, 255, 340, 510, 1020: 2, 3, 5, 17
HCF: 255; LCM: 1020
3 (a) 230, 200, 0 (b) 700, 700, 1000
(c) -350, -300, 0
4 Estimate: 1996; exact: 1906.7 to one dec. pl.
5 (a) -0.67 (b) 105.67 (c) -233.38 all to two dec. pl.
6 (a) 53.4, 53.35 (b) -9.05, -9.05
(c) 30.0, 29.95
7 (a) 6 (b) 3.02×10^{-7} to three sig. fig.

8 (a) 10.1, 101, 1010: 0.101, 0.0101, 0.001 01
(b) 0.283, 2.83, 28.3: 0.002 83, 0.000 283, 0.000 028 3
(c) −1721.012, −17 210.12, −172 101.2: −17.210 12, −1.721 012, −0.172 101 2
9 (a) 3.2567×10^2 (b) 5.24×10^{-3}
10 Estimate: 7.1; exact: 7.324 to four sig. fig.
11 (a) 3/7 (b) 3/4 (c) 11/9
12 (a) 7/8 (b) −1/24 (c) 12/5 (d) 4
(e) 5/6 (f) 3/35 (g) 1/3 (h) 1/125
13 (a) $-0.\dot{4}2857\dot{1}$ (b) $0.\dot{4}\dot{5}$ (c) $2.\dot{4}$
14 5:4:11
15 (a) 12 (b) 20%
16 By approximating them to rational numbers.
17 Irrational numbers.
18 (a) ±5j (b) ±25j (c) ±j/8 (d) j
(e) −1

Module 2

Unit 1: Test

1 644 m^3
2 (a) 0.01, 1% (b) −0.024, −2.4%
(c) 0.03, 3%
3 (a) -4.5×10^{-3}, −0.45% (b) −0.0153, −1.53% (c) 1.1×10^{-3}, 0.11%
(d) 1.5×10^{-2}, 1.5%
4 (a) 13.5% (b) 9%
5 (a) +1058 cm^3 or −1048 cm^3 (b) +0.0158 or −0.0157 (c) +1.58% or −1.57%
6 85.4 kph to one dec. pl., 9.997% error

Unit 1: Exercises

1 147.823 m^3 rounds up to 150 m^3.
2 (a) 0.35, 0.35/7, 0.05, 5% (b) −0.6, −0.6/26, −0.023, −2.3% (c) −0.4, −0.4/−12.8, 0.031, 3.1%
3 (a) 0.0115, 1.15% (b) −0.0305, −3.05%
(c) 0.005, 0.5% (d) −0.3, −30%
4 9%

Unit 2: Test

1 (a) 100.98 (b) 4.74 (c) 141.74 (d) 13.37 to four sig. fig. (e) 0.535 to three sig. fig. (f) 16.42 to four sig. fig.
2 (a) 0.06345 to four sig. fig. (b) 16.1604
(c) 1.054 to four sig. fig. (d) 5722.3 to one dec. pl.
3 (a) 2.10 to two dec. pl. (b) 0.474 to three dec. pl.
4 23.141
5 (a) 3.1536×10^9 (b) 70.4 (c) £623.99
6 17.67 g

Unit 2: Exercises

1 (a) 23.9 (b) 10.88 (c) 273.079
(d) 5.897 338 4 (e) 0.136 081 (f) 14.938 959
2 (a) 0.180 180 1 (b) 8.8804 (c) 5.047 969 9
(d) 19.756 878
3 (a) 3.616 766 5 (b) 1.139 208 5
(c) 85 320.892
4 9.87

Unit 3: Test

1 Frequencies: 9, 13, 18, 14, 6
2 Group frequencies: 2, 4, 5, 8, 4, 2

Unit 3: Exercises

1 Frequencies: 1, 7, 9, 8, 4, 1
2 Frequencies: 7, 10, 8, 8, 7

Unit 4: Test

1 1508 to nearest tonne
2 Mean 7.5, median 8, mode 9
3 Frequencies: 3, 5, 5, 4, 4, 3: mean 168.7 to one dec. pl.
4 Median: 169, modal groups: 150–159, 160–169

Unit 4: Exercises

1 Mean 35, median 32.5, mode 21
3 Mode 50.0, median 50.05

Unit 5: Test

1 342, 0.014, 0.002, 0.008, 1, 1
2 MAD: 1.0 to one dec. pl., standard deviation: 1.2 to one dec. pl.
3 MAD: 0.11 to two dec. pl., standard deviation: 0.14 to two dec. pl.
4 Alan: mean 65, standard deviation 2.55 to two dec. pl.
Beth: mean 65, standard deviation 3.35 to two dec. pl.

Unit 5: Exercises

2 MAD: 2.29 to two dec. pl., standard deviation 2.58 to two dec. pl.
3 Standard deviation 2.50 to two dec. pl.
4 A: mean 160.5, standard deviation 3.64
B: mean 160.5, standard deviation 13.43

Module 2: Further exercises

1 (a) 1.96 m, 3.0625 m, 0 m
2 $3\frac{1}{3}$ ohms
3 (a) 50, 51.5 (b) −3, −3.04 (c) 27, 24.7
(d) 0.000 06, 0.000 05
4 (a) 78 (b) 72.8 (c) 1.1 (d) 1.14

5 3%
6 3%
8 (a) 18.6°C (b) 18.3°C
9 Median 18°C, mode 18°C
10 16°C–20°C
11 MAD 2.8, standard deviation 3.64
12 0:05:41 to the nearest second
13 Standard deviation 40.1 seconds
14 1.5 standard deviations to one dec. pl.
15 3 times
16 Colin: Mean 3, standard deviation 1.41
Dorothy: mean 3, standard deviation 0.63

Module 3

Unit 1: Test

1 Km = (8/5) Miles
2 Amount = 6000 – 250 (Number paid)
3 (a) (i) u (ii) s (iii) T
(b) (i) u dependent; v independent
(ii) s dependent; u, t independent
(iii) T dependent, L independent

Unit 1: Exercises

1 $P = 0.025U$
2 $T = 25 + 5N$
3 (i) Subject and dependent variable r, independent variable s.
(ii) Subject and dependent variable u, independent variable v, w.
(iii) Subject and dependent variable d, independent variables x, y.

Unit 2: Test

1 $r = 0.017\dot{3}d$, (a) $0.78r$ interpolation
(b) 67.5° interpolation
(c) $3.4\dot{6}r$ extrapolation
(d) 7.5° extrapolation
2 $T = 125 + 53U$, (a) 496 interpolation
(b) 22 units interpolation
(c) 178 extrapolation
(d) 40 units extrapolation

Unit 2: Exercises

1 Ratio = Radius/3 (a) Ratio = 6, interpolation (b) Radius = 24 cm, interpolation (c) Ratio = 4, extrapolation (d) Radius = 45 cm, extrapolation
2 $U = 1000 + 20D$ (a) $D = 7.5\%$, interpolation (b) $U = 1200$, interpolation (c) $D = 25\%$, extrapolation (d) $U = 1020$, extrapolation

Module 3: Further exercises

1 ounces = 0.035 grams
2 pay = 7.25 hours
3 (a) Subject and dependent variable p, independent variable q.
(b) Subject and dependent variable z, independent variables x, y.
(c) Subject and dependent variable s, independent variables u, t.
4 $V = 1.4$ M (a) 21.7 cm^3, interpolation (b) 25 g, interpolation (c) 3.5 cm^3, extrapolation (d) 44.3 g to one dec. pl., interpolation

Module 4

Unit 1: Test

2 [gradient, intercept] (a) 2, –1 (b) –4, 2 (c) –1/2, 17/6 (d) 1, –4
3 (a) $y = 2x - 16$ (b) $y = -x - 1$
(c) $x + 4y = -19$ (d) $x - 5y = 35$
4 (a) $y = 8x - 2$ (b) $y = -3x + 3$
(c) $y = -7x + 11$ (d) $2y - 5x = -16$
5 (a) $q = 1$ (b) $x = 2$ (c) $v = -2$
(d) $s = 2$

Unit 1: Exercises

1 (a) Gradient 5, intercept –9. (b) Gradient –8/9, intercept 8.
2 [gradient, intercept] (a) 2, 3 (b) –6, –10 (c) –7/4, 11/4 (d) 5/3, –4
3 (a) $y = 8x - 13$ (b) $7x - 2y = 8$
4 (a) $y = 5x - 3$ (b) $y = -2x + 1$
(c) $y = 3x - 5$ (d) $y = (-8/9)x + 8$
5 (a) $v = 3$ (b) $t = -6$ (c) $y = 5$
(d) $m = 0$

Unit 2: Test

1 (a) (2, 1) (b) (6, –5)
2 (0, –1), (–5/2, –2), (–3/2, 0)
3 (–1, 0)

Unit 2: Exercises

(a) (2, –2) (b) (3, –1)

Module 4: Further exercises

2 [gradient, intercept] (a) 5, –3 (b) –2, 2
(c) –1/2, 4 (d) –1, 8
3 (a) $y = 3x - 4$ (b) $y = -2x + 1$
(c) $2x + 3y = -1$ (d) $4x - 2y = 6$
4 (a) $y = 6x + 4$ (b) $y = -5x + 17$
(c) $7x + 3y = 32$ (d) $3y - 4x = -12$

5 (a) (3, 10) (b) (−3, 1)
6 (5/2, 1/2), (−1, −3), (4/3, 5/3)
7 (3/2, 1/2)
8 (a) $r = 8/3$ (b) $x = 2$ (c) $v = 4$
(d) $n = -3$
9 (−3, −10)
10 A: $13\frac{1}{3}$ tonnes; B: $6\frac{2}{3}$ tonnes. The plant would close down.

Chapter 1: Miscellaneous exercises

1 (a) 25 (b) ± 3
3 12
5 15% decrease
7 Average: $1 + 1/n^2$, standard deviation: $\sqrt{(n-1)}/n^2$
9 37.5
11 4/11
13 6

CHAPTER 2

Module 5

Unit 1: Test

1 (a) $5x - 20y - 60z + 360$ (b) $2x - 5y - 23$
(c) $56x + 112y - 40z - 568$
2 No
3 No
4 (a) $(a^2 + b^2)/ab$ (b) a/b
(c) $(a^2b - ac + b^2d)/ab^2$

Unit 1: Exercises

1 (a) $-2x - 12y + 18z - 18$
4 (a) $(a^2 + bc)/ab$ (c) $(a^2 + b^2 + c^2)/abc$

Unit 2: Test

1 (a) $x + 7yz$ (b) $(s - 3u)(t - p)$
(c) $a[b(c + 1) + 1]$ (d) $-2x + 62y - 308$
2 1.42 seconds, 2.01 seconds, 2.84 seconds to two dec. pl.
3 1.8 gauss to two significant figures.
4 (a) a^2 (b) $(u + at)^2$

Unit 2: Exercises

1 (a) $3(4ab - 3x)$ (b) $l^2 - 2ml - m^2$
(c) $(a + b)(b + c)$ (d) $(x + y)/\sqrt{(xy)}$
2 35, 60, 85, 110
3 1.033
4 (a) a^2 (b) 1/2

Unit 3: Test

1 (a) $2(3x^2 + 5x - 12)$
(b) $2a^3 + a^2 - 4a + 15$
2 (a) $x - 5$ (b) $x + \dfrac{1}{x + 1}$ (c) $a - b$
(d) $p - 1 + (2p + 2)/(p^2 + p + 1)$

Unit 3: Exercises

1 (a) $2(6p^2 - 23p - 18)$
(b) $12x^3 + 16x^2 - 51x - 40$
2 (a) $x + 3$ (b) $(2x - 5) + 9/(x + 2)$
(c) $a^2 - ab + b^2$
(d) $x - (x + 1)/(3x^2 + 2x + 1)$

Unit 4: Test

1 (b)
2 Plot more points and join with a smooth curve.

Unit 4: Exercises

1 (a) a conditional equation (b) an identity
(c) an identity

Module 5: Further exercises

1 (a) $-3p + 21q + 168r - 504$
(b) $3t - 10s + 25$
(c) $-36x + 216y + 54z - 414$
2 (a) $(ab - c)/c$ (b) $(a/c)^2$
(c) $a^2(b^2 + c^2)/b^2c^2$
3 (a) $7u + 2v$
(b) $48w - 30u + 6uw - 12vw - 96$
(c) $x/(x + y)$
4 2.34 amps, 2.15 amps to two decimal places.
5 6.37 s, 7.36 s, 12.75 s to two decimal places.
6 (a) $(ac)^2$ (b) $1 - 2s$ or $2r - 1$ or $r - s$
7 (a) $24 - 10p - 6p^2$
(b) $15x^3 + 32x^2 - 17x + 2$
8 (a) $x + 6$ (b) $x + 1/(x - 1)$
(c) $a^2 - 2ab + 4b^2 - 520b^3/(a + 2b)$
(d) $-x + (11x - 1)/(3x^2 + 2)$
9 (b)
10 Plot more points and join them with a smooth curve.

Module 6

Unit 1: Test

1 (a) 4: 8, −3, 0, 7, −9
(b) 6: 9, 0, 0, 0, 2, −4, 8
2 (a) $a = -3$; $b = -2$; $c = 11$ (b) $a = 12$; $b = c = d = 0$; $e = 13$ (c) $a = -7$; $b = 3$; $c = 2$
3 (a) $x^2 + x - 12$ (b) $3x^2 + 2x - 8$
(c) $16x^2 - 40x + 25$
4 (a) $6x^3 - x^2 - 46x - 15$
(b) $9x^3 + 51x^2 - 80x + 28$
(c) $27x^3 - 54x^2 + 36x - 8$

5 (a) $(x+2)(x+3)$ (b) $(x-3)^2$
(c) $(2x+5)(2x-5)$
(d) $(3x+2)(9x^2-6x+4)$
(e) $(4x-3)(16x^2+12x+9)$

Unit 1: Exercises

1 (a) 4: 9, 8, 2, −3, 5 (b) 8: 100, 0, 0, 0, 0, 0, 0, 0, −13
2 (a) $a=12$; $b=-10$; $c=8$ (b) $a=8$; $b=0$; $c=0$; $d=-16$ (c) $a=13$; $b=5$; $c=2$
3 (a) $(x+4)(x+3)$ (b) $(x-5)^2$
(c) $(4x-6)(4x+6)$
(d) $(x-2a)(x^2+2ax+4a^2)$
(e) $(x+1/2)(x^2-x/2+1/4)$

Unit 2: Test

1 (a) x^2+6x+9
(b) $x^3-15x^2+75x-125$
(c) $216x^3+108x^2+18x+1$
(d) $4096x^4-6144x^3+3456x^2-864x+81$
2 (a) $x^8+8x^7+28x^6+56x^5+70x^4+56x^3+28x^2+8x+1$
(b) $x^4-16x^3+96x^2-256x+256$
(c) $125x^3+75x^2+15x+1$
(d) $6561x^4-17\,496x^3+17\,496x^2-7776x+1296$
3 (a) 5040 (b) 132 (c) 1 513 512 000
(d) $1/n$
4 (a) $23!/20!$ (b) $87!/84!$ (c) $(n+4)!/n!$
5 (a) $(55)6!$ (b) $n!(n+2)^2$
6 (a) See answer 2(a).
(b) $4\,782\,969x^7+3\,720\,087x^6+1\,240\,029x^5+229\,635x^4+25\,515x^3+1701x^2+63x+1$
(c) $-x^5/16\,807+5x^4/2401-10x^3/343+10x^2/49-5x/7+1$
(d) $x^4-32x^3+384x^2-2048x+4096$
(e) $x^6/7\,529\,536+3x^5/38\,416+15x^4/784+5x^3/2+735x^2/4+7203x+117\,649$
7 (a) $8008x^{10}/59\,049$
(b) $-(17\,006\,112)(792)x^5$
8 (a) $(1001)5^8$ (b) $(252)20^5$

Unit 2: Exercises

1 (a) x^3-3x^2+3x-1 (b) $4x^2+16x+16$
(c) $27x^3-108x^2+144x-64$
2 (a) $x^7+7x^6+21x^5+35x^4+35x^3+21x^2+7x+1$
(b) $x^4-8x^3+24x^2-32x+16$
(c) $125x^3-450x^2+540x-216$
3 (a) 120 (b) 336 (c) 10 080
(d) n^2+3n+2
4 (a) $16!/13!$ (b) $32!/29!$
(c) $(n+1)!/(n-2)!$
5 (a) $(29)4!$ (b) $(n^2+n-1)(n-1)!$
6 (b) $x^8/65\,536+x^7/2048+7x^6/1024+7x^5/128+35x^4/128+7x^3/8+7x^2/4+2x+1$
(c) $x^6-18x^5+135x^4-540x^3+1215x^2-1458x+729$
(d) $16x^4+192x^3+864x^2+1728x+1296$
7 (a) $(165)7^8x^8$ (b) $(630)15^4x^4$
8 (a) $-91/25$ (b) $156(110)^7$

Module 6: Further exercises

1 (a) $x^2+8x+16$ (b) $x^3-18x^2+108x-216$
(c) $125x^3+225x^2+135x+27$
(d) $1296x^4-1728x^3+864x^2-192x+16$
2 (a) $x^5+5x^4+10x^3+10x^2+5x+1$
(b) $x^6-36x^5+540x^4-4320x^3+19\,440x^2-46\,656x+46\,656$
(c) $243x^5+810x^4+1080x^3+720x^2+240x+32$
(d) $117\,649x^6-403\,368x^5+576\,240x^4-439\,040x^3+188\,160x^2-430\,08x+4096$
3 (a) 362 880 (b) 0.0625 (c) 39 916 800
(d) $n(n^3+2n^2-n-2)$
4 (a) $49!/46!$ (b) $17!/14!$
(c) $(n+3)!/(n-1)!$
5 (a) $(11\,879)8!$ (b) $n!(n^2+2n+1)$
6 (a) $x^5+5x^4+10x^3+10x^2+5x+1$
(b) $-x^7/2187+7x^6/729-7x^5/81+35x^4/81-35x^3/27+7x^2/3-7x/3+1$
(c) $x^7+42x^6+756x^5+7560x^4+45\,360x^3+163\,296x^2+326\,592x+279\,936$
(d) $-x^5/1024+5x^4/32-10x^3+320x^2-5120x+32\,768$
7 (a) $-101\,376x^7/78\,125$ (b) $3\,306\,744x^6/128$
8 (a) $(165)8^8$ (b) $(2)21^6$

Module 7

Unit 1: Test

1 (a) $x=3/2$ or -5 (b) $x=3/5$ (c) $x=1$, 5 or -3
2 (a) direct, 1/2 (b) direct, −1/6
(c) inverse, 5 (d) inverse, 60
3 16
4 243

Unit 1: Exercises

1 (a) $x=-4/3$ (b) $x=2$ or $-1/2$ (c) $x=1$, 1/2 or 1/3
2 (a) direct, 1/3 (b) direct, −1/4 (c) inverse, 1000 (d) direct, −2
3 6
4 6.5

Unit 2: Test

1 (a) (5, −3) (b) (2, 1) and (3/4, 7/4)
2 (a) $x = 1$ or 2 (b) $x = 2$ (c) $x = 4 \pm \sqrt{19}$ (d) $x = -7 \pm \sqrt{10}$ (e) $x = 0$ or 3 (f) $x = 1$ or 1/2 (g) $x = (-7 \pm \sqrt{93})/2$ (h) $x = 1$
3 (a) $x = -5/2 \pm \sqrt{3}j/2$ (b) $x = 3/4 \pm j/4$
4 (a) $x = (-1 \pm \sqrt{5})/2$ (b) $x = 1$

Unit 2: Exercises

1 (a) (1, −2) and (2, −1) (b) (3, −1)
2 (a) $x = 1$ or 2 (b) $x = 1$ or -3 (c) $x = 0$ or 2 (d) $x = 3$ or -2 (e) $x = (3 \pm \sqrt{5})/2$ (f) $x = 3$ (g) $x = (7 \pm \sqrt{89})/2$ (h) $x = 4$
3 (a) $x = 1/2 \pm \sqrt{3}j/2$ (b) $x = 1/5 \pm 2j/5$
4 (a) $x = 0$ (b) $x = (1 \pm \sqrt{57})/4$
5 $-b/c$, a/c
6 $k = 0$ or 8

Unit 3: Test

(a) $(x + 1)(x + 2)(x + 3)$ (b) $x(x - 3)(x + 5)$ (c) $(x - 1)(x - 2)(x + 2)$ (d) $(x - 1)$

Unit 3: Exercises

(a) $(x - 1)(x - 2)(x - 3)$ (b) $x(x + 2)(x + 4)$ (c) $(x + 1)(x - 1)(x - 5)$ (d) $(x + 1)(x^2 + x + 1)$

Unit 4: Test

(a) $2/(x - 1) + 3/(x + 2)$
(b) $-7/(x + 1) + (7x + 4)/(x^2 + x + 1)$
(c) $1/[18(x - 5)] + 5/[3(x - 5)^2] - 1/[18(x + 1)]$
(d) $x + 5 - 9/(x - 2) + 28/(x - 3)$

Unit 4: Exercises

(a) $5/[2(x - 5)] + 1/[2(x + 1)]$
(b) $-3/x + 3/[2(x - 1)] + 3/[2(x + 1)]$
(c) $-5/[4(x + 1)] - 5/[2(x + 1)^2] + 5/[4(x - 1)]$
(d) $1 + 1/(x + 1) - 3(x + 2)$

Module 7: Further exercises

1 (a) $x = -5/2$ or 3/2 (b) $x = -3/2$ (c) $x = -1$ or $+1$
2 (a) direct, 1/3 (b) direct, −1/5 (c) inverse, −4 (d) inverse, 5
3 7
4 324
5 (a) (13/4, −1/4) (b) (2, 0), (0, 2)
6 (a) $x = -2$ or 8 (b) $x = -1$ (c) $x = 0$ or 4 (d) $x = 0$ or -1 (e) $x = 2$ or -5 (f) $x = -5 \pm \sqrt{13}$ (g) $x = 1/3$ or 1 (h) $x = -1$ or -2
7 (a) $x = -2 \pm j/\sqrt{2}$ (b) $x = 1 \pm j/\sqrt{5}$
8 (a) $x = -3 \pm \sqrt{3}$ (b) $x = 0$
11 (a) $(x - 2)(x - 4)(x - 8)$ (b) $2x(x + 7)(x - 8)$ (c) $(x + 1)(x - 1)(x + \sqrt{2})(x - \sqrt{2})$ (d) $(x + 2)^3$
12 (a) $5/[2(x + 1)] + 1/[2(x - 1)]$ (b) $3/(x - 1) - 11/(x - 2) + 8/(x - 3)$ (c) $-1/[3(x + 2)] + 2/(x + 2)^2 + 1/[3(x - 1)]$ (d) $8(x - 1) + (56/5)[1/(x - 2) + 4/(x + 3)]$

Module 8

Unit 1: Test

1 (a) 1 × 3 row vector (b) 3 × 2 matrix (c) 2 × 2 unit matrix (d) 3 × 3 square matrix
2 (a) $\begin{bmatrix} 8 & 14 \\ 5 & -4 \end{bmatrix}$ (b) $\begin{bmatrix} 8 & 4 \\ -7 & 10 \end{bmatrix}$ (c) $\begin{bmatrix} 0 & 30 \\ 36 & -42 \end{bmatrix}$ (d) $\begin{bmatrix} -5 & 15 \\ 55 & 33 \end{bmatrix}$ (e) $\begin{bmatrix} 0 \\ 12 \end{bmatrix}$ (f) $[-1 \quad 3]$
3 No
4 (a) $\begin{bmatrix} 36 & -16 \\ -16 & 40 \end{bmatrix}$ (b) $\begin{bmatrix} 1 & 0 & -7 & -1 \\ 0 & 0 & 0 & 0 \\ -7 & 0 & 49 & 7 \\ -1 & 0 & 7 & 1 \end{bmatrix}$ (c) $\begin{bmatrix} 29 & -15 & -45 \\ -15 & 10 & 19 \\ -45 & 19 & 145 \end{bmatrix}$

Unit 1: Exercises

1 (a) 3 × 1 column vector (b) 2 × 2 symmetric matrix (c) 2 × 2 square matrix (d) 3 × 3 unit matrix
2 (a) $\begin{bmatrix} 6 & -1 \\ -2 & 3 \end{bmatrix}$ (b) $\begin{bmatrix} -6 & 5 \\ -4 & -5 \end{bmatrix}$

(c) $\begin{bmatrix} -36 & 18 \\ -6 & -24 \end{bmatrix}$ (d) $\begin{bmatrix} 9 & 15 \\ -12 & -2 \end{bmatrix}$

(e) $\begin{bmatrix} 24 \\ 22 \end{bmatrix}$ (f) $[6 \quad 14]$

3 No

4 (a) $\begin{bmatrix} 134 & 64 \\ 64 & 53 \end{bmatrix}$ (c) $\begin{bmatrix} 19 & 11 & -30 \\ 11 & 41 & 24 \\ -30 & 24 & 100 \end{bmatrix}$

Unit 2: Test

1 (a) $(1/6)\begin{bmatrix} 1 & 3 \\ 0 & 6 \end{bmatrix}$ (b) not possible

2 (a) $\begin{bmatrix} 4 & -5 \\ 1 & 3 \end{bmatrix}\begin{bmatrix} x \\ y \end{bmatrix} = \begin{bmatrix} 2 \\ 9 \end{bmatrix}$

(b) $\begin{bmatrix} 7 & 3 \\ -7 & 9 \end{bmatrix}\begin{bmatrix} x \\ y \end{bmatrix} = \begin{bmatrix} 14 \\ -2 \end{bmatrix}$

(c) $\begin{bmatrix} -2 & -3 \\ 4 & -3 \end{bmatrix}\begin{bmatrix} x \\ y \end{bmatrix} = \begin{bmatrix} 9 \\ 6 \end{bmatrix}$

(d) $\begin{bmatrix} 7 & -2 \\ 0 & 2 \end{bmatrix}\begin{bmatrix} x \\ y \end{bmatrix} = \begin{bmatrix} 15 \\ -2 \end{bmatrix}$

3 (a) $x = 3, y = 2$ (b) $x = 11/7, y = 1$
(c) $x = -1/2, y = -8/3$ (d) $x = 13/7, y = -1$

Unit 2: Exercises

1 (a) $(-1/17)\begin{bmatrix} -2 & -5 \\ -3 & 1 \end{bmatrix}$ (b) not possible

2 (a) $\begin{bmatrix} 1 & -1 \\ 2 & 5 \end{bmatrix}\begin{bmatrix} x \\ y \end{bmatrix} = \begin{bmatrix} 8 \\ 9 \end{bmatrix}$

(b) $\begin{bmatrix} 2 & -4 \\ -5 & 1 \end{bmatrix}\begin{bmatrix} x \\ y \end{bmatrix} = \begin{bmatrix} 12 \\ 6 \end{bmatrix}$

(c) $\begin{bmatrix} 2 & 2 \\ 7 & 5 \end{bmatrix}\begin{bmatrix} x \\ y \end{bmatrix} = \begin{bmatrix} 4 \\ 8 \end{bmatrix}$

(d) $\begin{bmatrix} 9 & -2 \\ 4 & 0 \end{bmatrix}\begin{bmatrix} x \\ y \end{bmatrix} = \begin{bmatrix} 8 \\ 2 \end{bmatrix}$

3 (a) $x = 7, y = -1$ (b) $x = -2, y = -4$
(c) $x = -1, y = 3$ (d) $x = 1/2, y = -7/4$

Unit 3: Test

1 speed $5\sqrt{2}$ kph, direction 45° below the horizontal, $x = 1000$ m
2 2 newtons

Unit 3: Exercises

1 (a) $\mathbf{u} = 50\hat{\mathbf{h}} + 100\hat{\mathbf{v}}$ (b) $\mathbf{u} = 50\hat{\mathbf{h}} + 75\hat{\mathbf{v}}$
(c) $\mathbf{u} = 50\hat{\mathbf{h}}$
2 Heading 30° west of north with a northerly speed of $100\sqrt{3}$ kph

Unit 4: Test

1 (b) (i) $-2\mathbf{i} - 3\mathbf{j} + 5\mathbf{k}$ (ii) $19\mathbf{i} + 6\mathbf{j} - 16\mathbf{k}$
(iii) 10 (iv) -3 (v) $2\mathbf{i} + 7\mathbf{j} + 5\mathbf{k}$
2 1
4 2

Unit 4: Exercises

1 (b) (i) $-8\mathbf{i} - 6\mathbf{j} - 10\mathbf{k}$ (ii) -13
(iii) $\mathbf{i} + 7\mathbf{j} - 5\mathbf{k}$
2 7
4 -3

Module 8: Further exercises

1 (a) 2×1 column matrix (b) 1×4 row matrix (c) 2×2 square matrix (d) 2×3 matrix

2 (a) $\begin{bmatrix} 1 & -2 \\ 2 & 2 \end{bmatrix}$ (b) $\begin{bmatrix} 15 & 4 \\ -2 & -8 \end{bmatrix}$

(c) $\begin{bmatrix} 21 & 9 \\ -6 & -15 \end{bmatrix}$ (d) $\begin{bmatrix} -56 & 2 \\ 16 & -13 \end{bmatrix}$

(e) $\begin{bmatrix} -75 \\ 38 \end{bmatrix}$ (f) $[-16 \quad -14]$

3 No

4 (a) $\begin{bmatrix} 130 & -99 \\ -99 & 265 \end{bmatrix}$

(b) $\begin{bmatrix} 100 & 40 & 0 & -90 \\ 40 & 16 & 0 & -36 \\ 0 & 0 & 0 & 0 \\ -90 & -36 & 0 & 81 \end{bmatrix}$

(c) $\begin{bmatrix} 59 & 18 & -97 \\ 18 & 36 & -72 \\ -97 & -72 & 229 \end{bmatrix}$

5 (a) not possible (b) $(-1/11)\begin{bmatrix} 3 & -2 \\ -4 & -1 \end{bmatrix}$

6 (a) $\begin{bmatrix} -1 & -2 \\ 1 & 6 \end{bmatrix}\begin{bmatrix} x \\ y \end{bmatrix} = \begin{bmatrix} 4 \\ -8 \end{bmatrix}$

(b) $\begin{bmatrix} 8 & 1 \\ 2 & -1 \end{bmatrix}\begin{bmatrix} x \\ y \end{bmatrix} = \begin{bmatrix} 24 \\ 6 \end{bmatrix}$

(c) $\begin{bmatrix} -9 & 2 \\ 3 & -4 \end{bmatrix}\begin{bmatrix} x \\ y \end{bmatrix} = \begin{bmatrix} -14 \\ -2 \end{bmatrix}$

(d) $\begin{bmatrix} -2 & 3 \\ 0 & 4 \end{bmatrix}\begin{bmatrix} x \\ y \end{bmatrix} = \begin{bmatrix} 7 \\ 20 \end{bmatrix}$

7 (a) $x = -2, y = -1$ (b) $x = 3, y = 0$ (c) $x = 2, y = 2$ (d) $x = 4, y = 5$

9 Swimmer's speed is $\sqrt{5/2}$ metres per second in a direction inclined at 18.4° to the river bank. Start 450 m from a point opposite the destination. Time taken is 5 minutes.

10 (b) (i) 14i − 36j + 64k (ii) 36 (iii) 4i − 2j − 2k

11 8

13 1, volume of parallelepiped.

Module 9

Unit 1: Test

1 (a) $3\pi/4$ (b) $\pi/12$ (c) $5\pi/4$ (d) $5\pi/12$ (e) $3\pi/5$

2 (a) 20° (b) 157.50° (c) 30.77° to two dec. pl. (d) 141.52° to two dec. pl.

3 (a) 102°, 40° (b) 50° (c) 60°, 50°, 70°

4 (a) isosceles (b) equilateral (c) acute-angled

Unit 1: Exercises

1 (a) $\pi/4$ (b) $\pi/3$ (c) $3\pi/2$ (d) $\pi/5$ (e) $17\pi/15$

2 (a) 22.5° (b) 108° (c) 64.46° to two dec. pl. (d) 1355.05° to two dec. pl.

3 (a) 90°, 60° (b) 80°, 80° (c) 60°, 60°, 60°

4 (a) scalene (b) isosceles (c) right-angled

Unit 2: Test

1 5.29 cm to two dec. pl.

2 4.58 m^2 to two dec. pl.

4 $2\sqrt{3}$ cm = 3.46 cm to two dec. pl.

5 150 m^2

Unit 2: Exercises

1 10.91 cm to two dec. pl.

4 6 cm

5 6.67 m to two dec. pl.

Unit 3: Test

1 (a) 0.5735... 0.8191... 0.7002... (b) 0.7660... 0.6427... 1.1917... (c) 0.3420... 0.9396... 0.3639...
(d) 0.0871... 0.9961... 0.0874...

2 (a) 2.3047... 1.1099... 2.0765... (b) 1.7013... 1.2360... 1.3763... (c) 1.5557... 1.3054... 1.1917...
(d) 2.1300... 1.1325... 1.8807...

3 $\sin \pi/3 = 0.8660\ldots$, $\cos \pi/3 = 0.5$, $\tan \pi/3 = 1.7320\ldots$
$\sin \pi/6 = 0.5$, $\cos \pi/6 = 0.8660\ldots$, $\tan \pi/6 = 0.5773\ldots$

4 $\sqrt{2}, \sqrt{2}, 1$

5 (a) 0.5414... 1.8470...
0.8407... 1.1894...
0.6439... 1.5528...
(b) 0.2393... 4.1785...
0.9709... 1.0299...
0.2464... 4.0571...

Unit 3: Exercises

1 (a) $1/2, \sqrt{3}/2, 1/\sqrt{3}$ (b) 0.8191..., 0.5735..., 1.4281... (c) $\sqrt{3}/2, 1/2, \sqrt{3}$ (d) 0.2079..., 0.9781..., 0.2126...

2 (a) $2, 2/\sqrt{3}, \sqrt{3}$ (b) 2.6131..., 1.0823..., 2.4142... (c) $2/\sqrt{3}, 2, 1/\sqrt{3}$

3 12/13, 5/13, 12/5, 5/12

4 5/4, 5/3, 3/4, 4/3

5 (a) 0.6242... 1.6018...
0.7811... 1.2800...
0.7991... 1.2513...
(b) 0.9981... 1.0018...
0.0604... 16.5377...
16.5074... 0.0605...

Unit 4: Test

1 (a) $(1/2)(\sqrt{3}\cos\beta - \sin\beta)$
(b) $(\sqrt{3}\tan\beta + 1)/(\sqrt{3} - \tan\beta)$
(c) $(1/\sqrt{2})(\sin\beta - \cos\beta)$

2 (a) $(\sqrt{3} - 1)/2\sqrt{2}$ (b) $(\sqrt{3} + 1)/(\sqrt{3} - 1)$

3 $\cos^4\beta - 6\cos^2\beta\sin^2\beta + \sin^4\beta$

Unit 4: Exercises

1 (a) $(1/2)(\sqrt{3}\sin\beta + \cos\beta)$
(b) $(\tan\beta - 1)/(\tan\beta + 1)$
(c) $(1/2)(\cos\beta + \sqrt{3}\sin\beta)$

2 (a) $(\sqrt{3} + 1)/2\sqrt{2}$ (b) $(\sqrt{3} - 1)/(\sqrt{3} + 1)$

3 $4\sin\beta\cos\beta(1 - 2\sin^2\beta)$

4 (a) $\sqrt{(3/2)}$ (b) $1/\sqrt{2}$
(c) $1 - (1/\sqrt{3})\tan 15°$

Unit 5: Test

1 37.6° to one dec. pl.

2 15.9 to one dec. pl.
3 (a) $A = 51.0°$, $B = 41.8°$, $C = 87.2°$, all to one dec. pl.
(b) $a = 14.1$ cm, $B = 53.9°$, $C = 101.1°$, all to one dec. pl.
(c) $C = 96°$, $b = 2.5$ to one dec. pl., $c = 14.5$ to one dec. pl.
(d) $A = 91°$, $b = 28.3$ to one dec. pl., $c = 16.5$ to one dec. pl.
4 $A = 29.0°$, $B = 46.6°$, $C = 104.4°$, all to one dec. pl.

Unit 5: Exercises

1 53.6° to one dec. pl.
2 7.09 cm to two dec. pl.
3 (a) $C = 81°$, $b = 29.8$ cm to one dec. pl., $c = 32.7$ cm to one dec. pl.
(b) $a = 16.1$ cm, $B = 77.5°$, $C = 26.5°$, all to one dec. pl.
4 $A = 51.3°$, $B = 69.5°$, $C = 59.2°$, all to one dec. pl.

Module 9: Further exercises

1 (a) $7\pi/4$ (b) $2\pi/3$ (c) $4\pi/3$ (d) $\pi/36$ (e) $17\pi/36$
2 (a) 15° (b) 144° (c) 180.02° to two dec. pl. (d) 1.38° to two dec. pl.
3 (a) 0.9396… 0.3420… 2.7474… (b) 0.9961… 0.0871… 11.4300… (c) 0.3862… 0.9238… 0.4142…
(d) 0.3090… 0.9510… 0.3249…
4 (a) 2.9238… 1.0641… 2.7474… (b) 2.9238… 1.0641… 2.7474…
5 9/15, 12/15, 9/12, 12/9
6 15/9, 15/12, 12/9, 9/12
7 (a) 0.6644… 1.5050…
0.7473… 1.3380…
0.8890… 1.1247…
(b) 0.2079… 4.8097…
0.9781… 1.0223…
0.2125… 4.7046…
8 (a) $(1/2)(\cos\beta + \sqrt{3}\sin\beta)$
(b) $(\tan\beta + 1)/(1 - \tan\beta)$
(c) $(1/2)(\sin\beta - \sqrt{3}\cos\beta)$
9 $3\sin\beta - 4\sin^3\beta$
10 $4\cos^3\beta - 3\cos\beta$
11 (a) $\pi/3$ radians (b) 35.26° to two dec. pl. (c) $\pi/4$ radians (d) 15°
15 $x^2/64 + y^2/36 = 1$
16 51.5° to one dec. pl.
17 4.3 to one dec. pl.
18 (a) 80.5°, 45.7°, 53.8° to one dec. pl.
(b) $a = 17.1$ cm, $B = 34.3°$, $C = 105.7°$ to one dec. pl.
(c) $C = 95°$, $b = 36.9$ cm, $c = 42.4$ cm to one dec. pl.
(d) $A = 112°$, $b = 10.3$, $c = 11.4$
19 41.4°, 55.8°, 82.8° to one dec. pl.

Module 10

Unit 1: Test

1 (a) 1, −2 (b) −3, 0 (c) 0, 6
3 (a) 1, $3\pi/4$ radians (b) 1, $-\pi/3$ radians
(c) 7.35, −101.77° to two dec. pl.

Unit 1: Exercises

1 (a) 6, −7 (b) −8, 0 (c) $j = 0 + j$
3 (a) 2, $\pi/3$ radians (b) 10, $3\pi/4$ radians
(c) 6.09°, 122.125°

Unit 2: Test

1 (b) $3 - 5j$, $1 + 7j$, -1, $-j$, $-7/25 - j/25$
2 $(1 + j)^2 = 2j$; argument $\pi/2$

Unit 2: Exercises

1 (b) $-2 - 11j$, $-2 + j$

Module 10: Further exercises

1 (a) 8, −4 (b) −4, 0 (c) 0, −9
3 (a) 1, $2\pi/3$ radians (b) $\sqrt{2}$, $-3\pi/4$ radians
(c) 7.43, 34.70° to two dec. pl.
4 (b) $-50 + 51j$, $-2 + 59j$, $-6 - 7j$, $-58/41 + 11j/41$

Chapter 2: Miscellaneous exercises

1 π
3 40 minutes
5 x^4
9 $k = 4$
11 5
13 30
15 All three
17 −14
19 $\pm\sqrt{2}$
21 $x \neq \pm a$
23 $x = 33$, $y = 12$
25 $-b/c$
29 75.0°, 43.1°, 61.9° to one dec. pl.

CHAPTER 3

Module 11

Unit 1: Test

1 (a) add 2/3 (b) multiply by 12.5 (c) raise to power −3 (d) multiply by 12 and then add 4 (e) square, multiply by 5 and subtract 2
2 (a) subtract 3 (b) divide by 2/3 (c) raise to power −4/3
3 (a) and (c)
4 (a) $1+x^{-2}$ (b) $(1+x)^{-2}$
(c) $1+(3x/4)^{-2}$ (d) $[(3/4)(1+x)]^{-2}$
(e) $2+x$
5 (a) multiply by −2, add 1, raise to power 5
(b) divide by −4, add 9, raise to power 1/3
(c) square, multiply by 2, add 5
6 (a)

Unit 1: Exercises

1 (a) subtract 2 (b) divide by 4 (c) raise to power 1/3 (d) divide by 5 then add 6
(e) raise to power −2 then multiply by 7
2 (a) add 2 (b) multiply by 3 (c) raise to power 4
3 (a) and (c)
4 (a) x^3-6 (b) $(x-6)^3$ (c) $(x/2)^3-6$
(d) $[(x-6)/2]^3$ (e) $x-12$
5 (a) multiply by 4, subtract 1, cube
(b) divide by π, subtract 2, square root
(c) $(x-2)(x+2) \equiv x^2-4$; square then subtract 4
6 (a) and (c)

Unit 2: Test

1 (a) domain $-\infty < x < 5$ and $5 < x < \infty$; range $0 < f(x) < \infty$
(b) domain $-1 \leqslant x < 1$; range $0 \leqslant f(x) \leqslant 4$
(c) domain $0 < x \leqslant 5$; range $f(x) = 5$
2 (a) $-x^2+3x-1$ and x^2+x-1, domain $-1 \leqslant x \leqslant 8$
$x(2x-1)(1-x)$, domain $-1 \leqslant x \leqslant 8$
$(2x-1)/x(1-x)$, domain $-1 \leqslant x < 0$ and $0 < x < 1$ and $1 < x \leqslant 8$
(b) $2x^2+x+3$ and $1-x$, domain $0 \leqslant x \leqslant 4$
$x^4+x^3+3x^2+2x+2$, domain $0 \leqslant x \leqslant 4$
$(2+x^2)/(1+x+x^2)$, domain $0 \leqslant x \leqslant 4$
3 (a) one to one (b) many to one (c) many to one (d) many to one

Unit 2: Exercises

1 (a) domain $-\infty < x < -1$ and $-1 < x < \infty$; range $-\infty < f(x) < \infty$
(b) domain $-3 < x < 3$; range $0 < f(x) < 9$
(c) domain $-5 \leqslant x \leqslant -4$; range $f(x) = -2$
2 (a) x^2+4x+2 and $-x^2+4x+2$ domain $-2 \leqslant x \leqslant 2$
$4x^3+2x^2$, domain $-2 \leqslant x \leqslant 2$
$4/x+2/x^2$, domain $-2 \leqslant x < 0$ and $0 < x \leqslant 2$
(b) $1+1/x-4x^2$ and $1-1/x-4x^2$, domain $-2 \leqslant x < 0$ and $0 < x \leqslant 6$
$1/x-4x$, domain $-2 \leqslant x < 0$ and $0 < x \leqslant 6$
$x-4x^3$, domain $-2 \leqslant x < 0$ and $0 < x \leqslant 6$
3 (a) one to one (b) one to one (c) many to one (d) many to one

Module 11: Further exercises

1 (a) subtract π (b) multiply by −7
(c) raise to power −2 (d) multiply by −3 and add 6 (e) square, multiply by 5, subtract 1
2 (a) add π (b) divide by −6 (c) raise to power −1/7
3 (a) and (b)
4 (a) $1/x^2-5$ (b) $1/(x-5)^2$
(c) $1/9x^2-5$ (d) $1/(15-3x)^2$
(e) $x-10$
5 (a) multiply by −3, subtract π, square
(b) square, divide by 4, add 5, cube
(c) raise to power −4, add 6, raise to power −2, multiply by 8
6 None
7 (a) domain $0 < x < 5$; range $0.024 < f(x) < \infty$
(b) domain $x > -2$; range $0 < f(x) < \infty$
(c) domain $-\infty < x < \infty$; range $0 \leqslant f(x) < \infty$
8 (a) $15+3x-2x^2$ and $-2x^2-3x-3$, domain $-1 \leqslant x \leqslant 1$
$54+18x-18x^2-6x^3$, domain $-1 \leqslant x \leqslant 1$
$(6-2x^2)/(9+3x)$, domain $-1 \leqslant x \leqslant 1$
(b) $6-8/x+7x^2$ and $6+8/x+7x^2$, domain $-8 \leqslant x < 0$ and $0 < x \leqslant 1$
$-48/x-56x$, domain $-8 \leqslant x < 0$ and $0 < x \leqslant 1$
$-(x/8)(6+7x^2)$, domain $-8 \leqslant x < 0$ and $0 < x \leqslant 1$
9 (a) many to one (b) many to one (c) one to one (d) one to one

Module 12

Unit 1: Test

1 (a) (0, −1), (−1, −2), (1, 0), (10, 9)
(b) (0, 1), (1, 3), (−1, 1), (5, 31)
(c) (0, 0), (1, 0), (−1, 8), (8, 784)

Unit 1: Exercises

1 (a) (−1, −11), (0, −6), (2, 4), (6/5, 0)
(b) (−2, −16), (0, −8), (2, 0), (1/2, −63/8)
(c) (−1, 1/3), (0, 0), (1, 1), (−1/2, 1/4)

Unit 2: Test

1 $g(x) = 2x + 9$
2 $g(x) = 16x^2 + 4x$
3 $g(x) = x^2 - 3x - 1$
4 $g(x) = f(x - 2)$

Unit 2: Exercises

1 $g(x) = 5 - x$
2 $g(x) = 8x^3$
3 $g(x) = x^2 - 4x + 6$
4 $g(x) = f(x + 8/3)$

Unit 3: Test

1 (a) odd (b) even (c) neither
(d) neither: $f(-x)$ cannot be defined
2 (a) 4 (b) x^4
3 (a) $-3x$ (b) $\frac{1}{x}$

Unit 3: Exercises

1 (a) odd (b) odd (c) neither (d) even
2 (a) 1 (b) $-x^2$
3 (a) $2x$ (b) x^3

Module 12: Further exercises

1 (a) (0, 7), (1, 2), (−1, 12), (10, −43)
(b) (0, 5), (1, −2), (−1, 6), (−5, −50)
(c) (1, 0), (−1, 0), $(2, \surd[3/5])$, $(3, 2/\surd 5)$
3 $g(x) = 3x - 14$
4 $g(x) = 32x^4$
5 $g(x) = x^2 - 6x + 10$
6 $g(x) = f(x + 3) + 1$
7 $g(x) = f(x + 10/9)$
8 (a) even (b) neither (c) neither
(d) odd
9 (a) −2 (b) 0
10 (a) $2x$ (b) $x/(x^2 - 1)$

Module 13

Unit 1: Test

1 (a) 0.9961... (b) −0.9510...
(c) −1.7320... (d) −0.7071...
3 (a) 6 (b) 3 (c) 1
4 (a) $\pi - 2$ (b) −7 (c) 0

Unit 1: Exercises

1 (a) −19.0811... (b) 0.9510... (c) 0
(d) −0.5773...
3 (a) 4 (b) 5 (c) 2
4 (a) 3 (b) $-\frac{1}{2}$ (c) $-\frac{3}{4}$

Unit 2: Test

1 (a) $11\pi/24 \pm 2n\pi$ and $35\pi/24 \pm 2n\pi$
(b) $11\pi/24 \pm 2n\pi$ and $35\pi/24 \pm 2n\pi$
(c) $A = 58.0° \pm n\pi$, $B = 21.8° \pm n\pi$ to one dec. pl.
2 −0.5
3 (a) $\pi/4 \pm n\pi$ and $71.6° \pm n\pi$ to one dec. pl.
(b) $\pm 120° \pm 2n\pi$ and $\pm 109.5° \pm 2n\pi$

Unit 2: Exercises

1 (a) $16.8° \pm 2n\pi$ and $163.2° \pm 2n\pi$
(b) any value of x for which $\tan x$ exists
(c) $\pm 67.5° \pm 2n\pi$ to one dec. pl.
2 $\surd 3/2$
3 (a) $\pm 30° \pm n\pi$ (b) $\pm 78.5° \pm 2n\pi$, $\pm 80.4° \pm 2n\pi$ to one dec. pl.
(c) $\pm 35.3° \pm 2n\pi$ and $\pm 30° \pm 2n\pi$ to one dec. pl. and $\pm 150° \pm 2n\pi$ and $\pm 144.7° \pm 2n\pi$.

Module 13: Further exercises

1 (a) 1 (b) 1 (c) 0.7313... (d) 0.9335...
3 (a) 7 (b) 2 (c) 5/2
4 (a) 2 (b) 3/2 (c) 1/4
6 (a) $14.5° \pm 2n\pi$ and $165.5° \pm 2n\pi$ (b) $\pm 13\pi/24 \pm n\pi$
(c) $A = 68.2° \pm n\pi$, $B = 78.7° \pm n\pi$ to one dec. pl.
7 $(1 - \surd 3)2/\surd 3$
8 (a) $\pi/4 \pm n\pi$ and $-18.4° \pm n\pi$ to one dec. pl.
(b) $\pm 80.4° \pm 2n\pi$, $\pm 109° \pm 2n\pi$
(c) $26.5° \pm n\pi$ and $-63.4° \pm n\pi$

Module 14

Unit 1: Test

1 (a) 0.1053... (b) 0.4723...
2 (a) 0 (b) 1/2 (c) 3.8714... (d) 6.2146...
3 (a) 2 (b) 0.6931..., 1.0986... (c) 0

Unit 1: Exercises

1 (a) 1.4953... (b) 0.1337...
2 (a) -1 (b) 2 (c) 3.8437... (d) -6.2146...
3 (a) 3 (b) 0, 1 (c) $-1, -2$

Unit 2: Test

1 (a) -0.2876... (b) 2.0794...
2 (a) 32/81 (b) 1/9 (c) 1024 (d) 6 (e) 2
3 $y = x^{9/8}$
4 8.5481...

Unit 2: Exercises

1 (a) -0.4771... (b) 0.4771...
2 (a) $1/2\sqrt{2}$ (b) 1/4 (c) 81 (d) 5 (e) 3
3 $y = x^{14/3}$
4 ± 3.1201...

Module 14: Further exercises

1 (a) 0.4054... (b) 1.6582...
2 (a) 0 (b) $-2/3$ (c) 0.5910... (d) -0.2171...
3 (a) ± 3 (b) 0.5848... (c) 125 (d) 2 (e) 2
4 $y = x^{2/3}$
5 1.4422...
6 (a) 87.8467... (b) 32.2926...
7 (a) 7 (b) 5
8 (a) 2 (b) 1, 3 (c) 0, 2

Module 15

Unit 1: Test

1 (a) $\text{arc}\, f(x) = x + 3$ (b) $x/2$ (c) $x - 1/3$ (d) $8x$ (e) $\sqrt{x}$
2 (a) $\arctan 4x$ (b) $(\arcsin x)/3$
3 (a) $\text{arc}\, f(x) = 3(x + 4)/2$
(b) $\text{arc}\, f(x) = (x^2 + 8)^{1/3}$
(c) $\text{arc}\, f(x) = \sqrt{[(\arccos x) + 1]}$
(d) $\text{arc}\, f(x) = (1/2)(10^x - 1)$
(e) $\text{arc}\, f(x) = [\ln(x)]/2$
4 (a) and (c)
6 $\text{arc}\, f(x) = (\arctan x)/2$
$-\pi/2 < \text{arc}\, f(x) < \pi/2$

Unit 1: Exercises

1 (a) $\text{arc}\, f(x) = 4x/3$ (b) $x + 2/3$ (c) x^4
2 (a) $\arccos x$ (b) $\exp(x)$
3 (a) $\text{arc}\, f(x) = (x + 1)/6$
(b) $\text{arc}\, f(x) = (x^{-1} - 3)/2$
(c) $\text{arc}\, f(x) = [\arctan x + 2]/5$
(d) $\text{arc}\, f(x) = (2^x + 1)/3$
4 (a) and (c)
6 $(\arcsin x)/2$; $-\pi/2 \leqslant \text{arc}\, f(x) \leqslant \pi/2$

Unit 2: Test

(a) $\pm 75.5° \pm 2n\pi$ to one dec. pl.
(b) $6.1° \pm n\pi/3$ and $23.9° \pm n\pi/3$ to one dec. pl.
(c) 82.73° to two dec. pl.

Unit 2: Exercises

(a) $63.4° \pm n\pi$ to one dec. pl.
(b) $21.1° \pm n\pi/3$ to one dec. pl.
(c) 116.2° and 200.2° to one dec. pl.

Module 15: Further exercises

1 (a) $\text{arc}\, f(x) = x + 1$
(b) $\text{arc}\, f(x) = x - \frac{2}{5}$
(c) $\text{arc}\, f(x) = \frac{6x}{5}$
(d) $\text{arc}\, f(x) = x^6$
2 (a) $\text{arc}\, f(x) = \text{arccosec}\, x$
$-\pi/2 \leqslant \text{arc}\, f(x) \leqslant \pi/2$
(b) $\text{arc}\, f(x) = \log_5 x$
3 (a) $\text{arc}\, f(x) = (x + 7)/9$
(b) $\text{arc}\, f(x) = [(x - 1)/2]^{1/2}$
(c) $\text{arc}\, f(x) = (\text{arcsec}\, x + 1)/2$
(d) $\text{arc}\, f(x) = (e^x + 4)/3$
(e) $\text{arc}\, f(x) = [\ln(x)]^{1/2}$
4 (b) and (c)
6 (a) $\text{arc}\, f(x) = |\sqrt{x}|$
(b) $\text{arc}\, f(x) = \ln(x)$
7 (a) $48.6° \pm 2n\pi$, $131.4° \pm 2n\pi$ to one dec. pl.
(b) $\pm 15.1° \pm 2n\pi/5$ to one dec. pl.
(c) $79.1° \pm 2n\pi$, $219.0° \pm 2n\pi$ to one dec. pl.

Chapter 3: Miscellaneous exercises

1 $x = 2$
3 $g = 2f$
9 No

CHAPTER 4

Module 16

Unit 1: Test

(a) 20, 25: $f(n) = 5n$, $n \geqslant 0$
(b) 3125, 15 625: $f(n) = 5^n$, $n \geqslant 1$
(c) $-1/4$, $1/2$: $f(n) = -(-2)^{n-6}$, $n \geqslant 0$
(d) $f(n) = 4(-3)^n$, $n \geqslant 0$
(e) $f(n + 3) = f(n + 2) - f(n)$: $f(0) = 1$, $f(2) = 2$, $f(3) = 3$
(f) $f(n + 2) = f(n + 1) + f(n)$: $f(0) = -16$, $f(1) = 2$

Unit 1: Exercises

(a) 16, 32: $f(n) = 2^n$, $n \geqslant 0$
(b) -34, -38: $f(n) = 18 - 4n$, $n \geqslant 0$
(c) 5/81, 5/243: $f(n) = 5/(3^n)$, $n \geqslant 0$
(d) -2, 2: $f(n) = 2(-1)^n$, $n \geqslant 1$
(e) $f(n+3) = f(n+2) - f(n)$: $f(0) = 1$, $f(2) = 2$, $f(3) = 3$
(f) $f(n+2) = f(n+1) + f(n)$: $f(0) = -16$, $f(1) = 2$

Unit 2: Test

(a) $-\infty$ (b) ∞ (c) 0.999 (d) ∞ (e) -1
(f) $-1/3$ (g) -1 (h) 54.7356...
(i) -0.22... (j) 1 (k) undefined

Unit 2: Exercises

(a) ∞ (b) 0 (c) ∞ (d) 0 (e) 0.75 (f) 0.4
(g) 1 (h) 45° (i) 1 (j) 1.6487...
(k) undefined

Unit 3: Test

(a) converges to 0 (b) diverges (c) converges to 0.4 (d) converges to 1 (e) converges to 0

Unit 3: Exercises

(a) converges to 1000 (b) diverges
(c) converges to $-5/3$ (d) converges to -1
(e) diverges

Module 16: Further exercises

1 (a) 21, 26: $f(n) = 1 + 5n$, $n \geqslant 0$
(b) -93, -91: $f(n) = -101 + 2n$, $n \geqslant 0$
(c) 0.001, 0.0001: $f(n) = 10^{1-n}$, $n \geqslant 0$
(d) 1.2345, 0.123 45: $f(n) = (12\,345)10^{-n}$, $n \geqslant 1$
(e) $f(n) = f(n-2) - f(n-1)$: $f(-2) = 2$, $f(-1) = 4$
(f) $f(n) = f(n-1) + f(n-2)$: $f(-2) = 5$, $f(-1) = 8$

2 (a) ∞ (b) ∞ (c) undefined (d) 0 (e) 1
(f) -0.2 (g) $-\infty$ (h) 30° (i) ∞
(j) 0 (k) 1

3 (a) diverge (b) diverge
(c) converges to $-1/8$
(d) converges to $2/\sqrt{3}$ (e) converges to 0

Module 17

Unit 1: Test

1 (a) 1, 2, 5/2, 8/3 (b) 0, 1/2, 7/6, 23/12
2 (a) 140 (b) -160 (c) 236 192 (d) 6.67 to three sig. fig.
3 £58 905.06

Unit 1: Exercises

1 (a) 1, 3/2, 11/6, 25/12 (b) 1, 2, 4, 10
2 (a) -85 (b) 20 (c) 1 398 100
(d) 1.0101...

Unit 2: Test

1 (a) 20/3 (b) $6.\dot{0}0\dot{6}$
2 (a) convergent by comparison test
(b) convergent by comparison test
(c) convergent by alternating sign test
(d) convergent by ratio test
(e) convergent by ratio test

Unit 2: Exercises

1 (a) 4 (b) $-\infty$
2 (a) convergent (b) divergent
(c) convergent (d) divergent
(e) convergent

Unit 3: Test

1 (a) $1 - 12x + 96x^2 - 640x^3 + \ldots$
(b) $3^{2/3}(1 - 2x/9 - x^2/81 - 4x^3/2187 - \ldots)$
(c) $9^{-1/3}[1 + 1/(27x) + 2/(729x^2) + 14/(59\,049x^3) + \ldots]$
2 $-0.5 < x < 0.5$

Unit 3: Exercises

1 (a) $1 + 6x + 27x^2 + 108x^3 + \ldots$
(b) $7^{1/3}(1 + 2x/21 - 4x^2/441 + 40x^3/27\,783 - \ldots)$
(c) $(-4)^{-1/5}[1 + 1/(20x) + 3/(400x^2) + 11/(8000x^3) + \ldots]$

Unit 4: Test

2 (a) £3273.31 (b) £3802.58 (c) £3943.10
(d) £4017.11

Module 17: Further exercises

1 (a) 0, 1, 5, 23 (b) -1, -1, 2, 10
2 (a) 295 (b) -195 (c) 0.6249
(d) 4.6621...
3 £151.87
4 (a) 3 (b) ∞
5 (a) convergent – geometric series with common ratio less than 1
(b) divergent by the comparison test
(c) convergent by alternating sign test
(d) convergent by ratio test
(e) divergent by ratio test
6 (a) $1 + 20x + 250x^2 + 2500x^3 + \ldots$
(b) $8^{3/4}(1 + 9x/32 - 27x^2/2048 + 135x^3/65\,536 - \ldots)$
(c) $1 - 1/x + 3/(2x^2) - 5/(2x^3) + \ldots$

7 $-5 < x < 5$
9 (a) £34 339.95 (b) £44 265.02
(c) £47 086.71 (d) £48 604.17

Module 18

Unit 1: Test

(a) -10 (b) -2 (c) -2

Unit 2: Test

(a) Infinite break discontinuities at every integer multiple of π.
(b) Finite break discontinuity at $x = 0$.
(c) Hole discontinuity at $x = 5$.
(d) Continuous everywhere.

Unit 2: Exercises

(a) Infinite break discontinuities at every integer multiple of π.
(b) Finite break discontinuity at $x = \pi/2$.
(c) Hole discontinuity at $x = -3/2$
(d) Continuous everywhere.

Module 18: Further exercises

1 (a) 8 (b) 2.4 (c) 2
2 (a) Infinite break discontinuities every odd integer multiple of $\pi/10$.
(b) Finite break discontinuity at $x = \pi/2$.
(c) Hole discontinuity at $x = -4/3$.
(d) Continuous everywhere.

Chapter 4: Miscellaneous exercises

1 1, 1/2, 2/3, 3/5, 5/8, 8/13,...: numerators and denominators form a Fibonacci sequence.
3 Yes
5 $-1/2 < x < 1/2$
7 0
9 1

CHAPTER 5

Module 19

Unit 1: Test

1 42 mph
2 $5/3$ cm s^{-2}
3 5
4 15

Unit 1: Exercises

3 -28
4 -3

Unit 2: Test

1 (a) $8x$ (b) $-9(1-3x)^2$
2 -32
3 (a) $9x^8$ (b) 0 (c) $-3/x^4$ (d) $2x^{-3}$

Unit 2: Exercises

1 $6x^2$
2 -405
3 (a) $-1.5x^{-2.5}$ (b) 0 (c) $-\sin x$ (d) $1/x$

Module 19: Further exercises

1 47.5 mph
2 -1 cm s^{-2}
3 -3
4 $-984\,375$
5 48
6 (a) $15x^2$ (b) $8(1+4x)$
7 (a) $(-1/3)x^{-2/3}$ (b) 0 (c) $(2/3)(1/x^{5/3})$
(d) $12x^{11}$

Module 20

Unit 1: Test

(a) $1 - 4x^3$ (b) $48x^3$ (c) $3x^2 \cos x - x^3 \sin x$
(d) $[x^3 - 4x^3 \ln(x)]/x^8$ (e) $2 \sec^2 x \tan x$
(f) $3 \sin(1-3x)$ (g) $(6x^2 - 2x)\exp(2x^3 - x^2)$
(h) $4e^{4x}$ (i) $-4 \sin 4x$

Unit 1: Exercises

1 (a) $-3x^{-4} - x^{-2}$ (b) $4x/3$
(c) $e^x \sin x + e^x \cos x$ (d) $[1 - \ln(x)]/x^2$
(e) $-\operatorname{cosec}^2 x$ (f) $6x(1+x^2)^2$
(g) $(4x-1)/(2x^2-x)$ (h) $3e^{3x}$ (i) $4 \sec^2 4x$

Unit 2: Test

1 (a) Rising curve $-\infty < x < 0$, falling curve $0 < x < 2/3$, rising curve $2/3 < x < \infty$, local maximum at $x = 0$; local minimum at $x = 2/3$.
(b) Rising curve for all x except at $x = 2$; vertical point of inflexion at $x = 2$.
2 Local maximum at $x = 1/3$; local minimum at $x = 1$.

Unit 2: Exercises

1 (a) Falling curve $-\infty < x < -3$, rising curve $-3 < x < 3$, falling curve $3 < x < \infty$, local minimum at $x = -3$, local maximum at $x = 3$.
(b) Rising curve for all x, vertical point of inflexion $x = 2$.
2 Rising curve $-\infty < x < -4$, falling curve $-4 < x < 1$, rising curve $1 < x < \infty$; local

maximum at $x = -4$, local minimum at $x = 1$.

Unit 3: Test

1 (a) $12x^2$, $24x$ (b) $-\sin x$, $-\cos x$ (c) $9e^{3x}$, $27e^{3x}$ (d) $-x^{-2}$, $2x^{-3}$
2 (a) Negative for $x < -1/3$, positive for $x > -1/3$
(b) Negative for all x.
3 (a) $x = -1/3$ (b) None

Unit 3: Exercises

1 (a) $36x$, 36 (b) $-2\cos 2x$, $4\sin 2x$ (c) $(x^2 + 4x + 3)e^x$, $(x^2 + 6x + 7)e^x$ (d) $1/x$, $-1/x^2$
2 (a) Positive for $x < 1/12$, negative for $x > 1/12$.
(b) Positive for $x < -2$, negative for $x > -2$.
3 (a) $x = 1/12$ (b) $x = -2$

Unit 4: Test

(a) Minimum at $x = 0$.
(b) No stationary points.
(c) Stationary points of inflexion at $x = 0$, local maximum at $x = 3$, points of inflexion at $x = 3 \pm \sqrt{3}$.

Unit 4: Exercises

1 (a) $x = 2$ (b) $x = n\pi/2 + \pi/4; n = 0, \pm 1, \ldots$ (c) $x = -1$
2 (a) Local maximum at $x = 0$.
(b) Local minimum at $x = 1/e$.
(c) Local minimum at $x = 0$, local maximum at $x = -2$.

Module 20: Further exercises

1 (a) $-2x^{-3} - 5x^{-6}$ (b) $27x^2/5$
(c) $e^x \cos x - e^x \sin x$
(d) $(3x^2 \sin x - x^3 \cos x)/\sin^2 x$
(e) $-2\,\text{cosec}^2 x \cot x$ (f) $-6x^2(1 - x^3)$
(g) $(12x - 2)/(4x^2 - x)$ (h) $-5e^{-5x}$
(i) $-6\,\text{cosec}^2 6x$
2 (a) Falling curve $-\infty < x < -1$, rising curve $-1 < x < 0$, falling curve $0 < x < 1$, rising curve $1 < x < \infty$; local minima at $x = \pm 1$, local maximum at $x = 0$.
(b) Rising curve for all x, vertical point of inflexion at $x = -2$.
3 Local maximum at $x = -2/3$, local minimum at $x = 4$.
5 (a) $6x^{-4}$, $-24x^{-5}$
(b) $4\sec^2 x \tan^2 x + 2\sec^4 x$
(c) $(4x - 4)e^{-2x}$, $(12 - 8x)e^{-2x}$
(d) $-2(x^2 + 1)/(x^2 - 1)^2$, $(4x^3 + 12x)/(x^2 - 1)^3$
6 (a) Negative for $x < -3/10$, positive for $x > -3/10$.
(b) Positive for $x < -2$, negative for $x > -2$.
7 (a) Local minimum at $x = 0$, local maximum at $x = -2/5$, point of inflexion at $x = -3/10$.
(b) No stationary points but vertical point of inflexion at $x = -2$.
8 (a) $x = 0$, $x = 2$ (b) all multiples of $\pi/2$ (c) no values
9 (a) Local maximum at $x = 0$.
(b) Minimum at $x = e^{-1/2}$.
(c) Local minimum at $x = 0$, local maximum at $x = -2$.

Module 21

Unit 1: Test

1 (a) $2x\,dx$ (b) $\cos x\,dx$ (c) $-\text{cosec}^2 x\,dx$ (d) $-2e^{-2x}\,dx$ (e) $(-1/x)\,dx$ (f) $(4x^3 - 2x)\,dx$
2 (a) $x^4/4 + c$ (b) $-x^{-5}/35 + c$ (c) $x^2/2 - 2x^5/5 + c$ (d) $(\sin 4x)/4 + c$ (e) $3e^{-3x} + c$ (f) $16\ln(x) + c$

Unit 1: Exercises

1 (a) $-x^{-2}\,dx$ (b) $2\cos 2x\,dx$ (c) $-3\,\text{cosec}^2 3x\,dx$ (d) $-e^{-x}\,dx$ (e) $(2/x)\,dx$ (f) $(2x - 2x^{-3})\,dx$
2 (a) $-x^{-1} + c$ (b) $4x^5/5 + c$ (c) $5x^4/4 - x^2/2 + c$ (d) $(\tan 3x)/3 + c$ (e) $-5e^{-2x}/2 + c$ (f) $12\ln(x) + c$

Unit 2: Test

1 1.44
2 1.73

Unit 2: Exercises

1 1.32
2 1.41

Unit 3: Test

1 1.442
2 1.732

Unit 3: Exercises

1 1.316
2 1.414

Module 21: Further exercises

1 (a) $-4x^{-5}\,dx$ (b) $-4\sin 4x\,dx$
(c) $6\sec^2 dx$ (d) $3e^{3x}\,dx$ (e) $(3/x)\,dx$
(f) $(1+x^{-2})\,dx$
2 (a) $-x^{-4}/4+c$ (b) $-2x^{-3}/9+c$
(c) $7x^2/2+9x^{-1}+c$ (d) $(-\cot 5x)/5+c$
(e) $3e^{4x}+c$ (f) $-3\ln(x)+c$
3 2.04
4 1.73
5 2.040
6 1.732

Chapter 5: Miscellaneous exercises

1 Derivative is zero in both (a) and (b) except at the origin where neither the function nor its derivative are defined.
3 $f'(x)=2x$ for $-\infty<x<-1$ and $1<x<\infty$
$f'(x)=-2x$ for $-1<x<1$ and $f'(x)$ undefined at $x=\pm 1$
5 $A=1$, $B=-1$
11 5% decrease

CHAPTER 6

Module 22

Unit 1: Test

1 (a) 0.994 (b) 1.290
(c) 7.328 to three dec. pl.
2 (a) 1.000 (b) 1.296
(c) 7.300 to three dec. pl.

Unit 1: Exercises

1 (a) 0.348 (b) 0.262
(c) 0.740 to three dec. pl.
2 (a) 0.347 (b) 0.264
(c) 0.748 to three dec. pl.

Unit 2: Test

(a) 5/2 (b) 40/3

Unit 2: Exercises

(a) 55
(b) $\beta^4/4$

Module 22: Further exercises

1 (a) 0.348 (b) 35.836
(c) 1.507 to three dec. pl.
2 (a) 0.347 (b) 35.679
(c) 1.369 to three dec. pl.
3 (a) 16 (b) 32 768/5

Module 23

Unit 1: Test

(a) 33/5 (b) 1 (c) $1-1/e$ (d) 1

Unit 2: Test

(a) $63\pi^6/6^7+(\sqrt{3}-1)/2$ (b) 22 099/5
(c) $1+\pi^2/32$

Unit 2: Exercises

(a) $\pi^5/1215-\sqrt{3}$ (b) 16/3 (c) $e-1$

Module 23: Further exercises

1 (a) 2/3 (b) $(\sqrt{3}-1)/\sqrt{3}$ (c) 15
2 (a) $6752/(5\pi^5)+(1-\sqrt{3})/\sqrt{3}$
(b) $9(8^8-3^8)/8$ (c) $3/8+\ln(2/3)$

Module 24

Unit 1: Test

(a) 211/5 (b) 0 (c) $(e^4-1)/2e$
(d) 2863/6 (e) 3

Unit 1: Exercises

(c) $(e-1)/2e^2$ (d) $(\sqrt{2}-1)/\sqrt{2}$

Unit 2: Test

(a) $3\pi/2$ (b) $8/\sqrt{3}$ (c) 1/4

Unit 2: Exercises

(a) $2(\sqrt{3}-1)$ (b) $26/9\sqrt{3}$

Unit 3: Test

(a) $\ln(2)$ (b) $\ln(\sqrt{3})$ (c) $\ln[1+\ln(2)]$

Unit 3: Exercises

(a) $\ln(2)$ (b) $\ln(\sqrt{2})$ (c) 1

Unit 4: Test

(a) $\ln(9/5)$ (b) $20+12\ln(6)-2\ln(2)$
(c) $(1/36)[\ln(8/7)-37\ln(2)-105]$

Unit 5: Test

(a) $5\pi/6\sqrt{3}-\ln(\sqrt{3})$ (b) $40\ln(2)-56/9$
(c) π

Unit 5: Exercises

(b) $2\ln(2)-3/4$

Module 24: Further exercises

1 (a) 1701 (b) 0 (c) $5/2\sqrt{3}$
2 (a) $8\pi/3$ (b) $\pi/6$ (c) $1/6 - \sqrt{3}/8$
3 (a) $(5/3)\ln(26/7)$ (b) 0
(c) $\ln(2) + \ln[1 + \ln(2)]$
4 (a) $1/2 \ln(343/500)$
(b) $(1/6)[9\ln(3) + 8\ln(2)] + 3$
(c) $(1/36)\ln(3/2) - 1/216$
(d) $(1/\sqrt{5})\ln(9/5)$
5 (a) $(1/3\sqrt{2})[\pi/2 - (2/3)(1 + \sqrt{2})]$
(b) 1/16 (c) $(3e^2/400)[1 - 2\ln(2)] + 3/100$

Chapter 6: Miscellaneous exercises

1 (a) 1 (b) 0
11 (a) $1/2(1 - e^{-k^2})$
(b) $1/2[1 - (k^2 + 2)e^{-k^2}]$
(c) $1/2[2 - (k^4 + 2k^2 + 1)e^{-k^2}]$

CHAPTER 7

Module 25

Unit 1: Test

1 (a) {black, blue, grey} (b) {x: x is a page number and $1 \leqslant x \leqslant 200$}
2 (a) {x: $2x$ is an integer and $2 \leqslant 2x \leqslant 7$}
(b) {Mon, Tue, Wed, Thu, Fri, Sat, Sun}
3 (a) True
4 ϕ, {A}, {B}, {1}, {2}, {3}, {A, B }, {A, 1}, {A, 2}, {A, 3}, {B, 1}, {B, 2}, {B, 3}, {1, 2}, {1, 3}, {2, 3}, {A, B, 1}, {A, B, 2}, {A, B, 3}, {A, 1, 2}, {A, 1, 3}, {A, 2, 3}, {B, 1, 2}, {B, 1, 3}, {B, 2, 3}, {1, 2, 3}, {A, B, 1, 2}, {A, B, 1, 3}{A, B, 2, 3}, {A, 1, 2, 3}, {B, 1, 2, 3}, {A, B, 1, 2, 3}
5 (a) True
6 (a) {Mon, Tue, Wed, Thu, Fri, Sat, Sun}
(b) {R, O, Y, G, B, I, V}

Unit 1: Exercises

1 (a) {Hearts, Spades, Diamonds, Clubs}
(b) {Jan, Feb, Mar, Apr, May, Jun, Jul, Aug, Sep, Oct, Nov, Dec}
2 (a) {x: x is even and $-4 \leqslant x \leqslant 4$}
(b) {R, O, Y, G, B, I, V}
3 (a) True (b) True
4 ϕ, {1}, {{1}}, {1, {1}}
5 (b) True
6 (a) {Defective, Non-defective}
(b) {June, July, August}

Unit 2: Test

1 (a) {1}, {2}, {3}
(b) {king}, {spade}, {diamond}, {club}
2 (a) {R}, {O}, {Y}, {G}, {B}, {I}, {V}
(b) {u}, {v}, {w}, {x}, {y}, {z}
4 (a) {salt and vinegar, salt and vinegar}, {salt and vinegar, cheese and onion}, {cheese and onion, cheese and onion}
(b) No {cheese and onion, cheese and onion}

Unit 2: Exercises

1 (a) {a}, {b}, {c}
(b) {red}, {blue}, {yellow}, {green}
2 (a) {4}, {6}, {8}, {10}
(b) {a}, {e}, {i}, {o}, {u}
4 (a) {black, black}, {black, white}
(b) {black, black}, {black, white}, {white, white}

Unit 3:Test

1 Union: {−1, 0, 1, 2, 3, 4}, intersection: {1}
{1, 2, 3, 4}′ = {x: x is an integer and $x > 4$ or $x < 1$}
{−1, 0, 1}′ = {x: x is an integer and $x(x^2 - 1) \neq 0$}
3 $E1 \cup E2 = \{2, 3, 5\}$, $E1 \cap E2 = f$,
$E1' = \{1, 2, 4, 6\}$, $E2' = \{1, 3, 4, 5, 6\}$
4 $E = \{1, 2, 4, 6\}$, $E' = \{3, 5\}$

Unit 3: Exercises

1 Union: {a, b, c, d, e, f, g, h}, Intersection: {e, f}
{a, c, e, f, h}′ = {b, d, g, i, j},
{d, e, f, g}′ = {a, b, c, h, i, j}
3 $E1 \cup E2$ = {red ball or number $\leqslant 2$},
$E1 \cap E2$ = {red ball numbered $\leqslant 2$}
$E1'$ = {blue or green ball},
$E2'$ = {number > 2}
4 $E1 \cup [E3 \cap E5]$ where $E1$ = {Toyota},
$E3$ = {Ford}, $E5$ = {white}

Unit 4: Test

2 Both are {d, e}
3 18

Unit 4: Exercises

2 Both are {a, b, c, f, g, h}
3 4,500

Module 25: Further Exercises

1 (a) {Volvo, Toyota, Ford}
(b) {cash, cheques}
2 (a) $\{x: x(x-1)=0\}$
(b) {2, 3}
3 (a), (b), (c) all true
4 ϕ, $\{\phi\}$, $\{\{\phi\}\}$, $\{\phi, \{\phi\}\}$
5 (a), (b) both true
6 (a) {on, on, on}, {on, on, off}, {on, off, off}, {off, off, off}
(b) $\{x: x$ is a chapter number and $1 \leqslant x \leqslant 6\}$
7 (a) {W}, {X}, {Y}, {Z}
(b) {Monday}, {Wednesday}, {Friday}, {Sunday}
8 (a) {Red}, {White}, {Blue}
(b) {1}, {2}, {3}, {4}, {5}, {6}, {7}, {8}
10 (a) {Red, A}, {Red, B}, {Red, C}, {Blue, A}, {Blue, B}, {Blue, C}
(b) No {Red, C}
11 Union: {R, O, Y, G, B, I, V, W}, Intersection: {R, B}, Complements: $\{x: x$ is a colour but not a colour of the rainbow$\}$ and $\{x: x$ is a colour but not red, white or blue$\}$
13 $E1 \cup E2 =$ {male or in management}, $E1 \cap E2 =$ {Male and in management} $E1' =$ {female}, $E2' =$ {not in management}
14 $E1 =$ {black}, $E2 =$ {court card}, $E1 \cup E2 =$ {black or court card}
16 Both {h}
17 290

Module 26

Unit 1: Test

1 (a) 5/13 (b) 1/26 (c) 1/52
2 (a) 1/60 (b) 1/31 (c) $1/N$ where N is the number of words in the dictionary
3 (a) 2/13 (b) 4/33 (c) 5/100
4 (a) 7/24 (b) 17/24 (c) 18/24
5 1/720 (6!/10!)

Unit 1: Exercises

1 (a) 12/52 (b) 2/52 (c) 1/52
2 (a) 1/10 (b) 1/120 (c) $1/N$ where N is the number of discs in the cabinet
3 (a) 8/12 (b) 1/10 (c) 17/100
4 (a) 9/19 (b) 11/19 (c) 17/19
5 1/5!

Unit 2: Test

1 (a) 9/15 (b) 7/15
2 (a) 28/52 (b) 36/52 (c) 26/52
3 (a) 15/16 (b) 12/16 (c) 1/16 (d) 7/16

Unit 2: Exercises

1 (a) 8/11 (b) 8/11
2 (a) 11/18 (b) 12/36 (c) 8/36
3 (a) 3/23 (b) 14/23 (c) 7/23 (d) 12/23

Unit 3: Test

1 (a) 36/225 (b) 49/225 (c) 16/225
2 (a) 144/2652 (b) 650/2652 (c) 144/2704, 676/2704
3 (a) 1/2 (b) 1/4 (c) 13/20 (d) 1/240

Unit 3: Exercises

1 (a) 60/121 (b) 25/121 (c) 60/110, 20/110
2 (a) 1/12 (b) 1/4
3 (a) 12 784/270 348 (b) 79 960/270 348 (c) 28 776/270 348

Module 26: Further Exercises

1 (a) 10/52 (b) 2/52 (c) 1/52
2 (a) 1/150 (b) 1/7 (c) $1/N$ where N is the number of hairs
3 (a) 7/1000 (b) 39/100
4 (a) 3/14 (b) 11/14 (c) 12/14
5 1/6840 (17!/20!)
6 (a) 24/44 (b) 14/44
7 (a) 32/44 (b) 8/44 (c) 20/44
8 (a) 9/12 (b) 6/12 (c) 3/12 (d) 4/12
9 (a) 0.03 (b) 0.63 (c) 0.07 (d) 0.34
10 (a) 280/1892 (b) 182/1892
11 (a) 20/132 (b) 2/132 (c) 6/132

Module 27

Unit 1: Test

1 Pr({Red}) = 0.4, Pr({Blue}) = 0.3, Pr({Yellow}) = 0.2, Pr({Green}) = 0.1
2 Assign numbers to events {Red} := 1, {Blue} := 2, {Yellow} := 3, {Green} := 4
Mean $\mu = 2.0$, Standard deviation $\sigma = 1.0$

3 (a) $\Pr(1) = 0.50$, $\Pr(2) = \Pr(3) = 0.25$
(b) Mean $\mu = 1.75$, Standard deviation $\sigma = 0.83$
(c) Expectation $E(x) = \mu = 1.75$

4 (a) 0.2304 (b) 0.077 76 (c) 0.66 304

Unit 1: Exercises

1 $\Pr(0.00 \leqslant x \leqslant 14.99) = 5/32$,
$\Pr(15.00 \leqslant x \leqslant 29.99) = 7/32$,
$\Pr(30.00 \leqslant x \leqslant 44.99) = 12/32$,
$\Pr(50.00 \leqslant x \leqslant 60.00) = 8/32$

3 Mean $\mu = 4.33$, Standard deviation $\sigma = 1.49$
(a) 12/21 (b) 14/21

4 (a) 0.034 to 3 dec. pl. (b) Same as (a)
(c) 0.65

Unit 2: Test

1 (a) (i) 0.0044, 0.0330, 0.1318 (ii) 0.8306

Unit 2: Exercises

1 (a) (i) 0.262, 0.082, 0.002 (ii) 0.01696

Unit 3: Test

1 (a) 0.0045 (b) 0.1255

2 (a) 0.0097 (b) 0.0005

Unit 3: Exercises

1 (a) 0.0077 (b) 0.0146

2 (a) 0.0905 (b) 0.0047

Unit 4: Test

1 (a) 0.642 (b) 0.654 (c) 0.988

2 (a) 1.2 (b) -1.6 (c) $-2.0 \leqslant z < 0.80$

3 (a) 0.387 (b) 0.099 (c) 0.174 (d) 0.448
(e) 0.135 (f) 0.295 (g) 0.905

4 (a) 0.606, 0.618, 0.385
(b) 152, 155, 96

Unit 4: Exercises

1 (a) 0.691 (b) 0.696 (c) 0.995

2 (a) 0.44 (b) -0.28 (c) $-1.01 \leqslant z < 0.61$

3 (a) 0.487 (b) 0.184 (c) 0.295 (d) 0.137
(e) 0.301 (f) 0.456 (g) 0.896

4 (a) 0.702, 0.702, 0.261
(b) 95, 95, 35

Module 27: Further Exercises

1 {fiction} := 1, {novel} := 2, {reference} := 3, {text} := 4
$\Pr(1) = 10/50$, $\Pr(2) = 24/50$, $\Pr(3) = 4/50$, $\Pr(4) = 12/50$.

4 (a) 0.075 (b) 0.263 (c) 0.666

6 (a) 0.005 (b) 0.361 (c) 0.001

8 (a) 0.295 (b) 0.518 (c) 0.777

11 (a) 0.692, 0.977, 0.787
(b) 996, 1407, 1133

Chapter 7: Miscellaneous Exercises

2 (a) $\{6, 7, 8\}$ (b) $\{x$: x is a number and $5 \leqslant x \leqslant 9\}$ (c) $\{x$: x is a dog$\}$
(d) {a, e, i, o, u}

4 (c)

6 (a) {(blue, head), (blue, tail), (grey, head), (grey, tail)}
(b) {(head, blue), (tail, blue), (head, grey), (tail, grey)}

8 (a) Union: $\{1, 2, 3, 4, 5\}$, Intersection: $\{3, 5\}$
(b) Union: $\{x\colon -1 \leqslant x \leqslant 3\}$, Intersection: $\{x\colon 0 \leqslant x < 2\}$
(c) Union: {C, H, S, D, H, T}, Intersection: ϕ

10 (a) 0.60 (b) 0.55 (c) 0.15 (d) 0.85
(e) 0.15

12 (a) 16/2704 (b) 1/4 (c) 2304/2704

14 (a) 0.126 (b) 0.809 (c) 0.191 (d) 0.223

16 (a) 0.663 (b) 0.663 (c) 0.326

■ ■ ■ Index